Springer

*Berlin
Heidelberg
New York
Barcelona
Hong Kong
London
Milan
Paris
Singapore
Tokyo*

Tamejiro Hiyama

Organofluorine Compounds

Chemistry and Applications

With contributions by
Tamejiro Hiyama, Kiyoshi Kanie, Tetsuo Kusumoto,
Yoshitomi Morizawa, Masaki Shimizu

With 96 Figures and 37 Tables

 Springer

Leading Author
Professor Tamejiro Hiyama
Department of Material Chemistry
Graduate School of Engineering
Kyoto University
Yoshida, Sakyo-ku
Kyoto 606-8501, Japan

Editor
Professor Hisashi Yamamoto
Nagoya University
Graduate School of Engineering
Chikusa
Nagoya 464-8603, Japan

ISBN 3-540-66689-3 Springer-Verlag Berlin Heidelberg New York

Library of Congress Cataloging-in-Publication Data
Hiyama, Tamejiro, 1946–
  Organofluorine compouds : chemistry and applications / Tamejiro Hiyama ; with contributions from Kiyoshi Kanie ... [et al.].
    p.cm.
  Includes index.
  ISBN 3540666893 (alk. paper)
  1. Organofluorine compounds. I. Title.

QD412.F1 H59 2000
547'.02–dc21   99.087168

This work is subject to copyright. All rights are reserved, whether the whole or part of the material is concerned, specifically the rights of translation, reprinting reuse of illustrations, recitation, broadcasting, reproduction on microfilm or in any other way, and storage in data banks. Duplication of this publication or parts thereof is permitted only under the provisions of the German Copyright Law of September 9, 1965, in its current version, and permission for use must always be obtained from Springer-Verlag. Violations are liable for prosecution under the German Copyright Law.

Springer-Verlag is a company in the BertelsmannSpringer publishing group
© Springer-Verlag Berlin Heidelberg 2000
Printed in Germany

The use of general descriptive names, registered names, trademarks, etc. in this publication does not imply, even in the absence of a specific statement, that such names are exempt from the relevant protective laws and regulations and therefore free for general use.

Product liability: The publishers cannot guarantee the accuracy of any information about dosage and application contained in this book. In every individual case the user must check such information by consulting the relevant literature.

Typesetting: Fotosatz-Service Köhler GmbH, Würzburg
Coverdesign: design & production, Heidelberg

Printed on acid-free paper   SPIN 10668915   2/3020 mh – 5 4 3 2 1 0

List of Contributors

Professor Tamejiro Hiyama
Department of Material Chemistry, Graduate School of Engineering,
Kyoto University, Yoshida, Sakyo-ku, Kyoto 606-8501, Japan

Kiyoshi Kanie
Department of Chemistry and Biotechnology,
Graduate School of Engineering, The University of Tokyo, Hongo,
Bunkyo-ku, Tokyo 113–0023, Japan

Dr. Tetsuo Kusumoto
Dainippon Ink and Chemicals, Inc., 4472-1 Komuro,
Ina-machi, Kita-adachi-gun, Saitama 362-8577, Japan

Dr. Yoshitomi Morizawa
Research Center, Asahi Glass Ltd., 1150 Hazawa, Kanagawa-ku,
Yokohama 221–0863, Japan

Dr. Masaki Shimizu
Department of Material Chemistry, Graduate School of Engineering,
Kyoto University, Yoshida, Sakyo-ku, Kyoto 606–8501, Japan

Preface

Recently, fluorine has become a key element in the remarkable progress of materials and biologically active agents. Since fluorine has quite unique reactivities and properties, fluorine chemistry is recognized as being totally different from the chemistry of the other halogens. Accordingly, many books and reviews have been published. Most of them, however, were written by experts in specialized fields. Although many excellent monographs on organofluorine chemistry exist, this new textbook was undertaken with the aim of overviewing organofluorine chemistry on the basis of organic synthesis.

Accordingly, the present work will focus on the organic chemistry of organofluorine compounds, starting with general discussions on the unique properties of the fluorine atom and carbon–fluorine bonds, natural resources, and the salient features of organofluorine compounds in addition to their toxicity. This will be followed by chapters on reagents and methods for the introduction of fluorine into organic compounds, organofluorine building blocks, reactions of C–F bonds, biologically active organofluorine compounds, fluorine-containing materials, fluorous media, and some synthetic reagents containing fluorine. The book focuses as much as possible on information that will be useful for synthetic chemists who have experienced or inexperienced organofluorine chemistry.

Kiyoshi Kanie contributed to Chapters 1, 2 and 7, Tetsuo Kusumoto to Chapters 2, 5, and 6, Yoshitomi Morizawa to Chapters 5 and 6, and Masaki Shimizu to Chapters 3, 4, and 8; all the coauthors worked very closely both with each other and with the main author Tamejiro Hiyama, who wishes to emphasize that this collaboration was enjoyed by them all. We also thank Dr. Joe Richmond both for his help and his warm encouragement. Finally, the authors thank Professor Hisashi Yamamoto, Nagoya University, Japan, for suggesting that we produce this book.

We wish to dedicate this book to Professor Hitoshi Nozaki, Professor Emeritus, Kyoto University, Japan, on the occasion of his 77th birthday. He introduced all of us either directly or indirectly to the fascination of organic chemistry.

Kyoto, January 2000   Tamejiro Hiyama

Table of Contents

Chapter 1 General Introduction 1
1.1 Nature of Organofluorine Compounds 1
1.1.1 Brief History 1
1.1.2 Properties of the Fluorine Atom 2
1.1.3 Fluorine Effects in Organic Compounds 3
1.1.4 Properties of Fluoroorganic Compounds 10
1.1.5 Properties of Perfluoroorganic Compounds 13
1.1.6 Spectroscopic Properties 14
1.2 Source of Fluorine 18
1.2.1 Hydrogen Fluoride 18
1.2.2 Fluorine Gas 18
1.3 Toxicity of Fluorinating Reagents 19
1.3.1 Hydrogen Fluoride and Fluorine Gas 19
1.3.2 First-Aid Treatment 20
1.3.3 Fluoroacetic Acid 21

Chapter 2 Reagents for Fluorination 25
2.1 Electrophilic Flurorinating Reagents 25
2.1.1 Fluorine Gas 25
2.1.2 Xenon Difluoride 28
2.1.3 Electrophilic Reagents Containing an O–F Bond 29
2.1.4 Electrophilic Reagents Containing an N–F Bond 34
2.2 Nucelophilic Fluorinating Reagents 39
2.2.1 Hydrogen Fluoride and Derivatives 39
2.2.2 Functional Group Transformation 43
2.2.3 Fluoride Reagents 48
2.3 Combination of an Electrophile and a Fluoride Reagent 56
2.3.1 Halofluorination of Olefins and Acetylenes 56
2.3.2 Thiofluorination and Selenofluorination of Olefins 58
2.3.3 Nitrofluorination 59
2.3.4 Oxidative Fluorination 58
2.3.5 Oxidative Desulfurization-Fluorination 61
2.3.6 Oxidative Fluorination of Amines 65
2.4 Electrochemical Fluorination 66

Chapter 3	**Organofluorine Building Blocks**	77
3.1	Fluorine-Substituted Nucleophilic Reagents	77
3.1.1	Alkylmetals	77
3.1.2	Alkenylmetals	84
3.1.3	Alkynylmetals	91
3.1.4	Metal Enolates	93
3.2	Fluorine-Substituted Electrophilic Reagents	99
3.3	Fluorine-Substituted Radicals	103
3.4	Fluorine-Substituted Carbenes	107
3.5	Electrophilic Perfluoroalkylating Reagents	111
3.5.1	(Perfluoroalkyl)aryliodonium Salts	111
3.5.2	(Polyfluoroalkyl)aryliodonium Salts	112
3.5.3	(Trifluoromethyl)chalcogenium Salts	113

Chapter 4	**Reactions of C–F Bonds**	119
4.1	Fluorine Leaving Group	119
4.1.1	1-Fluoro Sugars	119
4.1.2	Aromatic Nucleophilic Substitution	121
4.2	C–F Bond Activation by Metal Complexes	125
4.2.1	Activation of an Aliphatic C–F Bond	125
4.2.2	Activation of an Aromatic C–F Bond	126
4.3	Interaction of Fluorine with a Proton or Metal	128
4.3.1	Fluorine-Hydrogen Interaction	128
4.3.2	Fluorine-Metal Interaction	129

Chapter 5	**Biologically Active Organofluorine Compounds**	137
5.1	Fluorine Effect in Biological Activity	137
5.2	Strategies for Design and Synthesis	141
5.2.1	Structure-Activity Relationship	141
5.2.2	Commercially Available Fluorinated Materials	143
5.3	Fluorinated Amino Acids and Carbohydrates	144
5.3.1	Amino Acids	144
5.3.2	Protease Inhibitors	148
5.3.3	Carbohydrates	150
5.3.4	Nucleosides	151
5.4	Fluorine-Containing Pharmaceuticals	154
5.4.1	Prostanoids	154
5.4.2	Vitamin D_3	157
5.4.3	Central Nervous System Agents	160
5.4.4	Antibacterials and Antifungals	161
5.4.5	β-Lactam Antibiotics	164
5.4.6	Anesthetics	164
5.4.7	Artificial Blood Substitutes	165
5.4.8	^{18}F-Labeled Tracers for Positron Emission Tomography	166
5.5	Fluorine-Containing Agrochemicals	166

5.5.1	Insecticides	167
5.5.2	Herbicides	173
5.5.3	Fungicides	177

Chapter 6 Fluorine-Containing Materials 183

6.1	Fluorine Effect in Materials	183
6.1.1	Boiling Points and Melting Points	184
6.1.2	Solubility	186
6.1.3	Surface Tension	186
6.1.4	Refractive Index	187
6.1.5	Viscosity	187
6.2	Chlorofluorocarbons, Hydrochlorofluorocarbons, Hydrofluorocarbons, and Alternatives	188
6.2.1	Brief History	188
6.2.2	Production of Chlorofluorocarbons and Hydrochlorofluorocarbons	191
6.2.3	Syntheses of CFC Alternatives	192
6.2.4	Evaluation of Safety and Environmental Effects	195
6.2.5	Alternatives to the Third Generation	196
6.3	Fluorine-Containing Liquid Crystals	196
6.3.1	Nematic Liquid Crystals	197
6.3.2	Ferroelectric Liquid Crystals	202
6.3.3	Antiferroelectric Liquid Crystals	209
6.4	Fluorine-Containing Polymers	212
6.4.1	Brief History	212
6.4.2	Monomer Synthesis	214
6.4.3	Fluoroplastics	216
6.4.4	Fluoroelastomers	220
6.4.5	Fluoropolymer Coatings	224
6.4.6	Fluorosurfactants	225
6.4.7	Fluorinated Membranes	228

Chapter 7 Fluorous Media 235

7.1	Organic Reactions in Perfluorocarbons	235
7.2	Fluorous Biphase Reactions	237
7.2.1	Hydroformylation	237
7.2.2	Oxidation	239
7.3	Purification and Isolation by Phase Separation	243

Chapter 8 Organic Reactions with Fluorinated Reagents 249

8.1	Fluoride Ion in Organic Synthesis	249
8.1.1	Fluoride Base	249
8.1.2	Desilylative Elimination and Deprotection	250
8.1.3	Naked Anions and Fluorosilicates	252
8.2	Trifluoroacetic Acid and Trifluoroperacetic Acid	255
8.2.1	Trifluoroacetic Acid	255

8.2.2	Trifluoroperacetic Acid	256
8.3	Trifluoromethanesulfonic Acid and Derivatives	257
8.3.1	Trifluoromethanesulfonic Acid	257
8.3.2	Trimethylsilyl Trifluoromethanesulfonate	258
8.3.3	Metal Trifluoromethanesulfonates	259

Subject Index 265

CHAPTER 1

General Introduction

1.1
Nature of Organofluorine Compounds

1.1.1
Brief History

The name of fluorine was coined as *le fluor* by the French scientist Ampére in 1812 after its natural resource fluorspar. Fluorine, having been given the atomic number 9, is a second row element of Group 17 or the lightest element of the halogens.

Fluorine exists in nature as fluorides in fluorite (CaF_2), cryolite ($Na_3[AlF_6]$), and phosphorite ($Ca_5[F, Cl][PO_4]_3$). Fluorite has traditionally been the main source of hydrofluoric acid but, because it is now in short supply, it is gradually being replaced by cryolite. The abundance of fluorine in the Earth's crust is 625 wt/ppm, ca. five times that of chlorine (130 wt/ppm).

In the 17th century, it was already known that exposure of fluorite to sulfuric acid liberated an acid that corroded glass. Although not well characterized, this acid was used for glass etching. The acid was considered to consist of hydrogen and the new element fluorine on the basis of the discovery of chlorine and iodine in 1774 and 1811, respectively. Isolation of fluorine was successfully achieved in 1886 by Moissan, who electrolyzed a melt mixture of potassium hydrogen difluoride and hydrogen fluoride at –23°C. This success earned him the Nobel Prize in 1906.

Remarkable progress in fluorine chemistry has been made during the 20th century with the development in synthetic methods for organofluorine compounds. An epoch-making method is the fluorination of organic halides by nucleophilic substitution with antimony(III) fluoride, as discovered by Swarts [1]. This reaction was employed for the synthesis of a variety of organofluorine compounds and was later improved to become an industrial process. Another important finding is the Schiemann reaction uncovered in 1927 [2]. Aromatic amines were first converted into diazonium salts, which were then decomposed in the presence of hydrofluoric acid. This reaction was also continuously improved and is still used for the manufacture of fluoroaromatics.

In areas other than organic synthesis, fluorine also played key roles. For example, uranium(VI) fluoride was used in the United States during World War II for the collection of the radioactive uranium that was essential for the development of atomic bombs. Using volatile uranium(VI) fluoride, a radioactive isotope of uranium was condensed by fractional distillation. For this purpose, novel equipment, tools and materials were invented that could tolerate molecular fluorine and hydrogen fluoride generated during the whole process.

A great impact on the chemical industry resulted from the discovery of poly(tetrafluoroethene), a highly heat- and chemical-resistant polymer given the tradename Teflon. This discovery led to remarkable progress in fluorine-containing materials and chemicals that now play important roles in everyday life throughout the world.

1.1.2
Properties of the Fluorine Atom

The atomic weight of naturally abundant fluorine is estimated to be 18.9984032. In addition to the natural and stable isotope ^{19}F, radioisotopes ^{17}F (half-life time of 1.08 min) and ^{18}F (half-life time of 110 min) have been prepared. These radioisotopes decompose to give finally ^{17}O and ^{18}O, respectively. In particular, ^{18}F is prepared by the reaction of neon with deuteron or ^{18}O with proton and it has recently found applications in positron emission tomography (PET).

The electronegativity (EN) of fluorine is the largest among all elements and its ionization potential (IP) is also the largest with the exception of helium and neon. The C–F bond length is the shortest next to a C–H bond but is much stronger in energy. The carbon of a C–F bond is slightly polarized to δ^+ in sharp contrast to the carbon of a C–H bond. Thus, replacement of hydrogen in an organic molecule by fluorine causes polarization with very little change in total bulk. Accordingly, organofluorine compounds demonstrate a variety of unique biological activities and physical properties.

There is some controversy about the similarity of fluorine and hydrogen. According to Bondi's van der Waals radii estimation, fluorine is quite a bit larger than hydrogen and close to oxygen in size [3], as suggested in Table 1.1. A CF$_3$ group is much larger in steric bulk than a CH$_3$ group as expressed by the steric effect constant E_s (Table 1.2) [4]. Accordingly, a CF$_3$ group is not necessarily the same size as a CH$_3$ group, although such similarity is often referred to for the understanding of the biological activity of trifluoromethylthymidine.

Table 1.1. Several properties of elements

Element	IP (kcal/mol)	EA (kcal/mol)	CH$_3$-X (Å)	EN (Pauling)	r_v(Å) (Pauling)	r_v(Å) (Bondi)	BE CH$_3$-X (kcal/mol)
H	313.6	17.7	1.09	2.1	1.20	1.20	99
F	401.8	79.5	1.39	4.0	1.35	1.47	116
Cl	299.0	83.3	1.77	3.0	1.80	1.75	81
Br	272.2	72.6	1.93	2.8	1.95	1.85	68
O(OH)	310.4	33.7	1.43	3.5	1.40	1.52	86
S(SH)	238.9	48.0	1.82	2.5	1.85	1.80	65

IP, Ionization potential; EA, electron affinity; EN, electronegativity; BE, bond energy.

1.1 Nature of Organofluorine Compounds

Table 1.2. E_s values of typical substituents

R	H	Me	n-Bu	c-Hex	CF$_3$	i-Bu	s-Bu	t-Bu	Ph
E_s	−1.24	0	0.39	0.79	1.16	0.93	1.13	1.54	–
E_s':	−1-12	0	0.31	0.69	0.78	0.93	1.00	1.43	2.31

E_s: Steric parameters suggested by K. W. Tafts.
E_s': Steric parameters suggested later by J.-E. Dubois.

1.1.3
Fluorine Effects in Organic Compounds

Due to the large electronegativity of a fluorine atom, functional groups such as F, CF$_3$, CF$_3$O, and CF$_3$SO$_2$ are all electron withdrawing and find many applications in materials and biologically active agents [5]. These groups are successfully employed as control elements for selective organic syntheses. Substituent constants of such fluorine-containing groups are compared in Table 1.3 [6]. Although fluorine has a σ_m value comparable with other halogen substituents, its σ_p is much smaller due probably to the fairly large p–p repulsion (+I_π effect) or resonance effect (+R effect) of fluorine.

The following are worthy of note: A CF$_3$ group has σ_m and σ_p values only slightly smaller than those of nitro and cyano groups, indicating an electron-withdrawing nature of a CF$_3$ group in sharp contrast to a CH$_3$ group. A C$_2$F$_5$ group has an electron-withdrawing effect similar to a CF$_3$ group. A CF$_3$S group is electron withdrawing, whereas a CH$_3$S group is electron donating. CF$_3$CO and CF$_3$SO$_2$ groups withdraw electrons much more than a NO$_2$ group.

The electronic effects of a C–F bond are understood in terms of inductive and resonance effects [7] (Fig. 1.1). Owing to its large electronegativity, fluorine induces a −I_σ effect to reduce the electron density at the carbon of a C–F bond. This effect is remarkable particularly at an sp^3-carbon. Unshared electron pairs of fluorine bonded to a sp^2-carbon interact with π-electrons repulsively to push

Table 1.3. Hammet σ constants of several substituents

Substituents	σ_m	σ_p	Substituents	σ_m	σ_p
H	0.00	0.00	OCH$_3$	0.12	−0.27
F	0.34	0.06	OCF$_3$	0.38	0.35
Cl	0.37	0.23	SCH$_3$	0.15	0.00
Br	0.39	0.23	SCF$_3$	0.40	0.50
OH	0.12	−0.37	C(O)CH$_3$	0.38	0.50
NO$_2$	0.72	0.78	C(O)CF$_3$	0.63	0.80
CN	0.56	0.66	SO$_2$F	0.80	0.91
CH$_3$	−0.07	−0.17	SO$_2$CH$_3$	0.60	0.72
CF$_3$	0.43	0.54	SO$_2$CF$_3$	0.83	0.96
CH$_3$CH$_2$	−0.07	−0.15	C$_6$H$_5$	0.06	−0.01
CF$_3$CF$_2$	0.47	0.52	C$_6$F$_5$	0.26	0.27

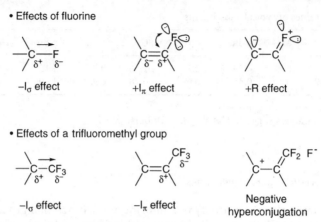

Fig. 1.1. Effects of fluorine-containing substituents

the π-electrons to a β-carbon ($+I_\pi$ effect). Like halogen atoms, fluorine can resonate with the π-electron system (+R effect). Thus, the electron density at the β-carbon or the *ortho-*(or *para-*)carbon of aromatic compounds increases.

A CF_3 group always withdraws electrons at the sp^3- and sp^2-carbons ($-I_\sigma$ effect). In particular, the electron-withdrawing effect by a CF_3 group on a sp^2-carbon is attributed to negative hyperconjugation, as illustrated in Fig. 1.1. Since the $-I_\sigma$ effect competes with the $+I_\pi$ and +R effects, fluorobenzene undergoes electrophilic substitution reactions more reluctantly than benzene but faster than chloro- or bromobenzene. In addition, fluorobenzene gives *p*-substituted products preferentially. These results are understood by a +R effect and especially by a $+I_\pi$ effect (Fig. 1.2) [8].

Substituent constants are divided into inductive and resonance factors and are listed in Table 1.4, which clearly shows that fluorine induces a resonance effect larger than other halogens [5, 6].

Because fluorine is an electron-withdrawing substituent, organic acids become much more acidic by fluorine substitution. Some examples are listed in Table 1.5. For example, the pK_a of polyfluoroacetic acid decreases according to the number of fluorine atoms and is ascribed to an inductive effect of fluorine. The effect is distinctive of fluorine as compared with chlorine. In contrast, the fact that pentafluorophenol is less acidic than pentachlorophenol is attributed to competing $-I_\sigma$ and $+I_\pi$ effects.

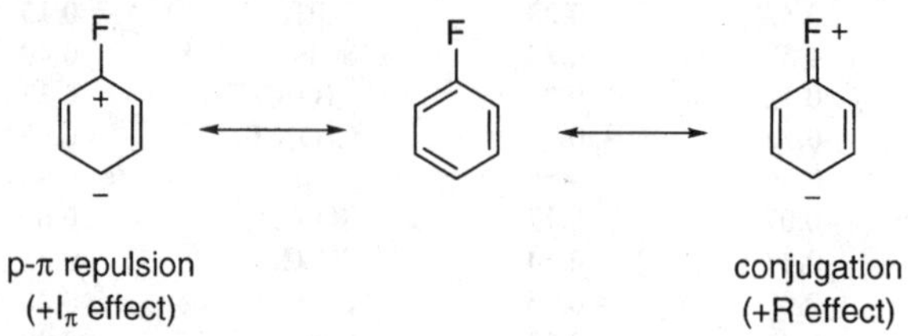

Fig. 1.2. Regioselectivity of aromatic electrophilic substitution

1.1 Nature of Organofluorine Compounds

Table 1.4. Modified Swain–Lupton constants of halogen substituents

Substituents	F	R
F	0.45	−0.39
Cl	0.42	−0.19
Br	0.45	−0.22
I	0.42	−0.24

F: field inductive effect.
R: resonance effect.

Table 1.5. Acidities of fluorine-containing organic acids

Acid	pK_a (25°C)	Acid	pK_a (25°C)
CH_3CO_2H	4.8	CH_3SO_3H	−1.9
CH_2FCO_2H	2.6	CF_3SO_3H	−5.1
CHF_2CO_2H	1.2	$C_6H_5CO_2H$	4.2
CF_3CO_2H	0.2	$C_6F_5CO_2H$	4.4
CH_2ClCO_2H	2.9	C_6H_5OH	9.9
$CHCl_2CO_2H$	1.3	C_6F_5OH	5.5
CCl_3CO_2H	0.6	C_6Cl_5OH	5.3

The high electron-withdrawing effect of fluorine allows the preparation of superacids that are more acidic than 100% sulfuric acid (Hammet acidity function, $H_0 = -11.3$). Typical examples are summarized in Table 1.6, along with values for H_0. As can easily be seen, FSO_3H and CF_3SO_3H are themselves superacids. A mixture of one of these with a Lewis acid like SbF_5 or SO_3 drastically enhances the acidity to dissolve and protonate hydrocarbons which are unreactive to most chemicals. These superacids are called magic acids. An example is FSO_3H/SbF_5 ($H_0 = -18.9$), invented by Olah [9]. The strongest magic acid is HF/SbF_5 whose H_0 reaches −24 [10].

Table 1.6. The Hammet acidity function (H_0) of acids

Acid	H_0	Acid	H_0
HCl (38 wt% in H_2O)	−4.4	CH_3SO_3H	−7.86
HCl (10 wt% in H_2O)	−1.0	CF_3SO_3H	−14.5
H_2SO_4 (100%)	−11.9	$C_4F_9SO_3H$	−13.2
H_2SO_4 (10 wt% in H_2O)	−0.4	$C_6F_{13}SO_3H$	−12.3
HF	−11.0	FSO_3H/SbF_5 (10 mol%)	−18.9
$ClSO_3H$	−13.8	FSO_3H/SbF_5 (0.3 mol%)	−21
FSO_3H	−15.1	$C_4F_9SO_3H/SbF_5$ (10 mol%)	−18.2

Table 1.7. Equilibrium acidities of halogenated carbon acids

Carbon acid	pK_{CsCHA}	Carbon acid	pK_{CsCHA}
CHF_3	30.5	CF_3CHF_2	28.2
$CHCl_3$	24.4	CF_3CHCl_2	24.4
$CHBr_3$	22.7	CF_3CHBr_2	23.7
CHI_3	22.5	CF_3CHI_2	24.1

pK_{CsCHA}: Cesium ion pair pK in cyclohexylamine.

Table 1.8. Effect of fluorine substitution on carbon acids

Compound	ΔH_{calcd}	$\Delta\Delta H$	$\Delta\Delta H'$	Compound	ΔH_{calcd}	$\Delta\Delta H$	$\Delta\Delta H'$
CH_4	416.8	0.0		CH_3NO_2	354.9	−61.9	0.0
CFH_3	406.3	−10.5	0.0	CFH_2NO_2	350.4	−66.4	−4.5
CF_2H_2	391.3	−25.5	−15.0	CF_2HNO_2	348.2	−68.6	−6.7
CF_3H	368.9	−47.9	−37.4	CH_3CHO	364.1	−52.7	0.0
CH_3NC	380.1	−36.7	0.0	CFH_2CHO	357.1	−59.7	−7.0
CFH_2NC	370.8	−46.0	−9.3	CF_2HCHO	360.8	−56.0	−3.3
CF_2HNC	357.1	−59.7	−23.0				
CH_3CN	373.5	−43.3	0.0				
CFH_2CN	366.8	−50.0	−6.7				
CF_2HCN	358.7	−58.1	−14.8				

ΔH_{calcd}: Enthalpy of ionization (kcal/mol).
$\Delta\Delta H$: Effect of substituents.
$\Delta\Delta H'$: Effect of fluorine substitution.

Introduction of fluorine substituents also enhances the acidity of a C–H bond. In general, hydrogen(s) on a carbon substituted by fluorine becomes acidic, but the effect is not as remarkable as for other halogens (Table 1.7) [11]. The mediocre results are explained by destabilization of a carbanion by fluorine owing to its $-I_\sigma$ effect.

As summarized in Table 1.8, the acidities of fluorine-substituted CH_3NC, CH_3CN, CH_3NO_2, and CH_3CHO, in addition to CH_4, depend on the number of fluorine substituents and the structure of the resulting anionic species [11]. Enthalpies of ionization calculated for each compound are listed in Table 1.8, which suggests that the acidities of fluoromethanes, fluoroisocyanomethanes and fluoroacetonitriles rise roughly in proportion to the number of fluorine atoms. In contrast, the number of fluorines does not explicitly reflect the acidity of fluoronitromethanes and fluoroacetonitriles. For example, difluoroacetaldehyde is less acidic than fluoroacetaldehyde.

Such a discrepancy is explained by the pyramidalization angle (θ) of a carbanion generated by proton abstraction. Fluorine prefers to be bonded to a carbanion carbon having an orbital with high p-character, and lone pairs prefer an orbital with high s-character. As Table 1.9 shows, substituted fluoromethanes

1.1 Nature of Organofluorine Compounds

Table 1.9. Pyramidalization angels of fluorine substituting anions

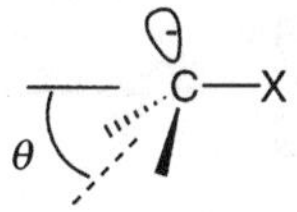

Anion	θ	Anion	θ
CH_2CHO^-	0.0	CH_2NC^-	33.6
$CFHCHO^-$	2.6	$CFHNC^-$	56.0
CF_2CHO^-	2.0	CF_2NC^-	67.7
$CH_2NO_2^-$	0.0	CH_2CN^-	52.7
$CFHNO_2^-$	38.1	$CFHCN^-$	66.9
$CF_2NO_2^-$	65.2	CF_2CN^-	72.4

that produce anions with large θ are more acidic. Since acetaldehyde gives an enolate in which the α-carbon has a high s-character, the anion form is not favored regarding difluoroacetaldehyde. Such an sp^2-enolate carbon also repulsively interacts with the unshared electron pair of fluorine to destabilize the enolate [12].

The unshared electron pairs in perfluoroalkylamines [13] and perfluoroalkyl ethers are not basic owing to the highly electron-withdrawing nature of the perfluoroalkyl group. Accordingly, these compounds find applications as fluorous media, perfluorocarbon solvents insoluble both in organic and aqueous phases. Details will be discussed in Chap. 7. Likewise trifluoromethylamines have oxidation potentials similar to ethers and thus tolerate the aerial conditions (Fig. 1.3).

Fluorine often exhibits distinctive effects on the stability of reactive species. In particular, the stabilization of an α-carbocation by the +R effect overrides the destabilization by the $-I_\sigma$ effect. Because the +R effect of fluorine is more pronounced than other halogens, the cation-stabilization effect is maximum with fluorine. The substituent effect of the cation-stabilization effect is as follows: $^+CH_3 < {}^+CF_3 < {}^+CH_2F < {}^+CHF_2$ and $^+CH_2CH_3 \ll {}^+CF_2CH_3 \approx {}^+CHFCH_3$ [14]. In contrast, fluorine destabilizes a β-carbocation owing to the $-I_\sigma$ effect and its incapability, unlike other halogens, to form a bridged ion (Fig. 1.4) [15]. Due to the destabilization effect by fluorine, or to the $-I_\sigma$ effect by a trifluoromethyl group, an S_N1 reaction at a trifluoromethyl-substituted carbon is not observed [16].

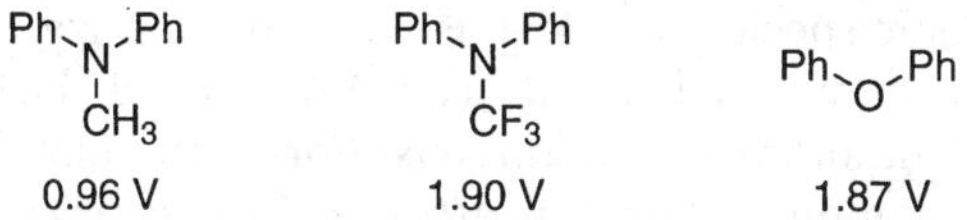

Ph–N(CH₃)–Ph   Ph–N(CF₃)–Ph   Ph–O–Ph
0.96 V   1.90 V   1.87 V

Fig. 1.3. Oxidation potential of trifluoromethylamines (vs. SCE)

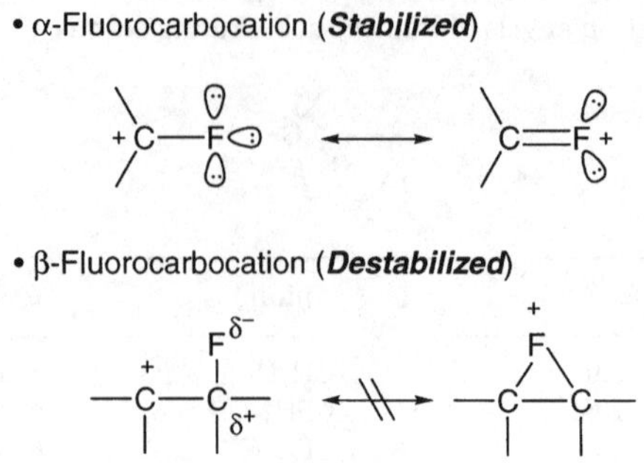

- α-Fluorocarbocation (**Stabilized**)

- β-Fluorocarbocation (**Destabilized**)

Fig. 1.4. Effects of fluorine on carbocation

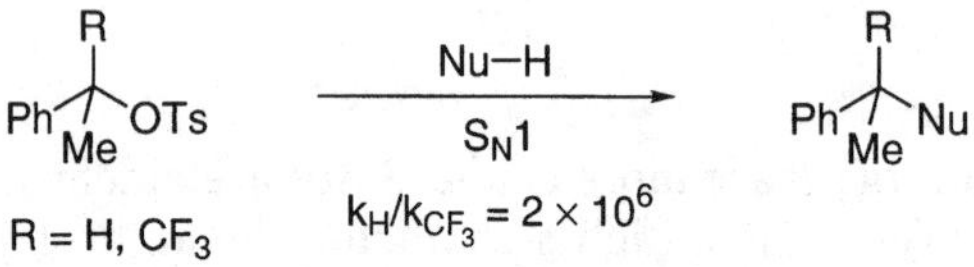

$$\text{Ph}\underset{\text{Me}}{\overset{R}{-}}\text{OTs} \xrightarrow[S_N1]{Nu-H} \text{Ph}\underset{\text{Me}}{\overset{R}{-}}\text{Nu} \quad (1.1)$$

R = H, CF$_3$

$k_H/k_{CF_3} = 2 \times 10^6$

In addition, an S_N2 reaction at a trifluoromethyl-substituted carbon is rare, also due to the destabilization by fluorine and repulsion between a nucleophile and unshared electron pairs of fluorine. With the aid of a heteroatom, a stereospecific reaction has recently been disclosed which proceeds with inversion of configuration [17].

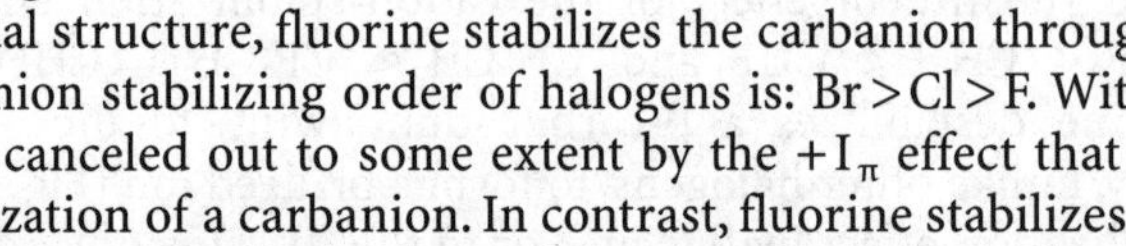

$$\text{Ph}\overset{\overset{CF_3}{|}}{\frown}\text{O}\frown\text{OTs} \xrightarrow[69\%]{LiAlEt_4} \text{Ph}\frown\text{O}\frown\overset{CF_3}{\frown} \quad (1.2)$$

100% ee   98% ee

As Tables 1.8 and 1.9 show, fluorine stabilizes or destabilizes an α-carbanion depending on the other substituents. When an anionic carbon can adopt a pyramidal structure, fluorine stabilizes the carbanion through a $-I_\sigma$ effect. The α-carbanion stabilizing order of halogens is: Br > Cl > F. With fluorine, the $-I_\sigma$ effect is canceled out to some extent by the $+I_\pi$ effect that contributes to the destabilization of a carbanion. In contrast, fluorine stabilizes a β-anionic center through a $-I_\sigma$ effect and negative hyperconjugation which is explained in terms of an orbital overlap of an anion orbital with a σ* orbital of a C-F bond (Fig. 1.5) [18]. Negative hyperconjugation appears to contribute to the stabilization of a trifluoromethoxy anion.

X-ray crystallography [19] and molecular orbital calculations [20] suggest such β-fluoroanionic species can exist as trifluoromethoxy and perfluoroalkyl anions with a soft counterion like TAS$^+$ (Fig. 1.6). Indeed, the facts that the C-O bond length of a trifluoromethoxyl anion is shorter and that the anionic carbon of a perfluoroalkyl anion is planar are attributed to the negative hyperconjugation.

1.1 Nature of Organofluorine Compounds

$$CF_3-\bar{X} \longleftrightarrow F^- \; CF_2=X$$

$$X = O, CF_2$$

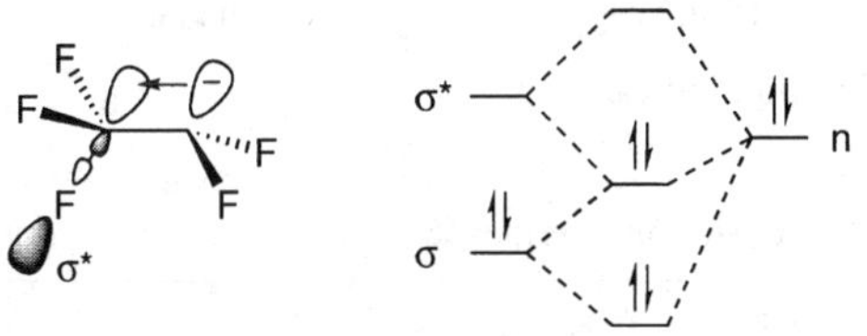

Fig. 1.5. Negative hyperconjugation of β-fluorocarbanions

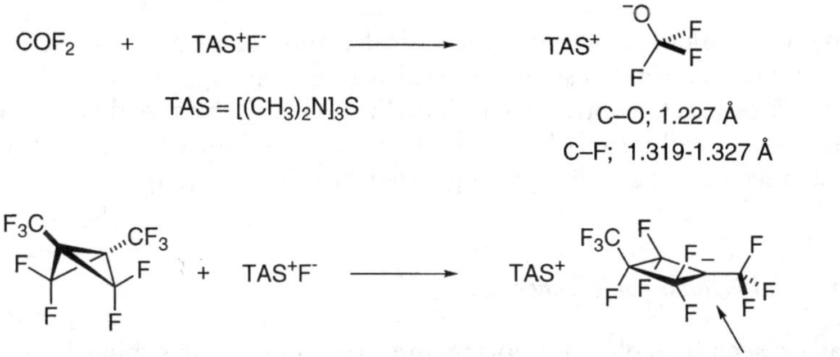

$$COF_2 + TAS^+F^- \longrightarrow TAS^+ \; \text{[}CF_3O^-\text{]}$$

$$TAS = [(CH_3)_2N]_3S$$

C–O; 1.227 Å
C–F; 1.319–1.327 Å

Fig. 1.6. Negative fluorine hyperconjugation of anions   planar carbanion

Calculations and experiments demonstrate that a fluorine at a β-carbon in a carbanion contributes to the stabilization more than one at an α-carbon. As Eqs. (1.3) and (1.4) indicate, fluorine at an α- or β-carbon brings stabilization of a carbanion as compared with a parent carbanion. In particular, three β-fluorines stabilize an anionic carbon by 13 kcal/mol more than two α-fluorines [12].

$$CF_3CH_2^- + CF_3CF_2H \longrightarrow CF_3CH_3 + CF_3CF_2^- \quad (1.3)$$

$$\Delta E = -21.9 \text{ kcal/mol}$$

$$CH_3CF_2^- + CF_3CF_2H \longrightarrow CH_3CF_2H + CF_3CF_2^- \quad (1.4)$$

$$\Delta E = -35.3 \text{ kcal/mol}$$

In sharp contrast to the planar configuration of a methyl radical, a trifluoromethyl radical [21] takes a pyramidal configuration (Fig. 1.7) owing to the repulsion between a half-filled radical orbital and unshared electron pairs on fluorine, as evidenced by calculation and experimental measurements [22].

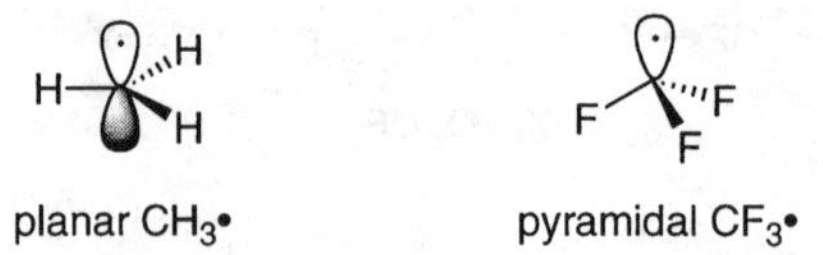

planar CH₃•   pyramidal CF₃•

Fig. 1.7. Structure of carbon radicals

Table 1.10. Calculated stabilization energies (SE) for substituted methyl radicals

Substituent	CH₃	(CH₃)₂	(CH₃)₃	F	F₂	F₃	FCH₂	F₂CH	CF₃
SE	3.3	5.8	8.0	1.6	0.6	–4.2	1.5	0.2	–1.3

One or two fluorine atoms on a radical carbon contribute to stabilization, whereas three fluorines destabilize a radical due probably to the inductive effect of three fluorines. In contrast to the stabilization effect by a methyl substituent, a mono-, di- or trifluoromethyl group does not contribute either to stabilization or destabilization of a radical, as suggested by Table 1.10 [23].

1.1.4
Properties of Fluoroorganic Compounds

As can be seen in Table 1.1, fluorine and oxygen are quite similar in their van der Waals radii and electronegativity. In addition, fluorine in fluoroalkanes and -alkenes has an electronic configuration of $2s^2 2p^6$, isoelectronic to oxygen. Thus, fluoroalkanes and -alkenes are called isosteres of alkanols and alkenols, as illustrated in Fig. 1.8. Indeed, such isosteric effect is successfully applied to the design of pharmaceuticals and agrochemicals. Very recently many potent protease inhibitors have appeared which contain such structural moieties [24].

The biological effects of organofluorine compounds are understood in terms of *block effect*, *mimic effect*, and *polar effect*. Details will be discussed in Chap. 5.

Recently, much attention has been focused on hydrogen bonding with fluorine as an acceptor base. Indeed, there is much evidence that fluorine forms a hydrogen bond (Fig. 1.9) [25]. In contrast to a hydroxyl group that works as a hydrogen donor and acceptor, fluorine behaves only as an acceptor.

Hydrogen bonding between fluorine and hydrogen is disclosed by molecular orbital calculations and X-ray crystallography. According to the Cambridge

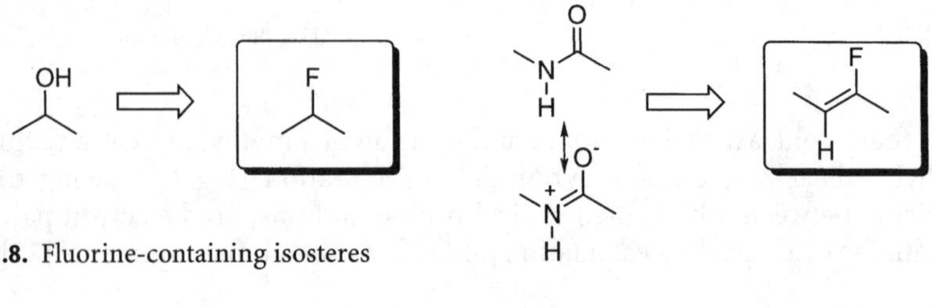

Fig. 1.8. Fluorine-containing isosteres

1.1 Nature of Organofluorine Compounds

Fig. 1.9. Hydrogen bonding of organofluorine compound

Structural Database Search (CSDS), there are 48 compounds that show a hydrogen bond with fluorine on an sp^3-carbon and 93 examples that exhibit a hydrogen bond with fluorine on an sp^2-carbon out of the registered 146,272 compounds [25b].

The hydrogen bond with fluorine is calculated to be 2 to 3 kcal/mol lower in energy than a hydrogen bond between H and O. Very recently, a stable H$\cdots$F bond of over 10 kcal/mol has been suggested by calculation [26] and demonstrated by X-ray crystallography [27].

The dissociation energies of ammonium salt clathrates with methane, fluoromethane, difluoromethane, and trifluoromethane were calculated to be 1.9, 13.5, 13.3, and 8.1 kcal/mol, respectively [26]. An example of $CH_3F\cdots H\text{-}NH_3^+$ is shown in Fig. 1.10.

The distance between F and N in an ammonium salt of a cryptand, shown in Fig. 1.11, is shortened by steric restraint, and coupling between fluorine and 15-methylene hydrogen has been observed by ^{19}F NMR [27].

The hydrogen bonding of fluorobenzene is a recent topic [28]. For example, 4,6-difluoro-m-xylene, an electronic and steric isostere of thymidine, is revealed to be able to form a hydrogen bond as strong as 60% of thymidine hydrogen bonding (Fig. 1.12) [29]. Accordingly, 4,6-difluoro-m-xylene is considered to be

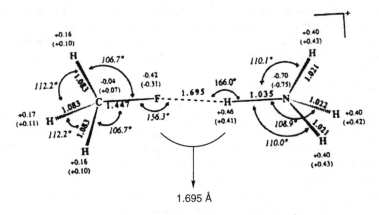

Fig. 1.10. Hydrogen bonding of CH_3F by ab initio calculations

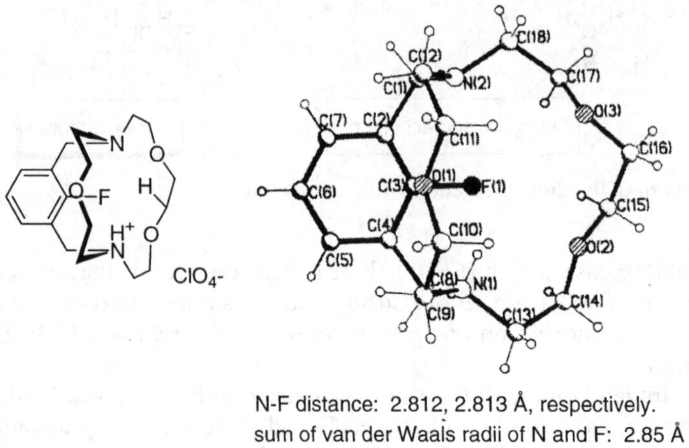

N-F distance: 2.812, 2.813 Å, respectively.
sum of van der Waals radii of N and F: 2.85 Å

Fig. 1.11. X-ray structure of fluorinated cryptand

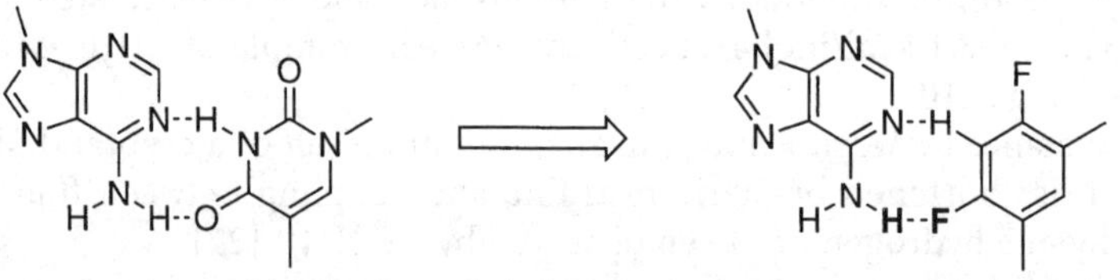

Fig. 1.12. Hydrogen bonding of fluorothymidine

a nonpolar nucleic acid base analog [30], although its hydrogen bond energy is much weaker. Such hydrogen bonding affects various biological activities. An example is the sympathomimetic neurotransmitter norepinephrin which, upon introduction of fluorine to its phenolic part, changes its activity drastically owing to the hydrogen bonding between fluorine and a phenolic hydroxyl group [31].

Fluorine interacts remarkably with an unshared elecon pair repulsively in a manner similar to the anomeric effect in sugars. For example, an *anti* conformation of fluoromethyl(dimethyl)amine has been demonstrated [32]. The orbital of the unshared electron pair (lone pair) is suggested to overlap with a σ*-orbital of a C-F bond to fix the antiperiplanar conformation. This is presented as lp(N)→σ*(C-F), being caused by a high level of HOMO of the unshared electrons and low-lying LUMO of a C-F bond. Gas electron diffraction has revealed the distance between F and CH_3 to be 2.9 Å (Fig. 1.13), a distance that corresponds well to the *anti*-conformer.

Ab initio calculations [HF/3-21G$^{(*)}$] reveal (Fig. 1.14) that the antiperiplanar conformer is ca. 4.5 kcal/mol more stable than a synclinal conformer. This means that over 95% of the molecule adopts the antiperiplanar conformation.

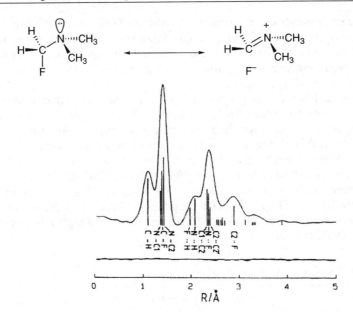

Fig. 1.13. Gas electron diffraction of FCH$_2$N(CH$_3$)$_2$

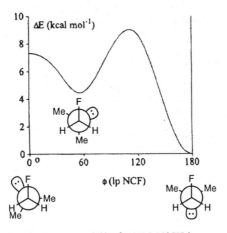

Fig. 1.14. Ab initio calculations (HF3-21G(*)) of FCHG$_2$N(CH$_3$)$_2$

1.1.5
Properties of Perfluoroorganic Compounds

When all of the hydrogen atoms in hydrocarbons, amines and ethers are replaced by fluorine, the resulting perfluorinated compounds are called perfluorocarbons or simply PFCs. They are stable against most chemicals, biologically inert, insoluble in both water and organic solvents, but dissolve some gaseous compounds like molecular oxygen [33]. These salient features have led to many applications of PFCs, including novel heat- and weather-resistant materials, artificial blood, new media in organic synthesis, and many key substances in modern science and technology. Details of such applications will be discussed in Chaps. 6 and 7.

Although PFCs are in general nonpolar, partially fluorinated compounds induce polarity due to the dipole moment of the C–F bond. Examples of linear C_6 compounds along with their dielectric constants (ε) are summarized in Table 1.11. Although the dielectric constant of perfluorohexane is lower than that of hexane, fluorohexane and heptafluorohexane induce a value of ε three times or more that of hexane. The dipole moment of each C–F bond is canceled in perfluorohexane resulting in a low ε. The same trend is also seen in the solvent polarity (P_s). The P_s's of PFCs are generally lower than the corresponding hydrocarbons; partially fluorinated hydrocarbons have high P_s's, as can easily be seen in Table 1.12.

Table 1.11. Dielectric constants of fluorinated hexanes

Hesanes	Dielectric constant (ε)
C_6F_{14}	1.69
$F(CF_2)_3(CH_2)_3H$	5.99
$C_6F_{13}F$	5.63
C_6H_{14}	1.89

Table 1.12. Solvent polarity (P_s) index

Solvent	Solvent polarity (P_s)
C_6F_{14}	0.00
C_6H_{14}	2.56
$C_4F_9CH_2CH_3$	4.01
C_8F_{18}	0.55
C_8H_{18}	2.86
$C_8F_{17}H$	4.33
C_6F_6	4.53
C_6H_6	6.95
C_6H_5F	7.52
$1,2-C_6H_4F_2$	7.86

1.1.6
Spectroscopic Properties

Fluorine in Nuclear Magnetic Resonance Spectroscopy. Since the spin quantum number of ^{19}F is 1/2, fluorine is detectable by nuclear magnetic resonance (NMR) spectroscopy. In addition, since the natural abundance of ^{19}F is 100% and fluorine has 83% of the sensitivity of hydrogen, ^{19}F NMR can be

1.1 Nature of Organofluorine Compounds

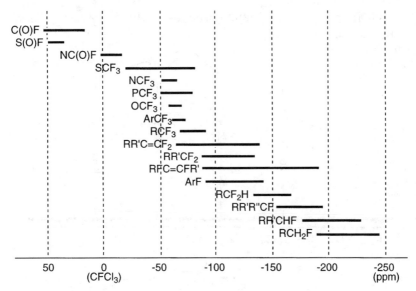

Fig. 1.15. ^{19}F-NMR chemical shifts of fluorine substituents

performed in a way equal to ^{1}H NMR. Chemical shifts of organofluorine compounds using CFCl$_3$ as a standard range from 50 to −250 ppm. A maximum range is as wide as 900 ppm, much wider than ^{1}H NMR, which ranges 10 to 20 ppm at best. ^{19}F NMR spectra are very much more sensitive to the structural and environmental changes of molecules. Accordingly, ^{19}F NMR is very often employed for the structural determination of compounds. To date, huge amounts of data are available [34]. Typical fluorine functional groups and their chemical shift ranges are illustrated in Fig. 1.15.

Fluorine, like hydrogen, gives characteristic coupling constants depending on the spacial displacement and the number of bonds between a coupling partner atom. In particular a long-range coupling of ^5J is observed in an olefinic system, as shown in Fig. 1.16.

A typical coupling of organofluorine compounds is observed in a geminal coupling ($^2J_{\text{H-F}}$) with a geminal hydrogen, being as large as 50 Hz. This coupling can also be observed by ^{1}H NMR. Coupling between *gem*-fluorines ($^2J_{\text{F-F}}$) also gives a large value of 250 to 300 Hz. Coupling between fluorine and carbon is also unique in ^{13}C NMR: $^1J_{\text{C-F}}$ ranges from 250 to 300 Hz. A typical example of a trifluoromethyl ether is shown in Fig. 1.17 [35]. A trifluoromethyl carbon splits into a quartet. To obtain such well-resolved spectra, high concentration of the

Fig. 1.16. Long-range F–F coupling

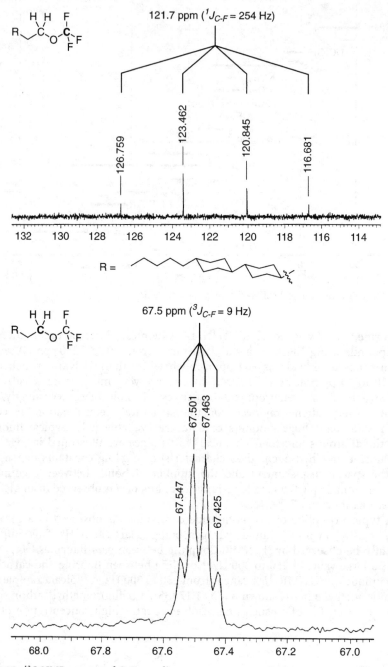

Fig. 1.17. ^{13}C-NMR spectra of C–F coupling

1.1 Nature of Organofluorine Compounds

- **Fluoroalkane**

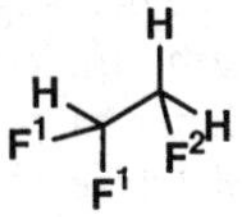

  F^1: -129.9 ppm ($^2J_{H\text{-}F}$ = 55 Hz, $^3J_{F\text{-}F}$ = 18 Hz, $^3J_{H\text{-}F}$ = 13 Hz)

  F^2: -288.9 ppm ($^2J_{H\text{-}F}$ = 46 Hz, $^3J_{F\text{-}F}$ = 18 Hz, $^3J_{H\text{-}F}$ = 6 Hz)

- **Fluoroalkene**

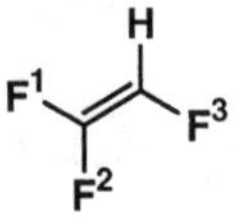

  F^1: -125.7 ppm ($^2J_{F\text{-}F}$ = 87 Hz, $^3J_{F\text{-}F\,(trans\text{-})}$ = 119 Hz, $^3J_{H\text{-}F\,(cis\text{-})}$ = 4 Hz)

  F^2: -99.7 ppm ($^2J_{F\text{-}F}$ = 87 Hz, $^3J_{F\text{-}F\,(cis\text{-})}$ = 33 Hz, $^3J_{H\text{-}F\,(trans\text{-})}$ = 13 Hz)

  F^3: -205.0 ppm ($^2J_{H\text{-}F}$ = 71 Hz, $^3J_{F\text{-}F\,(cis\text{-})}$ = 33 Hz, $^3J_{F\text{-}F\,(trans\text{-})}$ = 119 Hz)

Fig. 1.18. ^{19}F-NMR data of fluoroalkanes and -alkenes

sample and long-term accumulation are necessary. It is worth noting that $^3J_{C\text{-}F}$ can be as large as 9 Hz. Generally, coupling constants of $^1J_{C\text{-}F}$, $^2J_{C\text{-}F}$, $^3J_{C\text{-}F}$, and $^4J_{C\text{-}F}$ are, respectively, 160–370, 30–45, 5–25, and 1–5 Hz. The coupling constant $^3J_{F\text{-}F}$ of a *trans-vic*-difluoroolefin is larger than that of a *cis*-olefin like $^3J_{H\text{-}H}$. This fact is a reliable criterion for the determination of fluoroolefin configurations. Typical coupling constants of $^nJ_{H\text{-}F}$ and $^nJ_{F\text{-}F}$ of fluoroalkanes and -alkenes are summarized in Fig. 1.18.

Fluorine Compounds in High Resolution Mass Spectrometry. High resolution mass spectrometry is extremely effective for the determination of the elemental composition of organic compounds and their fragments. PFCs are used as standards for the calibration of the mass number particularly in the case of EI (electron impact) mass spectrometry. PFCs are stable compounds containing many fluorines within a molecule; the atomic weight of ^{19}F is slightly

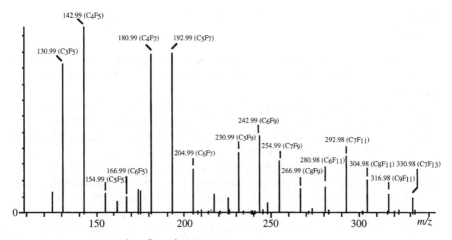

Fig. 1.19. Mass spectra of perfluorokerosene

lower than the round number 19 in sharp contrast to common elements: ^{1}H, 1.00894; ^{12}C, 12.0107; ^{14}N, 14.00674. Thus, PFCs give parent peaks slightly smaller than the corresponding hydrocarbons which have the same molecular weight. For example, if you consider a peak of m/z 338, a calculated parent peak of perfluorohexane (C_6F_{14}) is 337.9776, whereas the one of tetraeicosane ($C_{24}H_{30}$) is 338.3912. Thus the two peaks are easily resolved. In addition, use of a PFC allows estimation of an exact mass of an unknown compound by interpolation. A standard compound for EI mass spectrometry is perfluorokerosene [PFK, $CF_3(CF_2)_nCF_3$], whose mass spectrum is shown in Fig. 1.19.

1.2
Source of Fluorine

As described above, the main source of fluorine is fluorite. Fluorite reserves are estimated to be consumed at present: 60% for steel refinement, 20% for aluminum refinement, and 20% for HF production. This section briefly describes the manufacture of the most basic fluorinating reagents: hydrogen fluoride and fluorine gas.

1.2.1
Hydrogen Fluoride

Treatment of high grade (purity > 97%) fluorite with sulfuric acid gives hydrofluoric acid, an aqueous solution of hydrogen fluoride. Distillation of hydrofluoric acid affords anhydrous hydrogen fluoride (HF) which is then stored in a copper container. HF reacts with many metals to give metal fluorides and hydrogen gas. The reaction of HF with metal cyanides and sulfides gives hydrogen cyanide and hydrogen sulfide, respectively. HF forms intermolecular hydrogen bonds firmly and is thus relatively unreactive with organic compounds. To fluorinate various substrates easily, many reagents are prepared using HF or hydrofluoric acid. Details will be discussed in Chap. 2.

1.2.2
Fluorine Gas

Molecular fluorine is often called fluorine gas, an extremely pale yellow gas at ambient temperatures with a stinging smell. It boils at $-188\,°C$ and melts at $-218.6\,°C$.

The dissociation energy of an F–F bond is 158 kJ/mol, the smallest among all diatomic molecules. In sharp contrast to the very strong bond of fluorine with other elements, an F–F bond is easily cleaved. Thus fluorine gas readily reacts with other simple substances, even with a rare gas like xenon and krypton at high temperatures. However, it does not react with molecular nitrogen. It is easy to assume that fluorine gas spontaneously reacts with organic substances delivering an electrophilic fluorine atom. However, the reaction often takes place violently or explosively. Thus it is advised that reactions be carried out with great care under the supervision of an expert. This has hampered the use of fluorine gas in conventional laboratories for organic synthesis.

Fluorine gas is generated by anodic oxidation of a melt mixture of KF and HF, as discovered by Moissan. Nowadays fluorine gas is manufactured by electrochemical oxidation of KF·2HF (mp 70°C) at 90°C.

For reactions with fluorine gas, a metallic vessel, made of copper, bronze, or steel, is used. Since fluorine gas reacts with any trace of organic compounds with explosion or ignition, all vessels should be carefully cleaned. To effect handling and reaction with fluorine gas, a mixture of fluorine gas and molecular nitrogen is conveniently used in polyethylene or polyvinyl chloride containers according to a strict protocol.

Fluorine gas is converted into a variety of convenient electrophilic fluorination reagents. The preparations and reactions of these reagents will be discussed in Chap. 2.

1.3
Toxicity of Fluorinating Reagents

Before closing this chapter, the toxicity of HF and some organofluorine compounds, followed by first-aid treatment in case of exposure to the reagent, will be discussed briefly [36].

1.3.1
Hydrogen Fluoride and Fluorine Gas

The toxicity (50% lethal concentration, LC_{50}) of HF and fluorine gas is compared with other toxic gases in Table 1.13. As can easily be seen, HF and fluorine gas are extremely toxic upon inhalation. In addition, these agents cause various hazards upon exposure. HF, particularly in the moist form, is corrosive, causes burns to skin, penetrates relatively fast to necrotize deep tissues, and finally combines with a calcium ion to form calcium fluoride, which stays long in the body to interfere with the normal osteoplastic mechanism. Therefore, extreme care should always be paid when handling these fluorinating chemicals. Fluorine gas reacts spontaneously with atmospheric moisture to give HF and oxygen or ozone. In addition, fluorine gas reacts violently with organic substances. Accordingly, special vessels and protectors should be applied and the same special care taken as with HF.

Table 1.13. Lethal concentration (LC_{50}) of various gases against rats

Gas	LC_{50} (ppm)	Time for Effect
HF	1278	60 min
F_2	185	60 min
H_2S	712	60 min
HCN	544	5 min
HCl	5666	30 min
CO	1807	4 h

On dermal exposure to the human body, hydrofluoric acid, an aqueous solution of HF, causes pain and numbness and produces blisters depending on the concentration of HF and the time of the exposure.

(1) Contact with hydrofluoric acid of less than 20% concentration induces pain after several hours of an unconscious period, and the inflammation can heal in several days.
(2) Contact with hydrofluoric acid of 20–50% concentration immediately causes the affected skin to turn red with pain and then gradually to light grey, forming blisters. As HF readily penetrates skin, hypodermic tissues may be necrotized. However, prompt and appropriate first-aid treatment allows recovery without any aftereffects.
(3) Contact with hydrofluoric acid of over 50% or to anhydrous HF results in severe pain. HF penetrates to deep tissues; the affected skin turns grey purple, forming a dropsical swelling and then festers. The affected tissue will be recovered by granulation.

The cornea and the conjunctivae may be heavily damaged by contact with HF vapor. Highly concentrated hydrofluoric acid and anhydrous HF give serious and spontaneous damage to eyes, clouding the cornea and causing desquamation. It takes several days for recovery. Sometimes surgery on the eye is necessary.

Inhalation of HF vapor damages lungs often inducing an edema. Penetration of HF to lungs may continue over three weeks. Upon heavy inhalation, the liver and kidneys may also be damaged. In case of light inhalation, the damage will be limited to the bronchial passage.

According to reports so far, HF is not considered to be carcinogenic, as no carcinogenesis was observed in guinea pigs exposed to 18 ppm of HF for 18 months [37]. In contrast, mutation upon exposure to HF was observed with rats bred under an atmosphere containing 0.12 ppm of HF for a long period. However, this is not the case with mice [38]. There is a report on mutation in *Drosophila melanogaster* of Drosophilae when exposed to an atmosphere containing 2.9–4.3 ppm HF for 24 h [39].

1.3.2
First-Aid Treatment

Since HF penetrates skin rapidly, the person exposed to HF should move or be removed to a safe place, and the affected part of the body should be washed with copious volumes of running water for at least 15 min. If HF has come into contact with clothes and/or footwear, these should also be removed carefully and completely. Then the patient should consult a doctor for medical care. Regardless of the degree of the damage, first-aid treatment in laboratories is extremely important. Some treatments are described in the following.

(1) First-aid treatment for skin contact [36, 40]: The affected part is washed first with a large volume of water for at least 15 min, then dipped in 70% ethanol for at least 30 min. Alternatively, a poultice soaked in 70% ethanol is applied to the affected part, being exchanged at 2–3 min intervals, for a period of at least

30 min. Thus, HF may be extracted with ethanol. A pad containing benzethonium chloride solution (prepared by dissolving 10 ml of methyl salicylate, 56 g of phenol, and 20 ml of 20% aqueous benzethonium chloride solution in 30 l of ethanol) is also effective. The alcoholic treatment may be replaced by application of 2.5% calcium gluconate gel. The affected part should finally be examined by a medical doctor. In the case of serious inflammation, hypodermic injection of an aqueous calcium gluconate solution is effective.

(2) First-aid treatment for eye damage: The affected eye is first washed with a large volume of water and cooled with ice before examination by a medical doctor. An aqueous solution (1%) of calcium gluconate is used to wash the eye for a period of 5–10 min and then applied to the eye dropwise for a period of 2–3 h.

(3) First-aid treatment for inhalation of HF: The patient is immediately placed in a safe place and given oxygen inhalation. An aqueous solution (2.5–3%) of calcium gluconate is then administered with a nebulizer by a medical doctor.

1.3.3
Fluoroacetic Acid

Fluoroacetic acid is extremely toxic and induces systemic convulsions and cardiac lesion with $LDL_0 = 2-5$ mg/kg [41]. The mechanism of the toxicity is attributed to an inhibition of the TCA cycle, as shown in Scheme 1.1. Fluoroacetic

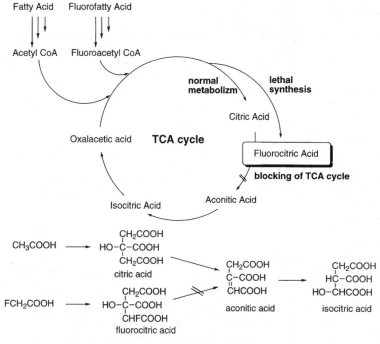

Scheme 1.1. TCA cycle of fatty acids and fluoro fatty acids

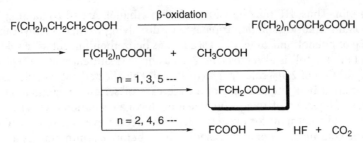

Scheme 1.2. Metabolic pathway of ω-fluoro fatty acids

acid is converted, in a way similar to acetic acid, into acetyl CoA (mimic effect), which is then transformed to fluorocitric acid. The resulting fluorocitric acid is not converted into aconitic acid due to a strong C–F bond (block effect). Since the enzyme performing the biological synthesis produces a suicide inhibitor, this is called lethal synthesis. Of the diastereomers of fluorocitric acid, only the (2R, 3R)-isomer is revealed to be toxic [42]. All organofluorine compounds, after a metabolic degradation such as the β-oxidation, give fluoroacetic acid (Scheme 1.2) [43].

References

1. A. L. Henne, *Org. React.*, **1944**, *2*, 49
2. A. Roe, *Org. React.*, **1949**, *5*, 193
3. A. Bondi, *J. Phys. Chem.*, **1964**, *68*, 441
4. (a) T. Nagai, G. Nishioka, M. Koyama, A. Ando, T. Miki, I. Kumadaki, *Chem. Pham. Bull.*, **1991**, *39*, 233; (b) I Kumadaki, *Fusso Yakugaku*, (Y. Kobayashi, I. Kumadaki, T. Taguchi, eds.), p 5, Hirokawa, Tokyo (1993); (c) K. W. Tafts Jr, *Steric Effects in Organic Chemistry*, (M. S. Newman, ed), p 556, John Wiley & Sons, New York (1956)
5. B. E. Smart, *Chemistry of Organic Fluorine Compounds II, A Critical Review*, (M. Hudlicky, A. E. Pavlath, eds.), ACS Monograph 187, p 979, American Chemical Society, Washington, DC (1995)
6. C. Hansch, A. Leo, R. W. Taft, *Chem. Rev.* **1991**, *91*, 165
7. T. Kitazume, T. Ishihara, T. Taguchi, *Fusso no Kagaku*, Chap 5, Kodansya, Tokyo (1993); (b) T. Kitazume, T. Yamazaki, *Experimental Methods in Organic Fluorine Chemistry*, Kodansha-Gordon & Breach Science, Tokyo (1998)
8. N. Ishikawa, *Fusso kagobutu no gosei to kinou*, (N. Ishikawa, ed.), Chap 4, CMC, Tokyo (1987)
9. (a) G. A. Olah, G. K. S. Prakash, J. Sommer, *Superacids*, John Wiley, New York (1985); (b) G. A. Olah, L. Heiliger, G. K. S. Prakash, *J. Am. Chem. Soc.* **1989**, *111*, 8020
10. R. J. Gillespie, J. Liang, *J. Am. Chem. Soc.* **1988**, *110*, 6053
11. A. Streitwieser, Jr., D. Holtz, G. R. Ziegler, J. O. Stoffer, M. L. Brokaw, F. Guibe, *J. Am. Chem. Soc.* **1976**, *98*, 5229
12. H. J. Castejon, K. B. Wiberg, *J. Org. Chem.* **1998**, *63*, 3937
13. K. Kanie, K. Mizuno, M. Kuroboshi, T. Hiyama, *Bull. Chem. Soc. Jpn.* **1998**, *71*, 1973
14. (a) R. J. Blint, T. B. McMahon, J. L. Beauchamp, *J. Am. Chem. Soc.* **1974**, *96*, 1269; (b) A. D. Williamson, P. R. LeBreton, J. L. Beauchamp, *J. Am. Chem. Soc.* **1976**, *98*, 2705
15. G. A. Olah, G. K. S. Prakash, V. V. Krishnamurthy, *J. Org. Chem.* **1983**, *48*, 5116
16. A. D. Allen, F. Shahidi, T. T. Tidwell, *J. Am. Chem. Soc.* **1982**, *104*, 2516
17. H. Matsutani, H. Poras, T. Kusumoto, T. Hiyama, *Chem. Commun.*, **1998**, 1259
18. D. A. Dixon, T. Fukunaga, B. E. Smart, *J. Am. Chem. Soc.* **1986**, *108*, 4027

19. (a) W. B. Farnham, B. E. Smart, W. J. Middleton, J. C. Calabrese, D. A. Dixon, *J. Am. Chem. Soc.* **1985**, *107*, 4565; (b) W. B. Farnham, D. A. Dixon, J. C. Calabrese, *J. Am. Chem. Soc.* **1988**, *110*, 2607; (c) W. B. Farnham, W. J. Middleton, W. C. Fultz, B. E. Smart, *J. Am. Chem. Soc.* **1986**, *108*, 3125
20. G. Raabe, H.-J. Gais, J. Fleischhauer, *J. Am. Chem. Soc.* **1996**, *118*, 4622
21. (a) W. R. Dolbier, Jr., *Chem. Rev.* **1996**, *96*, 1557; (b) P. J. Krusic, R. C. Bingham, *J. Am. Chem. Soc.* **1976**, *98*, 230; (c) D. V. Deardon, J. W. Hudgens, R. D. Johnson III, B. P. Tsai, S. A. Kafafi, *J. Phys. Chem.* **1992**, *96*, 585; (d) C. Yamada, E. Hirota, *J. Chem. Phys.* **1983**, *78*, 1703; (e) K. Morokuma, L. Pendersen, M. Karplus, *J. Chem. Phys.* **1968**, *48*, 4801
23. (a) D. J. Pasto, R. Krasnansky, C. Zercher, *J. Org. Chem.* **1987**, *52*, 3062; (b) D. J. Pasto, *J. Am. Chem. Soc.* **1988**, *110*, 8164
24. (a) J. T. Welch, *Tetrahedron*, **1987**, *43*, 3123; (b) T. Allmendinger, P. Furet, E. Hungerbuhler, *Tetrahedron Lett.*, **1990**, *31*, 7297; (c) D. O'Hagan, H. S. Rzepa, *Chem. Commun.*, **1997**, 645; (d) R. Filler, Y. Kobayashi, *Biomedical Aspects of Fluorine Chemistry*, Kodansha Ltd. and Elsevier Biochemical, Tokyo and Amsterdam (1982)
25. (a) H. Plenio, *Chem. Rev.*, **1997**, *97*, 3363; (b) J. A. K. Howard, V. J. Hoy, D. O'Hagan, G. T. Smith, *Tetrahedron*, **1996**, *52*, 12613
26. L. Shimoni, J. P. Glusker, C. W. Bock, *J. Phys. Chem.*, **1995**, *99*, 1194
27. H. Plenio, R. Diodone, *Chem. Ber.*, **1997**, *130*, 633
28. V. R. Thalladi, H.-C. Weiss, D. Bläser, R. Boese, A. Nangia, G. R. Desiraju, *J. Am. Chem. Soc.*, **1998**, *120*, 8702
29. T. A. Evans, K. R. Seddon, *Chem. Commun.*, **1997**, 2023
30. (a) B. A. Schweitzer, E. T. Kool, *J. Am. Chem. Soc.*, **1995**, *117*, 1863; (b) X.-F. Ren, B. A. Schweitzer, C. J. Sheils, E. T. Kool, *Angew. Chem. Int. Ed. Engl.*, **1996**, *35*, 743; (c) S. Moran, R. X.-F. Ren, S. Rumney IV, E. T. Kool, *J. Am. Chem. Soc.*, **1997**, *119*, 2056; (d) K. M. Guckian, E. T. Kool, *Angew. Chem. Int. Ed. Engl.*, **1997**, *36*, 2825
31. D. Cantacuzene, K. L. Kirk, D. H. McCulloh, C. R. Creveling, *Science*, **1979**, *204*, 1217
32. D. Christen, H.-G. Mack, S. Rüdiger, H. Oberhammer, *J. Am. Chem. Soc.*, **1996**, *118*, 3720
33. B. E. Smart, *Chemistry of Organic Fluorine Compounds II, A Critical Review* (M. Hudlicky, A. E. Pavlath, eds), ACS Monograph, p 980, American Chemical Society, Washington, DC (1995)
34. (a) T. S. Everett, *Chemistry of Organic Fluorine Compounds II, A Critical Review* (M. Hudlicky, A. E. Pavlath, eds), ACS Monograph, p 1037, American Chemical Society, Washington, DC (1995); (b) J. W. Emsley, L. Phillips, V. Wray, *Prog. NMR Spectr.*, **1971**, *7*, 1; (c) J. W. Emsley, L. Phillips, V. Wray, *Prog. NMR Spectr.*, **1976**, *10*, 83; (d) V. Wray, *Ann. Rep. NMR Spectr.*, **1980**, *10B*, 1; (e) V. Wray, *Ann. Rep. NMR Spectr.*, **1983**, *14*, 1; (f) C. A. Jameson, *Multinuclear NMR* (J. Mason, ed.), Plenum Press, New York (1987); (g) *Proton and Fluorine Nuclear Magnetic Resonance Spectral Data*, Varian Instruments/Japan, Halon, Tokyo (1988)
35. K. Kanie, T. Hiyama, unpublished results
36. (a) A. J. Finkel, *Adv. Fluorine Chem.*, **1973**, *7*, 199; (b) J. M. Wetherhold, F. P. Shepherd, *J. Occ. Med.*, **1965**, *7*, 193; (c) P. B. Teppermann, *J. Occ. Med.*, **1980**, *22*, 691; (d) T. D. Browne, *J. Soc. Occ. Med.*, **1974**, *24*, 80; (e) R. E. Gossetin, H. C. Hodge, R. P. Smith, M. N. Gleason, *Clinical Toxicology of Commercial Products: Acute Poisoning*, 4th Ed., p 159, Williams and Wilkins, Baltimore (1976)
37. P. Bourbon, *Pollut. Atmos.*, **1979**, *83*, 239
38. S. I. Voroshilin, *Tsitol. Genet.*, **1975**, *9*, 42
39. (a) A. H. Mohamaid, P. A. Kemmer, *Fluoride*, **1970**, *3*, 192; (b) R. A. Gerdes, *Atmos. Environ.*, **1971**, *5*, 117; (c) R. A. Gerdes, *Fluoride*, **1971**, *4*, 25
40. M. A. Trevino, G. H. Herrman, W. L. Sprout, *J. Occ. Med.*, **1983**, *25*, 861
41. P. Goldman, *Science*, **1969**, *164*, 1123
42. (a) R. Peters, Ciba Foundation Symposium: Carbon Fluorine Compounds, p 55, Elsevier Excerpta Medica, Amsterdam (1972); (b) R. J. Card, W. D. Hitz, *J. Am. Chem. Soc.*, **1984**, *106*, 5348
43. E. Kun, R. J. Dummel, *Methods in Enzymology*, Vol. XIII, p 623, Academic Press, New York (1969)

CHAPTER 2

Reagents for Fluorination

Since they are extremely rare in nature, organofluorine compounds are only accessible artificially by fluorination of organic compounds. The most basic fluorinating agents are fluorine gas (F_2) and hydrogen fluoride (HF). Starting from these, various kinds of electrophilic and nucleophilic fluorinating reagents have been invented, respectively, and their reactivities have been studied in detail. Although some of the reactions may be carried out in standard laboratories using common glassware, handling any fluorinating reagents still needs special care, because corrosive HF may be released. The toxicity of many organofluorine compounds is not yet fully understood, although most of them are nontoxic. Therefore, great care should be paid when dealing with any fluorinating reagents. Here, electrophlilic reagents, and subsequently nucleophilic ones, are described, followed by electrochemical fluorination.

2.1
Electrophilic Fluorinating Reagents

2.1.1
Fluorine Gas

A basic electrophilic fluorinating reagent is fluorine gas (molecular fluorine, F_2) that delivers a fluorine atom in an electrophilic manner. Therefore, fluorine gas is often called elemental fluorine. Fluorine gas is prepared by electrolysis of a melt mixture of HF and KF and is stored in a stainless steel cylinder. A dilute mixture with an inert gas like helium or molecular nitrogen is now commercially available.

As discussed in Chap. 1, fluorine gas is extremely toxic and corrosive and reacts explosively with any trace of organic compounds. The extremely high reactivity of fluorine gas is attributed to its high enthalpy of formation and oxidation potential. The oxidation potential of molecular fluorine is 2.87 V, much higher than chlorine (1.358 V), bromine (1.065 V), or iodine (0.535 V). In addition, the enthalpy of formation of a C–F bond from a C–H bond is 105 kcal/mol, much higher than that of a C–Cl bond (25 kcal/mol), a C–Br bond (9 kcal/mol), or a C–I bond (–6 kcal/mol). Accordingly, fluorine gas is far too reactive to selectively fluorinate highly functionalized substrates [1].

Alkanes are fluorinated by fluorine gas to give perfluorinated alkanes, but the yields are not necessarily high [2–4]. The industrial process for the production of perfluorinated organic compounds is called the La Mar process and is carried out by careful control of reaction temperatures and concentration of fluorine gas.

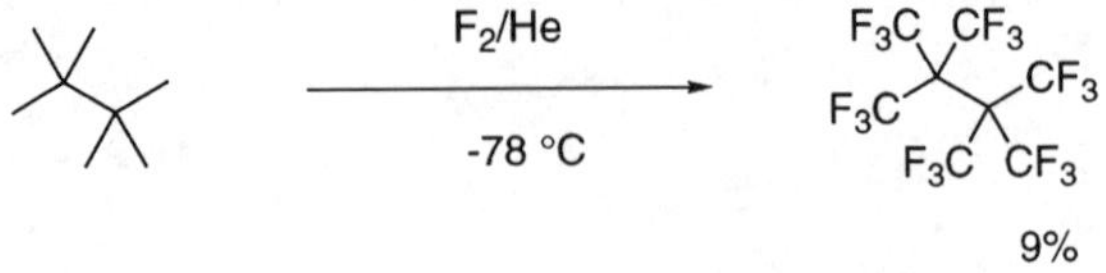

$$\text{(CH}_3\text{)}_3\text{C-C(CH}_3\text{)}_3 \xrightarrow[-78\,°C]{F_2/He} \text{(CF}_3\text{)}_3\text{C-C(CF}_3\text{)}_3 \quad 9\% \tag{2.1}$$

Because a tertiary C–H bond is a σ-bond of the highest electron density and thus is first attacked by an electrophilic reagent, selective fluorination at a tertiary carbon [5–11] is achieved using a chloroform or nitrobenzene solvent to suppress polyfluorination. Tertiary carbons of adamantane and steroid derivatives are successfully fluorinated by this method. Some examples are shown in Table 2.1.

Carbonyl compounds that easily produce enol forms are smoothly fluorinated by fluorine gas. For example, ethyl phenylpyruvate is fluorinated at an α-carbon of the keto group [12]. A ketone substrate whose enol form is extremely minor may be first converted into its enol silyl ether, which easily reacts

Table 2.1. Fluorination of tertiary carbons

Substrate	Products	Yield
 adamantane	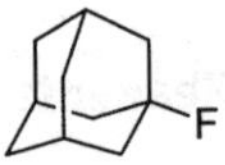 1-fluoroadamantane	71-90%
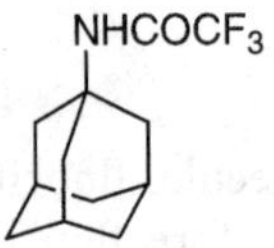 N-trifluoroacetyl aminoadamantane	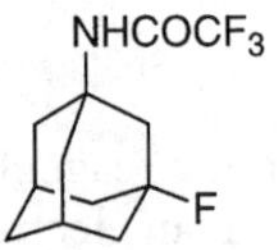 fluorinated N-trifluoroacetyl aminoadamantane	83%
 isopropylcyclohexane	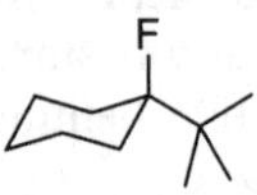 1-fluoro-1-isopropylcyclohexane	70%
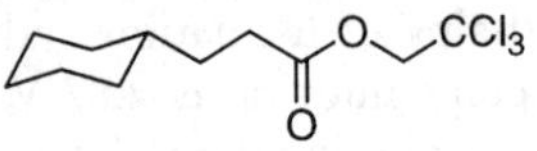 trichloroethyl cyclohexylbutanoate	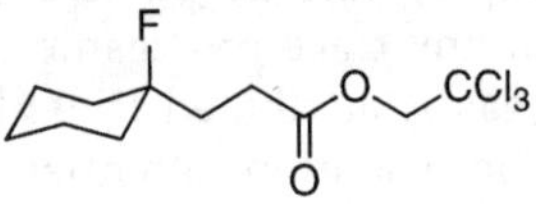 α-fluoro trichloroethyl cyclohexylbutanoate	60%
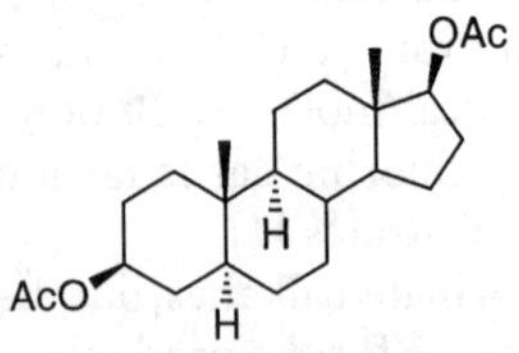 3,17-diacetoxy steroid	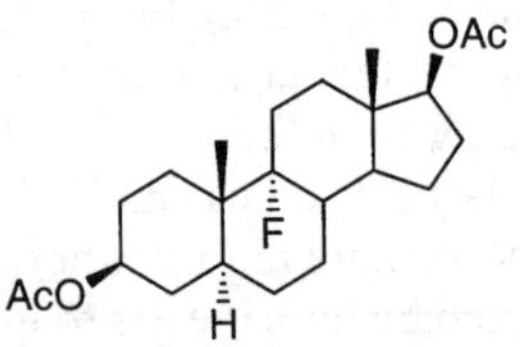 5-fluoro-3,17-diacetoxy steroid	50%

2.1 Electrophilic Fluorinating Reagents

with fluorine gas to give an α-fluoro ketone [13]. A six-membered transition state is proposed for the high yielding fluorination reaction.

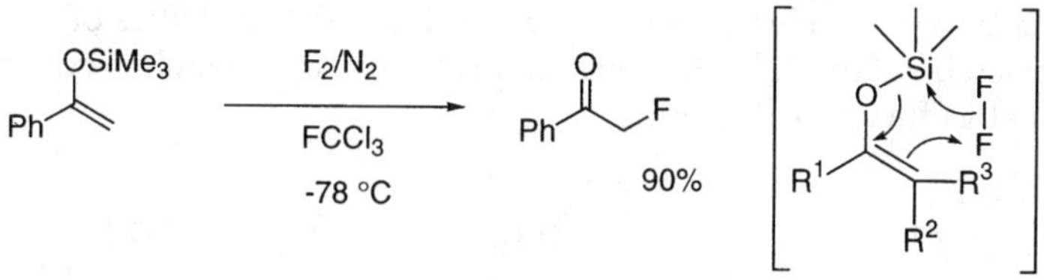

$$\text{Ph-C(OSiMe}_3\text{)=CH}_2 \xrightarrow[\text{-78 °C}]{\text{F}_2/\text{N}_2, \text{FCCl}_3} \text{PhCOCH}_2\text{F} \quad 90\% \qquad (2.2)$$

Fluorination of olefins with fluorine gas was studied ca. 30 years ago. Typical examples are listed in Table 2.2 [14–18]. Although bromination and chlorination of olefins proceed through a halonium ion intermediate to give *trans*-adducts, fluorination with fluorine gas proceeds preferentially in a *cis* fashion. Initially, a radical mechanism was proposed. However, a tight ion pair intermediate is now considered to be involved, as suggested in Scheme 2.1.

Table 2.2. Addition of fluorine to alkenes

Substrate	Products/Yield			
Ph₂C=CHPh	Ph₂CF–CHFPh (14%)	PhCH=CFPh (78%)	Ph₂CF–CF₂Ph (8%)	
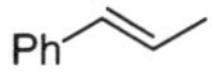 PhCH=CHCH₃	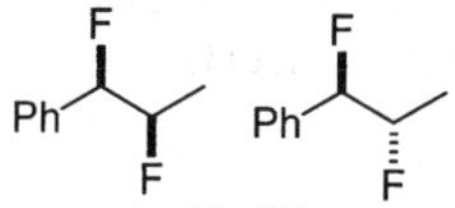 PhCHF–CHFCH₃ (69:31)		80–90%	
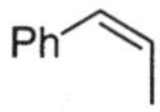 PhCH=CHCH=CH₂ (styryl)	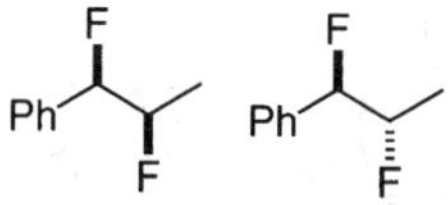 PhCHF–CHFR (22:78)		80–90%	
acenaphthylene	*cis*-difluoroacenaphthene (35%)		*trans*-difluoroacenaphthene (11%)	

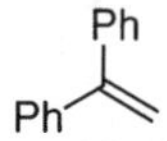

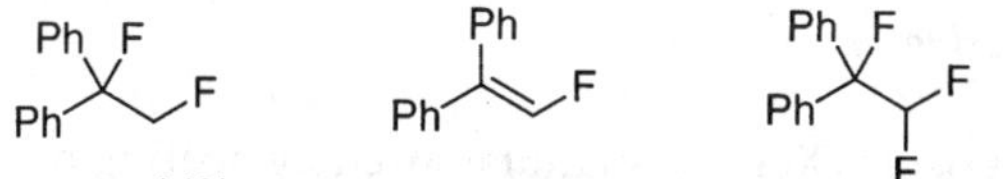

$$R^1R^2C=CHR^3 \xrightarrow{F_2} [R^2R^1C^+ \cdots F^- \cdots CHFR^3] \longrightarrow R^2R^1CF–CHFR^3 + R^1R^2C=CFR^3$$

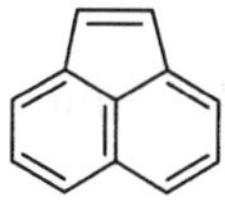

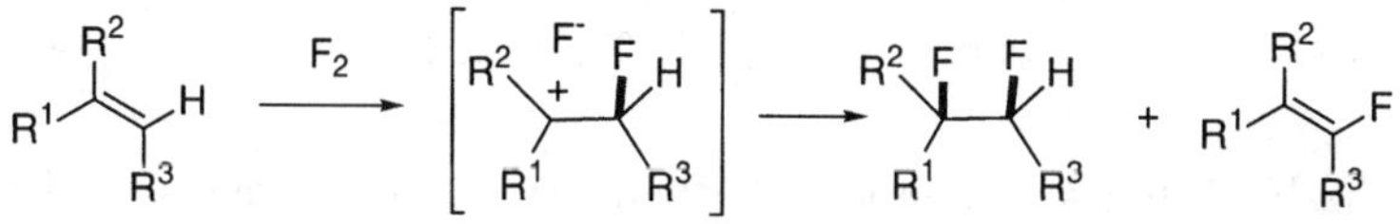

Scheme 2.1

Whereras electrophilic halogenation of aromatic compounds are common with chlorine or bromine, the reaction with fluorine gas is not popular because side reactions, like fluorine addition, polymerization, and a C–C bond cleavage, take place to a considerable extent, and yields and selectivities of the products are miserable [19–22]. However, fluorination of phenols takes place exceptionally efficiently [23, 24].

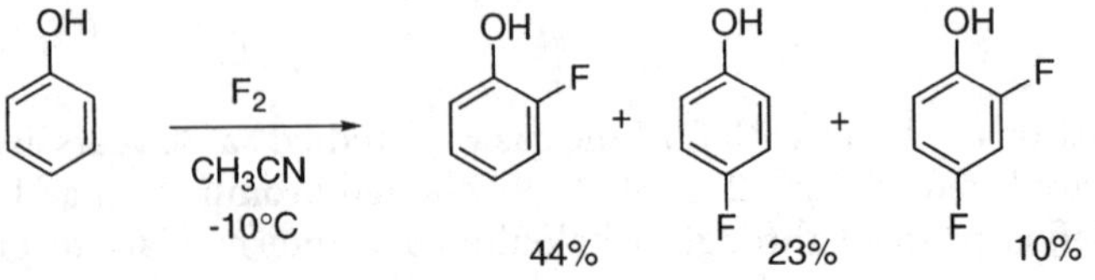

$$\text{PhOH} \xrightarrow[\text{-10°C}]{\text{F}_2, \text{CH}_3\text{CN}} \text{2-F-PhOH (44\%)} + \text{4-F-PhOH (23\%)} + \text{2,4-F}_2\text{-PhOH (10\%)} \quad (2.3)$$

Fluorine gas is converted into various electrophilic fluorinating reagents. ClF_3 and ClF_5 are prepared by the reaction of F_2 and Cl_2 and undergo fluorination reactions in a manner similar to fluorine gas but sometimes accompanied by chlorination [25–27].

2.1.2
Xenon Difluoride

Xenon difluoride (XeF_2) is manufactured by heating or electrically discharging a mixture of fluorine gas and xenon in a nickel vessel. Hereby, XeF_4 and XeF_6 also are prepared. In particular, XeF_2, a stable, colorless and commercially available crystalline reagent [28], selectively converts a tertiary C–H bond of hydrocarbons into a C–F bond [29].

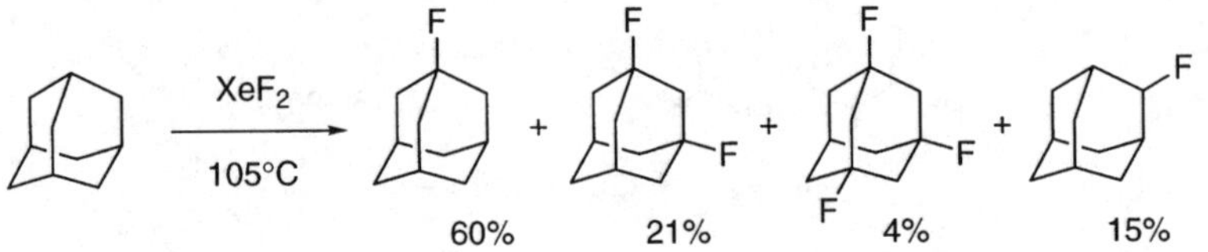

$$\text{adamantane} \xrightarrow[105°C]{XeF_2} \text{1-F (60\%)} + \text{1,3-F}_2 \text{(21\%)} + \text{1,3,5-F}_3 \text{(4\%)} + \text{2-F (15\%)} \quad (2.4)$$

Sulfides are fluorinated at an α-carbon, this reaction is the so-called fluoro-Pummerer reaction [30–32].

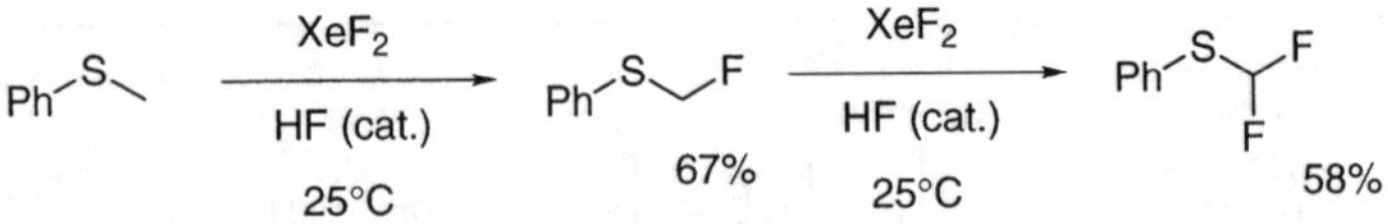

$$\text{Ph-S-CH}_3 \xrightarrow[\text{HF (cat.)}, 25°C]{XeF_2} \text{Ph-S-CH}_2\text{F (67\%)} \xrightarrow[\text{HF (cat.)}, 25°C]{XeF_2} \text{Ph-S-CHF}_2 \text{(58\%)} \quad (2.5)$$

Enolizable ketones and enol silyl ethers are selectively fluorinated to give α-fluoro ketones [33–36].

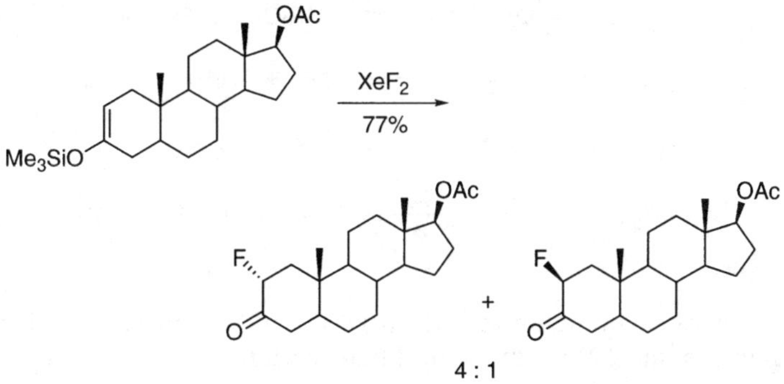

2.1 Electrophilic Fluorinating Reagents

[Scheme (2.6): Me₃SiO-substituted steroid with OAc group → XeF₂, 77% → two α-fluoro ketone products in 4:1 ratio]

Electrophilic fluorination of aromatic compounds with XeF_2 proceeds with yields better than with fluorine gas [37].

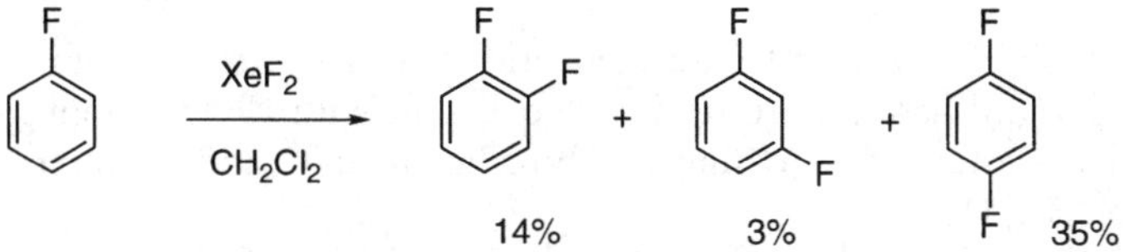

[Scheme (2.7): Fluorobenzene + XeF_2 / CH_2Cl_2 → 1,2-difluorobenzene (14%) + 1,3-difluorobenzene (3%) + 1,4-difluorobenzene (35%)]

When XeF_2 is reacted with a carboxylic acid it undergoes decarboxylation-fluorination [38, 39]. When this reaction is applied to (R)-(+)-α-methoxy-α-trifluoromethylphenylacetic acid, a decarboxylation-fluorination product is obtained in 95% yield but with total racemization. The initially produced RCOOXeF decomposes into R•, CO_2, and XeF to give finally R-F, CO_2 and Xe.

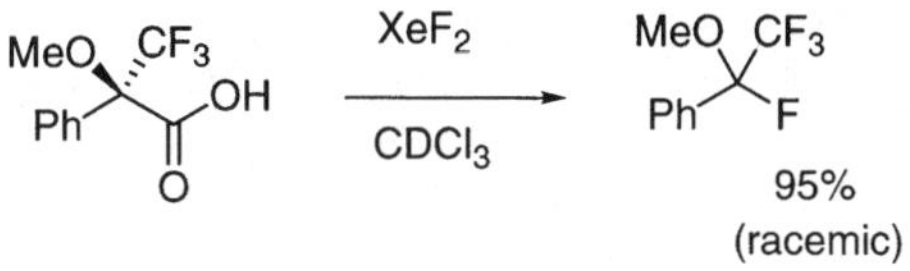

[Scheme (2.8): MeO,CF₃,Ph,COOH + XeF₂ / CDCl₃ → MeO,CF₃,Ph,F, 95% (racemic)]

2.1.3
Electrophilic Reagents Containing an O–F Bond

Reagents containing an O–F bond are extremely electrophilic. Well known are fluoroxytrifluoromethane (CF_3OF) and bis(fluoroxy)difluoromethane [$CF_2(OF)_2$]. These gaseous reagents (with bp –97 and –64 °C, respectively) are prepared by the reaction of F_2 with CO or CO_2 in the presence of CsF and used without purification [40]. In particular, CF_3OF is as reactive as fluorine gas and thus should be handled with equal care.

The reagent fluorinates a bridgehead carbon of adamantane in a manner similar to fluorine gas to give 1-fluoroadamantane [5]. Selective monofluorination of alanine at C-3 takes place under UV irradiation. The fluorination reaction is understood in terms of a fluorine radical and trifluoromethoxy

Scheme 2.2

$$CF_3OF \longrightarrow CF_3O\bullet + F\bullet$$

$$RH + F\bullet \longrightarrow R\bullet + HF$$

$$R\bullet + CF_3OF \longrightarrow RF + CF_3O\bullet$$

$$CF_3O\bullet + RH \longrightarrow R\bullet + COF_2 + HF$$

radical generated by homolysis of an O–F bond. A mechanism is proposed (shown in Scheme 2.2) for the radical fluorination.

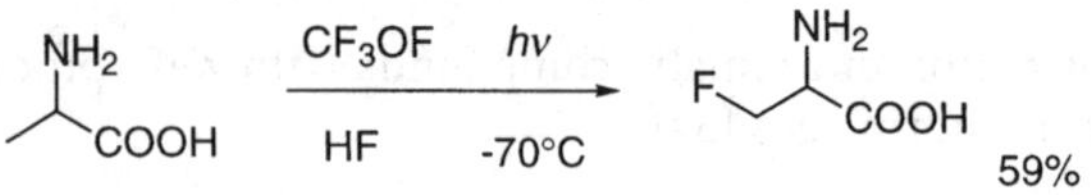

$$\underset{\text{COOH}}{\overset{\text{NH}_2}{\diagdown}} \xrightarrow[\text{HF} \quad -70°C]{CF_3OF \quad h\nu} \underset{\substack{\text{COOH}\\59\%}}{\overset{\text{NH}_2}{F\diagdown}} \tag{2.9}$$

Ketene silyl acetals and enol silyl ethers derived from esters (or carboxylic acids) and ketones are readily fluorinated with CF_3OF to give α-fluorocarboxylate esters (or α-fluorocarboxylic acids) and α-fluoro ketones (Scheme 2.3) [41].

Aromatic compounds are fluorinated with CF_3OF or $CF_2(OF)_2$. In particular, N-methylsulfonylaniline is fluorinated selectively at an *ortho* carbon probably through hydrogen-bonded delivery of an electrophilic fluorine (Scheme 2.4) [42].

Addition of CF_3OF to olefins [43–47] proceeds mainly with *cis* stereochemistry of the F and OCF_3 groups accompanied by difluorination. The mechanism of the *cis* addition is suggested to involve radical species F• and $CF_3O\bullet$ [45, 47].

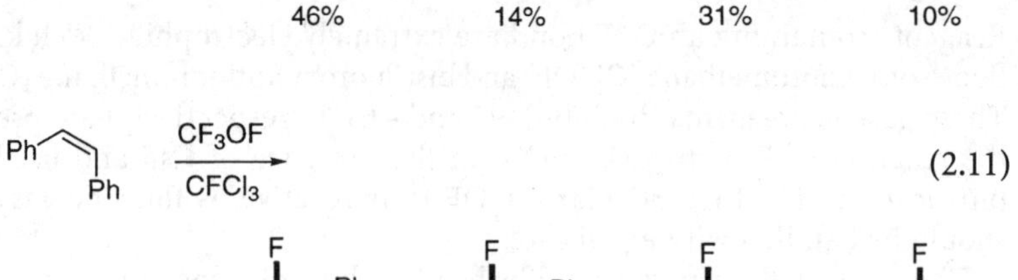

$$Ph\diagup\hspace{-0.3em}=\hspace{-0.3em}\diagdown Ph \xrightarrow[CFCl_3]{CF_3OF} \tag{2.10}$$

Ph–CHF–CH(OCF₃)–Ph (46%) + Ph–CHF–CH(OCF₃)–Ph (14%) + Ph–CHF–CHF–Ph (31%) + Ph–CHF–CHF–Ph (10%)

$$Ph\diagup\hspace{-0.3em}=\hspace{-0.3em}\diagup Ph \xrightarrow[CFCl_3]{CF_3OF} \tag{2.11}$$

Ph–CHF–CH(OCF₃)–Ph (25%) + Ph–CHF–CH(OCF₃)–Ph (62%) + Ph–CHF–CHF–Ph (3%) + Ph–CHF–CHF–Ph (9%)

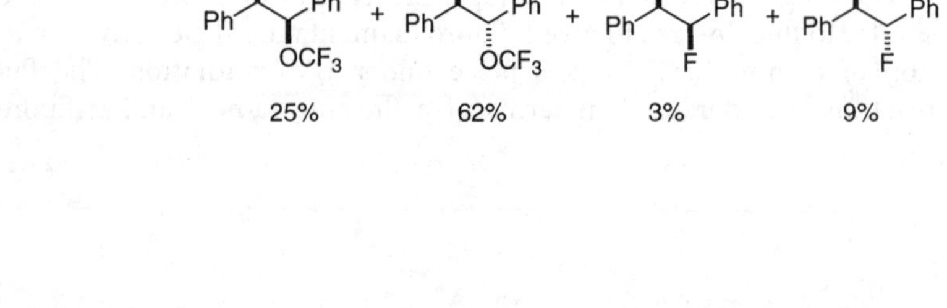

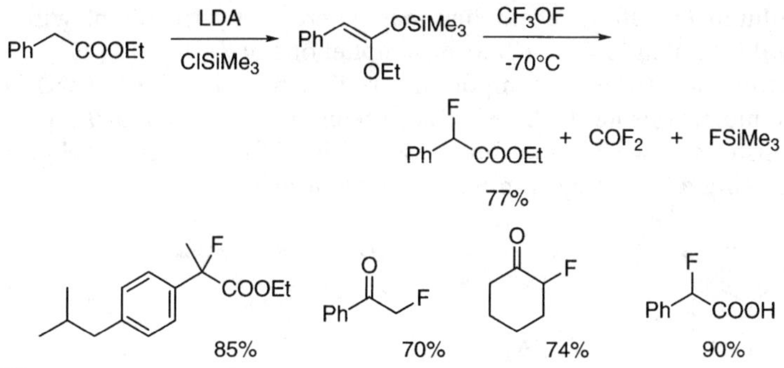

2.1 Electrophilic Fluorinating Reagents

Scheme 2.3

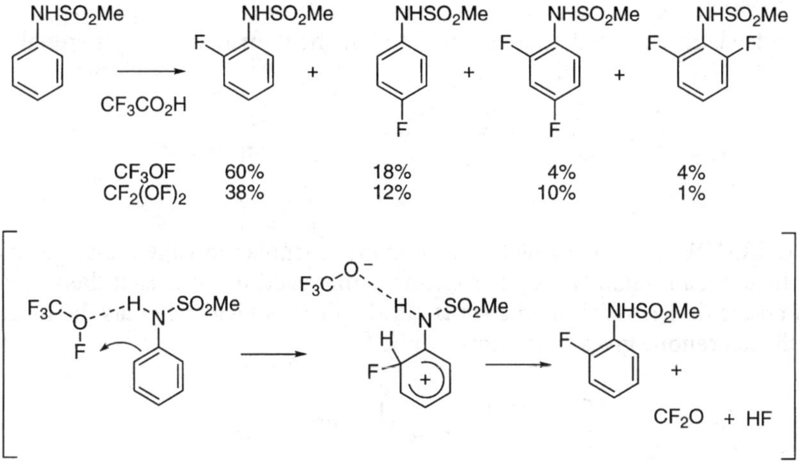

Scheme 2.4

Acetyl hypofluorite (CH_3COOF) is prepared by passing 10% F_2 in N_2 at −75 °C through a mixture of sodium or potassium acetate and acetic acid sometimes diluted with $CFCl_3$ and used directly at low temperatures [39]. This reagent is considered to be milder than fluorine gas and CF_3OF and fluorinates enolizable β-keto esters and β-diketones to give the corresponding monofluorination products. In particular, use of a base to generate an enolate form improves the yield remarkably [48].

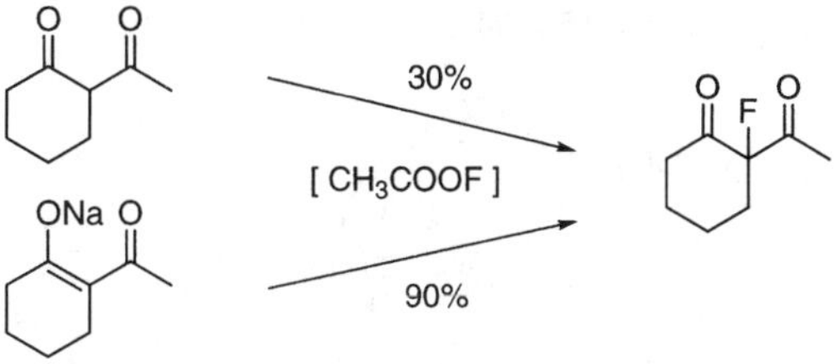

(2.12)

A lithium enolate of acetophenone, generated by treatment with LDA, is readily fluorinated to give α-fluoroacetophenone in 75% yield [49].

Electron-rich aromatic compounds are fluorinated with CH_3COOF [50, 51]. For example, anisole is fluorinated preferentially at an *ortho*-position. A mechanism is suggested that involves addition of CH_3COO and F groups to a benzene ring followed by elimination of acetic acid.

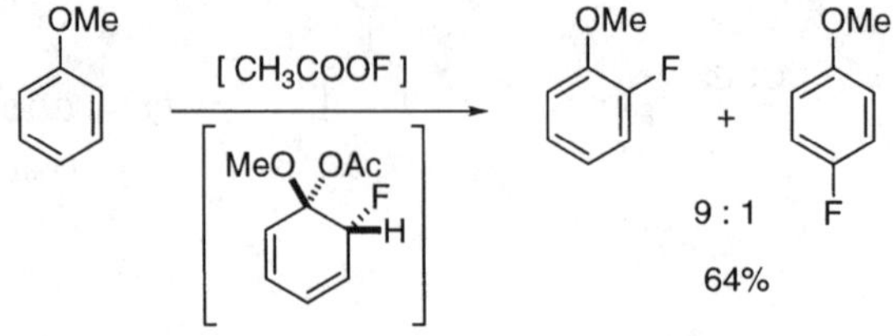

(2.13)

Indeed, the assumed adduct is isolated by the fluorination of piperonal.

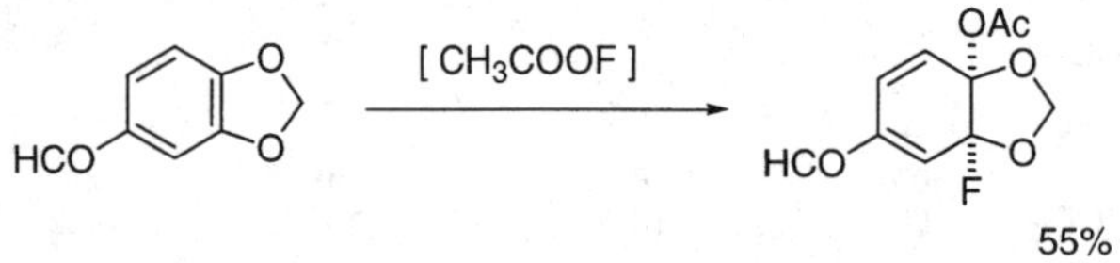

(2.14)

CH_3COOF reacts with olefins in a manner similar to CF_3OF and gives *cis*-adducts predominantly [52]. For example, the reaction with *cis*-stilbene gives a *cis*-adduct in 51% yield and a *trans*-adduct in 11% yield, whereas the reaction with suberenone gives a *cis*-adduct only.

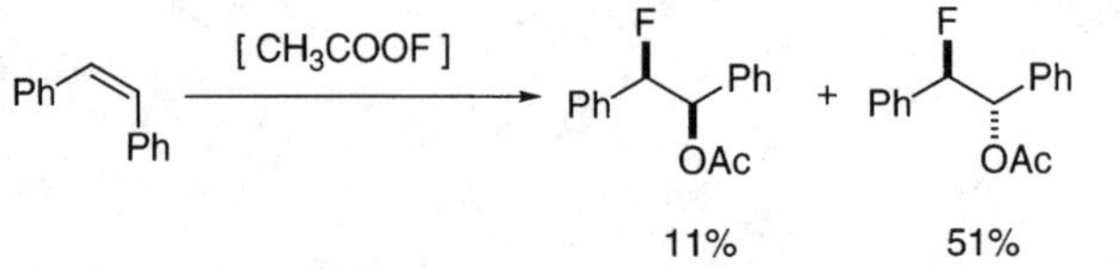

(2.15)

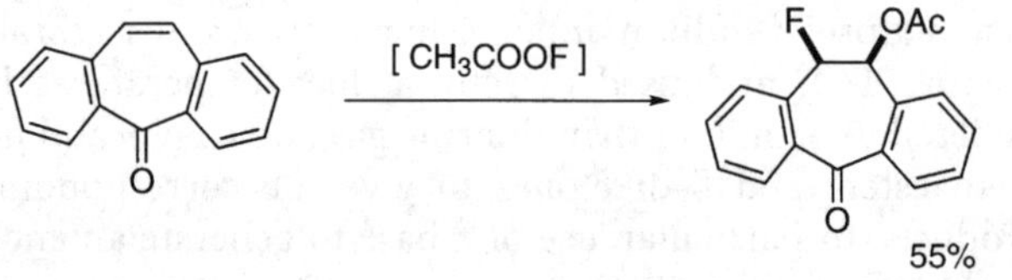

(2.16)

The fluorination is particularly effective for the fluorinaton of glycals and 2-fluoro sugars are readily prepared [53, 54].

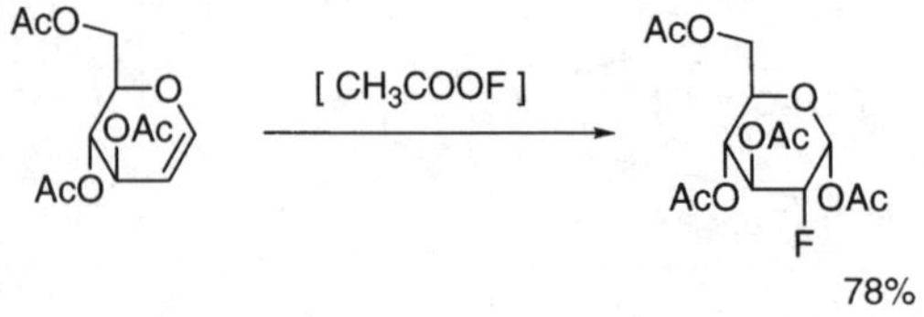

(2.17)

Enol acetates are convenient precursors of α-fluoro ketones [55, 56].

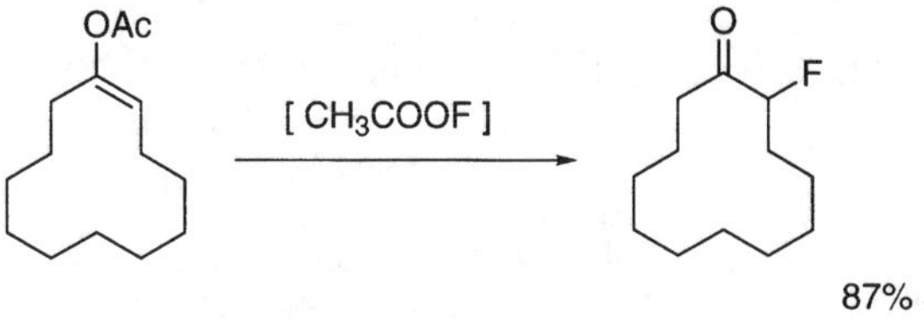

(2.18)

Earlier, perchloryl fluoride (ClO$_3$F), an explosive gas reagent with a bp of −46.7 °C, was used for the fluorination of enolates, enol esters, and enol ethers. Today, however, it is only rarely used due to its difficult handling [57–62]. Synthesis of a difluoroprostaglandin derivative (shown in Scheme 2.5) is an example.

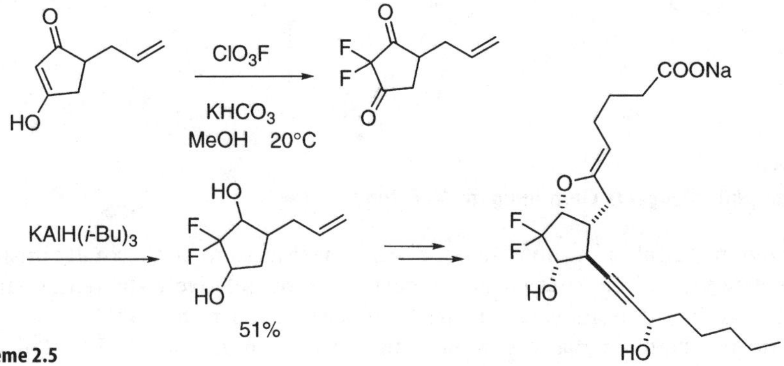

Scheme 2.5

Cesium fluorosulfate (CsSO$_4$F), a crystalline reagent prepared by the reaction of fluorine gas with cesium sulfate, fluorinates various substrates [63–68]. The reaction of norbornene with CsSO$_4$F gives two monofluoro products formed via skeletal rearrangement [63].

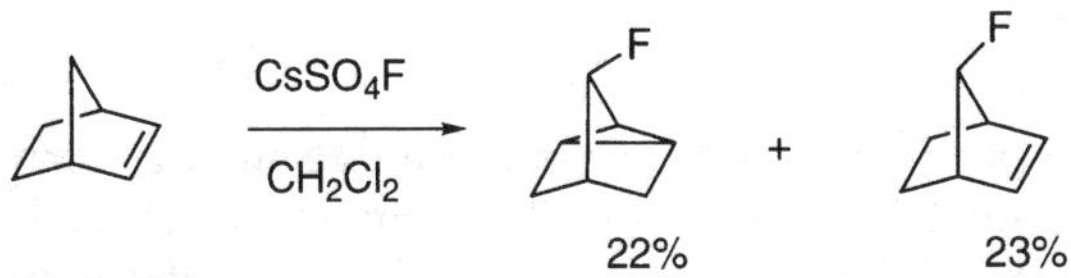

(2.19)

Although cesium fluorosulfate does not react with 1,3-dimethyluracil in acetonitrile, the reaction in methanol gives selectively a *cis*-adduct of F and MeO. Subsequent treatment with triethylamine gives 1,3-dimethyl-5-fluorouracil [64].

Phenol is fluorinated with CsSO$_4$F selectively at an *ortho*-carbon. However, phenyl ethers like anisole lose the *ortho*-selectivity [65].

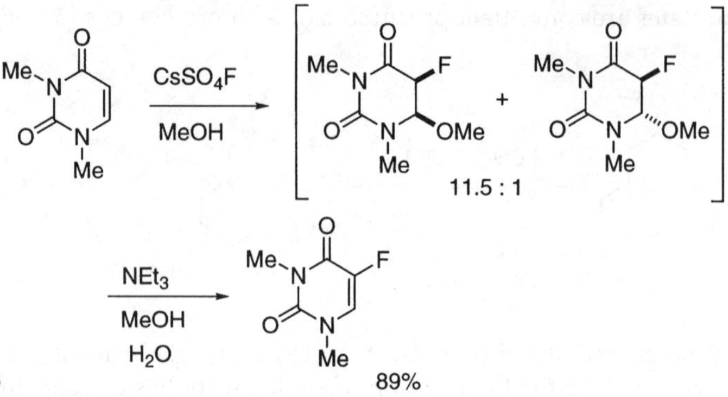

Scheme 2.6

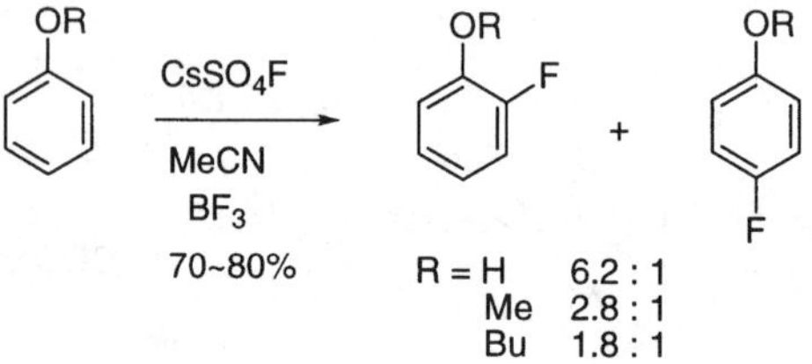

2.1.4
Electrophilic Reagents Containing an N–F Bond

Compounds containing an N–F bond were considered to be thermally unstable and were very rarely studied in the past. Recently, relatively stable derivatives have been found to fluorinate a variety of organic substrates [69].

The first example of a reagent with an N–F bond might be $NF_4^+ BF_4^-$ [70]. This reagent fluorinates hexafluorobenzene in HF to give perfluorocyclohexane in high yields [71].

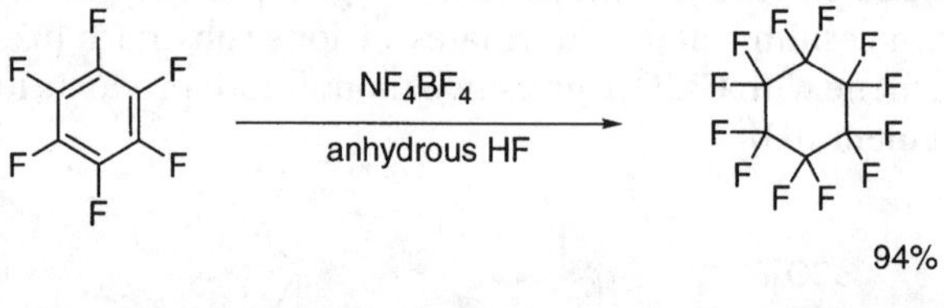

Although it was known that *N*-fluoropyridinium fluoride was formed by the reaction of fluorine gas with pyridine, in 1986 Umemoto succeeded in isolating its triflate salt (mp 185–187 °C) by treatment of labile *N*-fluoropyridinium fluoride with trifluoromethanesulfonic acid, its sodium salt, or trimethylsilyl triflate [72].

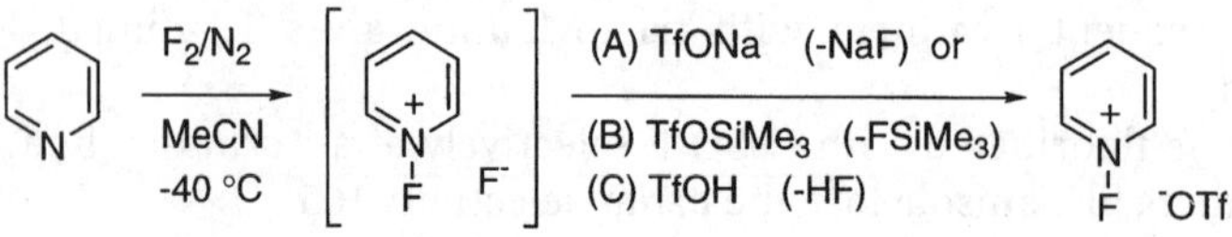

2.1 Electrophilic Fluorinating Reagents

By introducing an appropriate substituent and changing the counterion to BF_4^-, PF_4^-, or ClO_4^-, non-hygroscopic stable fluorinating reagents were prepared and have now been developed further [73–75]. Depending on the substrate, it is possible to use an appropriate reagent with suitable fluorinating power. In general, the reactivity of an N-fluoropyridinium reagent is enhanced when the electron density of the pyridinium nitrogen is decreased. For example, phenol is fluorinated with an electron-rich reagent only at 100 °C but with an electron-deficient reagent at lower temperatures.

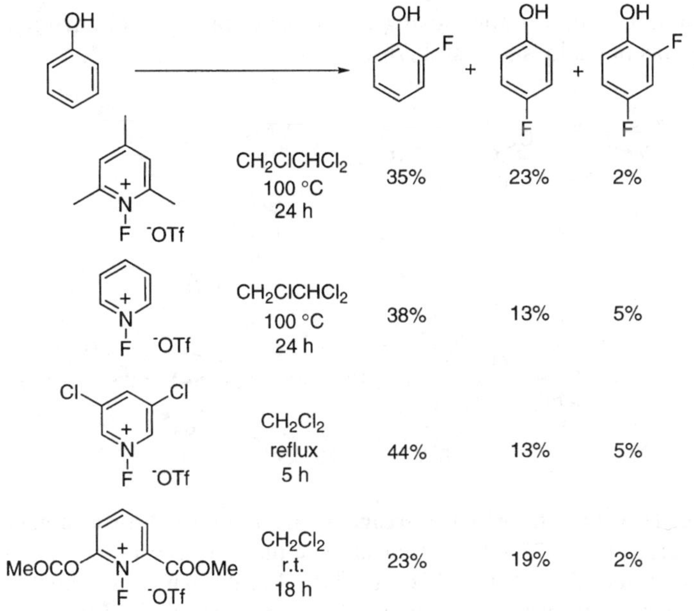

$$\text{PhOH} \longrightarrow \text{o-F-C}_6\text{H}_4\text{OH} + \text{p-F-C}_6\text{H}_4\text{OH} + \text{2,4-F}_2\text{-C}_6\text{H}_3\text{OH} \quad (2.23)$$

Reagent	Conditions	ortho	para	2,4
N-fluoro-2,4,6-trimethylpyridinium OTf	CH₂ClCHCl₂, 100 °C, 24 h	35%	23%	2%
N-fluoropyridinium OTf	CH₂ClCHCl₂, 100 °C, 24 h	38%	13%	5%
N-fluoro-2,6-dichloropyridinium OTf	CH₂Cl₂, reflux, 5 h	44%	13%	5%
N-fluoro-2,6-bis(methoxycarbonyl)pyridinium OTf	CH₂Cl₂, r.t., 18 h	23%	19%	2%

Grignard reagents are fluorinated easily. Thus, fluorobenzene is isolated in 58% yield from phenylmagnesium bromide and N-fluoro-2,4,6-lutidinium triflate.

Enolizable ketones and ketene silyl ethers are readily fluorinated to give α-fluoro ketones and esters.

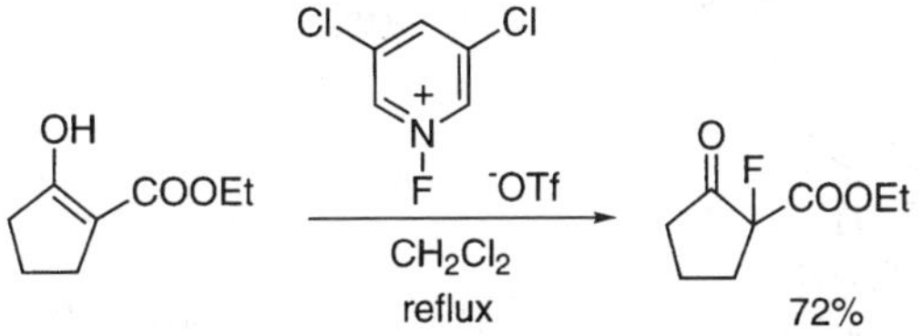

$$\text{2-hydroxy-1-cyclopentenyl-COOEt} \xrightarrow[\text{CH}_2\text{Cl}_2, \text{reflux}]{\text{N-fluoro-3,5-dichloropyridinium OTf}} \text{2-oxo-1-fluorocyclopentyl-COOEt} \quad (2.24)$$

72%

Fluorination of sulfides with an N-fluoropyridinium reagent gives α-fluoro sulfides via fluoro-Pummerer rearrangement. As can easily be seen, selection of a suitable reagent gives high yields of products under mild conditions.

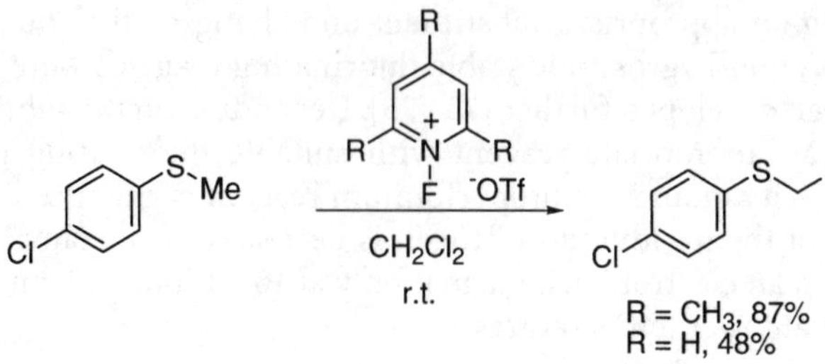

$$\text{(2.25)}$$

Asymmetric fluorination using a malonate having a phenylmenthyl ester moiety was partially successful [76].

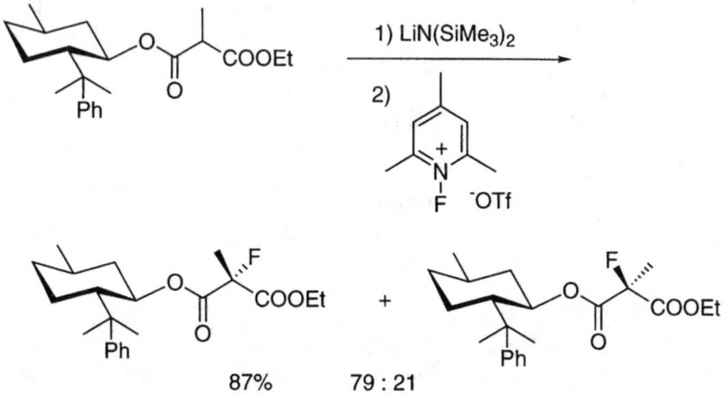

$$\text{(2.26)}$$

Other N–F type fluorination reagents are *N*-fluoro-1,4-diazabicyclo[2.2.2]-octane derivatives [77–80]. A typical example is 1-chloromethyl-4-fluoro-1,4-diazoniabicyclo[2.2.2]octane bis(tetrafluoroborate) which fluorinates electron-rich substrates under conditions milder than those with *N*-fluoropyridinium salts.

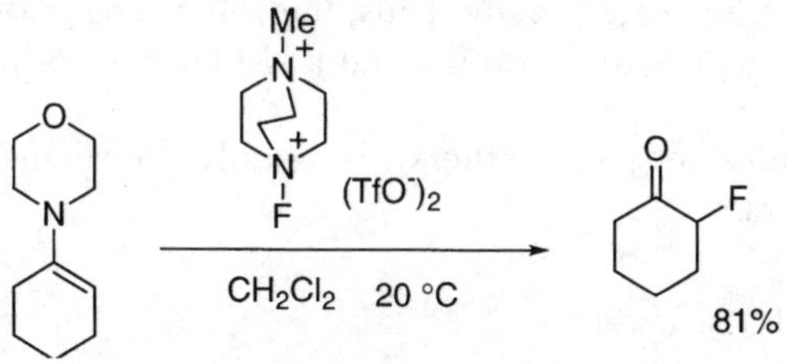

$$\text{(2.27)}$$

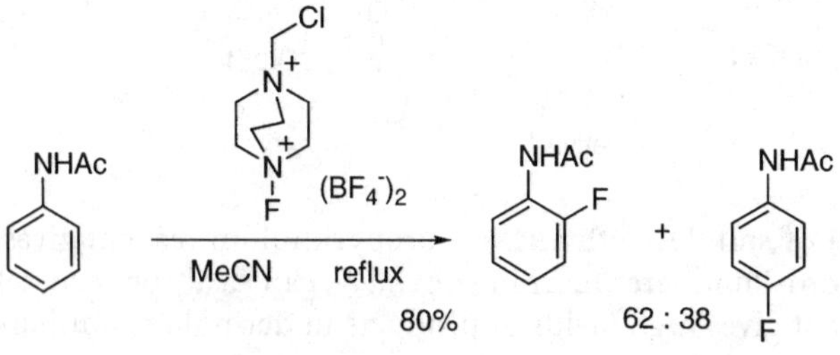

$$\text{(2.28)}$$

2.1 Electrophilic Fluorinating Reagents

The third category of N–F type fluorinating reagents are N-fluorosulfonamides and -imides. These reagents are prepared by the reaction of sulfonamides or -imides with fluorine gas [81]. A typical example is N-alkyl-N-fluoro-p-toluenesulfonamides and N-alkyl-N-fluorobenzenesulfonamides. These fluorinate anionic derivatives of malonates, nitroalkanes, organolithium compounds, and metal enolates.

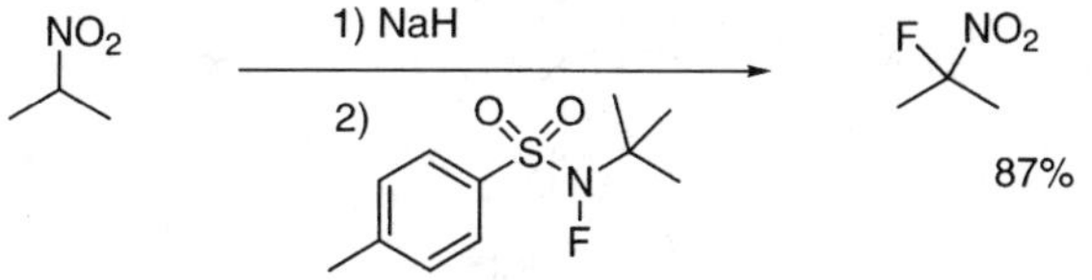

(2.29)

In particular, a vinyllithium is readily fluorinated to give a vinyl fluoride in high yield [82].

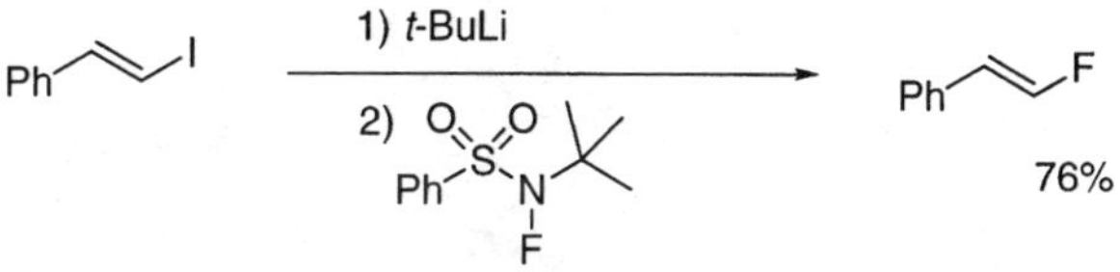

(2.30)

An N-fluorosultam reagent derived from saccharin is also effective for the fluorination of the enolates of ketones [83, 84]. For example, β-methyl-α-tetralone is converted in 97% yield into a β-fluorinated product through enolate formation.

In addition, use of an equimolar or 2 mole amount of a base and a fluorinating reagent gives rise to a monofluorinated product or a difluorinated product selectively.

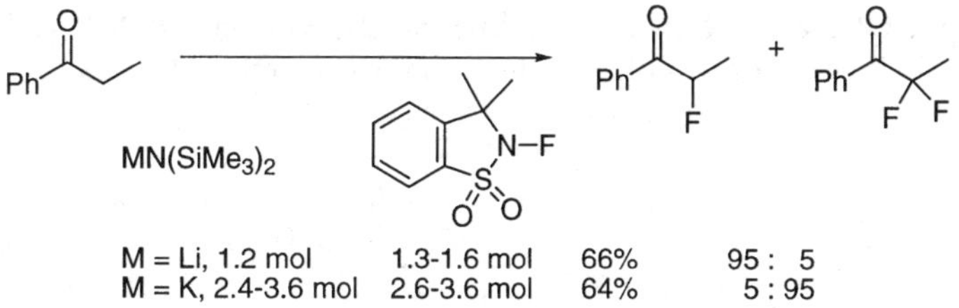

(2.31)

M = Li, 1.2 mol   1.3-1.6 mol   66%   95 : 5
M = K, 2.4-3.6 mol   2.6-3.6 mol   64%   5 : 95

Optically active sulfonamides serve as chiral fluorine carriers and allow asymmetric fluorination [84–88]. An early example used N-fluoro-2,10-camphorsultam, which fluorinated the enolate of 2-ethoxycarbonylcyclopentanone to give 2-ethoxycarbonyl-2-fluorocyclopentanone with 70% enantiomeric excess (ee). The reagent is not as effective with other substrates [85]. In order to improve the selectivity of asymmetric fluorination, chlorine-substituted N-fluoro-2,10-camphorsultam [86] and an N-fluorosaccharin analog were proved to show much improved results [88]. Over 80% ee and yields are attainable at present. In the near future, a reagent which can achieve a higher degree of asymmetric fluorination is expected to be put on the market.

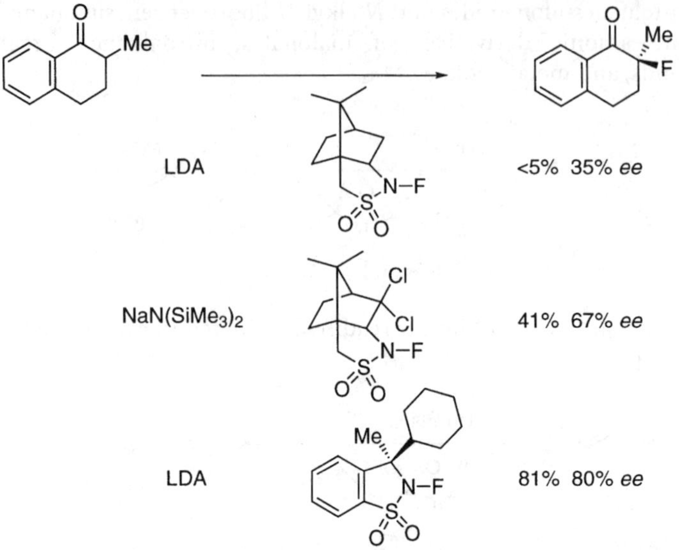

(2.32)

Base	Reagent	Yield	ee
LDA	(camphorsultam N–F)	<5%	35% ee
NaN(SiMe₃)₂	(dichlorocamphorsultam N–F)	41%	67% ee
LDA	(methyl-cyclohexyl benzisothiazole N–F)	81%	80% ee

Another class of electrophilic fluorinating reagent are N-fluorobis(sulfon)-imides [89–95] that are more reactive than N-fluorosulfonamides. In particular, N-fluorobis(trifluoromethylsulfon)imide, called the DesMarteu reagent (liquid, bp 90 °C), is one of the most electrophilic fluorinating reagents. The reagent fluorinates anisole at 22 °C in a quantitative yield to give mainly fluoroanisole [90], while $(PhSO_2)_2NF$ gives the same product only in 33% yield [89].

$(CF_3SO_2)_2NF$ fluorinates dimethyl phenylmalonate at 22 °C to give diethyl fluoro(phenyl)malonate in 92% yield [91, 92], whereas $(PhSO_2)_2NF$ needs a base [89]. Use of 1 or 2 moles of the DesMarteau reagent gives mono- or di-fluorinated products from benzoyl acetate [92].

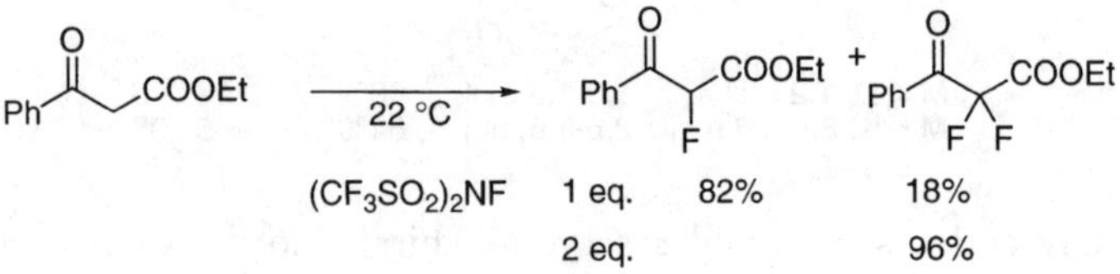

Ph–CO–CH₂–COOEt → Ph–CO–CHF–COOEt + Ph–CO–CF₂–COOEt (2.33)

22 °C
$(CF_3SO_2)_2NF$ 1 eq.   82%   18%
                    2 eq.   96%

An imide enolate having a chiral auxiliary is fluorinated with high diastereo-selectivity. Subsequent reductive or hydrolytic removal of the chiral auxiliary gives 2-fluoroalkanol or 2-fluoroalkanoic acid with very high ee [95].

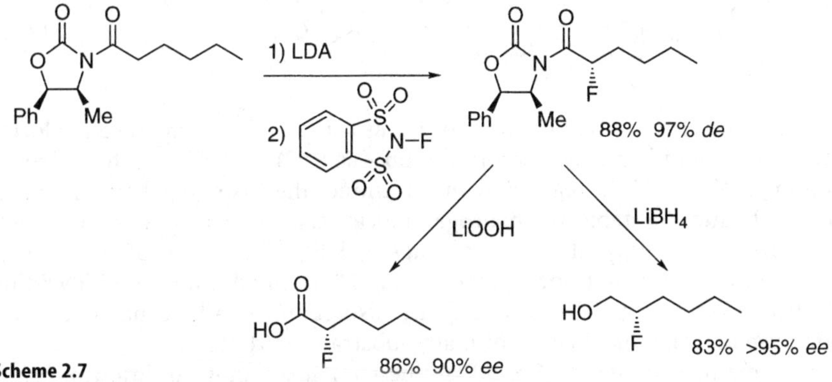

Scheme 2.7

2.2
Nucleophilic Fluorinating Reagents

2.2.1
Hydrogen Fluoride and Derivatives

The most basic nucleophilic fluorinating reagent is hydrogen fluoride (HF) and its aqueous solution, hydrofluoric acid. HF, a liquid with a boiling point of 19.5 °C, is employed for the manufacture of various organofluorine compounds, particularly chlorofluorocarbons [96, 97]. Details will be described in Chap. 6.2. HF undergoes a nucleophilic substitution reaction. However, as HF is highly corrosive, special equipment, as well as know-how, by talented experts are needed. To use HF conveniently in standard laboratories, many HF-derived reagents have been invented that are easily handled, less toxic, and more nucleophilic.

HF adds readily to a carbon–carbon double bond of olefins to give fluoroalkanes [98]. The reaction with propene proceeds at −45 °C to give exclusively a Markovnikov-type product 2-fluoropropane. Hydrofluoric acid is not capable of effecting the addition reaction.

Addition to cyclohexene takes place at −35 °C to give fluorocyclohexane. Cyclopropane is cleaved by HF to give mainly 1-fluoropropane.

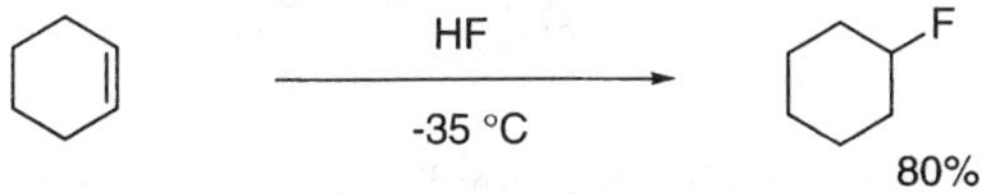

$$\text{cyclohexene} \xrightarrow[-35\,°\text{C}]{\text{HF}} \text{fluorocyclohexane} \quad 80\% \tag{2.34}$$

Benzotrichloride is readily fluorinated to give benzotrifluoride [99, 100]. Use of a Lewis acid catalyst like $FeCl_3$ or $SbF_3/SbCl_5$ accelerates the reaction.

A CCl_3 group is converted into a CF_3 group with chlorine on a tertiary carbon still intact.

$$\text{CCl}_3\text{-CHCl} \xrightarrow[130\,°C]{HF} \text{CF}_3\text{-CHCl} \quad 40\% \qquad (2.35)$$

The relatively high boiling point of HF as compared with hydrogen chloride (bp −85.0 °C) and hydrogen bromide (bp −66.8 °C) is attributed to hydrogen bonding. When HF is reacted with an amine, the hydrogen bond is partly cleaved to afford a stable polyhydrogenfluoride salt. For example, commercially available 70% HF/pyridine (Olah reagent, HF/pyridine = 70:30 w/w, or ca 9:1 mol ratio) does not liberate HF even at 50 °C. In addition to HF/pyridine, HF/triethylamine and HF/melamine are also commercially available and are often used for the fluorination of many substrates [101–103].

In a manner similar to HF, HF/amine complexes allow addition of HF to olefins. The reaction of cyclohexene is exemplified below [102]. The yields depend on the amine content as well as the solvent [103]. In THF, the addition of HF/pyridine and HF/melamine takes place smoothly at 0 °C.

$$\text{cyclohexene} \xrightarrow[THF \quad 0\,°C]{} \text{cyclohexyl-F} \qquad (2.36)$$

HF	5 min	71%
70% HF/pyridine	10 min	28%
"	1 h	80%
86% HF/melamine	10 min	88%
"	1 h	98%
"	5 min	77%
77% HF/melamine	5 min	46%
49% HF/NEt$_3$	10 min	12%
71% HF/BuNH$_2$	10 min	37%

HF/pyridine fluorinates secondary and tertiary alcohols to give the corresponding fluorides [101]. Likewise, protected glucoses with a free or acylated anomeric OH group are readily fluorinated to give 1-fluoro derivatives with an α-anomer being favored [104].

$$\text{BnO-glucose-OR} \xrightarrow[0\,°C]{HF\text{-pyridine}} \text{BnO-glucose-F} \qquad (2.37)$$

R = H   68%   α : β = 97 : 3
   Ac   89%   97 : 3

Oxiranes undergo a ring-opening fluorination by the action of an HF/amine complex [105–112]. In particular, 1,2-epoxyalkanes give preferentially 2-fluoro-1-alkanols [105] in sharp contrast to the ring-opening reaction with hydrogen chloride or bromide. Use of HF/HN(i-Pr)$_2$ at higher temperatures reverses the regioselectivity. The ring-opening reaction takes place stereospecifically with inversion of configuration. Thus, starting with optically active 1,2-epoxyoctane,

2.2 Nucleophilic Fluorinating Reagents

available by biological oxidation of 1-octene, gives optically active 2-fluoro-1-octanol [108], a requisite part of ferroelectric liquid crystals that respond extremely quickly to the change in an electric field. Details will be discussed in Chap. 6.3.

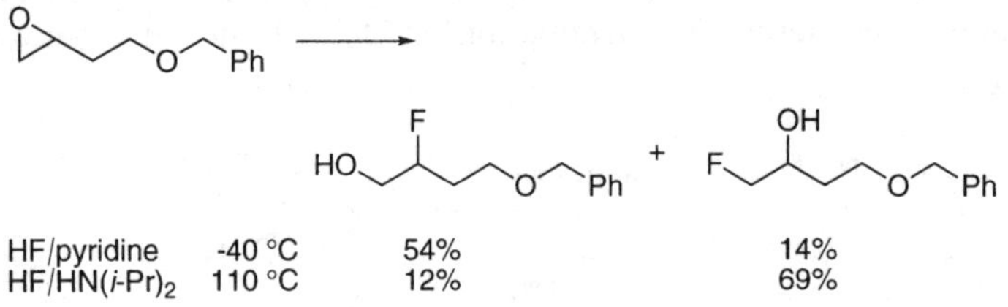

(2.38)

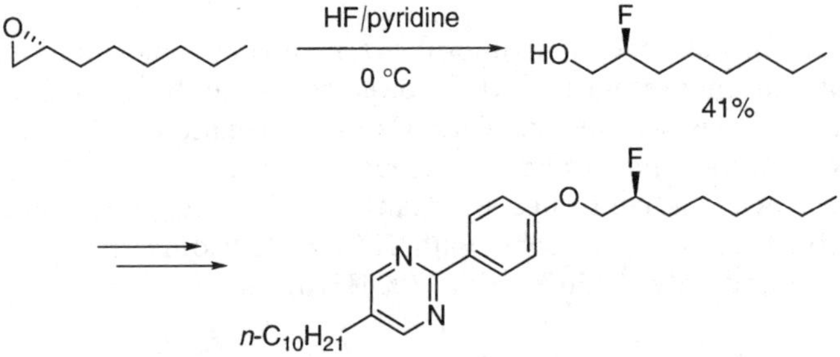

HF/pyridine	−40 °C	54%	14%
HF/HN(i-Pr)$_2$	110 °C	12%	69%

Scheme 2.8

HF/pyridine also reacts with α,β-epoxy nitriles and esters to give β-fluoro cyanohydrins and β-fluoro α-hydroxy esters selectively [109–112]. Similarly, aziridines undergo ring-opening fluorination [113–115].

For halogen-exchange reactions, HF/pyridine is rarely employed. In the presence of mercuric oxide, *gem*-dihaloalkanes and α-bromo ketones give the corresponding fluorine-substituted products [101].

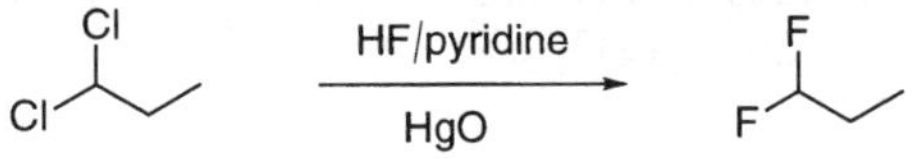

(2.39)

Although it is not easy to convert halobenzenes to fluorobenzenes with HF or an HF/amine complex, 2,4-dinitrochlorobenzene is fluorinated with HF at 180 °C in the presence of γ-collidine to give 2,4-dinitrofluorobenzene [116]. The use of HF and an amine base is also effective for the fluorine substitution of 2-chloropyridine. In contrast, 2-chloropyrimidine is fluorinated by HF at 50 °C in the absence of an amine base to give 2-fluoropyrimidine.

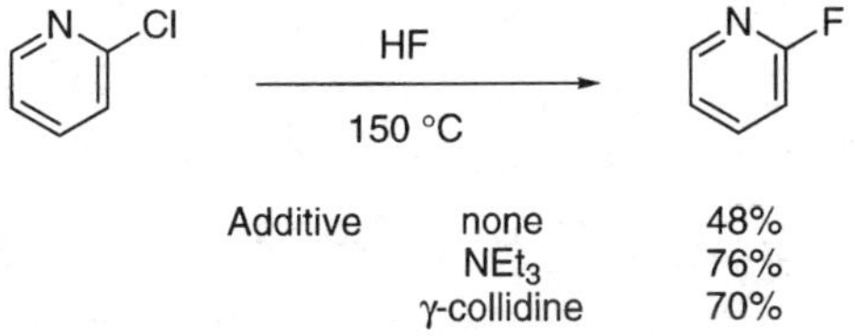

(2.40)

Additive	
none	48%
NEt$_3$	76%
γ-collidine	70%

Fluoroarenes with no electron-withdrawing substituents are not easily prepared from haloarenes. An effective method is the Schiemann reaction, a reaction of arenediazonium salts with HF or an HF/amine complex [117]. Upon heating, the reaction takes place, liberating molecular nitrogen and achieving fluorination, whereas substrates having an *ortho-* or *para-*substituent react sluggishly. Hereby UV irradiation accelerates the fluorination even at 12 °C [118].

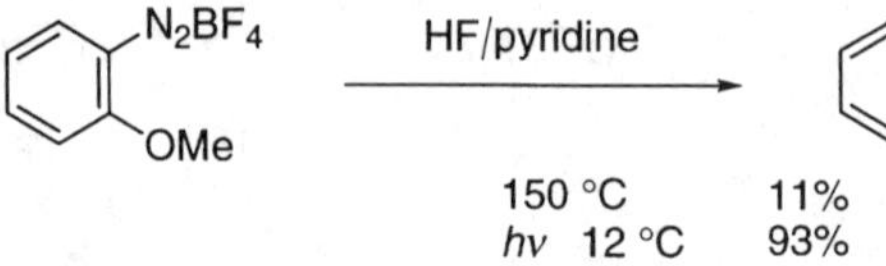

$$\text{o-MeO-C}_6\text{H}_4\text{-N}_2\text{BF}_4 \xrightarrow{\text{HF/pyridine}} \text{o-MeO-C}_6\text{H}_4\text{-F} \quad (2.41)$$

150 °C   11%
hν 12 °C   93%

A process for manufacturing fluorobenzene involves diazoniation of aniline with $NaNO_2$ in the presence of HF followed by thermolysis [119]. Use of 70% HF/pyridine allows the dediazoniation and fluorination to be performed at lower temperatures and extends the scope of the reaction [120]. The reactivity of the reagent is controlled by changing the ratio of HF and pyridine: the reaction takes place at 55 °C with 70% HF/pyridine, at 90 °C with 60% HF/pyridine, and only at 150 °C with 57% HF/pyridine.

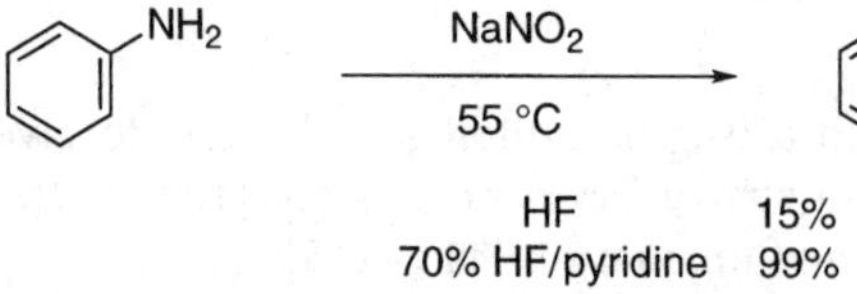

$$\text{PhNH}_2 \xrightarrow[55\ °\text{C}]{\text{NaNO}_2} \text{PhF} \quad (2.42)$$

HF   15%
70% HF/pyridine   99%

Diazoniation-fluorinations of some aniline derivatives are summarized in Table 2.3. Indeed, substrates having an *ortho-* or *para-*substituent need higher temperatures. In contrast, aminopyridines are fluorinated with HF or 60% HF/pyridine under milder conditions [121].

Table 2.3. Diazotization-fluorination of substituted anilines using 70% HF/pyridine

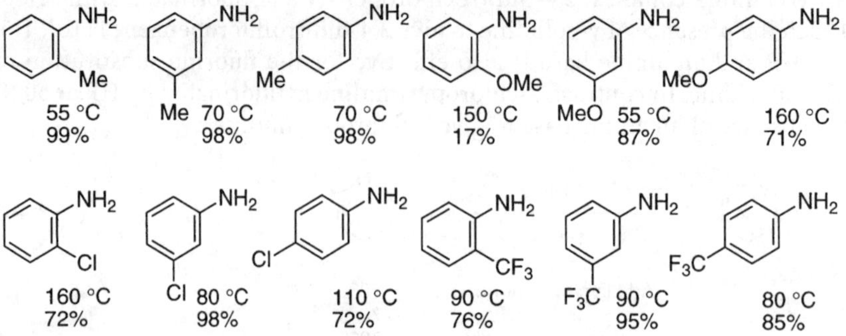

2-Me-C₆H₄NH₂	3-Me-C₆H₄NH₂	4-Me-C₆H₄NH₂	2-OMe-C₆H₄NH₂	3-MeO-C₆H₄NH₂	4-MeO-C₆H₄NH₂
55 °C	70 °C	70 °C	150 °C	55 °C	160 °C
99%	98%	98%	17%	87%	71%
2-Cl-C₆H₄NH₂	3-Cl-C₆H₄NH₂	4-Cl-C₆H₄NH₂	2-CF₃-C₆H₄NH₂	3-CF₃-C₆H₄NH₂	4-CF₃-C₆H₄NH₂
160 °C	80 °C	110 °C	90 °C	90 °C	80 °C
72%	98%	72%	76%	95%	85%

2.2 Nucleophilic Fluorinating Reagents

Because diazo compounds of aliphatic amines are unstable, diazoniation-fluorination is not popular with aliphatic and alicyclic amines except for the transformation of α-amino acids to α-fluorocarboxylic acids [111, 122, 123]. The reaction proceeds with almost complete retention of configuration and thus is a reliable method for the synthesis of optically active α-fluorocarboxylic acids, particularly useful for the synthesis of (S)-2-fluoropropanoic acid from L-alanine.

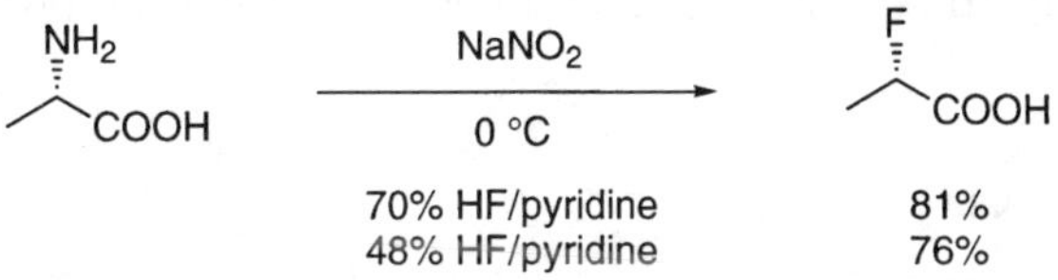

$$\underset{\text{COOH}}{\overset{\text{NH}_2}{\diagup}} \xrightarrow[\substack{0\,°C \\ 70\%\ \text{HF/pyridine} \\ 48\%\ \text{HF/pyridine}}]{\text{NaNO}_2} \underset{\text{COOH}}{\overset{F}{\diagup}} \quad\quad \substack{81\% \\ 76\%} \tag{2.43}$$

Phenylalanine and valine often give, in addition to the desired products, β-fluorine-substituted acids in fair amounts. Use of 48% HF/pyridine suppresses such by-product formation.

2.2.2
Functional Group Transformation

Typical functional groups like hydroxyl, carbonyl, and carboxyl groups are conveniently converted into fluorine functional groups with sulfur tetrafluoride (SF_4), diethylaminosulfur trifluoride (Et_2NSF_3) (abbreviated as DAST) or a fluoroalkyl amine (abbreviated as FAR).

A basic reagent is SF_4 that fluorinates various oxygeneous functional groups and thus is used very frequently [124, 125]. However, since it is an extremely toxic gas with a bp of −40 °C which on exposure to moisture is readily hydrolyzed to give HF, it is not so convenient for laboratory use. A typical transformation is replacement of hydroxyl to fluorine. Use of SF_4 alone needs heating for a long time [124], but the reaction in HF can be carried out cleanly at −78 °C with amino and carboxyl groups left intact [126, 127]. The substitution takes place roughly in an S_N2 manner with inversion of configuration.

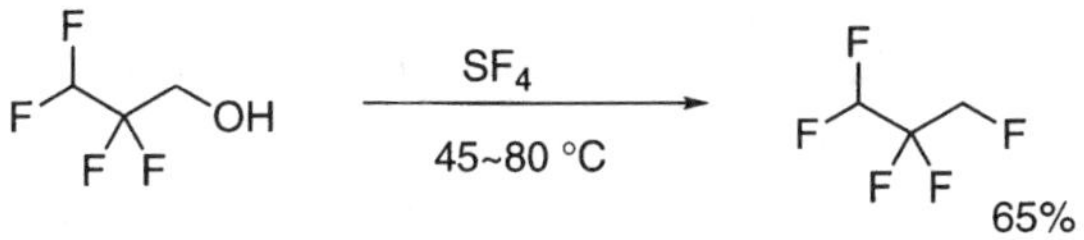

$$\text{(2.44)}$$

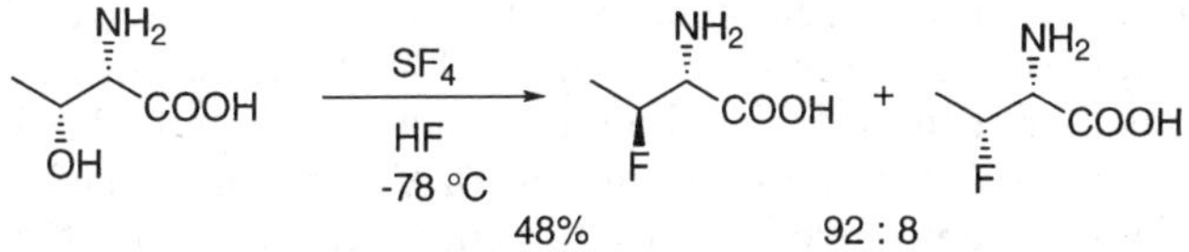

$$\text{(2.45)}$$

Aldehydes and ketones react with SF_4 above 100 °C, and the carbonyl group is transformed to a difluoromethylene functionality [124].

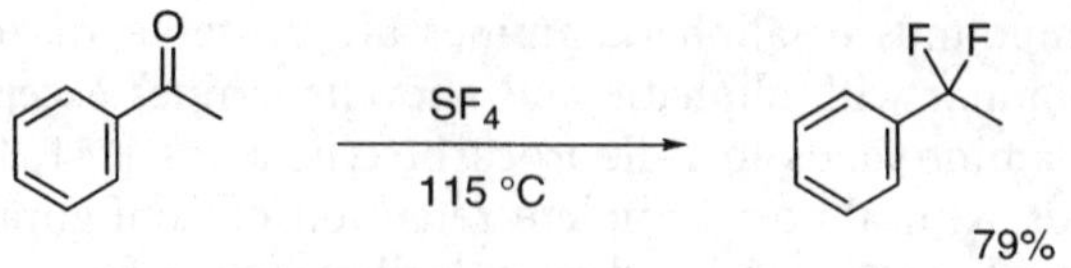

PhC(O)CH₃ →[SF₄, 115 °C] PhCF₂CH₃ (79%)   (2.46)

Because of the high reaction temperatures, the difluorination is sometimes overridden by dehydration to give fluoroolefins [128]. The side reaction may be suppressed by using HF as a coreagent and performing the reaction at room temperature.

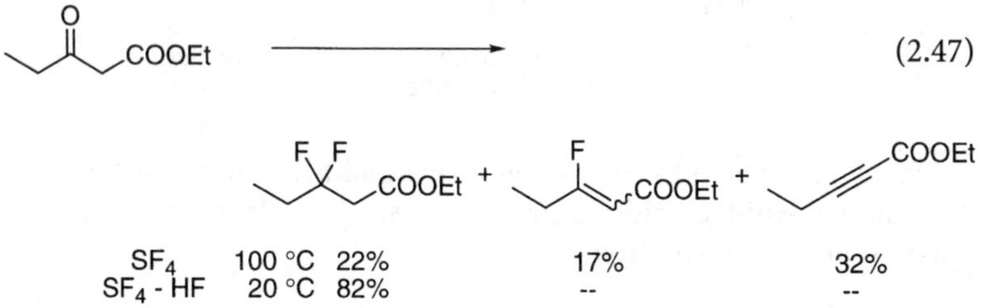

$$\text{CH}_3\text{C(O)CH}_2\text{COOEt} \longrightarrow \text{CH}_3\text{CF}_2\text{CH}_2\text{COOEt} + \text{CH}_3\text{CF=CHCOOEt} + \text{CH}_3\text{C≡CCOOEt-like}$$

		difluoro	vinyl F	alkyne
SF₄	100 °C	22%	17%	32%
SF₄–HF	20 °C	82%	--	--

Sulfur tetrafluoride can transform a carboxylic group to a trifluoromethyl group [124, 125]. The example of *p*-chlorobenzoic acid shows that heating at 160 °C is necessary with SF₄ alone [129], but a combination of SF₄/HF allows a lower reaction temperature and an improvement in the yield [130].

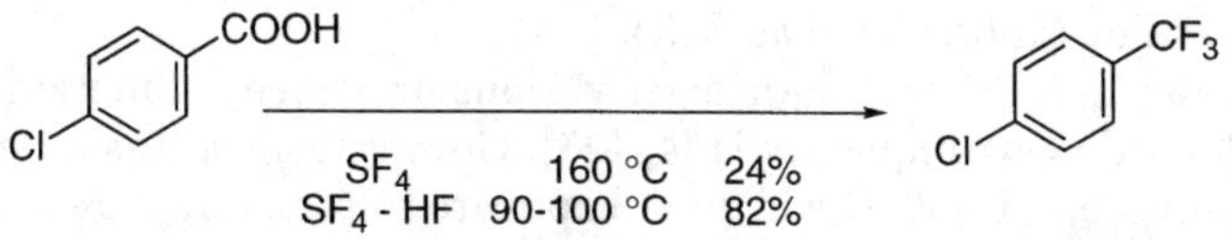

$$p\text{-Cl-C}_6\text{H}_4\text{-COOH} \longrightarrow p\text{-Cl-C}_6\text{H}_4\text{-CF}_3$$

SF₄	160 °C	24%
SF₄–HF	90–100 °C	82%

(2.48)

γ-Butyrolactone is fluorinated at 130 °C with SF₄ to give 4-fluorobutanoyl fluoride but, at 160 °C, the acid fluoride moiety is converted into a trifluoromethyl group [131] to give 1,1,1,4-tetrafluorobutane in 90% yield.

An oxirane ring is deoxygenated by SF₄ to give a *vic*-difluoro product [132].

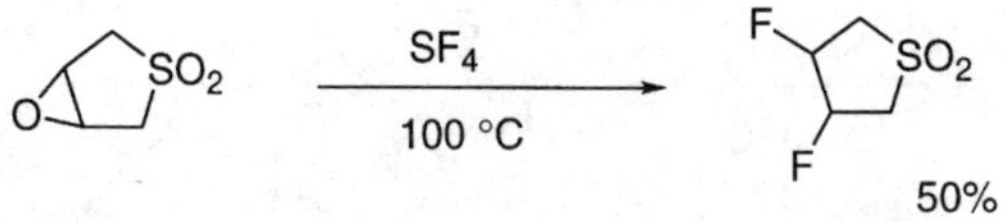

(2.49) — epoxide-sulfolane → *vic*-difluorosulfolane, SF₄, 100 °C, 50%

Reaction of SF₄ with an equimolar amount of dialkylamino(trialkyl)silane gives dialkylaminosulfur trifluoride and trialkylfluorosilane. One in particular, diethylaminosulfur trifluoride, is a liquid reagent called DAST. In addition to DAST, morpholinosulfur trifluoride (morph-DAST) is commercially available and is used for the conversion of alcohols into fluorides [133, 134]. Typical examples are shown below. Generally, the fluorination reaction takes place in high yields at room temperature or below with complete inversion of configuration [135]. Sometimes the fluorination reaction may be accompanied by dehydration or rearrangement [134]. Such side reactions may be suppressed using trimethylsilyl ethers as substrates [136].

$$SF_4 + R_2NSiMe_3 \longrightarrow R_2NSF_3 + FSiMe_3 \quad (2.50)$$

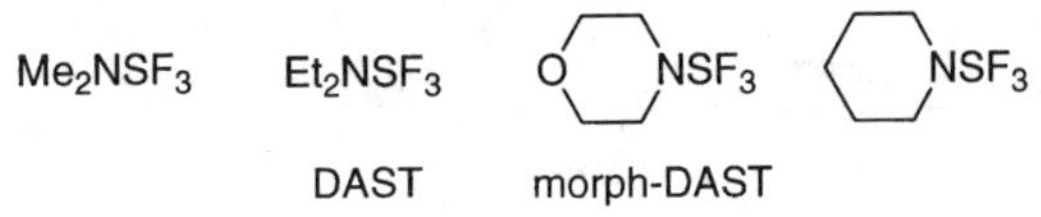

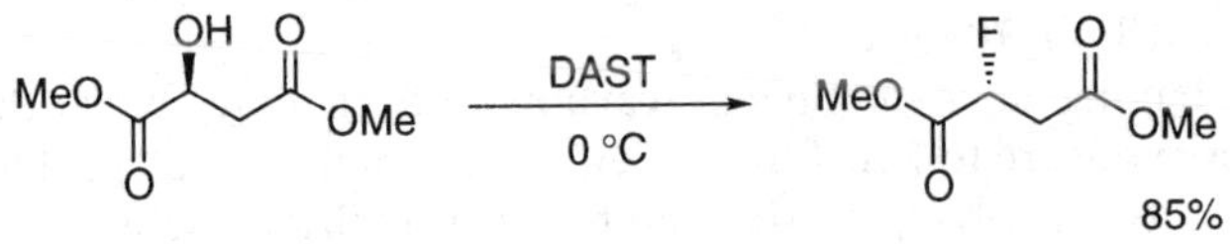

Me$_2$NSF$_3$ Et$_2$NSF$_3$ O(CH$_2$CH$_2$)$_2$NSF$_3$ (CH$_2$)$_5$NSF$_3$
               DAST   morph-DAST

(2.51)

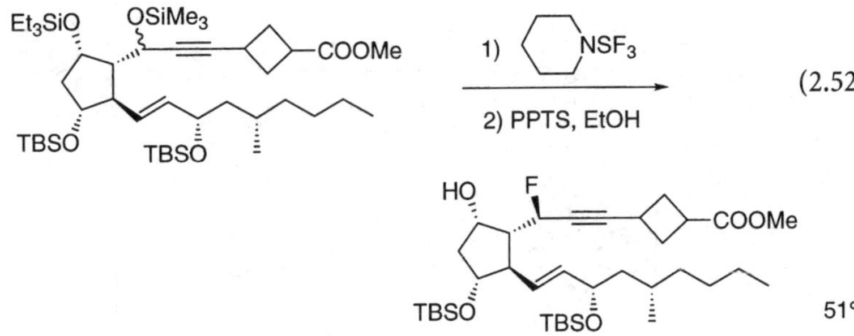

(2.52)

DAST is very often used for the synthesis of fluoro sugars, versatile precursors for glycosidation [137, 138], which will be discussed in Chap. 4.1.

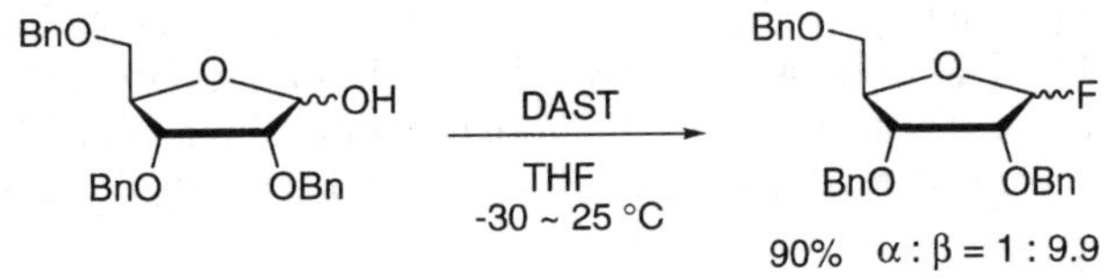

(2.53)

The carbonyl groups of aldehydes and ketones are conveniently converted by DAST into a difluoromethylene unit [134, 140]. gem-Difluorination of aldehydes, in particular, proceeds at room temperature, whereas that of ketones needs heating.

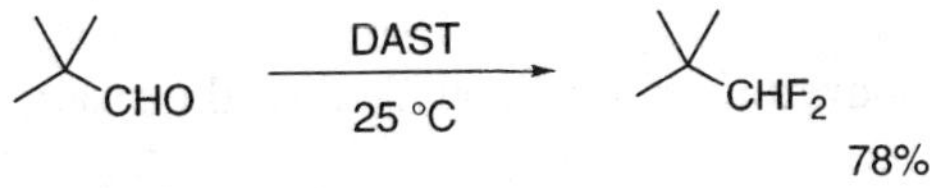

(2.54)

In contrast to SF_4, carboxylic acids are fluorinated to afford acid fluorides [139], which are also produced by the reaction of SF_4 with acid chlorides [140]. Trifluoromethyl formation is not observed in these reactions.

Thiocarbonyl esters are difluorinated by DAST to give α,α-difluoro ethers [141].

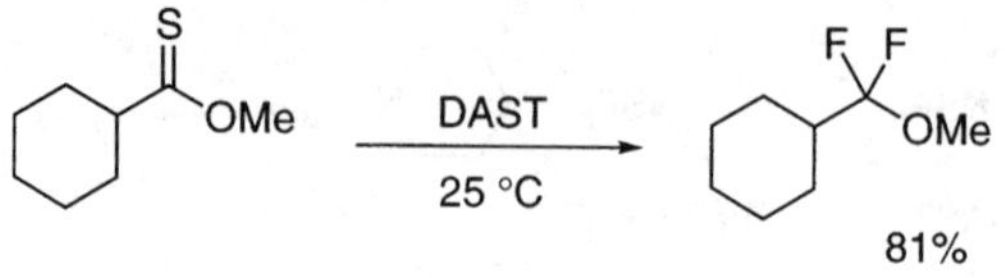

$$\text{Cyclohexyl-C(=S)-OMe} \xrightarrow{\text{DAST, 25 °C}} \text{Cyclohexyl-CF}_2\text{-OMe} \quad 81\% \tag{2.55}$$

The reaction of DAST with tetraethylthiuram disulfide [$Et_2NC(=S)S]_2$ at 20 °C gives Et_2NCF_3 in 70% yield [139].

Sulfoxides having an α-methylene carbon undergo the fluoro-Pummerer reaction upon exposure to DAST and a Lewis acid catalyst [142, 143]. A mechanism is proposed involving a six-membered transition state for α-proton elimination and a fluoride ion attack.

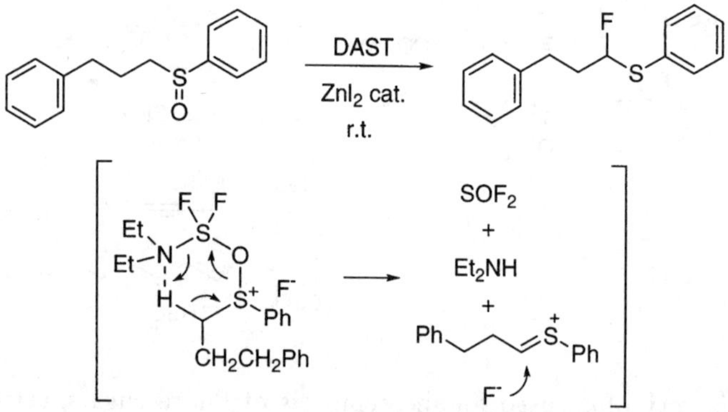

$$\text{Ph(CH}_2)_3\text{S(=O)Ph} \xrightarrow[\text{r.t.}]{\text{DAST, ZnI}_2 \text{ cat.}} \text{Ph(CH}_2)_2\text{CHF-SPh} \tag{2.56}$$

[transition state intermediate → SOF_2 + Et_2NH + Ph-CH=CH-S$^+$-Ph, F$^-$]

The reaction of diethylamine with chlorotrifluoroethene gives 2-chloro-1,1,2-trifluoroethyl(diethyl)amine, which converts steroidal alcohols to fluoro steroids [97, 144]. The reagent is simply called FAR, or the Yarovenko reagent after the inventor, and also converts 1-butanol to 1-fluorobutane. A drawback of the reagent is that it can only be stored for several days in the dark at low temperatures.

$$Et_2NH + CF_2=CFCl \longrightarrow Et_2NCF_2CHFCl \tag{2.57}$$

FAR, Yarovenko reagent

$$R\text{-OH} + Et_2NCF_2CHFCl \longrightarrow R\text{-F} + Et_2NCOCHFCl + HF \tag{2.58}$$

The Yarovenko reagent converts carboxylic acids and their salts into acid fluorides (Scheme 2.9).

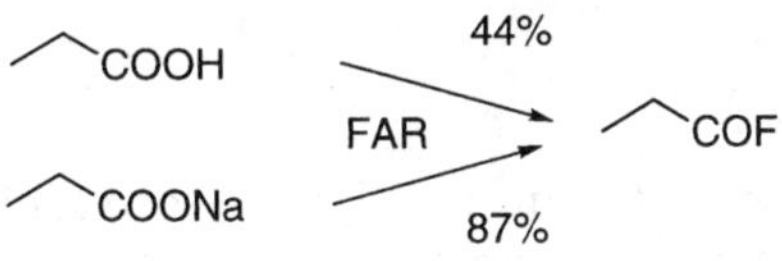

2.2 Nucleophilic Fluorinating Reagents

Scheme 2.9

$$\text{CH}_3\text{CH}_2\text{COOH} \xrightarrow{\text{FAR}, 44\%} \text{CH}_3\text{CH}_2\text{COF}$$
$$\text{CH}_3\text{CH}_2\text{COONa} \xrightarrow{87\%}$$

Use of hexafluoropropene in lieu of chlorotrifluoroethene gives a ca. 1:1 mixture (PPDA) of diethyl(1,1,2,3,3,3-hexafluoropropyl)amine and diethyl(hexafluoro-1-propenyl)amine [145]. The mixture, sometimes called the Ishikawa reagent, can be stored at room temperature and is now commercially available. Like FAR, the reagent fluorinates alcohols. Both components of PPDA are effective for the fluorination. Although primary alcohols are fluorinated in high yields, secondary alcohols often undergo dehydration.

$$\text{Et}_2\text{NH} + \text{CF}_2=\text{CFCF}_3 \longrightarrow \text{Et}_2\text{NCF}_2\text{CHFCF}_3 + \text{Et}_2\text{NCF}=\text{CFCF}_3 \quad (2.59)$$
$$\text{PPDA ca. 1:1}$$

$$\text{PhCH}_2\text{CH}_2\text{OH} \xrightarrow[\substack{\text{Et}_2\text{O} \\ 20\,°\text{C}}]{\text{PPDA}} \text{PhCH}_2\text{CH}_2\text{F} \quad 89\% \quad (2.60)$$

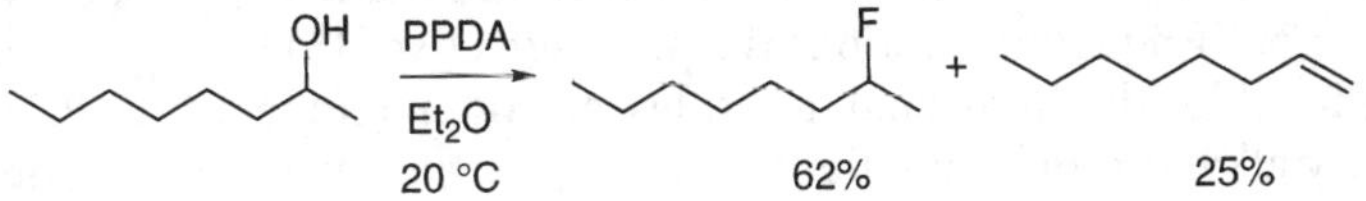

$$\text{sec-alcohol} \xrightarrow[\substack{\text{Et}_2\text{O} \\ 20\,°\text{C}}]{\text{PPDA}} \text{sec-fluoride} \; 62\% + \text{alkene} \; 25\% \quad (2.61)$$

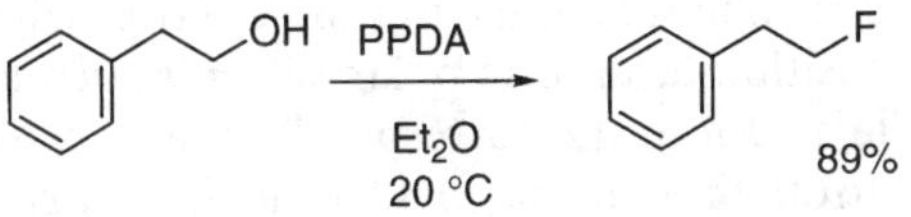

The substitution reaction proceeds mostly with inversion of configuration. Thus, ethyl (R)-mandelate is converted into ethyl (S)-2-fluoro(phenyl)acetate [146].

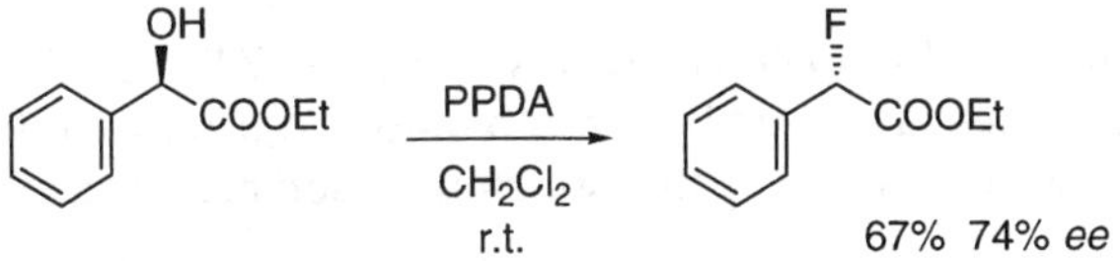

$$\text{PhCH(OH)COOEt} \xrightarrow[\substack{\text{CH}_2\text{Cl}_2 \\ \text{r.t.}}]{\text{PPDA}} \text{PhCHFCOOEt} \quad 67\% \; 74\% \; ee \quad (2.62)$$

The reactive species of FAR and PPDA are assumed to be α-fluoro enamines. Indeed, an α-fluoro enamine generated from DAST and a bulky tertiary amide converts nonanol at 20 °C into 1-fluorononane in high yields [147, 148]. However, a less bulky reagent does not give any fluorination product.

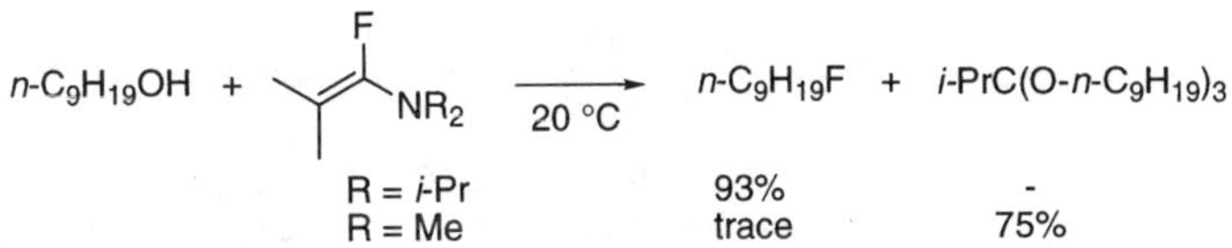

$$n\text{-C}_9\text{H}_{19}\text{OH} + \text{CF=C(NR}_2\text{)} \xrightarrow{20\,°\text{C}} n\text{-C}_9\text{H}_{19}\text{F} + i\text{-PrC}(\text{O-}n\text{-C}_9\text{H}_{19})_3 \quad (2.63)$$

R = i-Pr : 93%, 75%
R = Me : trace, –

2.2.3
Fluoride Reagents

The reagents discussed so far are mostly toxic or corrosive liquids or gases. Therefore, special care and equipment are necessary to perform the fluorination reactions safely. In addition, these reagents are usually costly. In contrast, fluoride reagents are generally much less expensive, stable, easy-to-handle, and thus attractive in view of manufacture. Hereafter typical fluoride reagents, KF, $Bu_4N^+F^-$ (TBAF), $(R_2N)_3S^+[F_2SiMe_3]^-$, and $Ph_4P^+F^-$ are discussed.

It has long been known that fluoride salts can substitute halogen and sulfonate groups in aromatic and aliphatic substrates. Typical examples are KF, CsF, and AgF, all highly hygroscopic. Powdered KF with particle size of 200–300 mm is commercially available [149]. Furthermore, a fine powder of 10–50 mm size can be made by spraying a 30% aqueous solution at between 300 and 500 °C. This powder is called spray-dried KF, is less hygroscopic and is now commercially available [150]. Spray-dried KF has a surface area much larger than conventional KF and thus is much more reactive. These KF reagents are barely soluble in organic solvents.

Spray-dried KF fluorinates halobenzenes having an electron-withdrawing group in the *ortho*- or *para*-position in aprotic polar solvents such as DMSO, DMF, sulfolane or benzonitrile, to afford the corresponding fluorobenzenes. *p*-Nitrochlorobenzene gives *p*-nitrofluorobenzene in 47% yield using commercial KF in DMSO at 250 to 300 °C, whereas use of spray-dried KF gives the fluorobenzene in 82% yield. Freeze-dried KF (ca. 40 mm diameter, prepared by freeze-drying a 5% KF aqueous solution) is equally effective [151]. Spray-dried KF converts acid chlorides into acid fluorides at room temperature. Primary and secondary alkyl bromides are fluorinated to give the corresponding fluorides.

$$\text{cyclohexyl-Br} \xrightarrow[\text{reflux}]{\text{spray-dried KF, MeCN}} \text{cyclohexyl-F} \quad 82\% \tag{2.64}$$

Heating a mixture of chloromalonate with spray-dried KF at 100 °C followed by reaction with an aldehyde or ketone gives α-fluoro-α,β-unsaturated esters. The transformation is understood by the decarboxylation mechanism shown in Scheme 2.10 [152].

To make KF soluble in organic solvents, use of a crown ether is effective [153–155]. For example, heating 1-bromooctane and KF at 83 °C in acetonitrile containing a small amount of 18-crown-6 gives 1-fluorooctane in 92% yield. The less polar solvent benzene is also equally applicable. Likewise, acid chlorides and sulfonyl chlorides are easily converted into the corresponding fluorides at room temperature.

$$\text{n-C}_8\text{H}_{17}\text{-Br} \xrightarrow[\text{18-crown-6}]{\text{KF}} \text{n-C}_8\text{H}_{17}\text{-F} \tag{2.65}$$

MeCN 83 °C 92%
benzene 90 °C 92%

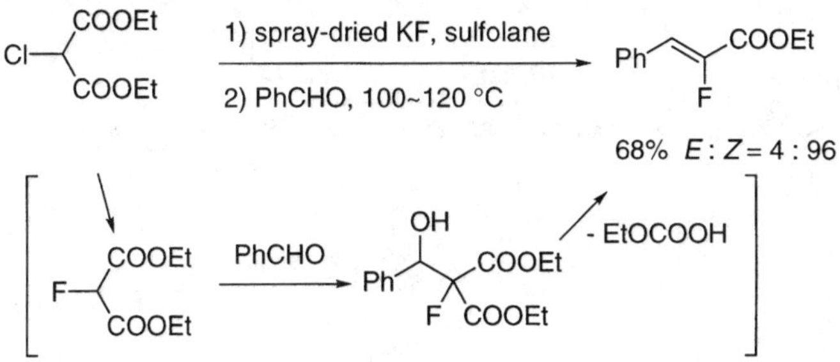

2.2 Nucleophilic Fluorinating Reagents

Scheme 2.10

In addition to 18-crown-6, a phase-transfer catalyst phosphonium salt is effective. Whereas 18-crown-6 does not accelerate the fluorination of *p*-chlorobenzaldehyde with KF, tetraphenylphosphonium bromide does to give *p*-fluorobenzaldehyde in 41% yield. Combined use of 18-crown-6 and the phosphonium salt further enhances the yield up to 74%. In addition to 18-crown-6, trimethyleneglycol dimethyl ether and polyethyleneglycol dimethyl ether also accelerate the nucleophilic substitution [156, 157].

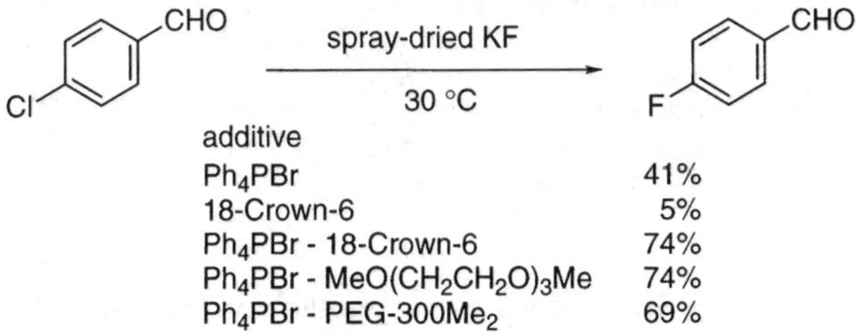

(2.66)

additive	
Ph_4PBr	41%
18-Crown-6	5%
Ph_4PBr - 18-Crown-6	74%
Ph_4PBr - $MeO(CH_2CH_2O)_3Me$	74%
Ph_4PBr - PEG-300Me_2	69%

A nitro group in 1,3-dinitrobenzene is substituted by fluorine upon reaction with KF particularly in the presence of phthaloyl dichloride. The reaction is especially accelerated by Ph_4PBr [158, 159].

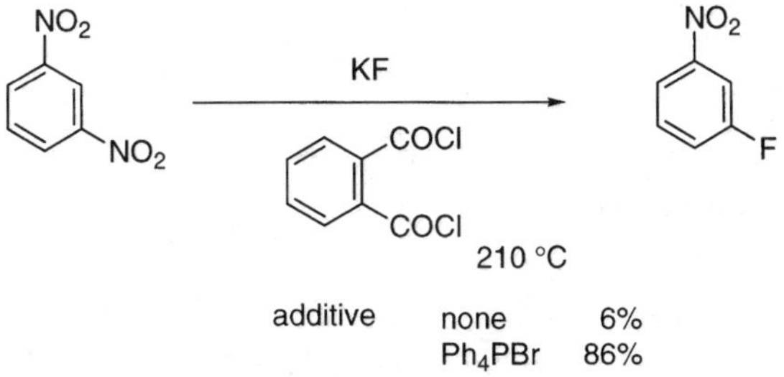

(2.67)

additive	none	6%
	Ph_4PBr	86%

Difluorination takes place with 4-chloro-3-nitrobenzonitrile by combination of phthaloyl dichloride and Ph_4PBr to give 3,4-difluorobenzonitrile. The phosphonium salt may be immobilized in a polystyrene resin [160].

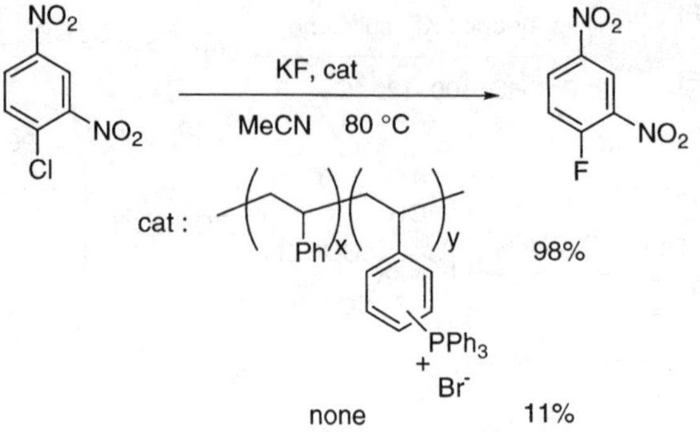

$$\text{(2.68)}$$

2,4-dinitrochlorobenzene + KF, cat / MeCN 80 °C → 2,4-dinitrofluorobenzene

cat: copolymer with Ph/x and y (pendant PPh$_3^+$ Br$^-$ arene) — 98%
none — 11%

Calcium fluoride (CaF$_2$), when used in combination with KF, is an equally effective reagent for nucleophilic fluorination [161, 162]. An example is fluorine substitution of benzyl bromide. Use of KF or CaF$_2$ alone is much less effective. The combined reagent is more active than spray-dried KF alone or a combined reagent of KF and 18-crown-6.

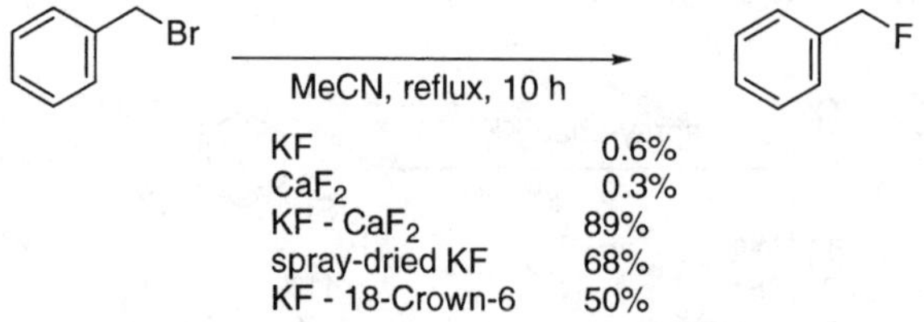

$$\text{(2.69)}$$

PhCH$_2$Br → PhCH$_2$F, MeCN, reflux, 10 h

Reagent	Yield
KF	0.6%
CaF$_2$	0.3%
KF - CaF$_2$	89%
spray-dried KF	68%
KF - 18-Crown-6	50%

At high temperatures, five chlorines in pentachlorobenzonitrile are fully fluorinated by KF to give pentafluorobenzonitrile, a synthetic intermediate of color film materials and artificial antibiotics [163].

Potassium hydrogen fluoride (potassium difluoride, KHF$_2$) also serves as a fluoride ion reagent and reacts with epoxides to give fluorohydrins. This reaction, when applied to epoxy sugars, gives such fluoro sugars [164] as fluoro-D-arabinofuranoside [165].

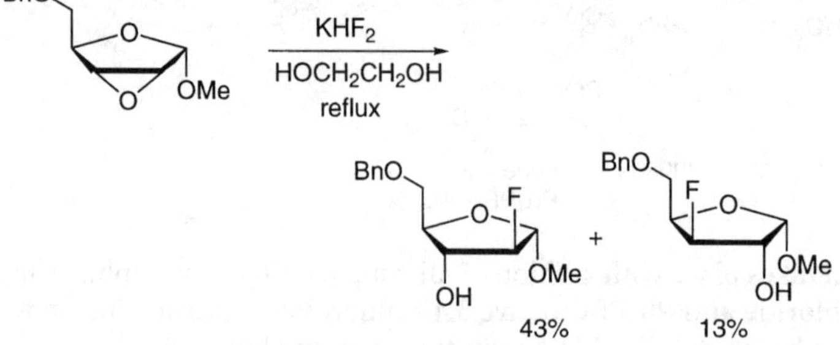

$$\text{(2.70)}$$

BnO-epoxy sugar-OMe + KHF$_2$ / HOCH$_2$CH$_2$OH, reflux → fluorohydrin products (43% + 13%)

2.2 Nucleophilic Fluorinating Reagents

Likewise, fluoro hydroxy vitamin D_3 is prepared starting from an epoxy steroid as shown below [166].

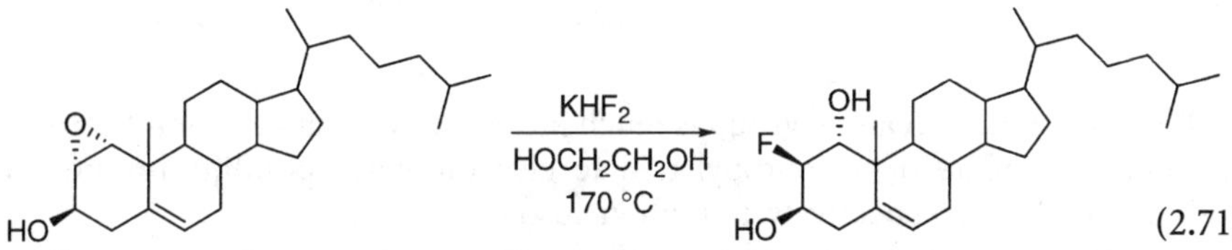

$$\text{epoxy steroid} \xrightarrow[\text{170 °C}]{\text{KHF}_2, \text{HOCH}_2\text{CH}_2\text{OH}} \text{fluoro hydroxy steroid} \tag{2.71}$$

Tetraalkylammonium fluorides are highly nucleophilic fluoride ion reagents soluble in organic solvents. In particular, $Bu_4N^+F^-$, prepared by halogen exchange of $Bu_4N^+Br^-$ or by neutralization of $Bu_4N^+OH^-$ with hydrofluoric acid, is very often used. This reagent is normally called TBAF. Its trihydrate form is stable and commercially available. However, dehydration under vacuum induces its decomposition via E2-elimination. To prevent this elimination reaction, a fluoride reagent is often employed that has a counterion like Me_4N^+, $PhCH_2(Me_3)N^+$, $Me_3(t\text{-}BuCH_2)N^+$ or 1,1,3,3,5,5-hexamethylpiperidinium. These fluorides appear less nucleophilic than TBAF. Anhydrous $Me_4N^+F^-$ is prepared by neutralization of its hydroxide with hydrofluoric acid, followed by azeotropic removal of water with isopropyl alcohol and recrystallization from the same solvent. This salt decomposes slowly at 170 °C [167].

Treatment of TBAF with active halides like benzyl bromide or benzoyl chloride without a solvent gives the corresponding fluorides [168]. Primary alkyl bromides, e.g. octyl bromide, are fluorinated in moderate yields; secondary bromides undergo dehydrobromination to give olefins as major products. Use of tosylates in these cases improves the yields remarkably.

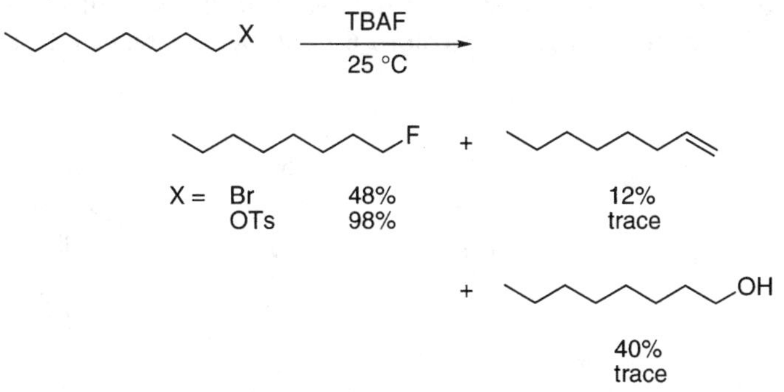

$$R\text{-}X \xrightarrow[25\,°C]{\text{TBAF}} R\text{-}F + \text{olefin} + R\text{-}OH \tag{2.72}$$

X =	R-F	olefin	R-OH
Br	48%	12%	40%
OTs	98%	trace	trace

A substrate having both primary and secondary hydroxyl groups is selectively fluorinated to give a primary alkyl fluoride by reaction with TBAF and tosyl fluoride [169]. A mechanism involving selective tosylation followed by nucleophilic substitution is suggested.

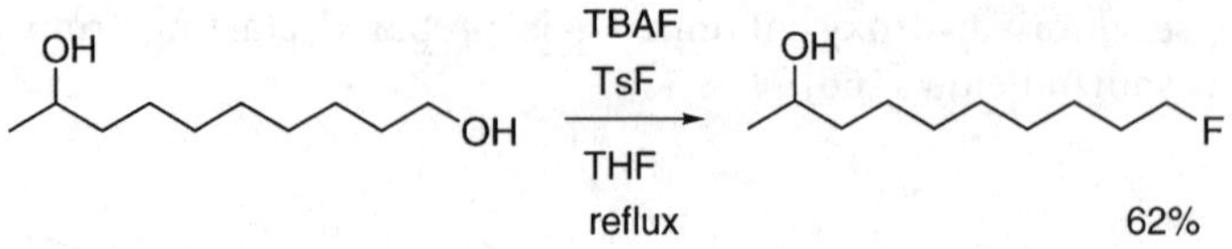

[Scheme 2.73: diol + TBAF/TsF in THF reflux → fluorohydrin, 62%]

Because a triflate leaving group is much more reactive than a bromide group, bromoethyl triflate or iodopropyl triflate give the corresponding bromo- or iodofluoroalkane, respectively, in high yields [170].

A sulfonate ester moiety of various alcohols is replaced by fluorine with TBAF [171–173].

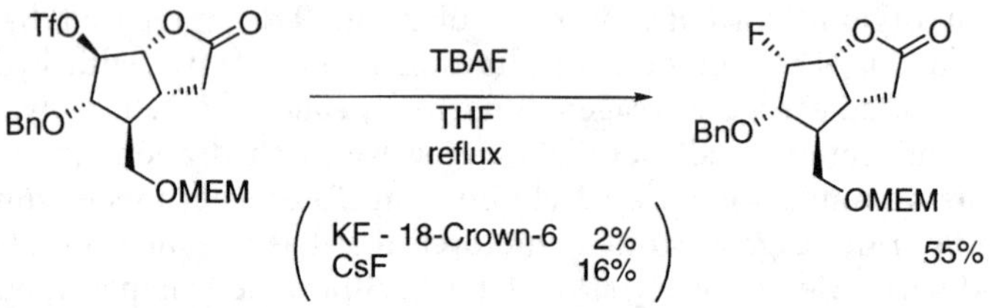

[Scheme 2.74: triflate bicyclic lactone + TBAF/THF reflux → fluoride product, 55% (KF-18-Crown-6 2%; CsF 16%)]

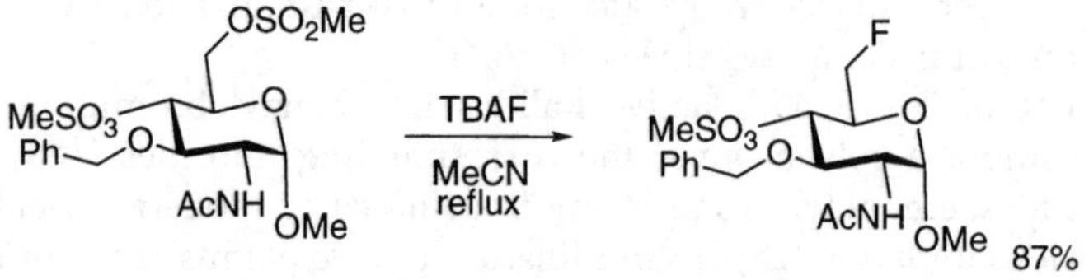

[Scheme 2.75: sugar mesylate + TBAF/MeCN reflux → fluoro sugar, 87%]

Denitrofluorination of nitrobenzenes is carried out with TBAF [174]. o-Dinitrobenzene is converted quantitatively into 2-fluoronitrobenzene, and 2-chloro-6-nitrobenzonitrile to 2-chloro-6-fluorobenzonitrile and 2,6-difluorobenzonitrile.

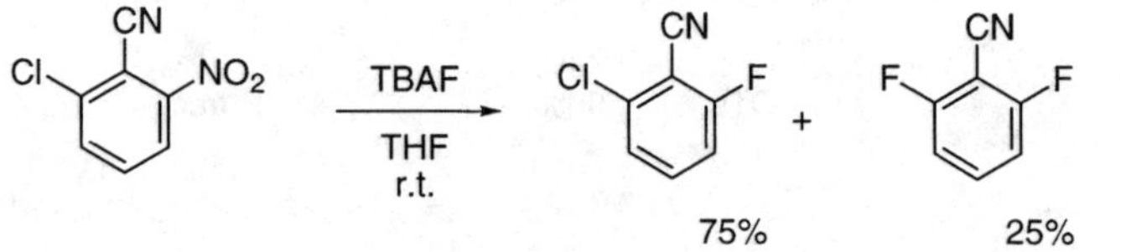

[Scheme 2.76: 2-chloro-6-nitrobenzonitrile + TBAF/THF r.t. → 2-chloro-6-fluorobenzonitrile (75%) + 2,6-difluorobenzonitrile (25%)]

Tetrabutylammonium hydrogen fluoride ($Bu_4N^+HF_2^-$; TBA·HF_2) is prepared by exchange of a counter anion of $Bu_4N^+HSO_3^-$ with $KHCO_3$ and then with KHF_2.

Like TBAF, TBA·HF_2 fluorinates alkyl halides and sulfonates in a 1:1 mixture of THF and HMPA at 95 °C, a temperature slightly higher than the reaction temperature of TBAF, to give fluorine-substituted products with much less olefin formation [175, 176]. Nucleophilic fluorination of *p*-chloronitrobenzene also takes place, but a longer reaction time is necessary.

$$n\text{-}C_{12}H_{25}\text{-}X \xrightarrow[\text{THF : HMPA (1 : 1)}]{\text{TBA·HF}_2} n\text{-}C_{12}H_{25}\text{-}F + n\text{-}C_{10}H_{21}\text{-}CH=CH_2 \quad (2.77)$$
$$95\,°C$$

X =				
	Cl	22 h	83%	5%
	OTs	14 h	96%	--
	OMs	6 h	~100%	--

Use of an equimolar amount of pyridine with TBA·HF$_2$ suppresses an elimination reaction to give the desired fluorine-substituted products in higher yields [177]. This modification is also effective for secondary bromides.

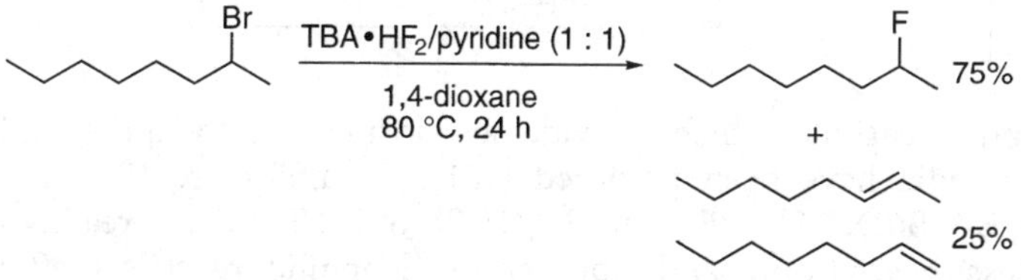

$$(2.78)$$

with conditions TBA·HF$_2$/pyridine (1:1), 1,4-dioxane, 80 °C, 24 h, giving fluoride product 75% and olefin 25%.

Tetrabutylammonium dihydrogen trifluoride (Bu$_4$N$^+$H$_2$F$_3^-$; TBA·H$_2$F$_3$) is readily prepared by treatment of TBAF with hydrofluoric acid and excess KHF$_2$ in 1,2-dichloroethane. The resulting TBA·H$_2$F$_3$ is dissolved in an organic solvent and is easily concentrated in vacuo by azeotropy to give the very nucleophilic fluoride reagent. Alternatively, the salt may be prepared using an ion-exchange resin like Amberlyst A26 or Amberlite IRA 900.

The reagent allows HF to add across an electron-deficient carbon–carbon triple bond [178, 179]. The stereochemistry of the addition is *trans* with dimethyl acetylenedicarboxylates and methyl phenylacetylenecarboxylate, whereas *cis* addition is not negligible with 2-alkynoates and 2-alkynenitriles.

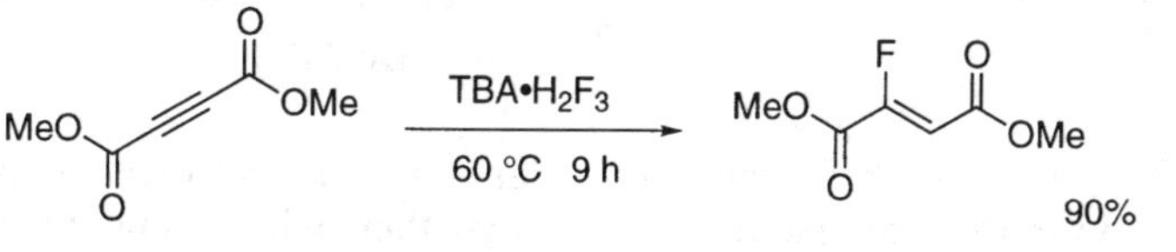

$$\text{MeO}_2\text{C-C}\equiv\text{C-CO}_2\text{Me} \xrightarrow[60\,°C\ 9\,h]{\text{TBA·H}_2\text{F}_3} \text{(F,CO}_2\text{Me) alkene, 90\%} \quad (2.79)$$

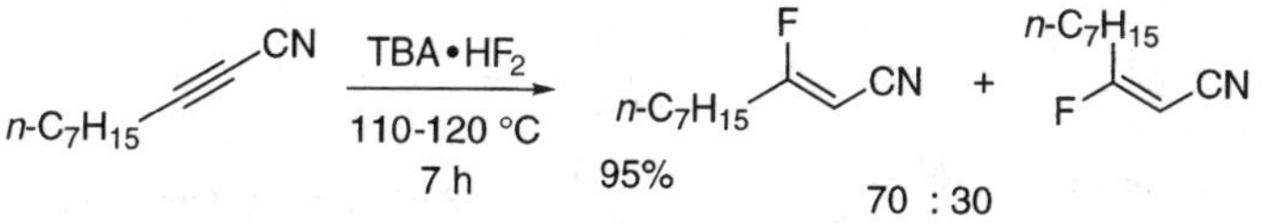

$$n\text{-}C_7H_{15}\text{-C}\equiv\text{C-CN} \xrightarrow[\substack{110\text{-}120\,°C \\ 7\,h}]{\text{TBA·HF}_2} \text{fluoroalkenes, 95\%, 70 : 30} \quad (2.80)$$

Ring-opening fluorination of oxiranes is readily carried out with TBA·H$_2$F$_3$ [180]. Although KHF$_2$ is barely soluble in organic solvents, use of 10 mol% of TBA·H$_2$F$_3$ and solid KHF$_2$ at 120 °C gives fluorohydrins in high yields. The coproduced reagent TBA·HF$_2$ is assumed to be readily converted back to TBA·H$_2$F$_3$ by KHF$_2$. Thus, TBA·H$_2$F$_3$ is considered to be a highly nucelophilic fluoride ion supplier.

$$R\overset{O}{\triangle} \xrightarrow[120\ °C]{\text{KHF}_2\ (\text{solid})\atop \text{TBA·H}_2\text{F}_3\ (10\ \text{mol\%})} R\overset{OH}{\underset{}{\diagdown}}F + R\overset{F}{\underset{}{\diagdown}}OH \quad (2.81)$$

R =			
Ph	8 h	74%	61 : 39
n-C$_{10}$H$_{21}$	48 h	84%	74 : 26
PhOCH$_2$	6 h	90%	100 : 0

$$\left[\ \overset{O}{\triangle} + \text{TBA·H}_2\text{F}_3 \xrightarrow{\text{KHF}_2\ (\text{solid})} \overset{OH}{\diagdown}F + \text{TBA·HF}_2\ \right]$$

Recently, reagents which provide a fluoride ion by an equilibrium in organic media have been prepared [181]. Examples are (R$_2$N)$_3$S$^+$ Me$_3$SiF$_2^-$ (R = Me, Et), Bu$_4$N$^+$ Ph$_3$SiF$_2^-$, and Bu$_4$N$^+$ Ph$_3$SnF$_2^-$. These are readily prepared, much less hygroscopic and undergo nucleophilic reactions efficiently in aprotic organic solvent. In particular, tris(dimethylamino)sulfonium difluorotrimethylsilicate (TASF) is prepared from Me$_3$SiNMe$_2$ and SF$_4$ and is commercially available. Its ethyl analog is also called TASF, and thus there is often confusion. Here we discriminate them by noting TASF (Me) and TASF (Et), respectively.

$$3\ \text{Me}_3\text{SiNMe}_2 + \text{SF}_4 \xrightarrow[25\ °C]{\text{Et}_2\text{O}} [(\text{Me}_2\text{N})_3\text{S}]^+(\text{F}_2\text{SiMe}_3)^- + 2\ \text{FSiMe}_3 \quad (2.82)$$
$$\text{TASF (Me)}$$

$$2\ \text{Me}_3\text{SiNEt}_2 + \text{Et}_2\text{NSF}_3 \xrightarrow[25\ °C]{\text{Et}_2\text{O}} [(\text{Et}_2\text{N})_3\text{S}]^+(\text{F}_2\text{SiMe}_3)^- + \text{FSiMe}_3 \quad (2.83)$$
$$\text{TASF (Et)}$$

TASF (Et) generates a fluoride ion under strictly anhydrous conditions and thus is used for an aldol reaction of enol silyl ethers with aldehydes [182, 183].

$$\overset{OSiMe_3}{\diagup\!\!\diagdown} + \text{PhCHO} \xrightarrow[\text{THF}\atop 78\ °C]{\text{TASF (Et)}} \overset{O\ \ OH}{\diagup\!\!\diagdown\!\!\diagup}\text{Ph} \quad (2.84)$$

(E : Z = 0 : 100)   65% erythro : threo = 86 : 14

TASF undergoes nucleophilic fluorination of triflates. This reaction is a highly efficient alternative of the fluorination of mesylates with TBAF or of alcohols with DAST (Scheme 2.11). Analogous reagents, Bu$_4$N$^+$ Ph$_3$SiF$_2^-$ [184,

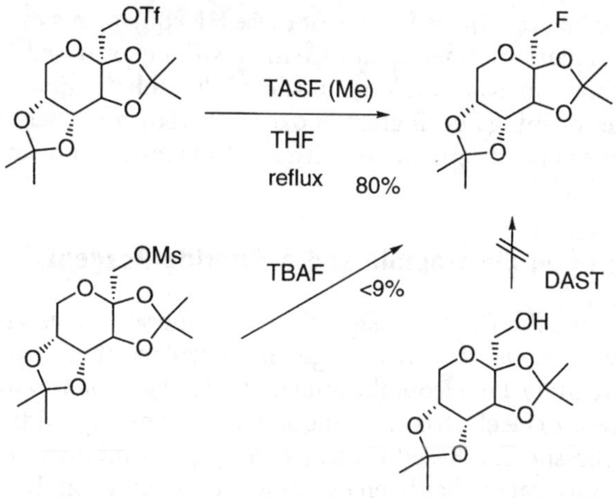

2.2 Nucleophilic Fluorinating Reagents

Scheme 2.11

185] and $Bu_4N^+ Ph_3SnF_2^-$ [186], are readily prepared from TBAF and triphenylsilyl fluoride and triphenylstannyl fluoride, respectively.

$$Bu_4NF + Ph_3SiF \xrightarrow[\text{r.t.}]{CH_2Cl_2} [Bu_4N]^+[Ph_3SiF_2]^- \qquad (2.85)$$

$$Bu_4NF + Ph_3SnF \xrightarrow[\text{r.t.}]{CH_2Cl_2} [Bu_4N]^+[Ph_3SnF_2]^- \qquad (2.86)$$

$Bu_4N^+ Ph_3SiF_2^-$ undergoes nucleophilic substitution in a manner similar to TBAF but in a longer reaction time. Formation of olefin by-products is not so prevalent.

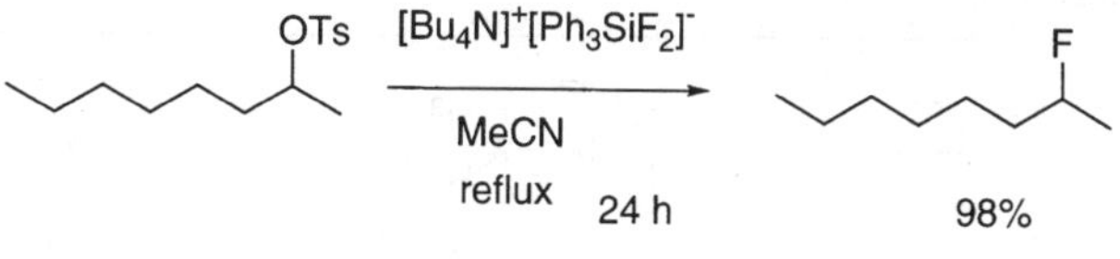

$$\text{R-OTs} \xrightarrow[\substack{\text{MeCN} \\ \text{reflux} \quad 24\,h}]{[Bu_4N]^+[Ph_3SiF_2]^-} \text{R-F} \quad 98\% \qquad (2.87)$$

The order of nucleophilicity of a fluoride ion is compared in a bromine-fluorine exhange reaction: anhydrous TBAF > $TBAHF_2$ > $TBAPh_3$, $SnF_2 \cdot TBAPh_3SiF_2$; selectivity order: $TBAPh_3SiF_2$ ~ $TBAHF_2$ > anhydrous TBAF.

Phosphonium analogs of the quaternary ammonium fluoride salts can be prepared with the exception of $Me_4P^+F^-$ [187]; the reactivities are disclosed to be parallel to the corresponding ammonium salts. Unique reagents are tetraarylphosphonium salts $Ar_4P^+F^-$, $Ar_4P^+HF_2^-$, and $Ar_4P^+H_2F_3^-$. These, when prepared by a standard procedure, give hydrates like $Ph_4P^+F^- \cdot 4H_2O$ and $Ph_4P^+HF_2^- \cdot H_2O$. $Ph_4P^+HF_2^- \cdot 2H_2O$ loses a crystal of water at 0.1 mmHg, 50 °C. The nucleophilicity order of these fluoride reagents is: $Ph_4P^+ HF_2^- > Ph_4P^+ H_2F_3^- > Ph_4P^+F^-$. For nucleophilic substitution with $Ph_4P^+HF_2^-$, two moles of the well-dried reagent are required. Some of these HF equivalents may allow the use of conventional glassware but great care must still be exercised.

The phosphonium salts undergo nucleophilic substitution at sp^3 carbons [188–190], ring-opening fluorination of oxiranes [188], and nucleophilic fluorine substitution of aromatic compounds with an electron-withdrawing group [191].

2.3
Combination of an Electrophile and a Fluoride Reagent

Since electrophilic fluorinating reagents are in general expensive and often too reactive to be selective, it is advantageous to replace them with fluoride ion reagents. To facilitate a nucleophilic attack of a fluoride ion towards a substrate, a substrate needs to be electrophilic enough or to be mildly oxidized to generate an elelctrophilic species. Based on this concept, a combination of an oxidant and a fluoride ion reagent has been examined. Depending on the substrate, both the oxidant and the fluoride ion reagent can be optimized to give the desired product in high yields.

2.3.1
Halofluorination of Olefins and Acetylenes

Originally, halofluorination of olefins was carried out with a halogen fluoride that had been prepared from molecular halogen and fluorine gas [192]. A more convenient reagent system is made by combining a positive halogen oxidant like *N*-haloimide and a fluoride ion reagent. Some examples are listed in Table 2.4 [101a, 193–200]. Every reagent system affords halofluorination products in good yields. The reaction is considered to proceed through a halonium ion intermediate and thus to give *trans*-adducts.

Table 2.4. Halofluorination of olefins

F^-	X^+	References
70% HF/py	NBS, DBH, NIS	[101a]
60% HF/PVP	NIS	[193]
33% HF/Et$_3$N	NBS	[194]
Bu$_4$NF·3H$_2$O	NBS	[195]
Bu$_4$NH$_2$F$_3$	DBH, NIS	[196]
NH$_4$HF$_2$-AlF$_3$	NIS	[197]
SiF$_4$	DBH	[198]
HFP/DA	DBH, NIS	[199]
Mtl$_m$(HF)$_n$	DBH, NIS	[200]

HFP/DA; hexafluoropropene-diethylamine complex.

2.3 Combination of an Electrophile and a Fluoride Reagent

Some reagents deserve comment. TBA·H_2F_3 allows the use of conventional glassware. 60% HF/PVP is readily removed after the reaction simply by filtration. The Olah reagent, 70% HF/pyridine, is very reactive to give adducts of electron-deficient olefins such as α,β-unsaturated esters and fluoroolefins. By tuning the ratio of HF and pyridine, labile olefins like styrene also undergo halofluorination. The product can be further converted into a difluorination product with the aid of silver fluoride [101a].

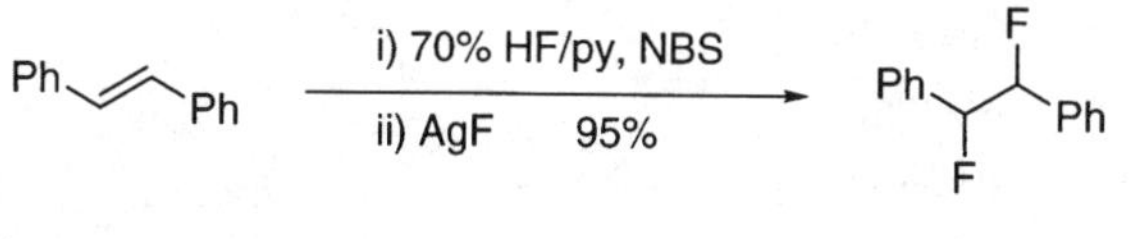

$$\text{Ph}\diagup\diagdown\text{Ph} \xrightarrow[\text{ii) AgF \quad 95\%}]{\text{i) 70\% HF/py, NBS}} \text{Ph-CF}_2\text{-CHF-Ph} \qquad (2.88)$$

The reaction may be carried out in a two-phase system, using NIS, TBAF, KHF_2, and 1M hydrofluoric acid in dichloromethane. Hereby TBA·H_2F_3 is considered to be generated and to move through two phases as a phase-transfer catalyst. Some examples are shown in Table 2.5 [201]. As dilute hydrofluoric acid is readily accessible, the phase-transfer reaction may find wide applications.

Alkynes also undergo the halofluorination reaction to give halofluoroalkenes [202].

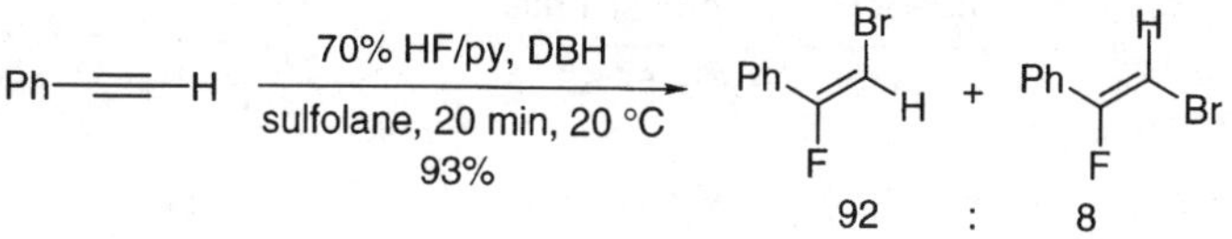

$$\text{Ph−≡−H} \xrightarrow[\text{sulfolane, 20 min, 20 °C}]{\text{70\% HF/py, DBH}} \underset{92}{\text{Ph(F)C=CH(Br)}} + \underset{8}{\text{Ph(F)C=C(H)Br}} \qquad (2.89)$$

93%

Table 2.5. Iodofluorination of olefins with NIS and TBAF/KHF_2

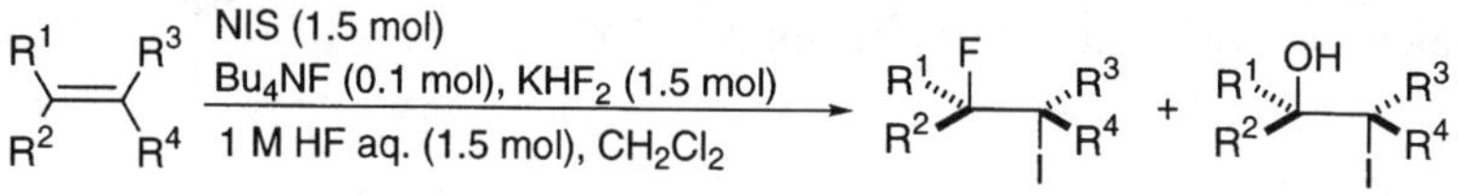

$$R^1R^2C=CR^3R^4 \xrightarrow[\text{1 M HF aq. (1.5 mol), CH}_2\text{Cl}_2]{\text{NIS (1.5 mol), Bu}_4\text{NF (0.1 mol), KHF}_2\text{ (1.5 mol)}} R^1R^2C(F)-CR^3R^4(I) + R^1R^2C(OH)-CR^3R^4(I)$$

Product	X = F	X = OH	Product	X = F	X = OH
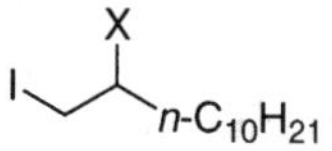 I-CH$_2$-CHX-n-C$_{10}$H$_{21}$	82%	0.2%	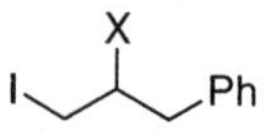 I-CH$_2$-CHX-Ph	78%	5%
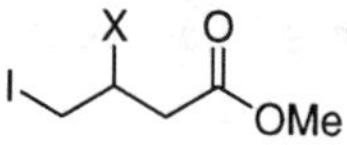 I-CH$_2$-CHX-C(O)OMe	77%	5%	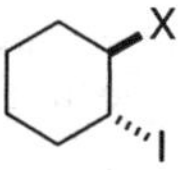 cyclohexane with X and I trans	75%	trace
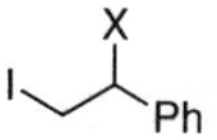 I-CH$_2$-CX(Ph)-CH$_3$	59%	6%	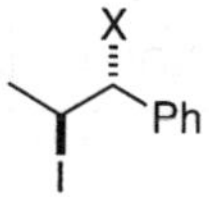 I-CH(CH$_3$)-CX(Ph)-CH$_3$	70%	trace

2.3.2
Thiofluorination and Selenofluorination of Olefins

Using a reagent that generates an electrophilic sulfur species in lieu of an N-haloimide, thiofluorination readily takes place selectively with Markovnikov selectivity and *anti* stereochemistry. An olefin is considered to first react with dimethyl(methylthio)sulfonium fluoroborate then with $3HF/NEt_3$ [203].

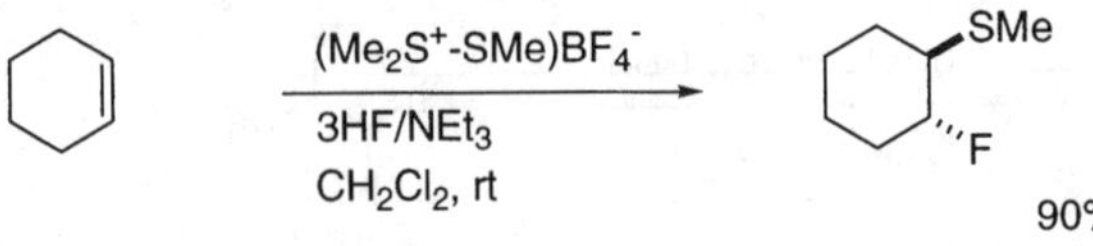

$$\text{cyclohexene} \xrightarrow[\text{CH}_2\text{Cl}_2,\text{ rt}]{\substack{(Me_2S^+\text{-}SMe)BF_4^- \\ 3HF/NEt_3}} \text{trans-2-(methylthio)fluorocyclohexane, 90\%} \quad (2.90)$$

N-Phenylthiophthalimide [204] or phenylsulfenyl chloride [205] in combination with 70% HF/pyridine can also achieve thiofluorination of olefins and acetylenes with *trans* stereochemistry.

In a similar manner, selenofluorination of olefins and acetylenes proceeds in good yields using N-phenylselenylphthalimide and 33% HF/NEt_3 [206]. It is possible to selenofluorinate an olefin using PhSeBr and AgF under sonication [207].

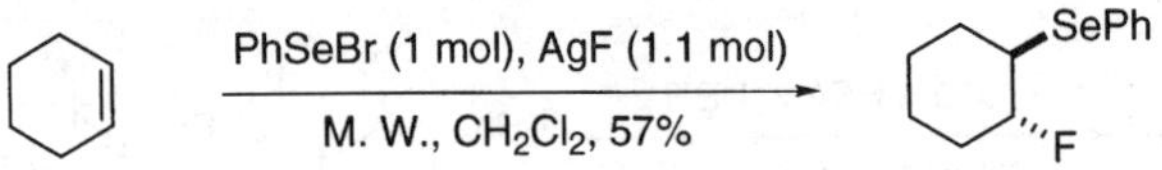

$$\text{cyclohexene} \xrightarrow[\text{M. W., CH}_2\text{Cl}_2,\text{ 57\%}]{\text{PhSeBr (1 mol), AgF (1.1 mol)}} \text{trans-2-(phenylseleno)fluorocyclohexane} \quad (2.91)$$

2.3.3
Nitrofluorination

Addition of F and NO_2 groups to an olefin in a *trans* manner proceeds using nitronium tetrafluoroborate and 70% HF/pyridine. The resulting β-fluoro(nitro)alkanes are reduced to β-fluoro amines [208].

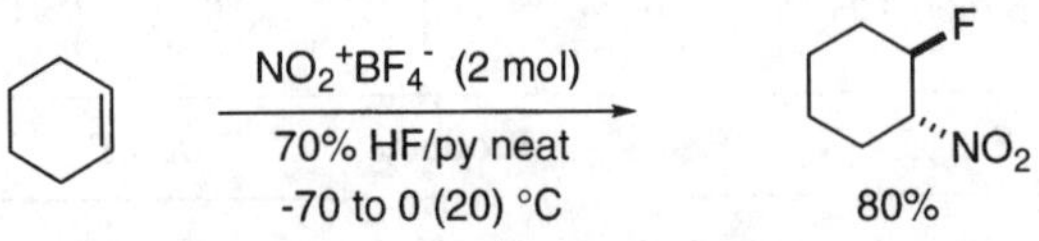

$$\text{cyclohexene} \xrightarrow[\substack{70\% \text{ HF/py neat} \\ -70 \text{ to } 0 \text{ (20) °C}}]{NO_2^+BF_4^- \text{ (2 mol)}} \text{trans-2-fluoronitrocyclohexane, 80\%} \quad (2.92)$$

In contrast, reaction of diarylacetylenes with nitrosonium tetrafluoroborate and 60% HF/pyridine gives 1,2-diaryl-substituted 1,1,2,2-tetrafluoroethanes [209], products equivalent to double fluorination of the acetylenes. The mechanism in Scheme 2.12 is suggested for this reaction.

$$C_6H_5C\equiv CC_6H_5 \xrightarrow[\text{CH}_2\text{Cl}_2,\text{ 0 °C to rt}]{NO^+BF_4^- \text{ (3 mol), 60\% HF/py}} C_6H_5CF_2CF_2C_6H_5 \quad (2.93)$$
$$75\%$$

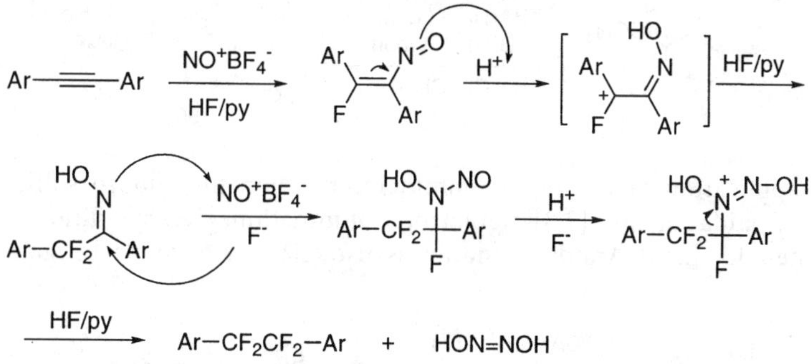

2.3 Combination of an Electrophile and a Fluoride Reagent

Scheme 2.12. Suggested mechanism of tetrafluorination of diarylacetylenes

2.3.4
Oxidative Fluorination

Since oxidation of organosulfur compounds easily affords electrophilic species, the reaction carried out in the presence of a fluoride ion induces fluorination of organosulfur compounds. Although this type of reaction has been achieved with HF/CF$_3$OF and HF/F$_2$, a facile method employs NBS and 70% HF/pyridine or NBS and DAST [210]. The reaction is particularly convenient for the synthesis of fluoro sugars, precursors of O-glycosidation.

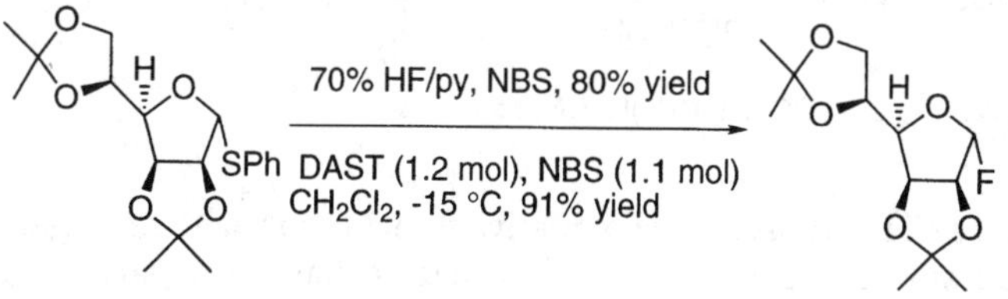

$$(2.94)$$

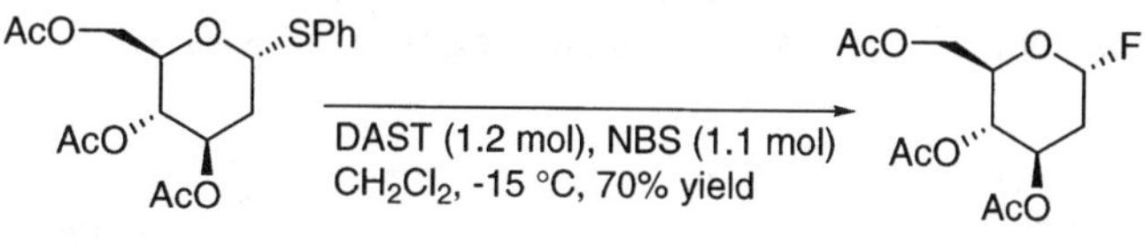

$$(2.95)$$

Very recently, ArIF$_2$ has been shown to be the reagent of choice for this transformation [211]. Hereby the stereochemistry of fluorination is mostly retention of configuration probably through an S$_N$i mechanism [212].

A combination of NOBF$_4$ and 60% HF/pyridine induces substitution of a phenylthio group with fluorine. An alternative reagent system of methyl fluorosulfate and CsF is also known [213].

When sulfides are treated with 3,3-dimethyl-1,3-dibromohydantoin (DBH) and Bu$_4$NH$_2$F$_3$, fluoro-Pummerer-type products are obtained [214].

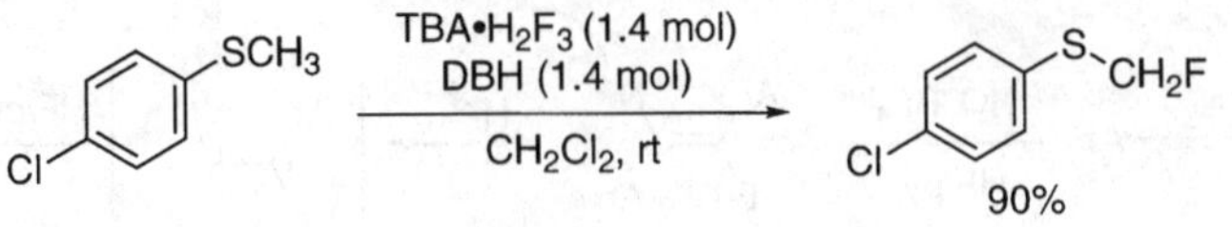

$$\text{(2.96)}$$

An application of the fluoro-Pummerer reaction to β-fluoro sulfides, followed by oxidation and [2,3]-sigmatropic elimination, gives *vic*-difluoroolefins (Scheme 2.13) [215]. Anodic oxidation is also effective to convert organosulfur

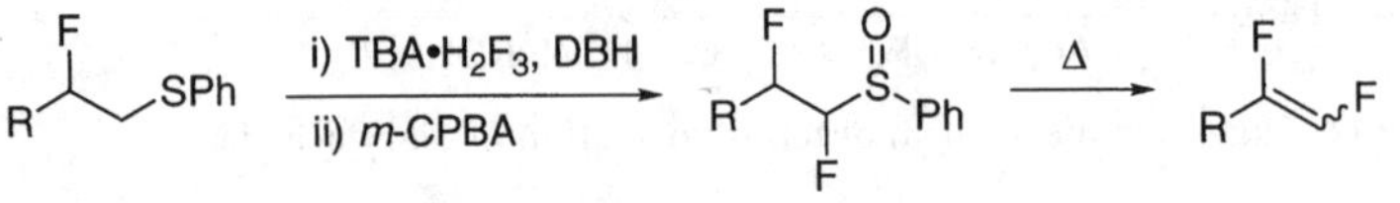

Scheme 2.13. Synthesis of *vic*-difluoroolefins

substrates into electrophilic species, and thus the fluoro-Pummerer reaction is easily achieved [216]. The reaction takes place with particular ease when organosulfur compounds have an electron-withdrawing group at an α-carbon. For this reaction 3HF/Et$_3$N is shown to be nucleophilic enough and tolerant of the oxidation conditions owing to its high oxidation potential. Substrates containing a lactam [217] or heteroaromatic ring [218] are also applicable. Organic selenides also undergo a similar fluorination reaction [219].

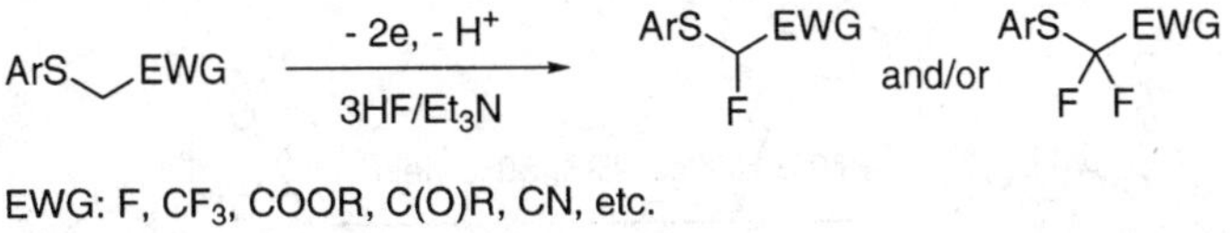

$$\text{(2.97)}$$

EWG: F, CF$_3$, COOR, C(O)R, CN, etc.

Although a diastereoselective oxidative fluorination with an organic oxidant is yet to be studied [220], anodic fluorination has been partially successful [221]. An example is shown in Eq. (2.98), wherein a phenyl group in a chiral auxiliary is said to protect an *Re* face of an electrophilic reaction center, thus allowing a nucleophilic attack of a fluoride ion from an *Si* face.

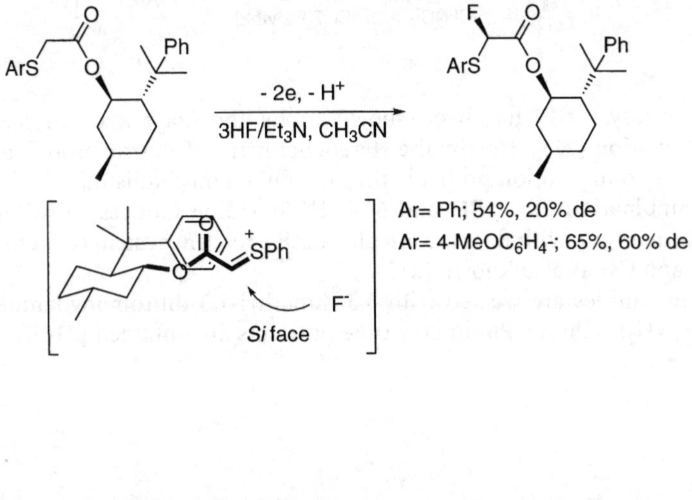

Ar= Ph; 54%, 20% de
Ar= 4-MeOC$_6$H$_4$-; 65%, 60% de

$$\text{(2.98)}$$

2.3 Combination of an Electrophile and a Fluoride Reagent

As already discussed in Chap. 2.1.4, an asymmetric fluorination with an electrophilic reagent proceeds with relatively high diastereoselectivity. In contrast, nucleophilic fluorination lacks such high selectivity, probably because a fluoride ion is too small to be affected by a steric bulk and/or an electron transfer mechanism may be involved.

2.3.5
Oxidative Desulfurization-Fluorination

Compounds containing a *gem*-difluoromethylene functionality at a specific position have been prepared by treatment of carbonyl substrates with the highly toxic reagent SF_4 at 100 °C [222]. However, protected forms, 1,3-dithiolanes or 1,3-dithianes, have proved to be much more convenient substrates. This transformation was first achieved with CF_3OF [223]. Later, the functional transformation was carried out with a combined reagent system of HF/pyridine and an oxidant like DBH [224], SO_2ClF [225], or $NOBF_4$ [212]. In particular, a reaction using HF/pyridine and a positive halogen oxidant is often accompanied by halogenation of an aromatic part of the substrates and/or side reactions caused by the acidic reagent.

Such side reactions can be avoided by use of $TBA \cdot H_2F_3$ as a fluoride ion reagent. Even an acid-labile oxirane ring tolerates the reaction conditions [226].

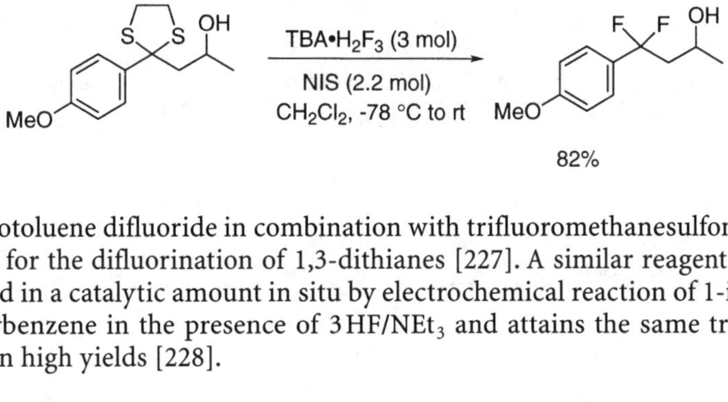

$$\text{(2.99)}$$

Iodosotoluene difluoride in combination with trifluoromethanesulfonic acid is useful for the difluorination of 1,3-dithianes [227]. A similar reagent can be generated in a catalytic amount in situ by electrochemical reaction of 1-iodo-4-methoxybenzene in the presence of $3\,HF/NEt_3$ and attains the same transformation in high yields [228].

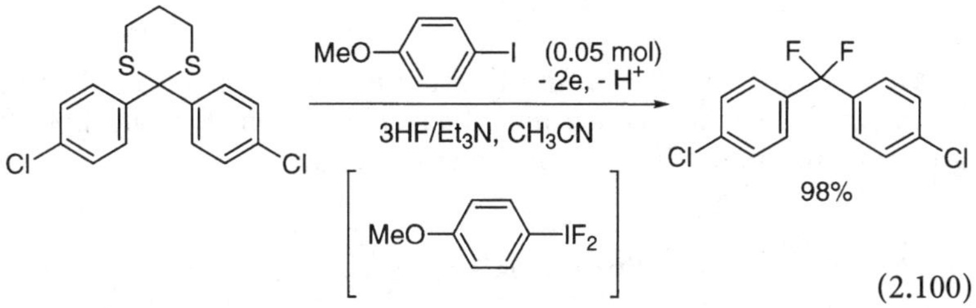

$$\text{(2.100)}$$

When the reaction is applied to dithioacetals of aryl perfluoroalkyl ketones, perfluoroalkyl-substituted aromatics are produced conveniently. Because a perfluoroalkanoyl group is attached to an aromatic moiety by the reaction of an arylmagnesium halide with a perfluoroalkanoate ester, an electrophilic perfluoroalkylation of an aromatic substrate is thus achieved [229]. For the fluorinating agent, 70% HF/pyridine or 80% HF/melamine is particularly effective.

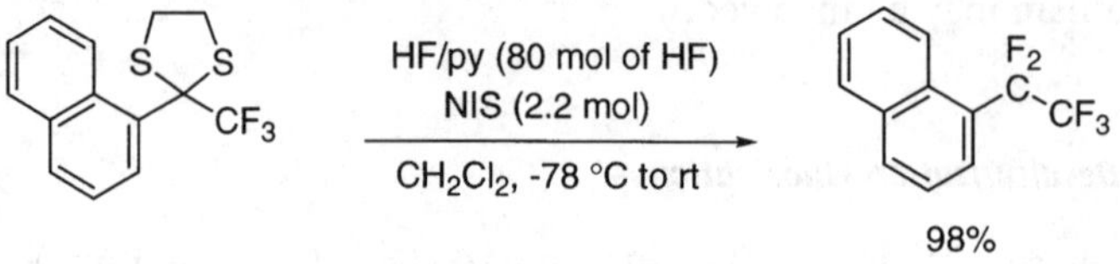

$$\text{1-napth-C(S-CH}_2\text{-CH}_2\text{-S)CF}_3 \xrightarrow[\text{CH}_2\text{Cl}_2, -78\,°\text{C to rt}]{\text{HF/py (80 mol of HF)}\atop \text{NIS (2.2 mol)}} \text{1-napth-CF}_2\text{CF}_3 \quad 98\% \quad (2.101)$$

Orthothioesters, ArC(SR)$_3$, are converted into trifluoromethyl-substituted products, ArCF$_3$, under similar conditions [230].

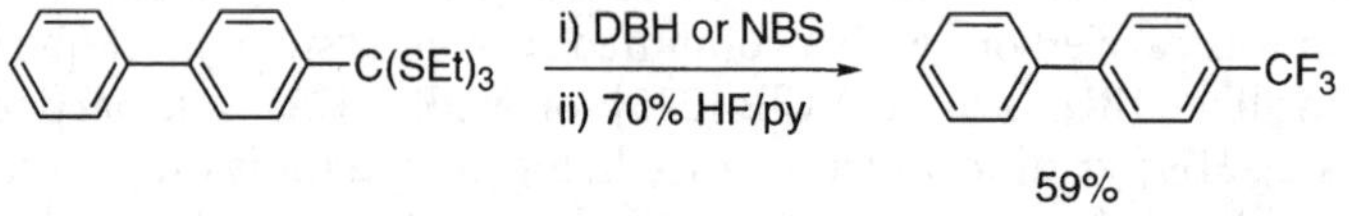

$$\text{Ph-C}_6\text{H}_4\text{-C(SEt)}_3 \xrightarrow[\text{ii) 70\% HF/py}]{\text{i) DBH or NBS}} \text{Ph-C}_6\text{H}_4\text{-CF}_3 \quad 59\% \quad (2.102)$$

In contrast, orthothioesters of alkanoic acids are not fully fluorinated. In general, difluorination accompanied by β-bromination takes place. The products, after oxidation to sulfoxides and thermolysis, give 2-bromo-1,1-difluoro-1-alkenes.

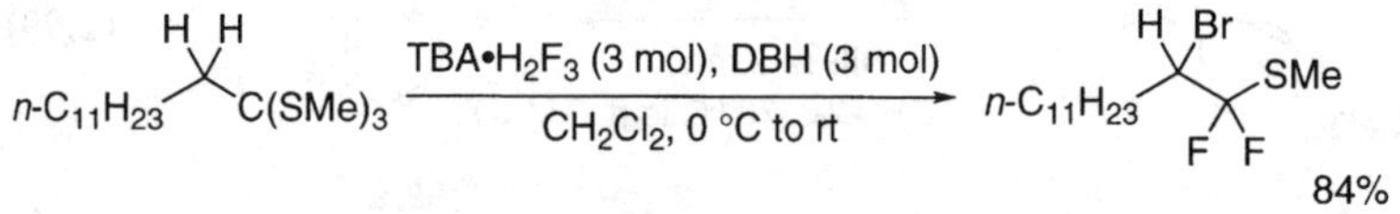

$$n\text{-C}_{11}\text{H}_{23}\text{CH}_2\text{C(SMe)}_3 \xrightarrow[\text{CH}_2\text{Cl}_2, 0\,°\text{C to rt}]{\text{TBA·H}_2\text{F}_3\text{ (3 mol), DBH (3 mol)}} n\text{-C}_{11}\text{H}_{23}\text{CHBrCF}_2\text{SMe} \quad 84\% \quad (2.103)$$

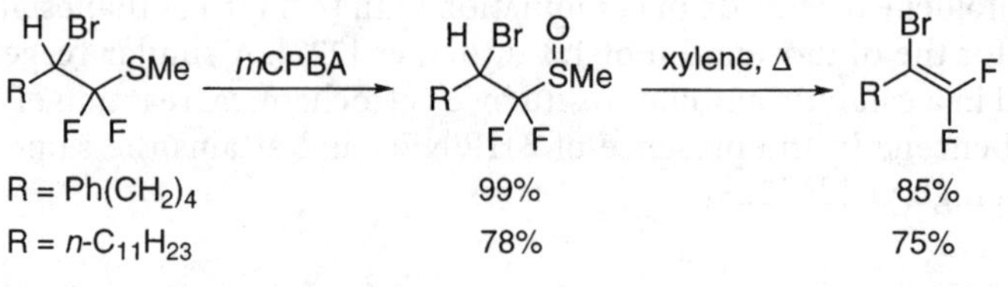

$$\text{R-CHBr-CF}_2\text{SMe} \xrightarrow{m\text{CPBA}} \text{R-CHBr-CF}_2\text{S(O)Me} \xrightarrow{\text{xylene, }\Delta} \text{R-CBr=CF}_2 \quad (2.104)$$

R = Ph(CH$_2$)$_4$ 99%   85%
R = n-C$_{11}$H$_{23}$ 78%   75%

An oxidative desulfurization-fluorination reaction of 1-substituted 2,2,2-tri(methylthio)ethanols is accompanied by oxidation of the alcoholic functional group to give difluoro(methylthio)methyl ketones [231].

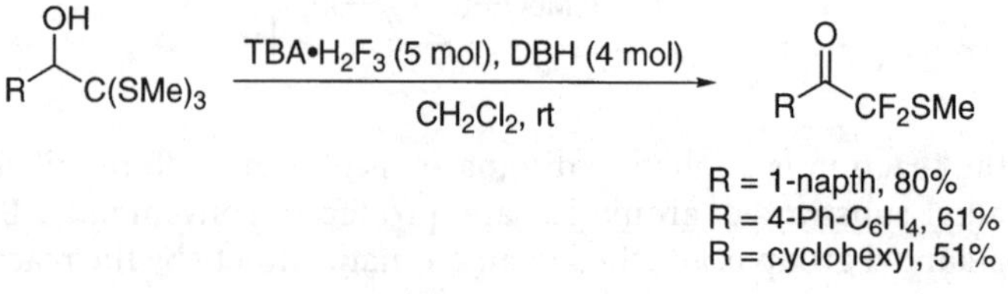

$$\text{R-CH(OH)-C(SMe)}_3 \xrightarrow[\text{CH}_2\text{Cl}_2, \text{rt}]{\text{TBA·H}_2\text{F}_3\text{ (5 mol), DBH (4 mol)}} \text{R-C(O)-CF}_2\text{SMe} \quad (2.105)$$

R = 1-napth, 80%
R = 4-Ph-C$_6$H$_4$, 61%
R = cyclohexyl, 51%

Oxidative desulfurization-fluorination of dithioesters is a convenient and straightforward method for the synthesis of trifluoromethyl-substituted compounds, Ar-CF_3, from arenedithiocarboxylates, $ArCS_2Me$. By tuning the oxidant and the fluoride ion reagent, either trifluorination or difluorination is achieved. For example, a combination of DBH and TBA·H_2F_3 or NIS and HF/pyridine gives a trifluorination product, whereas difluorination results with NIS and TBA·H_2F_3 (Scheme 2.14) [232].

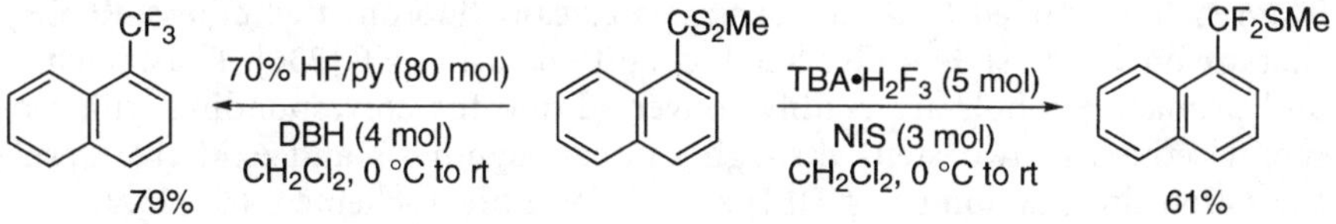
CF3
70% HF/py (80 mol)
DBH (4 mol)
CH2Cl2, 0 °C to rt
79%
CS2Me
TBA•H2F3 (5 mol)
NIS (3 mol)
CH2Cl2, 0 °C to rt
CF2SMe
61%

Scheme 2.14. Di- and trifluorination of dithiocarboxylates

Trifluoromethylation is achieved with α,β-unsaturated dithiocarboxylates, which give 1-substituted 3,3,3-trifluoropropenes [233].

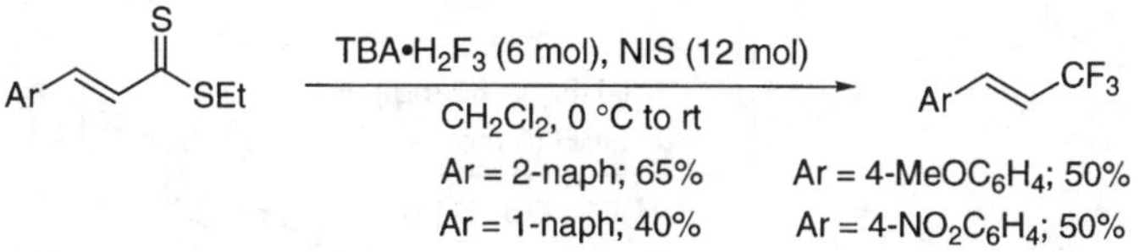
S
Ar
SEt
TBA•H2F3 (6 mol), NIS (12 mol)
CH2Cl2, 0 °C to rt
Ar
CF3
Ar = 2-naph; 65%
Ar = 1-naph; 40%
Ar = 4-MeOC6H4; 50%
Ar = 4-NO2C6H4; 50%

(2.106)

In contrast, dithioalkanoates are difluorinated to convert a -CS_2Me group into a -F_2SMe moiety. The resulting products, after oxidation and thermolysis, give 1,1-difluoro-1-alkenes (Scheme 2.15) [234].

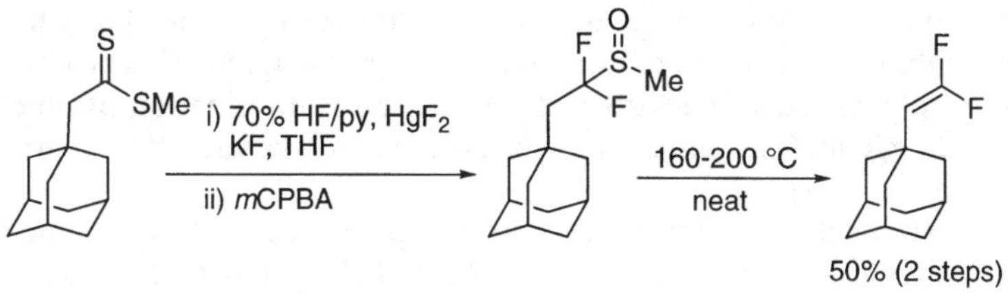
S
SMe
i) 70% HF/py, HgF2
KF, THF
ii) mCPBA
O
F
S
Me
F
160-200 °C
neat
F
F
50% (2 steps)

Scheme 2.15. *gem*-Difluoroolefin synthesis

Thione esters and thione carbonates are also difluorinated readily. The resulting difluoro ethers and difluoromethylenedioxy compounds, respectively, can be isolated except for those having an electron-donating group [235].

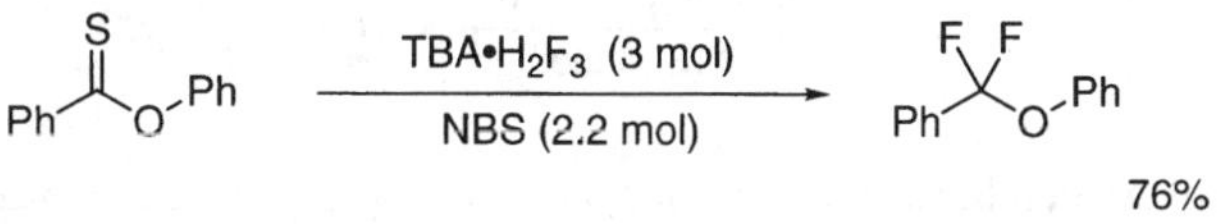
S
Ph
O
Ph
TBA•H2F3 (3 mol)
NBS (2.2 mol)
F F
Ph
O
Ph
76%

(2.107)

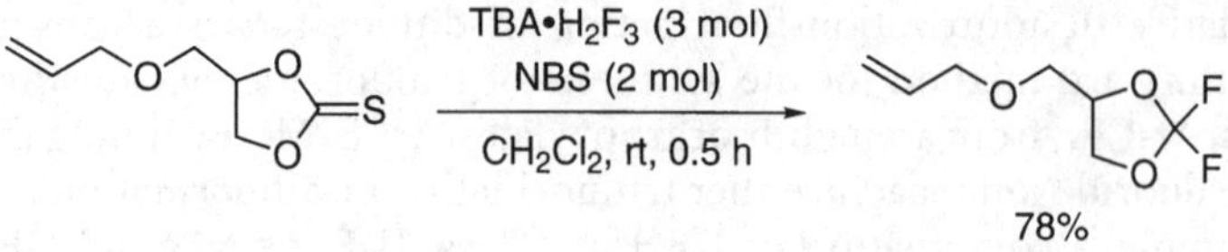

[Scheme showing allyl ether dioxolane-thione + TBA·H₂F₃ (3 mol), NBS (2 mol), CH₂Cl₂, rt, 0.5 h → difluoro dioxolane product, 78%] (2.108)

The characteristic features of the oxidative desulfurization-fluorination reaction are demonstrated by use of xanthates or dithiocarbonates, ROCS$_2$Me. The reaction applied to these substrates gives trifluoromethyl ethers, ROCF$_3$, that are only accessible under harsh conditions using HF [236]. Thus, phenols and primary alcohols are readily converted into the corresponding trifluoromethyl ethers in two steps through xanthate synthesis and oxidative desulfurization-fluorination using DBH and HF/pyridine (Scheme 2.16) [237].

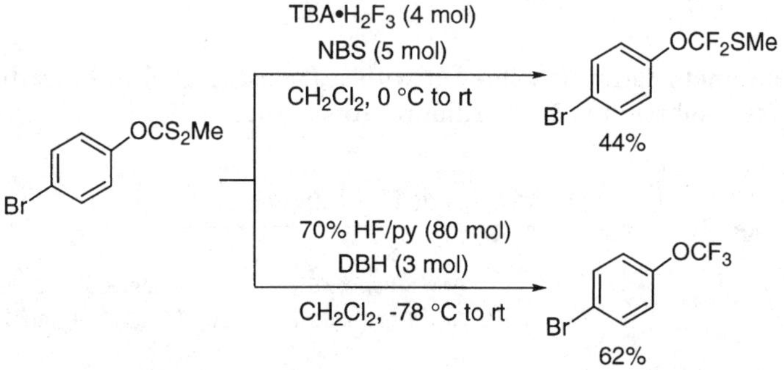

[Scheme: 4-Br-C₆H₄-OCS₂Me converted with TBA·H₂F₃ (4 mol), NBS (5 mol), CH₂Cl₂, 0 °C to rt → 4-Br-C₆H₄-OCF₂SMe, 44%; or with 70% HF/py (80 mol), DBH (3 mol), CH₂Cl₂, -78 °C to rt → 4-Br-C₆H₄-OCF₃, 62%]

Scheme 2.16. Di- and trifluorination of xanthates

Xanthates of secondary alcohols are converted into fluorination products under the original oxidative fluorination conditions, namely use of 70% HF/pyridine and NIS. However, use of NBS and 50% HF/pyridine allows the isolation of trifluoromethyl ethers of secondary alcohols [238]. This reaction contrasts sharply with the one with TolIF$_2$ which gives fluorination products only [239]. The reaction of xanthates of tertiary alcohols affords fluorination products (Scheme 2.17) [238].

Dithiocarbamates, readily available from amines, carbon disulfide and methyl iodide, are readily converted into trifluoromethyl amines under oxidative desulfurization-fluorination conditions [240]. Although the synthesis of trifluoro-

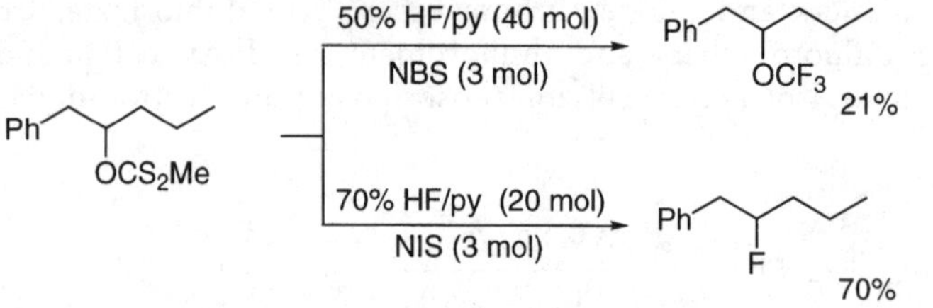

[Scheme: PhCH(OCS₂Me)CH₂CH₃ with 50% HF/py (40 mol), NBS (3 mol) → PhCH(OCF₃)CH₂CH₃, 21%; or with 70% HF/py (20 mol), NIS (3 mol) → PhCHF-CH₂CH₃, 70%]

Scheme 2.17. Oxidative desulfurization-fluorination of xanthates of secondary alcohols

methylamines has precedents [241], the present method is applicable to (hetero)-aromatic and aliphatic amines. Under forcing conditions, the (hetero)aromatic ring may be halogenated, but the halogen substituent may be removed or functionalized by metalation followed by a reaction with an electrophile.

$$\text{NO}_2\text{-C}_6\text{H}_4\text{-N(Me)(CS}_2\text{Me)} \xrightarrow[\text{CH}_2\text{Cl}_2, \text{rt, 1 h}]{\substack{\text{TBA·H}_2\text{F}_3 \text{ (5 mol)} \\ \text{DBH (4 mol)}}} \text{NO}_2\text{-C}_6\text{H}_4\text{-N(Me)(CF}_3\text{)} \quad 94\% \qquad (2.109)$$

When thioamides of perfluoroalkanoic acids are applied to oxidative desulfurization-fluorination, perfluoroalkylamines are produced [242].

$$\text{4-Cl-C}_6\text{H}_4\text{-N(CH}_2\text{Ph)(C(=S)}n\text{-C}_3\text{F}_7\text{)} \xrightarrow[\text{CH}_2\text{Cl}_2, 0\,°\text{C to rt, 0.5 h}]{\substack{\text{TBA·H}_2\text{F}_3 \text{ (3 mol)} \\ \text{NBS (2.2 mol)}}} \text{4-Cl-C}_6\text{H}_4\text{-N(CH}_2\text{Ph)(}n\text{-C}_4\text{F}_9\text{)} \quad 80\% \qquad (2.110)$$

2.3.6
Oxidative Fluorination of Amines

Amino acids are converted into α-fluorocarboxylic acids by diazoniation followed by substitution by a fluoride ion [243] with retention of configuration probably via an α-lactone intermediate. Esters of amino acids are also fluorinated similarly.

$$\text{R-CH(NH}_2\text{)-COOH} \xrightarrow[\text{rt, 1 h}]{\substack{\text{NaNO}_2 \text{ (1.5 mol)} \\ 70\%\text{ HF/py (2.5 mL/mol)}}} \text{R-CH(F)-COOH} \qquad (2.111)$$

R = H, 38%   R = Et, 80%
R = Me, 96% R = 4-HO-C$_6$H$_4$CH$_2$, 90%

Aliphatic diazo compounds decompose in the presence of HF/pyridine to give monofluoro compounds. When the reaction is carried out in the presence of a positive halogen oxidant and HF/pyridine, fluorine and halogen atoms are connected to the diazo carbon (Scheme 2.18) [244].

Under oxidative fluorination conditions, oximes are converted into *gem*-difluoromethylene compounds. Although similar transformations have been achieved with IF [245] and BrF$_3$ [246], combination of HF/pyridine and the oxidant NOBF$_4$ allows the easy implementation of the *gem*-difluorination in standard laboratories [247].

$$\text{adamantanone oxime (=NOH)} \xrightarrow[\text{neat, 0 °C to rt}]{\substack{60\%\text{ HF/py (excess)} \\ \text{NO}^+\text{BF}_4^- \text{ (1.2 mol)}}} \text{adamantane-CF}_2 \quad 95\% \qquad (2.112)$$

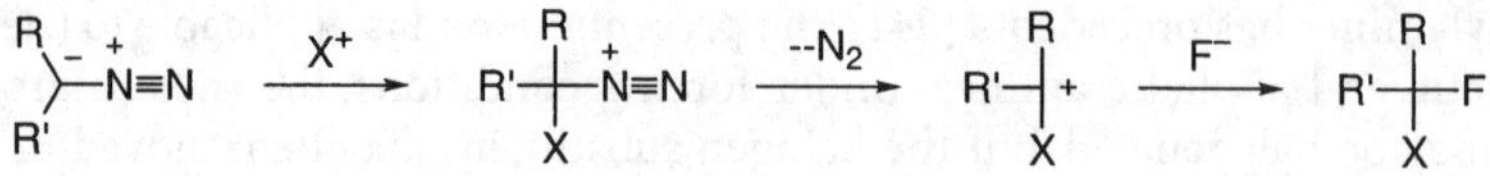

$$\underset{R'}{\overset{R}{\underset{|}{>}}}\!\!-\!\!\overset{+}{N}\!\!\equiv\!\!N \xrightarrow{X^+} R'\!\!-\!\!\underset{X}{\overset{R}{\underset{|}{\overset{|}{C}}}}\!\!-\!\!\overset{+}{N}\!\!\equiv\!\!N \xrightarrow{-N_2} R'\!\!-\!\!\underset{X}{\overset{R}{\underset{|}{\overset{|}{C}}}}\!\!+ \xrightarrow{F^-} R'\!\!-\!\!\underset{X}{\overset{R}{\underset{|}{\overset{|}{C}}}}\!\!-\!\!F$$

NXS (1.5 mol), 70% HF/py, 0 °C, 20 min, Et$_2$O

substrate	NXS	product	yield (%)
Ph–CO–CHN$_2$	----	Ph–CO–CH$_2$F	32
Ph–CO–CHN$_2$	NBS	Ph–CO–CHBrF	63
 ⌬–CO–CHN$_2$	NIS	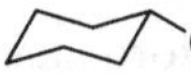 ⌬–CO–CHIF	80
PhCHN$_2$	----	PhCH$_2$F	70

Scheme 2.18. Fluorination of diazo compounds

2.4
Electrochemical Fluorination

The electrochemical reaction of organic substrates in HF or in a solvent containing HF or its equivalent gives fluorinated products. This process is called electrochemical fluorination. In particular, the one in HF is called the Simons method [248] and is still used today for the manufacture of various perfluoro compounds [249]. Although no special reagent is used in this process, some examples will be discussed in the following sections. Readers are advised to consult reference books and review articles [250–254].

The Simons method involves electrolysis of a solution of an organic substrate in HF using a nickel electrode for an anode to produce perfluoro products. At the cathode, molecular hydrogen is evolved. Since HF is commercially available and the reaction takes place in a single step, the reaction is of great industrial significance. Limitations of the reaction are: (1) fluorination of compounds barely soluble in HF is not satisfactory; (2) by-products due to ring opening, ring formation, and/or isomerization often accompany the reaction; and (3) control of the stereo- and regioselectivity of the fluorination is not easy.

Because hydrocarbons are not very soluble in HF, they are not good substrates for electrochemical fluorination. However, partly fluorinated substrates like 2,2-difluorohexane gives perfluorohexane and its lower homologues [255].

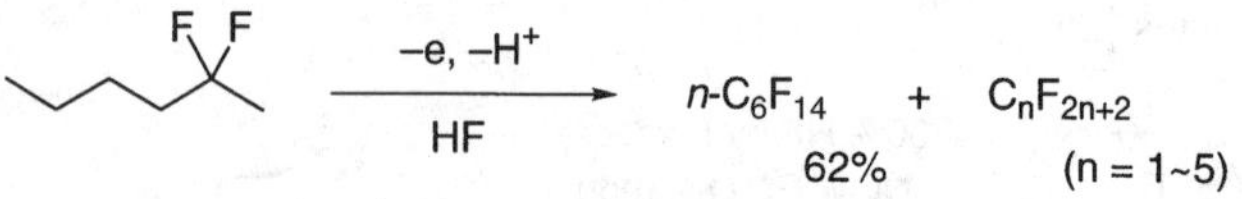

$$\text{CH}_3\text{CF}_2\text{C}_4\text{H}_9 \xrightarrow[\text{HF}]{-e,\, -H^+} n\text{-C}_6\text{F}_{14} + \text{C}_n\text{F}_{2n+2} \qquad (2.113)$$
$$ 62\% \quad\;\; (n = 1{\sim}5)$$

Electrochemical fluorination of 1,1-difluorocylcohexane gives a ring-contraction product in addition to perfluorocyclohexane.

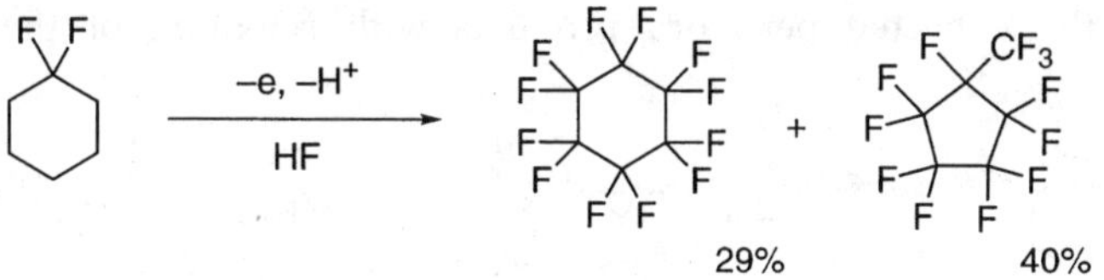

2.4 Electrochemical Fluorination

[Reaction scheme 2.114: cyclohexane + HF, −e, −H⁺ → perfluorocyclohexane (29%) + perfluoromethylcyclopentane (40%)] (2.114)

Use of NaF as a conductivity enhancer allows the fluorination of adamantane, which is insoluble in HF, to give perfluoroadamantane in 31% yield [256].

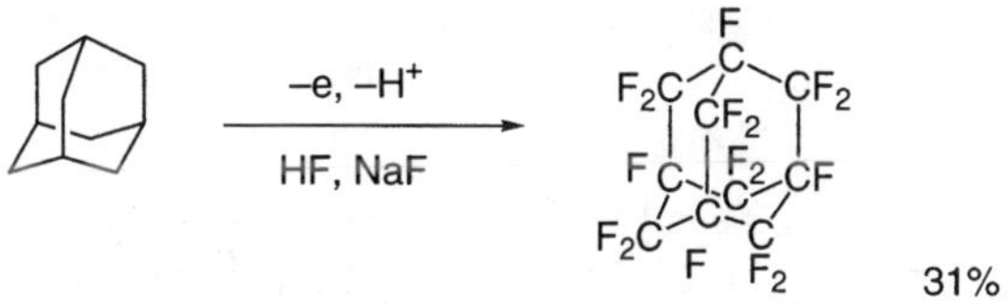

[Reaction scheme 2.115: adamantane, −e, −H⁺, HF, NaF → perfluoroadamantane (31%)] (2.115)

Electrochemical fluorination of carboxylic acids, acid halides, and esters gives perfluoroalkanoyl fluorides along with perfluoro cyclic ethers, which sometimes predominate [257].

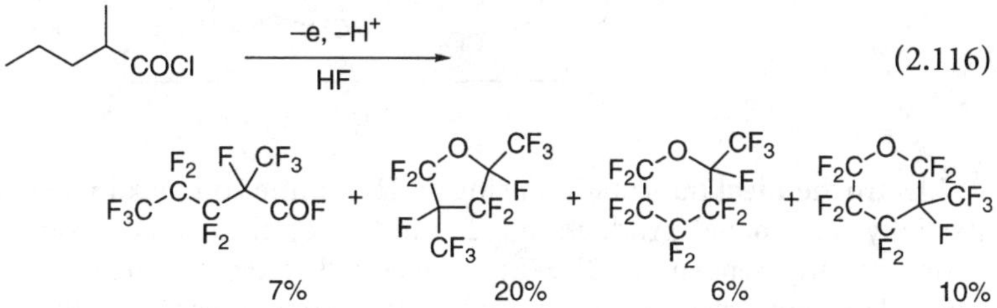

[Reaction scheme 2.116: (CH₃)₂CHCH₂COCl, −e, −H⁺, HF → $F_3C-CF_2-CF(CF_3)-COF$ (7%) + perfluoro cyclic ethers (20%, 6%, 10%)] (2.116)

In a similar manner, alkanesulfonic acids and derivatives are fluorinated to give perfluoroalkanesulfonyl fluorides that are used for the synthesis of perfluorinated surfactants and superacids [258, 259]. Although methane- and ethanesulfonyl chlorides give the corresponding perfluorosulfonyl fluorides, higher homologs are obtained in inferior yields.

$$H(CH_2)_nSO_2Cl \xrightarrow[HF]{-e, -H^+} F(CF_2)_nSO_2F$$ (2.117)

n =		n =	
1	87%	5	45%
2	79%	6	36%
3	68%	7	31%
4	58%	8	25%

Perfluorination of ethers [260] and tertiary amines [261] has been well studied. Again ring-opened and/or rearranged products are produced in addition to the expected perfluoro products with retention of the original structure.

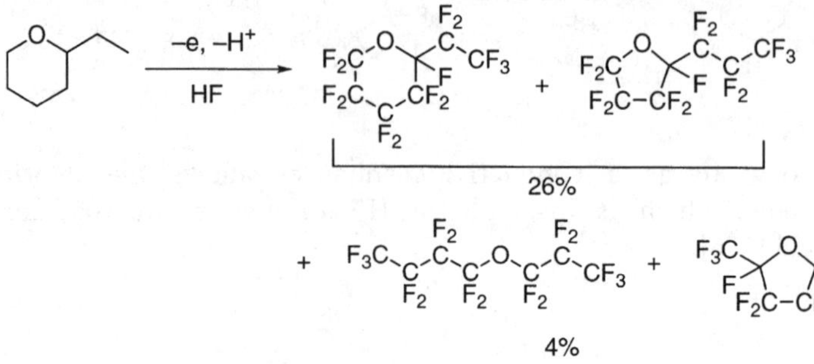

$$(2.118)$$

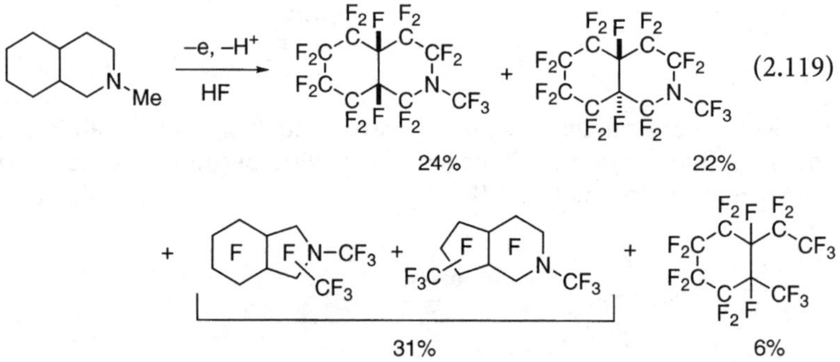

$$(2.119)$$

Electrochemical fluorination using an HF/amine complex in lieu of HF gives partially fluorinated products under milder conditions. For example, the reaction of ethylbenzene in 3HF/Et$_3$N gives 1-fluoroethylbenzene in 42% yield [262]. The same reaction when applied to styrene gives 1,2-difluoroethylbenzene in 51% yield [263].

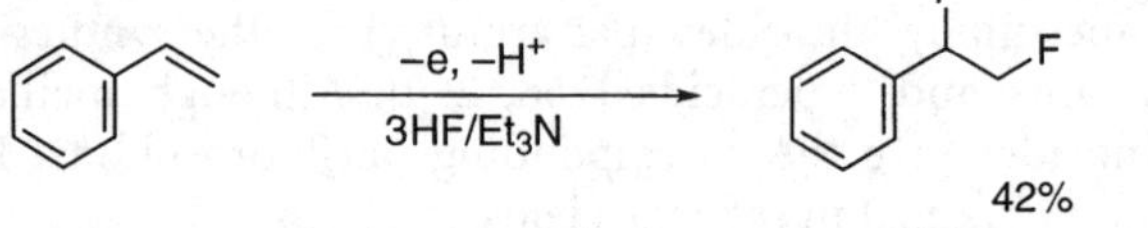

$$(2.120)$$

Organic solvents, such as acetonitrile, DMF and THF, can be used to control the reaction. An example is the electrochemical fluorination of the enol acetate of deoxybenzoin using 3HF/Et$_3$N in acetonitrile. The initial product is assumed to be a difluorination product as above, but this is converted into fluorodeoxybenzoin under the reaction conditions [264].

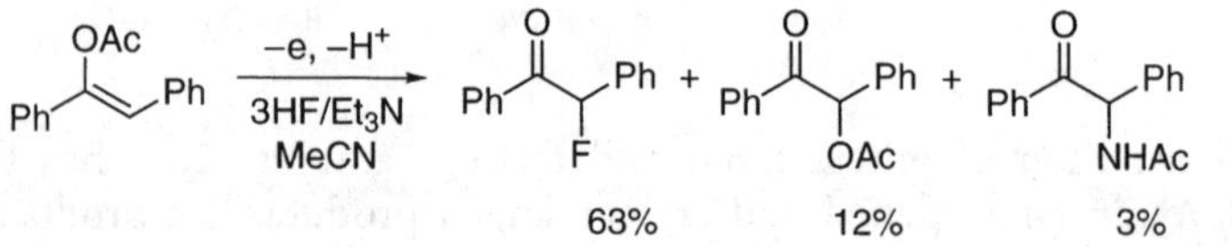

$$(2.121)$$

2.4 Electrochemical Fluorination

An optically active enol ester of α-tetralone allows isolation of diastereomeric difluorinated products, which upon hydrolysis give β-fluoro tetralones of high enantiomeric excess (Scheme 2.19) [265].

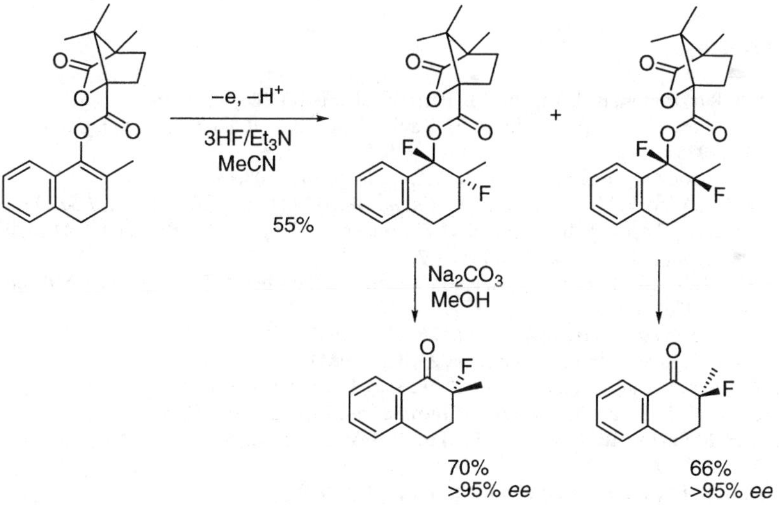

Scheme 2.19. Electrochemical asymmetric fluorination

Electrochemical fluorination of sulfides gives α-fluoro sulfides [266, 267]. Scheme 2.20 indicates that the degree of fluorination can be controlled by applied voltage and electric current.

Electrochemical fluorination conditions can be successfully applied to the synthesis of fluorine-containing β-lactams [268].

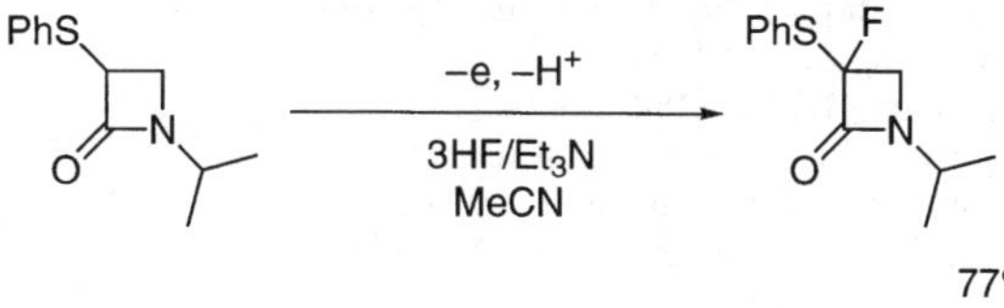

(2.122)

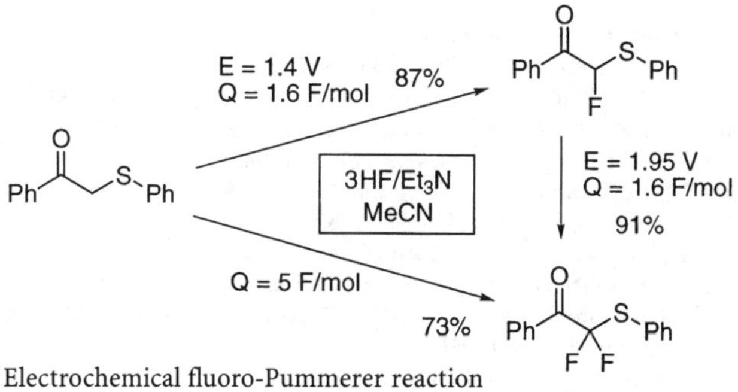

Scheme 2.20. Electrochemical fluoro-Pummerer reaction

Future studies on various substrates will demonstrate that electrochemical fluorination is one of the most promising and reliable methods for highly clean and selective fluorination.

References

1. S. T. Purrington, B. S. Kagen, T. B. Patrick, *Chem. Rev.*, **86**, 997 (1986)
2. N. J. Maraschin, B. D. Catsikis, L. H. Davis, G. Jarvinen, R. J. Lagow, *J. Am. Chem. Soc.*, **97**, 513 (1975)
3. J. L. Adcock, K. Horita, E. B. Renk, *J. Am. Chem. Soc.*, **103**, 6937 (1981)
4. (a) J. L. Adcock, M. L. Robin, *J. Org. Chem.*, **48**, 3128 (1983); (b) T.-Y. Lin, W.-H. Lin, W. D. Clark, R. J. Lagow, S. B. Larson, S. H. Simonsen, V. M. Lynch, J. S. Brodbelt, S. D. Maleknia, C.-C. Liou, *J. Am. Chem. Soc.*, **116**, 5172 (1994)
5. D. H. R. Barton, R. H. Hesse, R. E. Markwell, M. M. Pechet, H. T. Toh, *J. Am. Chem. Soc.*, **98**, 3034 (1976)
6. C. Gal, S. Rozen, *Tetrahedron Lett.*, **26**, 2793 (1985)
7. C. Gal, S. Rozen, *Tetrahedron Lett.*, **25**, 449 (1984)
8. S. Rozen, C. Gal, Y. Faust, *J. Am. Chem. Soc.*, **102**, 6860 (1980)
9. C. Gal, G. Ben-Shoshan, S. Rozen, *Tetrahedron Lett.*, **21**, 5067 (1980)
10. D. H. R. Barton, R. H. Hesse, R. E. Markwell, M. M. Pechet, S. Rozen, *J. Am. Chem. Soc.*, **98**, 3036 (1976)
11. D. H. R. Barton, *Pure Appl. Chem.*, **49**, 1241 (1977)
12. T. Tsushima, K. Kawada, T. Tsuji, S. Misaki, *J. Org. Chem.*, **47**, 1107 (1982)
13. S. T. Purrington, N. V. Lazaridis, C. L. Bumgardner, *Tetrahedron Lett.*, **27**, 2715 (1986)
14. R. F. Merritt, *J. Org. Chem.*, **31**, 3871 (1966)
15. R. F. Merritt, F. A. Johnson, *J. Org. Chem.*, **31**, 1859 (1966)
16. R. F. Merritt, T. E. Stevens, *J. Am. Chem. Soc.*, **88**, 1822 (1966)
17. D. H. R. Barton, J. Lister-James, R. H. Hesse, M. M. Pechet, S. Rozen, *J. Chem. Soc., Perkin Trans. 1*, **1982**, 1105
18. V. Grakauskas, *J. Org. Chem.*, **34**, 2835 (1969)
19. V. Grakauskas, *J. Org. Chem.*, **35**, 723 (1970)
20. F. Cacace, P. Giacomello, A. P. Wolf, *J. Am. Chem. Soc.*, **102**, 3511 (1980)
21. S. T. Purrington, D. L. Woodard, *J. Org. Chem.*, **56**, 142 (1991)
22. S. Misaki, *J. Fluorine Chem.*, **21**, 191 (1982)
23. S. Misaki, *J. Fluorine Chem.*, **17**, 159 (1981)
24. J. W. Grisard, H. A. Bernhardt, G. D. Oliver, *J. Am. Chem. Soc.*, **73**, 5725 (1951)
25. J. F. Fllis, W. K. R. Musgrave, *J. Chem. Soc.*, 3608 (1950)
26. M. M. Boudakian, G. A. Hyde, *J. Fluorine Chem.*, **25**, 435 (1984)
27. R. Filler, *Isr. J. Chem.*, **17**, 71 (1978)
28. (a) B. Zajc, M. Zupan, *Bull. Chem. Soc. Jpn.*, **59**, 1659 (1986); (b) M. A. Tius, *Tetrahedron*, **51**, 6605 (1995)
29. M. Zupan, *J. Fluorine Chem.*, **8**, 305 (1976)
30. R. K. Marat, A. F. Janzen, *Can. J. Chem.*, **55**, 3031 (1977)
31. A. F. Janzen, P. M. C. Wang, A. E. Lemire, *J. Fluorine Chem.*, **22**, 557 (1983)
32. T. Tsushima, K. Kawada, T. Tsuji, *Tetrahedron Lett.*, **23**, 1165 (1982)
33. G. L. Cantrell, R. Filler, *J. Fluorine Chem.*, **27**, 35 (1985)
34. B. Zajc, M. Zupan, *J. Chem. Soc., Chem. Commun.*, **1980**, 759
35. B. Zajc, M. Zupan, *J. Org. Chem.*, **47**, 573 (1982)
36. M. J. Shaw, H. H. Hyman, R. Filler, *J. Am. Chem. Soc.*, **92**, 6498 (1970)
37. T. B. Patrick, K. K. Johri, D. H. White, *J. Org. Chem.*, **48**, 4158 (1983)
38. T. B. Patrick, K. K. Johri, D. H. White, W. S. Bertrand, R. Mokhtar, M. R. Kilbourn, M. J. Welch, *Can. J. Chem.*, **64**, 138 (1986)
39. S. Rozen, *Chem. Rev.*, **96**, 1717 (1996)

40. J. Kollonitsch, L. Barash, *J. Am. Chem. Soc.*, **98**, 5591 (1976)
41. W. J. Middleton, E. M. Bingham, *J. Am. Chem. Soc.*, **102**, 4845 (1980)
42. M. J. Fifolt, R. T. Olczak, R. F. Mundhenke, J. F. Bieron, *J. Org. Chem.*, **50**, 4576 (1985)
43. D. H. R. Barton, R. H. Hesse, G. P. Jackman, L. Ogunkoya, M. M. Pechet, *J. Chem. Soc., Perkin Trans. 1*, **1974**, 739
44. T. B. Patrick, G. L. Cantrell, S. M. Inga, *J. Org. Chem.*, **45**, 1409 (1980)
45. J. B. Levy, D. M. Sterling, *J. Org. Chem.*, **50**, 5615 (1985)
46. K. Sekiya, K. Ueda, *Chem. Lett.*, **1990**, 609
47. W. Navarrini, A. Russo, V. Tortelli, *J. Org. Chem.*, **60**, 6441 (1995)
48. O. Lerman, S. Rozen, *J. Org. Chem.*, **48**, 724 (1983)
49. S. Rozen, M. Brand, *Synthesis*, **1982**, 665
50. O. Lerman, Y. Tor, S. Rozen, *J. Org. Chem.*, **46**, 4629 (1981)
51. O. Lerman, Y. Tor, D. Hebel, S. Rozen, *J. Org. Chem.*, **49**, 806 (1984)
52. S. Rozen, O. Lerman, M. Kol, D. Hebel, *J. Org. Chem.*, **50**, 4753 (1985)
53. M. J. Adam, *J. Chem. Soc., Chem. Commun.*, **1982**, 730
54. M. J. Adam, B. D. Pate, J.-R. Nesser, L. D. Hall, *Carbohydr. Res.*, **124**, 215 (1983)
55. S. Rozen, Y. Menahem, *Tetrahedron Lett.*, **1979**, 725
56. S. Rozen, Y. Menahem, *J. Fluorine Chem.*, **16**, 19 (1980)
57. J. Fried, D. K. Mitra, M. Nagarajan, and M. M. Mehrotra, *J. Med. Chem.*, **23**, 237 (1980)
58. C.-L. J. Wang, P. A. Grieco, F. J. Okuniewic, *J. Chem. Soc., Chem. Commun.*, **1976**, 468
59. J. S. Mills, *J. Am. Chem. Soc.*, **81**, 5515 (1959)
60. S. W. Djyric, R. B. Garland, L. N. Nysted, R. Pappo, G. Plume, L. Swenton, *J. Org. Chem.*, **52**, 978 (1987)
61. V. C. O. Njar, T. Arungavel, E. Caspi, *J. Org. Chem.*, **48**, 1007 (1983)
62. C. M. Sharts, W. A. Sheppard, *Org. React.*, **21**, 125 (1974)
63. S. Stavber, M. Zupan, *Tetrahedron*, **42**, 5035 (1986)
64. S. Stavber, M. Zupan, *J. Chem. Soc., Chem. Commun.*, **1983**, 563
65. S. Stavber, M. Zupan, *J. Chem. Soc., Chem. Commun.*, **1981**, 148
66. S. Stavber, M. Zupan, *J. Org. Chem.*, **50**, 3609 (1985)
67. D. P. Ip, C. D. Arthur, R. E. Winans, E. H. Appelman, *J. Am. Chem. Soc.*, **103**, 1964 (1981)
68. E. H. Appelman, L. J. Basile, R. C. Thompson, *J. Am. Chem. Soc.*, **101**, 3384 (1979)
69. G. S. Lal, G. P. Pez, R. G. Syvret, *Chem. Rev.*, **96**, 1737 (1996)
70. K. O. Christe, C. J. Schack, R. D. Wilson, *J. Fluorine Chem.*, **8**, 541 (1976)
71. C. J. Schack, K. O. Christe, *J. Fluorine Chem.*, **18**, 363 (1981)
72. T. Umemoto, K. Tomita, *Tetrahedron Lett.*, **27**, 3271 (1986)
73. T. Umemoto, K. Kawada, K. Tomita, *Tetrahedron Lett.*, **27**, 4465 (1986)
74. T. Umemoto, S. Fukami, G. Tomizawa, K. Harasawa, K. Kawada, K. Tomita, *J. Am. Chem. Soc.*, **112**, 8563 (1990)
75. (a) T. Umemoto, G. Tomizawa, *Bull. Chem. Soc. Jpn.*, **59**, 3625 (1986); (b) T. Umemoto, G. Tomizawa, *J. Org. Chem.*, **60**, 6563 (1995); (c) T. Umemoto, K. Harasawa, G. Tomizawa, K. Kawada, K. Tomita, *Bull. Chem. Soc. Jpn.*, **64**, 1081 (1991)
76. M. Ihara, T. Kai, N. Taniguchi, K. Fukumoto, *J. Chem. Soc., Perkin Trans. 1*, **1990**, 2357
77. A. G. Gilicinski, G. P. Pez, R. G. Syvret, G. S. Lal, *J. Fluorine Chem.*, **59**, 157 (1992)
78. R. E. Banks, S. N. Mohialdin-Khaffaf, G. S. Lal, I. Sharif, R. G. Syvret, *J. Chem. Soc., Chem. Commun.*, **1992**, 595
79. G. S. Lal, *J. Org. Chem.*, **58**, 2791 (1993)
80. R. E. Banks, N. J. Lawrence, A. L. Popplewell, *J. Chem. Soc., Chem. Commun.*, **1994**, 343
81. W. E. Barnette, *J. Am. Chem. Soc.*, **106**, 452 (1984)
82. S. H. Lee, J. Schwartz, *J. Am. Chem. Soc.*, **108**, 2445 (1986)
83. E. Differding, R. W. Lang, *Helv. Chim. Acta*, **72**, 1248 (1989)
84. E. Differding, G. M. Rüegg, R. W. Lang, *Tetrahedron Lett.*, **32**, 1779 (1991)
85. E. Differding, R. W. Lang, *Tetraheron Lett.*, **29**, 6087 (1988)
86. F. A. Davis, P. Zhou, C. K. Murphy, *Tetrahedron Lett.*, **34**, 3971 (1993)
87. Y. Takeuchi, A. Satoh, T. Suzuki, A. Kameda, M. Dohrin, T. Satoh, T. Koizumi, K. L. Kirk, *Chem. Pharm. Bull.*, **45**, 1085 (1997)

88. Japan Kokai Tokkyo Koho, JP 09-249653 (1997) [*Chem. Abstr.*, 127:262674 (1997)]; T. Suzuki, N. Shibata, A. Satoh, H. Kakuda, Y. Takeuchi, *15th Int. Symp. Fluorine Chem.*, Vancouver, 1997, Abstracts, FRx C-6.
89. E. Differding, H. Ofner, *Synlett*, **1991**, 187
90. S. Singh, D. D. DesMarteau, S. S. Zuberi, M. Witz, H.-N. Huang, *J. Am.Chem. Soc.*, **109**, 7194 (1987)
91. Z.-Q. Xu, D. D. DesMarteau, Y. Gotoh, *J. Fluorine Chem.*, **58**, 71 (1992)
92. G. Resnati, D. D. DesMarteau, *J. Org. Chem.*, **57**, 4281 (1992)
93. G. Resnati, D. D. DesMarteau, *J. Org. Chem.*, **56**, 4925 (1991)
94. (a) F. A. Davis, W. Han, *Tetrahedron Lett.*, **32**, 1631 (1991); (b) F. A. Davis, W. Han, C. K. Murphy, *J. Org. Chem.*, **60**, 4730 (1995)
95. (a) F. A. Davis, W. Han, *Tetrahedron Lett.*, **33**, 1153 (1992); (b) F. A. Davis, P. V. N. Kasu, G. Sundarababu, H. Qi, *J. Org. Chem.*, **62**, 7546 (1997)
96. M. Hudlicky, *Chemistry of Organic Fluorine Compounds*, Pergamon Press, p 88 (1961)
97. C. M. Sharts, W. A. Sheppard, *Org. React.*, **21**, 125 (1974)
98. A. V. Grosse, C. B. Linn, *J. Org. Chem.*, **3**, 26 (1939)
99. F. Smith, M. Stacey, J. C. Tatlow, I. K. Dowson, B. R. J. Thomas, *J. Appl. Chem.*, **2**, 97 (1952)
100. P. Tarrant, J. Atlaway, A. M. Lovelace, *J. Am. Chem. Soc.*, **76**, 2343 (1954)
101. (a) G. A. Olah, J. T. Welch, Y. D. Vankar, M. Nojima, I. Kerekes, J. A. Olah, *J. Org. Chem.*, **44**, 3872 (1979); (b) G. A. Olah, X.-Y. Li, Synthetic Fluorine Chemistry (G. A. Olah, R. D. Chambers, G. K. S. Prakash, eds), p 163, John Wiley & Sons, New York (1992); (c) N. Yoneda, *Tetrahedron*, **4 J**, 5329 (1991); (d) T. Fukuhara, N. Yoneda, T. Abe, S. Nagata, A. Suzuki, *Nippon Kagaku Kai Shi*, 1951 (1985)
102. N. Yoneda, T. Abe, T. Fukuhara, A. Suzuki, *Chem. Lett.*, **1983**, 1135
103. N. Yoneda, S. Nagata, T. Fukuhara, A. Suzuki, *Chem. Lett.*, **1984**, 1241
104. M. Hayashi, S. Hashimoto, R. Noyori, *Chem. Lett.*, **1984**, 1747
105. M. Muehlbacher, C. D. Poulter, *J. Org. Chem.*, **53**, 1026 (1988)
106. S. Takano, M. Yanase, K. Ogasawara, *Chem. Lett.*, **1989**, 1689
107. H. Amri, M. M. E. Gaieo, *J. Fluorine Chem.*, **46**, 75 (1990)
108. H. Nohira, S. Nakamura, M. Kamei, *Mol. Cryst. Liq. Cryst.*, **180B**, 379 (1990)
109. A. I. Ayi, M. Remli, R. Guedj, *Tetrahedron Lett.*, **22**, 1505 (1981)
110. N. Mongelli, F. Animati, R. D'Alessio, L. Zuliani, C. Gandolfi, *Synthesis*, **1988**, 310
111. A. I. Ayi, M. Remli, R. Condom, R. Guedj, *J. Fluorine Chem.*, **17**, 565 (1981)
112. A. Ourari, R. Condom, R. Guedj, *Can. J. Chem.*, **60**, 2707 (1982)
113. T. N. Wade, F. Gaymard, R. Guedi, *Tetrahedron Lett.*, **1979**, 2681
114. G. Alvernhe, E. Kozlowska-Gramsz, S. Lacombe-Ber, A. Laurent, *Tetrahedron Lett.*, **1978**, 5203
115. A. Barama, R. Condom, R. Guedj, *J. Fluorine Chem.*, **16**, 183 (1980)
116. T. Fukuhara, N. Yoneda, *Chem Lett.*, **1993**, 509
117. A. Roe, *Org. React.*, **5**, 193 (1949)
118. T. Fukuhara, N. Sekiguchi, N. Yoneda, *Chem. Lett.*, **1994**, 1011
119. R. D. Beaty, W. K. R. Musgrave, *J. Chem. Soc.*, **1952**, 875
120. T. Fukuhara, N. Yoneda, T. Sawada, A. Suzuki, *Synth. Commun.*, **17**, 685 (1987)
121. T. Fukuhara, N. Yoneda, A. Suzuki, *J. Fluorine Chem.*, **38**, 435 (1988)
122. F. Faustini, S. De Munari, A. Panzert, V. Villa, C. A. Gandolfi, *Tetrahedron Lett.*, **22**, 4533 (1981)
123. J. Barber, R. Keck, J. Rétey, *Tetrahedron Lett.*, **23**, 1549 (1982)
124. G. A. Boswell, Jr., W. C. Ripka, R. M. Scribner, C. W. Tullock, *Org. React.*, **21**, 1 (1974)
125. C.-L. J. Wang, *Org. React.*, **34**, 319 (1985)
126. J. Kollonitsch, S. Marburg, L. M. Perkins, *J. Org. Chem.*, **40**, 3808 (1975)
127. J. Kollonitsch, S. Marburg, L. M. Perkins, *J. Org. Chem.*, **44**, 771 (1979)
128. F. A. Bloshchitsa, A. I. Burmakov, B. V. Kunshenko, L. A. Alekseeva, L. M. Yagupol'skii, *J. Org. Chem. USSR*, **18**, 679 (1982)
129. A. I. Burmakov, L. A. Alekseeva, L. M. Yagupol'skii, *J. Org. Chem. USSR*, **8**, 156 (1972)
130. B. V. Kunshenko, A. I. Burmakov, L. A. Alekseeva, V. G. Lukmanov, L. M. Yagupol'skii, *J. Org. Chem. USSR*, **10**, 896 (1972)

131. N. N. Muratov, A. L. Burmakov, B. V. Kunshenko, L. A. Alekseeva, L. M. Yagupol'skii, *J. Org. Chem. USSR*, **18**, 1220 (1982)
132. V. I. Golikov, A. M. Aleksanfrov, L. A. Alekseeva, T. É. Bezmenova, L. M. Yagupol'skii, *J. Org. Chem. USSR*, **9**, 2446 (1972)
133. M. Hudlicky, *Org. React.*, **35**, 513 (1988)
134. W. J. Middleton, *J. Org. Chem.*, **40**, 574 (1975)
135. T. J. Tewson, M. J. Welch, *J. Org. Chem.*, **43**, 1090 (1978)
136. G. Lowe, B. V. L. Potter, *J. Chem. Soc., Perkin Trans. 1*, **1980**, 2029
137. G. H. Posner, S. R. Haines, *Tetrahedron Lett.*, **26**, 5 (1985)
138. K. C. Nicolaou, T. Ladduwahetty, J. L. Randall, A. Chuchlolowski, *J. Am. Chem. Soc.*, **108**, 2466 (1986)
139. L. N. Markovskij, V. E. Pashinnik, A. V. Kirsanov, *Synthesis*, **1973**, 787
140. L. N. Markovski, V. E. Pashinnik, *Synthesis*, **1975**, 801
141. W. H. Bunnelle, B. R. McKinnis, B. A. Narayanan, *J. Org. Chem.*, **55**, 768 (1990)
142. J. R. McCarthy, N. P. Peet, M. E. LeTourneau, M. Inbasekaran, *J. Am. Chem. Soc.*, **107**, 735 (1985)
143. M. J. Robins, S. F. Wnuk, *Tetrahedron Lett.*, **29**, 5729 (1988)
144. N. N. Yarovenko, M. A. Raksha, *J. Gen. Chem. USSR*, **29**, 2125 (1959)
145. (a) A. Takaoka, H. Iwakiri, N. Ishikawa, *Bull. Chem. Soc. Jpn.*, **52**, 3377 (1979); (b) S. Watanabe, T. Fijita, Y. Usui, T. Kitazume, *J. Fluorine Chem.*, **31**, 247 (1986)
146. T. Asai, Y. Morizawa, T. Shimada, T. Nakayama, M. Urushihara, Y. Matsumura, A. Yasuda, *Tetrahedron Lett.*, **36**, 273 (1995)
147. F. Munyemana, A.-M. Frisque-Hesbain, A. Devos, L. Ghosez, *Tetrahedron Lett.*, **30**, 3077 (1989)
148. B. Ernst, T. Winkler, *Tetrahedron Lett.*, **30**, 3081 (1989)
149. Y. Kimura, *Yuki Gosei Kagaku Kyokaishi*, **47**, 258 (1989)
150. N. Ishikawa, T. Kitazume, T. Yamazaki, Y. Mochida, T. Tatsuno, *Chem. Lett.*, **1981**, 761
151. Y. Kimura, H. Suzuki, *Tetrahedron Lett.*, **30**, 1271 (1989)
152. T. Kitazume, N. Ishikawa, *Chem. Lett.*, **1981**, 1259
153. C. L. Liotta, H. P. Harris, *J. Am. Chem. Soc.*, **96**, 2250 (1974)
154. J. Cuomo, R. A. Olofson, *J. Org. Chem.*, **44**, 1016 (1979)
155. T. A. Bianchi, L. A. Cate, *J. Org. Chem.*, **42**, 2031 (1977)
156. Y. Yoshida, Y. Kimura, *Chem. Lett.*, **1988**, 1355
157. Y. Yoshida, Y. Kimura, *J. Fluorine Chem.*, **44**, 291 (1989)
158. N. Yazawa, H. Suzuki, Y. Yoshida, O. Furusawa, Y. Kimura, *Chem. Lett.*, **1989**, 2213
159. H. Suzuki, N. Yazawa, Y. Yoshida, O. Furusawa, Y. Kimura, *Bull. Chem. Soc. Jpn.*, **63**, 2010 (1990)
160. Y. Yoshida, Y. Kimura, M. Tomoi, *Chem. Lett.*, **1990**, 769
161. J. Ichihara, T. Matsuo, T. Hanafusa, T. Ando, *J. Chem. Soc., Chem. Commun.*, **1986**, 793
162. J. H. Clark, A. J. Hyde, D. K. Smith, *J. Chem. Soc., Chem. Commun.*, **1986**, 791
163. T. Nakamura, O. Kaieda, *Yuki Gosei Kagaku Kyokaishi*, **47**, 20 (1989)
164. A. B. Foster, J. H. Westwood, *Pure Appl. Chem.*, **35**, 147 (1973)
165. J. A. Wright, N. F. Taylor, J. J. Fox, *J. Org. Chem.*, **34**, 2632 (1969)
166. (a) J. Oshida, M. Morisaki, N. Ikekawa, *Tetrahedron Lett.*, **21**, 1755 (1980); (b) E. Ohshima, H. Sai, S. Takatsuto, N. Lkekawa, Y. Kobayashi, Y. Tanaka, H. F. DeLuca, *Chem. Pharm. Bull.*, **32**, 3525 (1984)
167. K. O. Christe, W. W. Wilson, R. D. Wilson, R. Bau, J. Feng, *J. Am. Chem. Soc.*, **112**, 7619 (1990)
168. D. P. Cox, J. Terpinski, W. Lawrynowicz, *J. Org. Chem.*, **49**, 3216 (1984)
169. M. Shimizu, Y. Nakahara, H. Yoshioka, *Tetrahedron Lett.*, **26**, 4207 (1985)
170. D. Y. Chi, M. R. Kilbourn, J. A. Katzenellenbogen, M. J. Welch, *J. Org. Chem.*, **52**, 658 (1987)
171. P. A. Grieco, E. Williams, T. Sugahara, *J. Org. Chem.*, **44**, 2194 (1979)
172. D. O. Kiesewetter, J. A. Katzenellenbogen, M. R. Kilbourn, M. J. Welch, *J. Org. Chem.*, **49**, 4900 (1984)
173. L. Hough, A. A. E. Penglis, A. C. Richardson, *Can. J. Chem.*, **59**, 396 (1981)

174. J. H. Clark, D. K. Smith, *Tetrahedron Lett.*, **26**, 2233 (1985)
175. P. Bosch, F. Camps, E. Chamorro, V. Gasol, A. Guerrero, *Tetrahedron Lett.*, **28**, 4733 (1987)
176. M. B. Giudicelli, D. Picq, B. Veyron, *Tetrahedron Lett.*, **31**, 6527 (1990)
177. K. Moughamir, A. Atmani, H. Mestdagh, C. Rolando, C. Francesch, *Tetrahedron Lett.*, **39**, 7305 (1998)
178. P. Albert, J. Cousseau, *J. Chem. Soc., Chem. Commun.*, **1985**, 961
179. J. Cousseau, P. Albert, *Bull. Soc. Chim. Fr.*, **1986**, 910
180. D. Landini, M. Penso, *Tetrahedron Lett.*, **31**, 7209 (1990)
181. W. J. Middleton, *Org. Synth.*, **64**, 221 (1986)
182. R. Noyori, I. Nishida, J. Sakata, *J. Am. Chem. Soc.*, **103**, 2106 (1981)
183. R. Noyori, I. Nishida, J. Sakata, *Tetrahedron Lett.*, **21**, 2085 (1980)
184. A. S. Pilcher, H. L. Ammon, P. DeShong, *J. Am. Chem. Soc.*, **117**, 5166 (1995)
185. A. S. Pilcher, P. DeShomg, *J. Org. Chem.*, **61**, 6901 (1996)
186. M. Gingras, *Tetrahedron Lett.*, **32**, 7381 (1991)
187. (a) O. Gasser, H. Schmidbaur, *J. Am. Chem. Soc.*, **97**, 6281 (1975); (b) S. J. Brown, J. H. Clark, *J. Chem. Soc., Chem. Commun.*, **1983**, 1256; (c) S. J. Brown, J. H. Clark, D. J. Macquarrie, *J. Chem. Soc., Dalton Trans.*, **1988**, 277
188. (a) J. Bensoam, J. Leroy, F. Mathey, C. Wakselman, *Tetrahedron Lett.*, **1979**, 353; (b) H. Seto, Z. Qian, H. Yoshioka, Y. Uchibori, M. Umeno, *Chem. Lett.*, **1991**, 1185
189. S. J. Brown, J. H. Clark, *J. Chem. Soc., Chem. Commun.*, **1985**, 672
190. S. J. Brown, J. H. Clark,. *J. Fluorine Chem.*, **30**, 251 (1985)
191. Y. Uchibori, M. Umeno, H. Seto, Z. Qian, H. Yoshioka, *Synlett*, **1992**, 345
192. S. Rozen, M. Brand, *J. Org. Chem.*, **1985**, 50, 3342
193. (a) A. Gregorcic, M. Zupan, *Bull. Chem. Soc. Jpn.*, **60**, 3083 (1987); (b) G. A. Olah, X.-Y. Li, *Synlett*, **1990**, 267. (c) G. A. Olah, X.-Y. Li, Q. Wang, G. K. S. Prakash, *Synthesis*, **1993**, 693
194. (a) G. Alvernhe, A. Laurent, G. Haufe, *Synthesis*, **1987**, 562. (b) G. Haufe, G. Alvernhe, A. Laurent, *Tetrahedron Lett.*, **27**, 4449 (1986); (c) M. M. Kremlev, G. Haufe, *J. Fluorine Chem.*, **90**, 121 (1998)
195. M. Maeda, M. Abe, M. Kojima, *J. Fluorine Chem.*, **34**, 337 (1987)
196. M. Kuroboshi, T. Hiyama, *Tetrahedron Lett.*, **32**, 1215 (1991)
197. J. Ichihara, K. Funabiki, T. Hanafusa, *Tetrahedron Lett.*, **31**, 3167 (1990)
198. M. Shimizu, Y. Nakahara, H. Yoshioka, *J. Chem. Soc., Chem. Commun.*, **1989**, 1881
199. M. Shimizu, M. Okamura, T. Fujisawa, *Bull. Chem. Soc. Jpn.*, **64**, 2596 (1991)
200. M. Tamura, M. Shibakami, A. Sekiya, *Synthesis*, **1995**, 515
201. M. Kuroboshi, T. Hiyama, *Bull. Chem. Soc. Jpn.*, **68**, 1799 (1995)
202. (a) S. Eddarir, H. Mestdagh, C. Rolando, *Tetrahedron Lett.*, **32**, 69 (1991); (b) S. Eddarir, C. Francesch, H. Mestdagh, C. Rolando, *Bull. Soc. Chim. Fr.*, **134**, 741 (1997)
203. G. Haufe, G. Alvernhe, D. Anker, A. Laurent, C. Sluzzo, *Tetrahedron Lett.*, **29**, 2311 (1988)
204. C. Saluzzo, A.-M. L. Spina, D. Picq, G. Alvernhe, D. Anker, D. Wolf, G. Haufe, *Bull. Soc. Chim. Fr.*, **131**, 831 (1994)
205. C. Saluzzo, G. Alvernhe, D. Anker, *J Fluorine Chem.*, **47**, 467 (1990)
206. C. Saluzzo, G. Alvernhe, D. Anker, G. Haufe, *Tetrahedron Lett.*, **31**, 663 (1990)
207. S. Tomoda, Y. Usuki, *Chem. Lett.*, **1989**, 1235
208. G. A. Olah, M. Nojima, *Synthesis*, **1973**, 785
209. C. York, G. K. S. Prakash, G. A. Olah, *J. Org. Chem.*, **59**, 6493 (1994)
210. (a) K. C. Nicolaou, R. E. Dolle, D. P. Papahatjis, J. L. Randall, *J. Am. Chem. Soc.*, **106**, 4189 (1984); (b) R. E. Dolle, K. C. Nicolaou, *J. Am. Chem. Soc.*, **107**, 1691 (1985)
211. S. Caddick, L. Gazzard, W. B. Motherwell, J. A. Wilkinson, *Tetrahedron*, **52**, 149 (1996)
212. C. York, G. K. S. Prakash, G. A. Olah, *Tetrahedron*, **52**, 9 (1996)
213. J. Ichikawa, K. Sugimoto, T. Sonoda, H. Kobayashi, *Chem. Lett.*, **1987**, 1985
214. S. Furuta, M. Kuroboshi, T, Hiyama, *Tetrahedron Lett.*, **36**, 8243 (1995)
215. K. Kanie, Y. Tanaka, M. Shimizu, S. Takehara, T. Hiyama, *Chem. Lett.*, **1998**, 1169
216. (a) A. Konnno, K. Nakagawa, T. Fuchigami, *J. Chem. Soc., Chem. Commun.*, **1991**, 1027; (b) T. Fuchigami, M. Shimojo, A. Konno, K. Nakagawa, *J. Org. Chem.*, **55**, 6074 (1990);

References

(c) T. Fuchigami, A. Konno, K. Nakagawa, M. Shimojo, *J. Org. Chem.*, **59**, 5937 (1994); (d) T. Brigaud, E. Laurent, *Tetrahedron Lett.*, **31**, 2287 (1990)
217. (a) T. Fuchigami, S. Narizuka, A. Konno, *J. Org. Chem.*, **57**, 3755 (1992); (b) S. Narizuka, T. Fuchigami, *J. Org. Chem.*, **58**, 4200 (1993)
218. (a) A. W. Erian, A. Konnno, T. Fuchigami, *Tetrahedron Lett.*, **35**, 7245 (1994); (b) S. Higashiya, T. Sato, T. Fuchigami, *J. Fluorine Chem.*, **87**, 203 (1998)
219. T. Fuchigami, T. Hayashi, A. Konno, *Tetrahedron Lett.*, **33**, 3161 (1992)
220. K. Iseki, Y. Kobayashi, *Yuki Gosei Kagaku Kyokai Shi*, **52**, 40 (1994)
221. (a) L. Kabore, S. Chebli, R. Faure, E. Laurent, B. Marquet, *Tetrahedron Lett.*, **31**, 3137 (1990); (b) S. Narizuka, H. Koshiyama, A. Konno, T. Fuchigami, *J. Fluorine Chem.*, **73**, 121 (1995)
222. (a) S. Rozen, R. Filler, *Tetrahedron*, **41**, 1111 (1985); (b) W. R. Hasek, W. C. Smith, V. A. Engelhardt, *J. Am. Chem. Soc.*, **82**, 543 (1960); (c) G. A. Boswell, Jr., W. C. Ripka, R. M. Scribner, C. W. Tullock, *Org. React.*, **21**, 1 (1974); (d) W. J. Middleton, *J. Org. Chem.*, **40**, 574 (1975); (e) M. Hudlicky, *Org. React.*, **35**, 513 (1988)
223. J. Kollonitsch, S. Marburg, L. M. Perkins, *J. Org. Chem.*, **41**, 3107 (1976)
224. S. C. Sondej, J. A. Katzenellenbogen, *J. Org. Chem.*, **51**, 3508 (1986)
225. G. K. S. Prakash, D. Hoole, V. P. Reddy, G. A. Olah, *Synlett*, **1993**, 691
226. M. Kuroboshi, T. Hiyama, *Synlett*, **1991**, 909
227. W. B. Motherwell, J. A. Wilkinson, *Synlett*, **1991**, 191
228. (a) T. Fuchigami, T. Fujita, *J. Org. Chem.*, **59**, 7190 (1994); (b) T. Fujita, T. Fuchigami, *Tetrahedron Lett.*, **37**, 4725 (1996)
229. M. Kuroboshi, T. Hiyama, *J. Fluorine Chem.*, **69**, 127 (1994)
230. D. P. Matthews, J. P. Whitten, J. R. McCarthy, *Tetrahedron Lett.*, **27**, 4861 (1986)
231. (a) M. Kuroboshi, S. Furuta, T. Hiyama, *Tetrahedron Lett.*, **36**, 6121 (1995); (b) S. Furuta, M. Kuroboshi, T. Hiyama, *Bull. Chem. Soc. Jpn.*, **71**, 1939 (1998)
232. M. Kuroboshi, T. Hiyama, *Chem. Lett.*, **1992**, 827
233. S. Furuta, T. Hiyama, *Synlett*, **1996**, 1199
234. K.-I. Kim, J. R. McCarthy, *Tetrahedron Lett.*, **37**, 3223 (1996)
235. M. Kuroboshi, T. Hiyama, *Synlett*, **1994**, 251
236. (a) W. A. Sheppard, *J. Org. Chem.*, **29**, 1 (1964); (b) P. E. Aldrich, W. A. Sheppard, *J. Org. Chem.*, **29**, 11 (1964); (c) F. Mathey, J. Bensoam, *Tetrahedron Lett.*, **25**, 2253 (1973); (d) A. E. Feiring, *J. Org. Chem.*, **44**, 2907 (1979)
237. (a) M. Kuroboshi, K. Suzuki, T. Hiyama, *Tetrahedron Lett.*, **33**, 4173 (1992); (b) K. Kanie, K. Mizuno, M. Kuroboshi, S. Takehara, T. Hiyama, *Bull. Chem. Soc. Jpn.*, **72**, 2523 (1999)
238. K. Kanie, Y. Tanaka, M. Shimizu, M. Kuroboshi, T. Hiyama, *Chem. Commun.*, **1997**, 309
239. M. J. Koen, F. L. Guyader, W. B. Motherwell, *J. Chem. Soc., Chem. Commun.*, **1995**, 1241
240. (a) K. Kanie, K. Mizuno, M. Kuroboshi, T. Hiyama, *Bull. Chem. Soc. Jpn.*, **71**, 1973 (1998); (b) M. Kuroboshi, K. Mizuno, K. Kanie, T. Hiyama, *Tetrahedron Lett.*, **36**, 563 (1995); (c) M. Kuroboshi, T. Hiyama, *Tetrahedron Lett.*, **33**, 4177 (1992)
241. (a) W. Dmowski, M. Kamiński, *J. Fluorine Chem.*, **1983**, 23, 207; (b) W. Dmowski, M. Kamiński, *J. Fluorine Chem.*, **23**, 219 (1983); (c) L. N. Markovskij, V. E. Pashinnik, A. V. Kirsanov, *Synthesis*, **1973**, 787; (d) L. M. Yagupol'skii, N. V. Kondratenko, G. N. Timofeeva, M. I. Dronkina, Y. L. Yagupol'skii, *Zh. Org. Khim.*, **16**, 2508 (1980); (e) G. Pawelke, *J. Fluorine Chem.*, **52**, 229 (1991); (f) G. Wedler, *Angew. Chem. Int. Ed. Engl.*, **5**, 848 (1966)
242. M. Kuroboshi, T. Hiyama, *Tetrahedron Lett.*, **35**, 3983 (1994)
243. (a) G. A. Olah, J. Welch, *Synthesis*, **1974**, 652; (b) S. Hamman, C. G. Beguin, *Tetrahedron Lett.*, **24**, 57 (1983); (c) J. Barber, R. Keck, J. Rétey, *Tetrahedron Lett.*, **23**, 1549 (1982); (d) G. A. Olah, G. K. S. Prakash, Y. L. Chao, *Helv. Chim. Acta*, **64**, 2528 (1981)
244. G. A. Olah, J. Welch, *Synthesis*, **1974**, 896
245. S. Rozen, M. Brand, D. Zamir, D. Hebel, *J. Am. Chem. Soc.*, **109**, 896 (1987)
246. S. Rozen, E. Mishani, A. Bar-Haim, *J. Org. Chem.*, **59**, 2918 (1994)
247. C. York, G. K. S. Prakash, Q. Wang, G. A. Olah, *Synlett*, **1994**, 425
248. J. H. Simons, *J. Fluorine Chem.*, **32**, 7 (1986)

249. W. H. Person, *J. Fluorine Chem.*, **32**, 29 (1986)
250. S. Nagase, *Fluorine Chemistry Reviews*, Vol. 1, P. Tarrant (ed), Marcel Dekker, New York, p 77 (1967)
251. T. Abe, S. Nagase, *Preparation, Properties, and Industrial Application of Organofluorine Compounds*, R. E. Banks (ed), Ellis Horwood, Chichester, p 19 (1982)
252. J. Burdon, J. C. Tatlow, *Advances in Fluorine Chemistry*, Vol. 1, M. Stacey, J. C. Tatlow, A. G. Sharp (eds), Butterworths, London, p 166 (1960)
253. R. E. Banks, B. E. Smart, J. C. J. C. Tatlow (eds), *Organofluorine Chemistry*, Plenum, New York, p 121 (1994)
254. *Introduction to Fluorine Chemistry. Principles and Experiments*, Japan Society for Promotion of Science, The 155th Committee, Nikkan Kogyo Shinbun, Japan, p 233 (1997)
255. M. Sander, W. Blöchel, *Chem. Eng. Tech.*, **37**, 7 (1965)
256. H. Baba, E. Hayashi, T. Abe, H. Muramatsu, *Chem. Express*, **1**, 371 (1986)
257. T. Abe, K. Kodaira, H. Baba, S. Nagase, *J. Fluorine Chem.*, **12**, 1 (1977)
258. T. Gramstad, R. N. Haszeldine, *J. Chem. Soc.*, **1956**, 173
259. T. Gramstad, R. N. Haszeldine, *J. Chem. Soc.*, **1957**, 2640
260. T. Abe, E. Hayashi, H. Baba, K. Kodaira, S. Nagase, *J. Fluorine Chem.*, **15**, 353 (1980)
261. Y. Naito, Y. Inoue, T. Ono, Y. Arakawa, C. Fukaya, K. Yokoyama, Y. Kobayashi, K. Yamanouchi, *J. Fluorine Chem.*, **26**, 485 (1984)
262. J. H. H. Meurs, D. W. Sopher, W. Eilenberg, *Angew. Chem. Int. Ed. Engl.*, **28**, 927 (1989)
263. J. H. H. Meurs, W. Eilenberg, *Tetrahedron*, **47**, 705 (1991)
264. E. Laurent, B. Marquet, R. Tardive, H. Thiebault, *Bull. Soc. Chim. Fr.*, **1986**, 955
265. F. M. Ventalon, R. Faure, E. G. Laurent, B. S. Marquet, *Tetrahedron Asymmetry*, **5**, 1909 (1994)
266. T. Brigaud, E. Laurent, *Tetrahedron Lett.*, **31**, 2287 (1990)
267. T. Fuchigami, M. Shimojo, A. Konno, K. Nakagawa, *J. Org. Chem.*, **55**, 6074 (1990)
268. S. Narizuka, T. Fuchigami, *J. Org. Chem.*, **58**, 4200 (1993)

CHAPTER 3

Organofluorine Building Blocks

The synthetic methodology for organofluorine compounds is generally classified into two approaches: one involves C–F bond formation via transformation of a suitable functional group by a fluorinating reagent, as described in Chap. 2, and the other includes C–C bond formation starting from a relatively small fluorine-substituted molecule which is readily available and can be easily handled. This approach is called the building block approach and is familiar to synthetic chemists, because hazardous and toxic reagents and special equipment are not necessary. This chapter focuses on the synthetic transformation of fluorine-containing building blocks in the order of the type of reagent.

3.1
Fluorine-Substituted Nucleophilic Reagents

3.1.1
Alkylmetals

Perfluoroalkyllithiums are, in general, thermally unstable due to their rapid α- or β-elimination. In particular, trifluoromethyllithium is yet unknown. However, except for a trifluoromethyl group, a halogen-metal exchange of a perfluoroalkyl iodide with an organolithium reagent at low temperatures conveniently generates the corresponding perfluoroalkyllithium reagent which reacts with such a co-existing electrophile as an aldehyde, ketone, or ester to give a perfluoroalkylated product [1]. For example, pentafluoroethyl iodide undergoes a rapid exchange with methyllithium in the presence of benzaldehyde to produce 1-phenylpentafluoro-1-propanol. For the reaction with an imine, the use of $BF_3 \cdot OEt_2$ is essential to obtain a perfluoroalkylated amine.

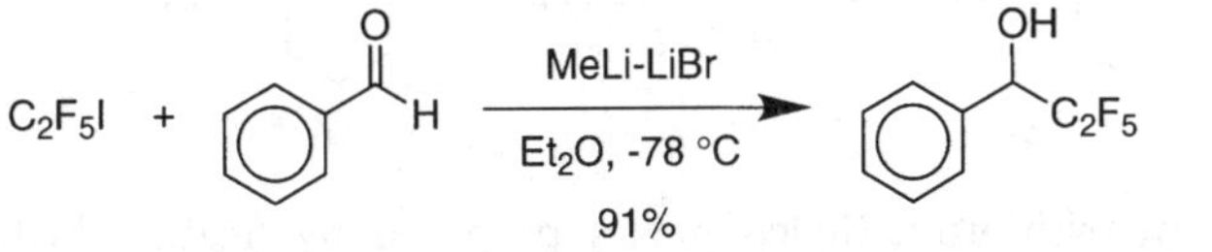

$$C_2F_5I + PhCHO \xrightarrow[\text{Et}_2\text{O, -78 °C}]{\text{MeLi-LiBr}} PhCH(OH)C_2F_5 \quad 91\% \quad (3.1)$$

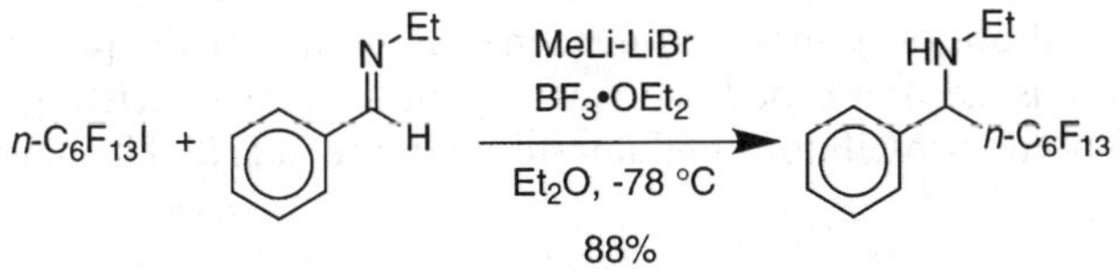

$$n\text{-}C_6F_{13}I + PhCH=NEt \xrightarrow[\text{Et}_2\text{O, -78 °C}]{\text{MeLi-LiBr, } BF_3 \cdot OEt_2} PhCH(NHEt)n\text{-}C_6F_{13} \quad 88\% \quad (3.2)$$

Perfluoroalkylmagnesium reagents are prepared by a conventional procedure from perfluoroalkyl iodides with magnesium metal, in contrast to labile perfluoroalkyllithium reagents [2]. Alternatively, an iodine-magnesium exchange reaction of a perfluoroalkyl iodide with a Grignard reagent gives the corresponding perfluoroalkylmagnesium reagent [3]. An example is the reaction of heptafluoro-1-iodopropane with cyclohexanone shown below.

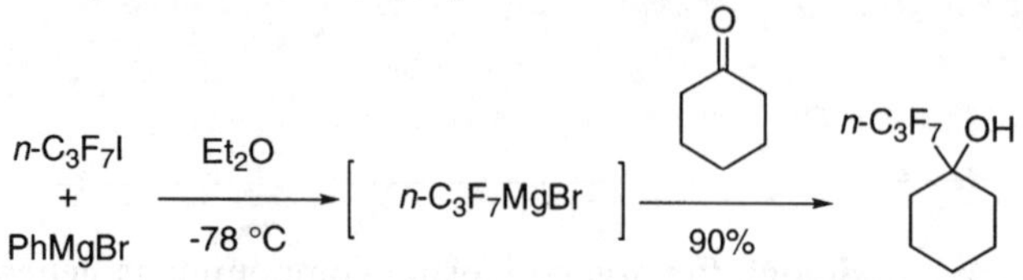

$$n\text{-}C_3F_7I + PhMgBr \xrightarrow[-78\ °C]{Et_2O} [n\text{-}C_3F_7MgBr] \xrightarrow{\text{cyclohexanone}} \underset{90\%}{n\text{-}C_3F_7\text{-cyclohexanol}} \quad (3.3)$$

Perfluoroalkylcopper reagents, easily prepared from perfluoroalkyl iodides and copper metal, are more stable than the corresponding lithium or magnesium reagents [4]. Thus, pentadecafluoroheptyl iodide and iodobenzene are smoothly coupled by means of copper metal.

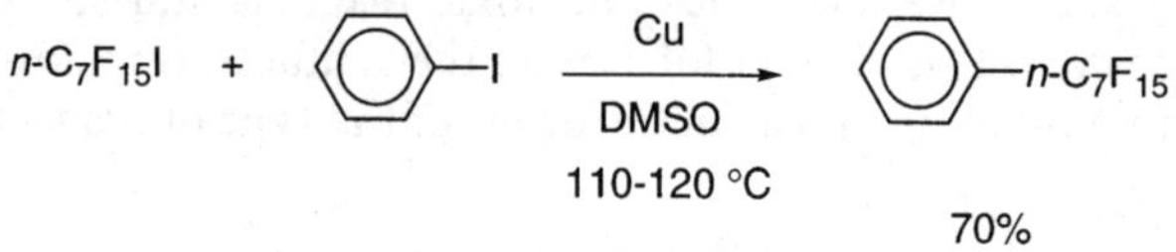

$$n\text{-}C_7F_{15}I + PhI \xrightarrow[\substack{DMSO \\ 110\text{-}120\ °C}]{Cu} \underset{70\%}{Ph\text{-}n\text{-}C_7F_{15}} \quad (3.4)$$

Trifluoromethylation is effectively achieved with trifluoromethylcopper produced by heating trifluoromethyl iodide and copper metal in an aprotic solvent. Thus, m-iodotoluene and alkenyl dibromide are converted into the corresponding trifluoromethylated products [5].

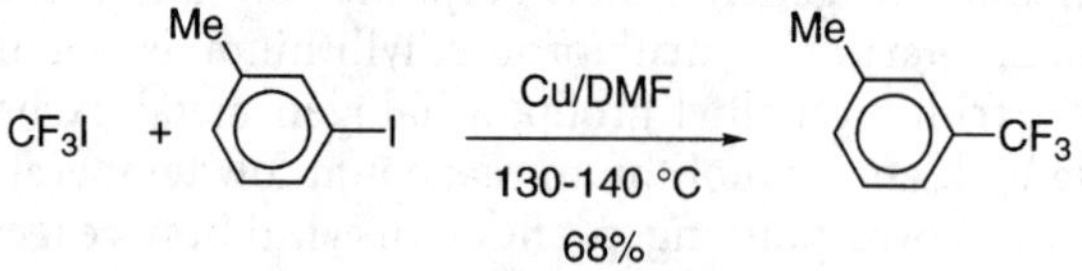

$$CF_3I + m\text{-}MeC_6H_4I \xrightarrow[\substack{130\text{-}140\ °C \\ 68\%}]{Cu/DMF} m\text{-}MeC_6H_4\text{-}CF_3 \quad (3.5)$$

$$\underset{\text{di-}t\text{-Bu-cyclohexylidene-CBr}_2}{} \xrightarrow[\substack{CF_3CuI \\ \text{rt - 50 °C, 81\%}}]{\substack{CF_3I + Cu \\ HMPA,\ 120\ °C}} \underset{\text{di-}t\text{-Bu-cyclohexylidene-C(CF}_3)_2}{} \quad (3.6)$$

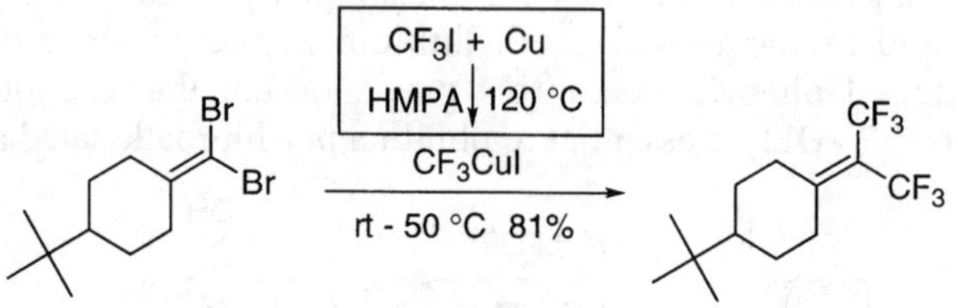

Alternatively, trifluoromethylcopper is prepared by heating bis(trifluoromethyl)mercury with copper powder and is allowed to undergo trifluoromethylation of p-iodonitrobenzene in good yields [6]. The same transformation is accomplished with a reagent system consisting of N-trifluoromethyl-N-nitrosotrifluoromethanesulfonamide (TNS-Tf) and copper (Scheme 3.1) [7].

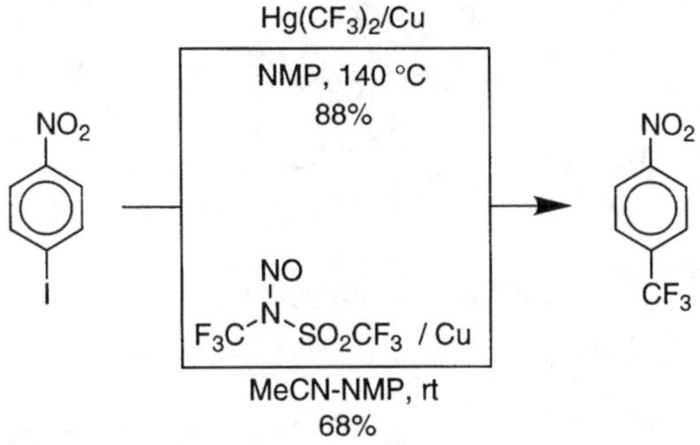

3.1 Fluorine-Substituted Nucleophilic Reagents

Scheme 3.1

The reaction of difluorodihalomethanes with cadmium or zinc produces thermally stable trifluoromethylcadmium or -zinc reagents, respectively, and both transfer a trifluoromethyl group to copper to generate a trifluoromethyl-copper reagent in excellent yields [8].

$$CF_2X_2 + 2M \xrightarrow[rt]{DMF} [CF_3MX + (CF_3)_2M] \xrightarrow[-80°C\ to\ rt]{CuX} [CF_3Cu] \quad (3.7)$$

X = Br, Cl
M = Cd, Zn
80-95%   90-100%

This protocol is the most convenient in view of the cheap and readily available precursors and mild conditions. Trifluoromethylcopper prepared in this manner reacts with substituted iodobenzenes smoothly irrespective of the nature of the substituent.

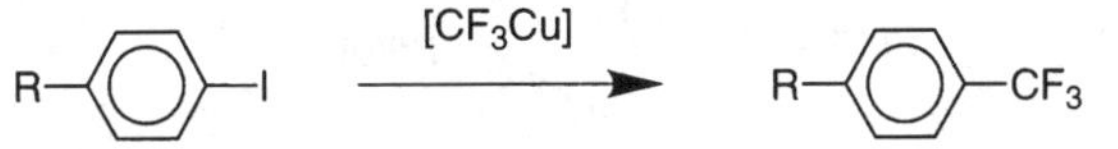

$$R-C_6H_4-I \xrightarrow{[CF_3Cu]} R-C_6H_4-CF_3 \quad (3.8)$$

R = H (100%); R = NO_2 (81%); R = OMe (78%); R = Br (95%)

A trifluoromethylzinc reagent, generated by treatment of a difluorodihalomethane with zinc, also gives a trifluoromethyl anionic reagent, as illustrated in Scheme 3.2 [9]. A mechanism is proposed that involves a difluorocarbene intermediate. Pathways are illustrated in Scheme 3.3.

Under ultrasonic irradiation, perfluoroalkyl iodides are converted into the corresponding zinc reagents which then undergo perfluoroalkylation of various halides, as summarized in Scheme 3.4 [10].

A fluorine-containing zinc carbenoid, 1,1-dichloro-2,2,2-trifluoroethylzinc chloride [11], is prepared simply by reduction of CFC-113a (CF_3CCl_3) with zinc and is thermally stable enough to react with aldehydes (Scheme 3.5). Use of a large excess of zinc in the presence of acetic anhydride or aluminum chloride leads to the stereoselective in situ formation of 2-chloro-1,1,1-trifluoro-2-

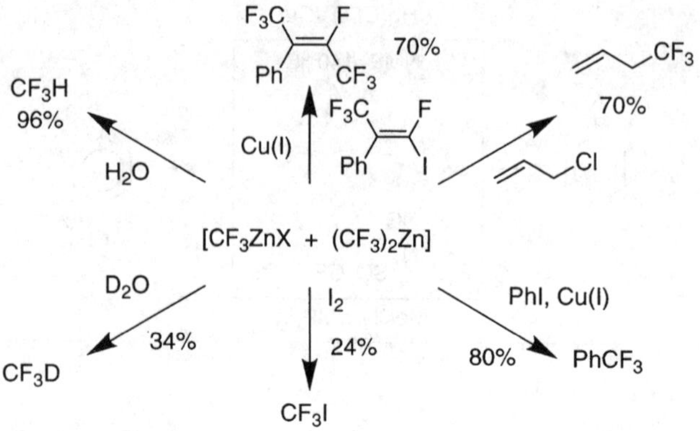

Scheme 3.2

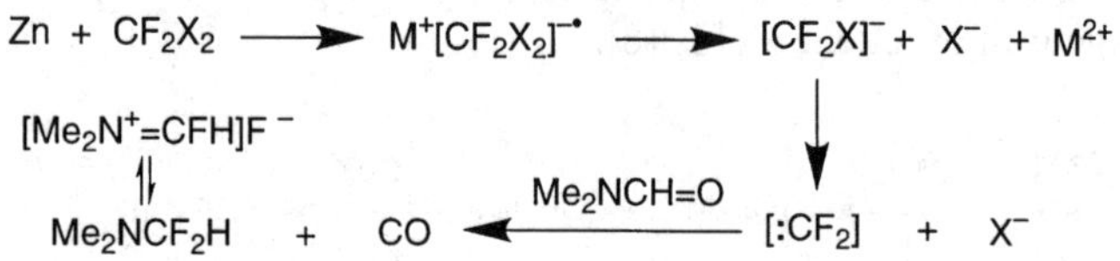

Scheme 3.3

$$Zn + CF_2X_2 \longrightarrow M^+[CF_2X_2]^{-\bullet} \longrightarrow [CF_2X]^- + X^- + M^{2+}$$

$$[Me_2N^+=CFH]F^- \rightleftharpoons Me_2NCF_2H$$

$$Me_2NCF_2H + CO \xleftarrow{Me_2NCH=O} [:CF_2] + X^-$$

$$F^- + [:CF_2] \longrightarrow [CF_3^-] \longrightarrow CF_3MX + (CF_3)_2M$$

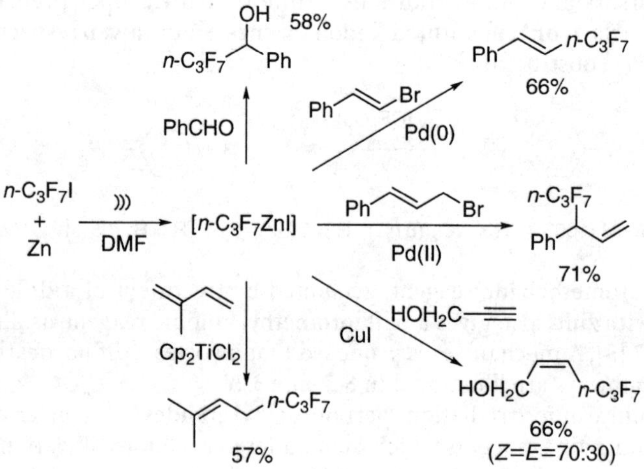

Scheme 3.4

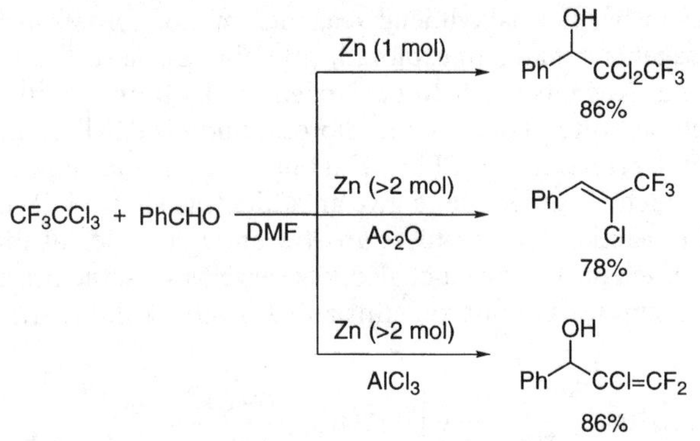

3.1 Fluorine-Substituted Nucleophilic Reagents

Scheme 3.5

CF$_3$CCl$_3$ + PhCHO

- Zn (1 mol), DMF → Ph-CH(OH)-CCl$_2$CF$_3$ (86%)
- Zn (>2 mol), Ac$_2$O, DMF → Ph-C(Cl)=C(CF$_3$) (78%)
- Zn (>2 mol), AlCl$_3$, DMF → Ph-CH(OH)-CCl=CF$_2$ (86%)

alkenes or 2-chloro-1,1-difluoro-1-alken-3-ols, respectively. This approach has been applied to a one-step synthesis of an artificial pyrethroid, bifenthrin [12].

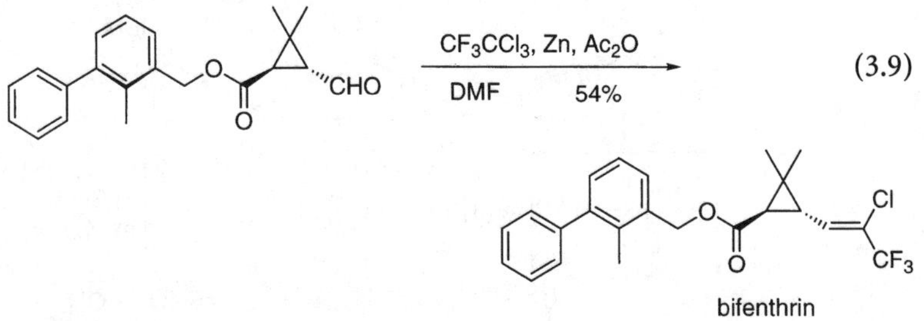

$$\text{(substrate)} \xrightarrow{\text{CF}_3\text{CCl}_3,\ \text{Zn},\ \text{Ac}_2\text{O},\ \text{DMF},\ 54\%} \text{bifenthrin} \quad (3.9)$$

Trifluoromethyl(trimethyl)silane, upon activation by a fluoride ion, reacts with aldehydes, ketones, or esters to give, respectively, trifluoromethylated alcohols or trifluoromethyl ketones (Scheme 3.6) [13].

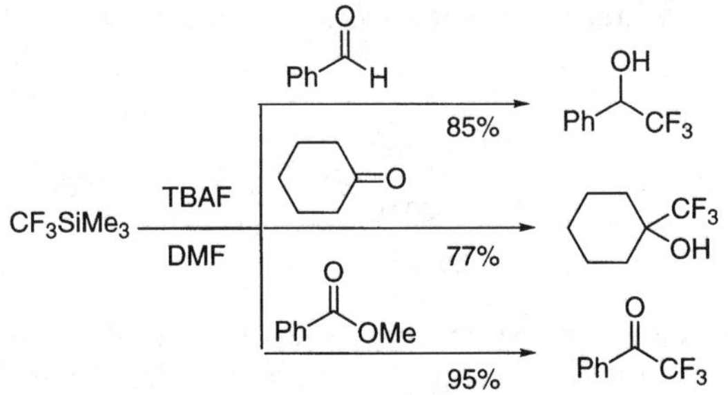

Scheme 3.6

CF$_3$SiMe$_3$ (TBAF, DMF):
- PhCHO → Ph-CH(OH)-CF$_3$ (85%)
- cyclohexanone → 1-CF$_3$-cyclohexan-1-ol (77%)
- PhCO$_2$Me → Ph-CO-CF$_3$ (95%)

In the presence of copper(I) iodide, the silane reagent undergoes a cross-coupling reaction with aryl and alkenyl iodides [14].

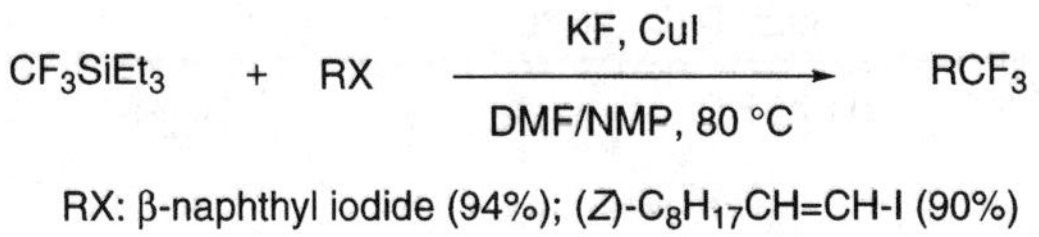

$$\text{CF}_3\text{SiEt}_3 + \text{RX} \xrightarrow[\text{DMF/NMP, 80 °C}]{\text{KF, CuI}} \text{RCF}_3 \quad (3.10)$$

RX: β-naphthyl iodide (94%); (Z)-C$_8$H$_{17}$CH=CH-I (90%)

In contrast to perfluoroalkylmetal reagents, monofluoroalkyl- and difluoroalkylmetal reagents remain unexplored, probably because the preparation of fluoroalkylmetal reagents needs to be carried out by bromine-lithium exchange at very low temperatures. For example, dibromofluoromethyllithium is generated only at −130 °C by treatment of $CFBr_3$ with butyllithium and allowed to react with a co-existing aldehyde or ketone to give an adduct, as shown in Scheme 3.7 [15]. As long as the reaction temperatures are low enough, yields of the adducts are moderate to excellent. The product alcohols serve as versatile precursors for the stereoselective synthesis of fluoroolefins and 2-bromo-2-fluoro-1,3-diols [16].

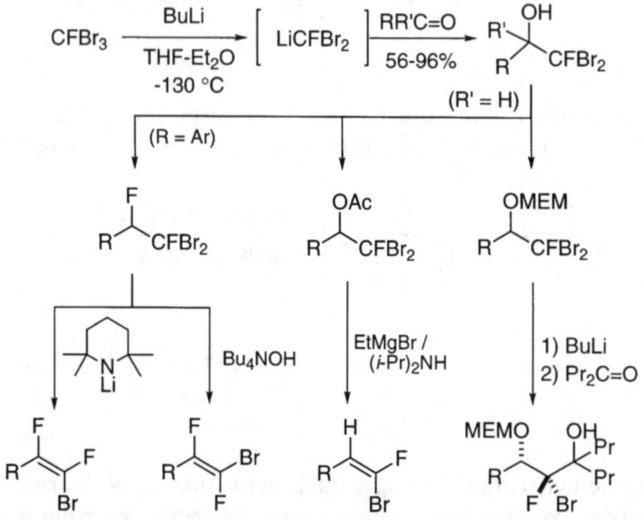

Scheme 3.7

An alternative zinc carbenoid reagent is prepared by the reduction of $CFBr_3$ with diethylzinc and reacts chemoselectively with aldehydes in preference to ketones [17].

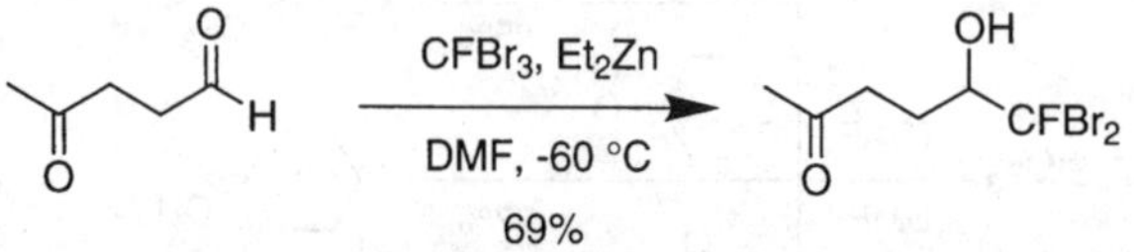

$$\text{(3.11)}$$

with CFBr$_3$, Et$_2$Zn, DMF, −60 °C, 69%.

Introduction of an anion-stabilizing group, such as R_3Si-, ArSO-, $ArSO_2$- or $(RO)_2P(O)$-, at an anionic carbon allows a thermally labile fluoroalkylmetal reagent to be handled at slightly higher temperatures of −98 to −70 °C [18]. Substituents introduced can be utilized for their further transformations. Some examples are presented in Schemes 3.8 and 3.9 and in Eq. (3.12).

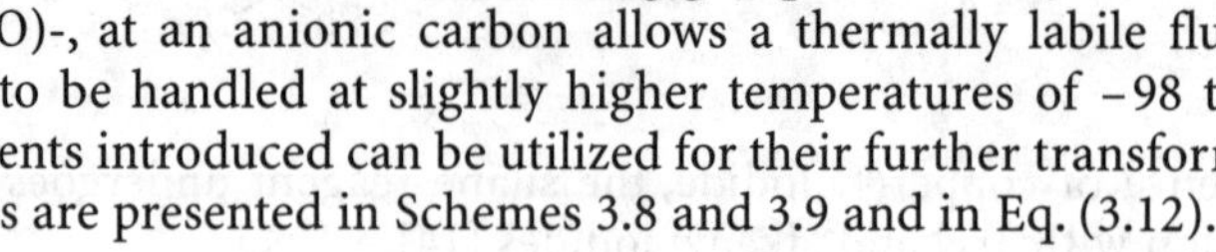

$$\text{(3.12)}$$

$(EtO)_2OP-CHF-CO_2H$ 1) BuLi, THF, −78 °C; 2) PhCHO → Ph-CHF-CO$_2$H, 91%.

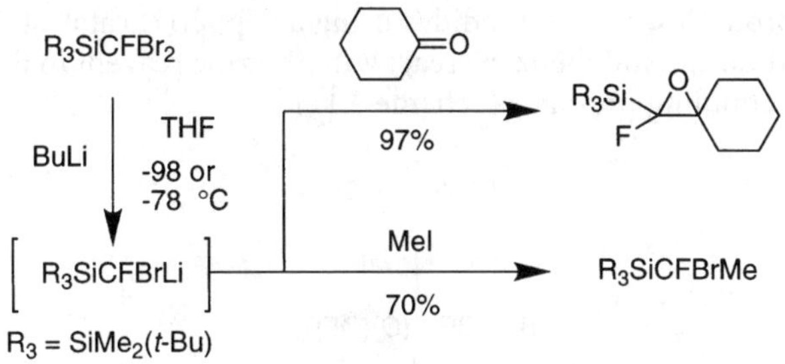

3.1 Fluorine-Substituted Nucleophilic Reagents

Scheme 3.8 $R_3 = SiMe_2(t\text{-Bu})$

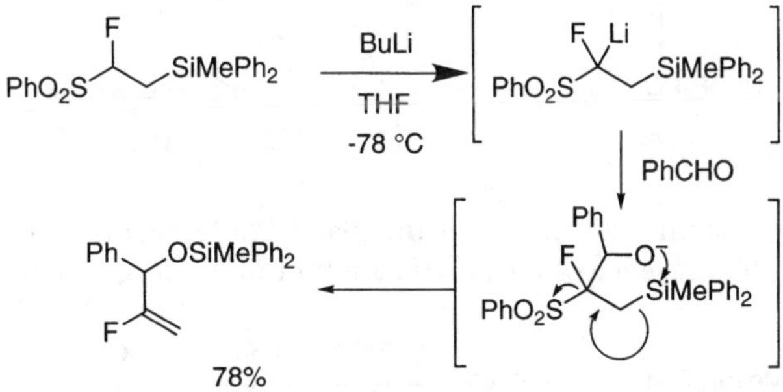

Scheme 3.9

Difluoroalkylmetals substituted by a $(RO)_2P(O)$- group are prepared by deprotonation of diethyl difluoromethylphosphonate with LDA (Scheme 3.10) [19]. The resulting reagent, $(EtO)_2P(O)CF_2Li$, undergoes alkylation as well as carbonyl addition. Reaction with esters and nitroalkenes is facilitated in the presence of $CeCl_3$ (Scheme 3.10) [20].

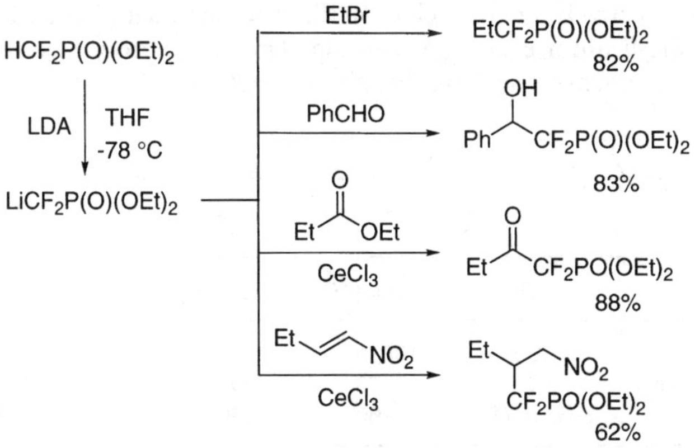

Scheme 3.10

Reduction of $(EtO)_2P(O)CF_2Br$ with zinc also produces a similar zinc reagent, $(EtO)_2P(O)CF_2ZnBr$, which reacts with electrophiles such as acyl chlorides, alkyl bromides, and aryl iodides using a copper(I) catalyst. For example, 1-hexenyl iodide and iodobenzene react with the zinc reagent in the presence of CuBr to give coupling products (Scheme 3.11) [21].

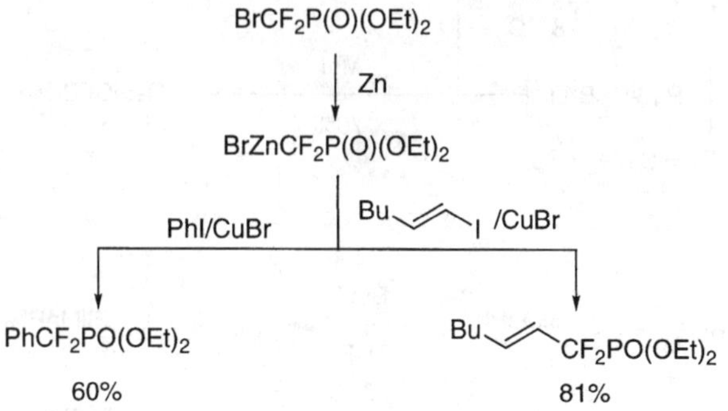

$$BrCF_2P(O)(OEt)_2$$
$$\downarrow Zn$$
$$BrZnCF_2P(O)(OEt)_2$$

PhI/CuBr → PhCF$_2$PO(OEt)$_2$ 60%

Bu-CH=CH-I /CuBr → Bu-CH=CH-CF$_2$PO(OEt)$_2$ 81%

Scheme 3.11

In a manner similar to trifluoromethylsilane, PhMe$_2$SiCF$_2$H reacts with benzaldehyde in the presence of a catalytic amount of KF to give PhCH(OH)CF$_2$H [22].

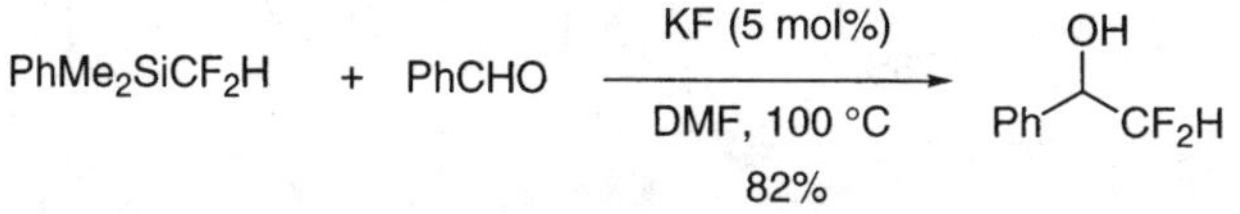

$$PhMe_2SiCF_2H + PhCHO \xrightarrow[\text{DMF, 100 °C}]{\text{KF (5 mol\%)}} Ph\underset{\text{82\%}}{\overset{OH}{\underset{|}{C}}}CF_2H \quad (3.13)$$

3.1.2
Alkenylmetals

Fluoroalkenyllithiums are conveniently prepared by a hydrogen-lithium and/or halogen-lithium exchange of fluoro(halo)alkenes with butyllithium [23]. For example, trifluorovinyllithium is generated from trifluoroethene or bromotrifluoroethene with butyllithium via proton abstraction or bromine-lithium exchange, respectively, or from chlorotrifluoroethene with s- or t-butyllithium via a chlorine-lithium exchange reaction. The latter procedure in particular provides an inexpensive and highly effective method.

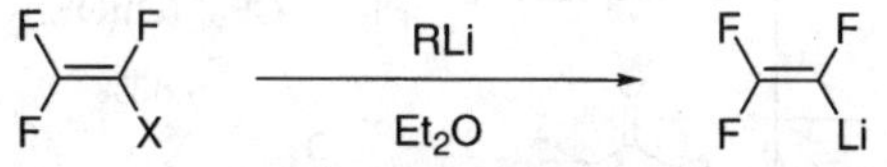

$$\underset{F}{\overset{F}{\diagup}}C=C\underset{X}{\overset{F}{\diagdown}} \xrightarrow[\text{Et}_2\text{O}]{\text{RLi}} \underset{F}{\overset{F}{\diagup}}C=C\underset{Li}{\overset{F}{\diagdown}} \quad (3.14)$$

X	R	Temp (°C)	Yield (%)
H	Bu	-100 to -78	>79
Br	Bu	-78	>73
Cl	s-Bu or t-Bu	-60	>95

Similarly, 2,2-difluorovinyllithium, 1,2-difluoroalkenyllithiums, and perfluoroalkenyllithiums are obtained by a deprotonation or halogen-lithium exchange of the corresponding fluoro(halo)alkenes. The stability of 1,2-difluoroalkenyllithiums depends on the olefinic geometry: (E)-RCF=CFLi is stable below $-5\,°C$ and above $-5\,°C$ cleanly undergoes dimerization and elimination to give an enyne, whereas (Z)-RCF=CFLi decomposes to give an intractable complex mixture of products even below $-80\,°C$, as shown in Scheme 3.12.

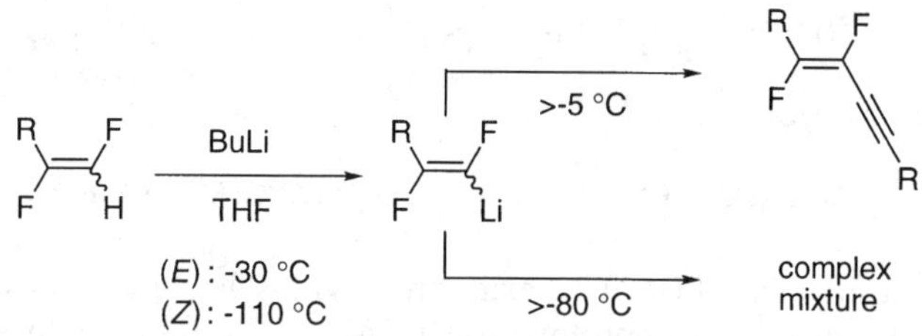

Scheme 3.12

Perfluoroalkenyllithiums react with carbonyl compounds in the normal fashion. In the presence of $BF_3 \cdot OEt_2$, the reagents add to oxiranes and oxetanes smoothly to give fluorinated homo- and bishomoallylic alcohols [24].

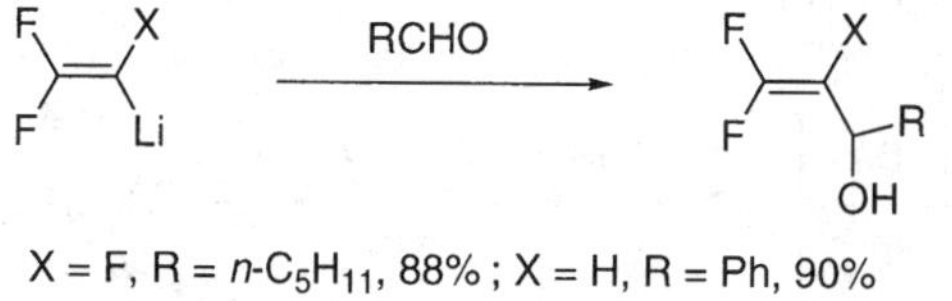

(3.15)

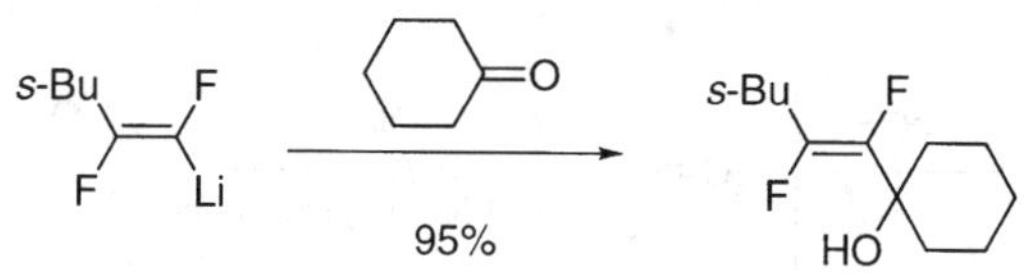

X = F, R = n-C$_5$H$_{11}$, 88% ; X = H, R = Ph, 90%

(3.16)

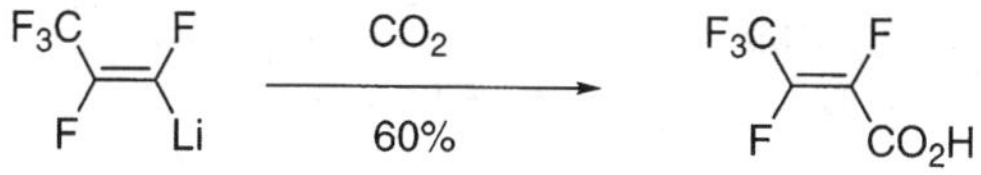

(3.17)

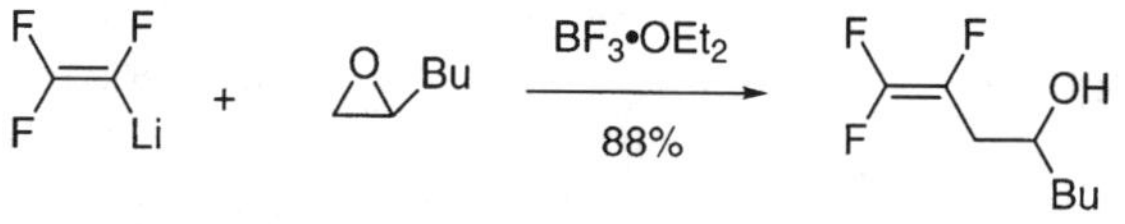

(3.18)

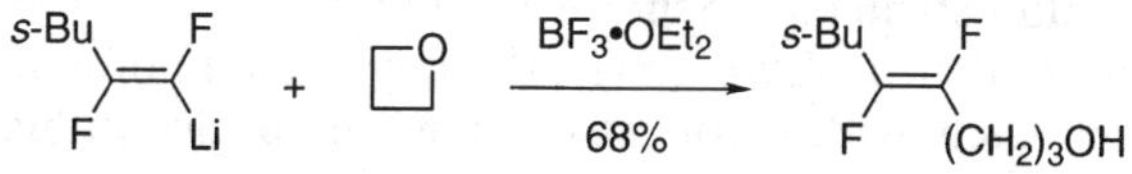

(3.19)

An iodine-lithium exchange reaction is applied to the generation of *N*-aryl-trifluoroacetimidoyllithiums, synthons of a trifluoroacetyl anion, which react with various electrophiles to give imines of trifluoromethyl ketones [25].

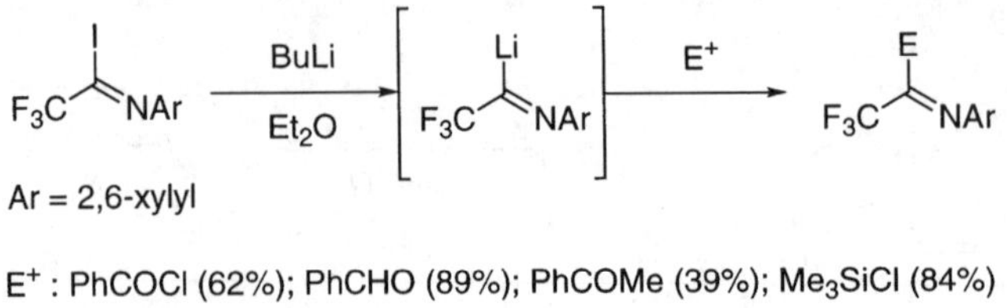

$$\underset{\text{Ar = 2,6-xylyl}}{F_3C\overset{I}{-}C=NAr} \xrightarrow[\text{Et}_2O]{\text{BuLi}} \left[F_3C\overset{Li}{-}C=NAr \right] \xrightarrow{E^+} F_3C\overset{E}{-}C=NAr \quad (3.20)$$

E^+ : PhCOCl (62%); PhCHO (89%); PhCOMe (39%); Me$_3$SiCl (84%)

Upon treatment with LDA, 2,2,2-trifluoroethyl tosylate undergoes sequential dehydrofluorination and deprotonation to give 2,2-difluoro-1-tosyloxyvinyl-lithium, a versatile and useful reagent for a building block of a CF$_2$=C< moiety. For example, its reaction with aldehydes and ketones provides the corresponding adducts (Scheme 3.13) [26]. Furthermore, trapping the lithium reagent with trialkylboranes affords 1-substituted 2,2-difluoroethenylboranes, which can then be converted into various types of *gem*-difluoroolefins by migration or a coupling reaction [27].

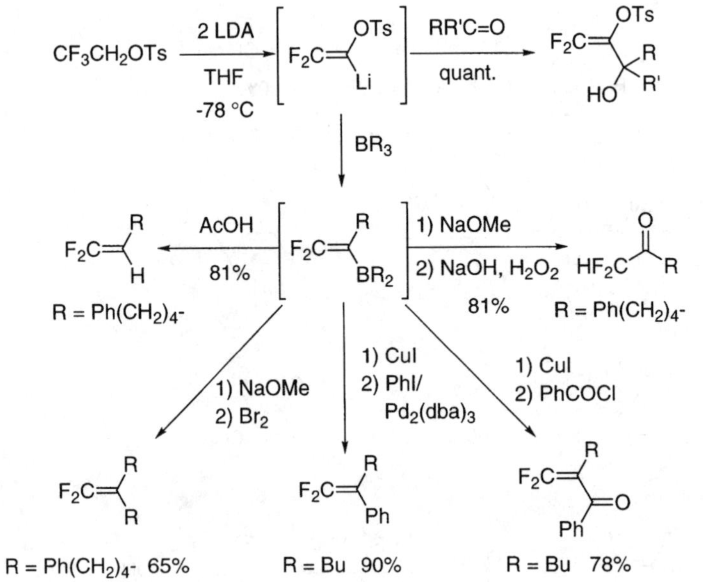

Scheme 3.13

In a similar manner, 2,2-difluoro-1-methoxyethoxymethoxyvinyllithium is also prepared. The reagent reacts with electrophiles such as aldehydes, ketones, chlorosilanes, and chlorostannanes (Scheme 3.14) [28].

It is worth noting that 1-*N*,*N*-diethylcarbamoyl-2,2-difluorovinyllithium, generated as above, reacts with carbonyl compounds to give difluoromethyl

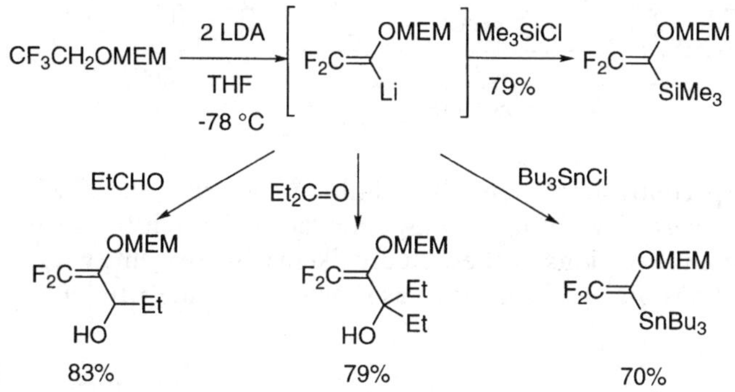

3.1 Fluorine-Substituted Nucleophilic Reagents

Scheme 3.14

ketones derived from 1,4-migration of a carbamoyl group (Scheme 3.15). The intermediary difluoroenolates then react with benzaldehyde to give three component coupled products.

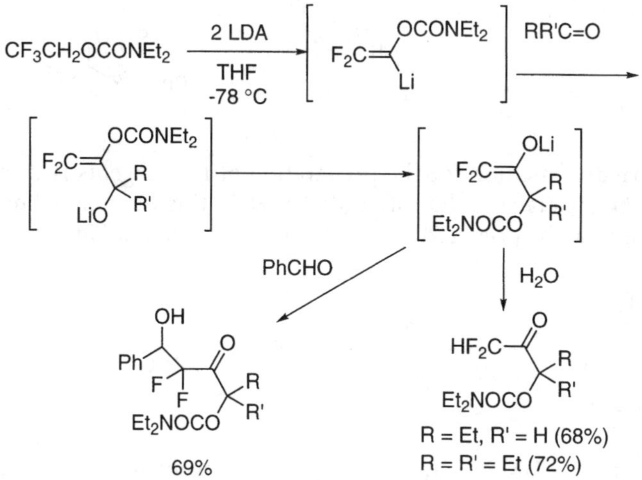

Scheme 3.15

Since fluoroalkenyllithiums are generally unstable and their applications are limited, stable fluoroalkenylmetals are prepared by transmetalation. In this way, fluoroalkenylzinc reagents are prepared at very low temperatures. Thus, 1,1,2-trifluorovinyl-, 1,2-difluoroalkenyl-, 2,2-difluorovinyl-, and 1-chloro-2-fluoro-alkenylzinc chlorides are prepared [29].

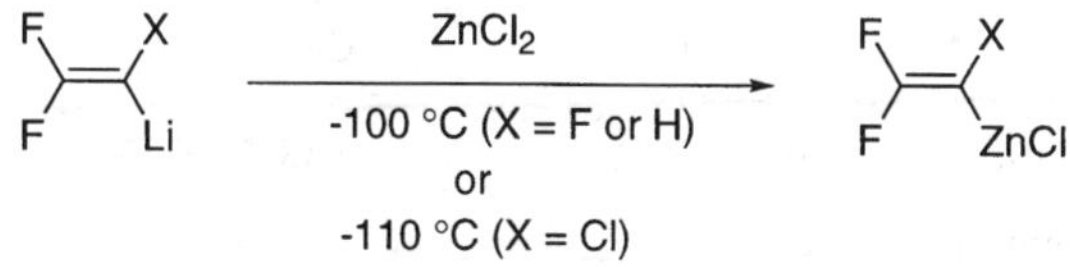

(3.21)

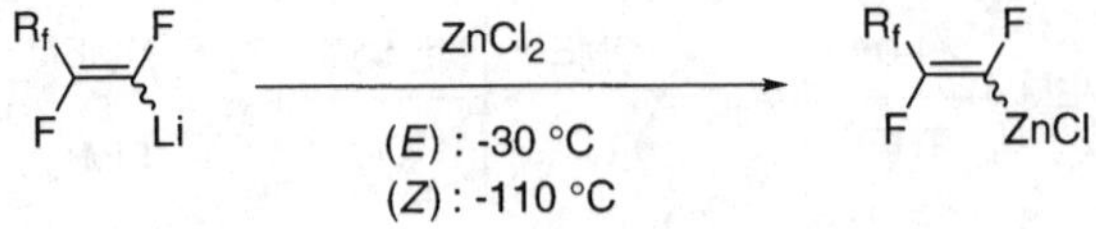

$$\underset{\text{F}}{\overset{R_f}{\diagdown}}\!\!=\!\!\underset{\text{Li}}{\overset{F}{\diagup}} \xrightarrow[\substack{(E):-30\ °C\\(Z):-110\ °C}]{ZnCl_2} \underset{\text{F}}{\overset{R_f}{\diagdown}}\!\!=\!\!\underset{\text{ZnCl}}{\overset{F}{\diagup}} \qquad (3.22)$$

In sharp contrast to fluoroalkenyllithium reagents, the zinc reagents are stable for several days even at room temperature. Thus, it is possible to achieve cross-coupling reactions of these fluoroalkenylzinc reagents with aryl iodides, alkenyl iodides, and acid chlorides, as shown in Scheme 3.16 [30].

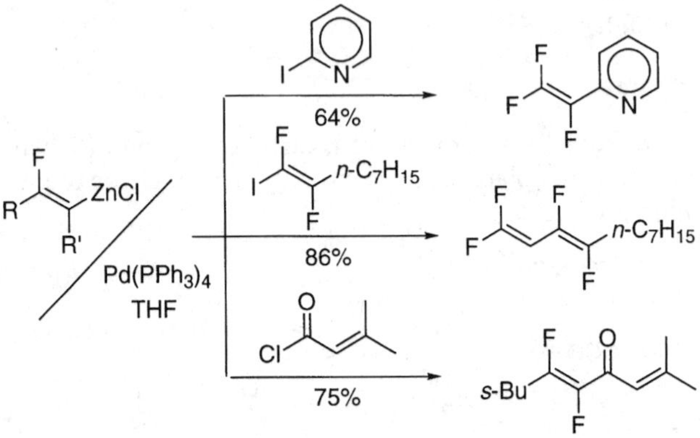

Scheme 3.16

An alternative method for the preparation of the reagents is direct metalation using zinc metal and fluoro(halo)alkenes. In this way, many fluoroalkenylzinc reagents can be prepared, some of which are listed in Fig. 3.1 [31].

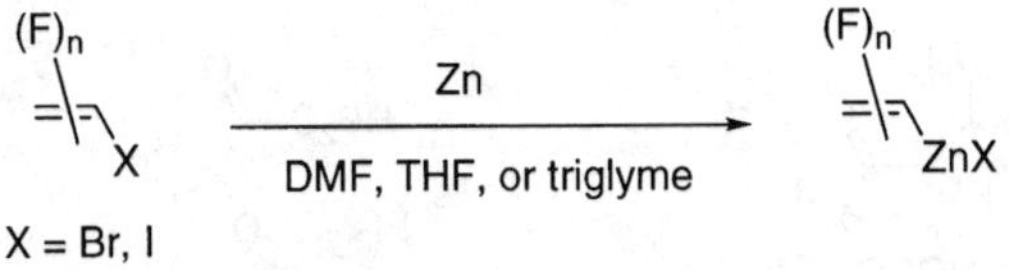

$$\underset{X=Br,\ I}{\overset{(F)_n}{=}\!\!\diagdown_X} \xrightarrow[\text{DMF, THF, or triglyme}]{Zn} \overset{(F)_n}{=}\!\!\diagdown_{ZnX} \qquad (3.23)$$

Solvent such as DMF, THF, or triglyme are conveniently used, whereas THF containing a small amount of tetramethylethylenediamine (TMEDA) is shown to be effective for the formation of a 2-trifluoromethylethenylzinc reagent.

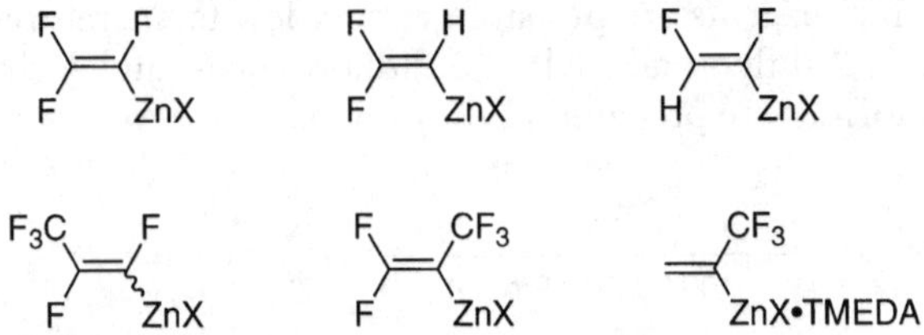

Fig. 3.1. Typical fluoroalkenylzinc reagents

3.1 Fluorine-Substituted Nucleophilic Reagents

These zinc reagents also undergo cross-coupling reactions with aryl iodides in the presence of a palladium (Pd) catalyst to give fluorinated styrene derivatives.

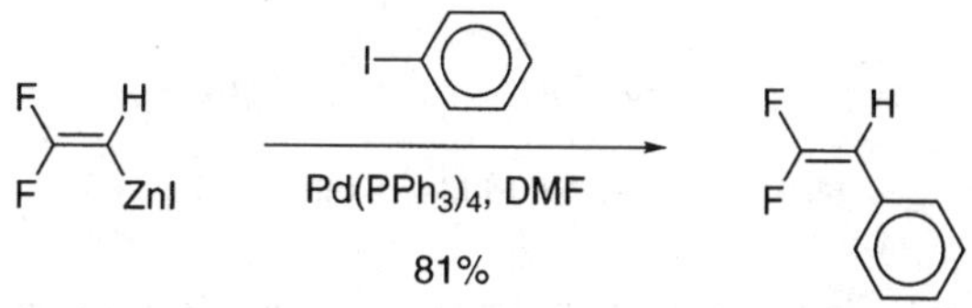

$$\underset{F}{\overset{F}{\diagup}}C=C\underset{ZnI}{\overset{H}{\diagdown}} \quad \xrightarrow[\text{81\%}]{\text{I-Ph} \quad \text{Pd(PPh}_3)_4, \text{ DMF}} \quad \underset{F}{\overset{F}{\diagup}}C=C\underset{Ph}{\overset{H}{\diagdown}} \tag{3.24}$$

Fluoroalkenylcopper reagents are another example of thermally stable fluoroalkenylmetals. The reagent is prepared by transmetalation of the corresponding fluoroalkenylzinc or -cadmium with a copper(I) halide [32]. The resulting vinylcopper reagent, (Z)-CF$_3$CF=CFCu, for example, participates in alkylation, acylation, and a coupling reaction in good yields, as shown in Scheme 3.17.

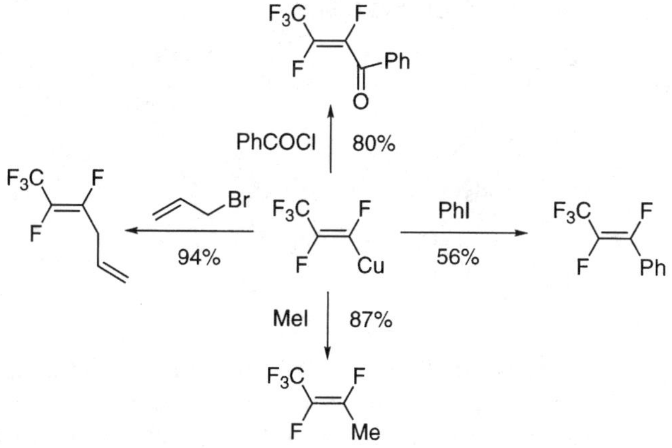

Scheme 3.17

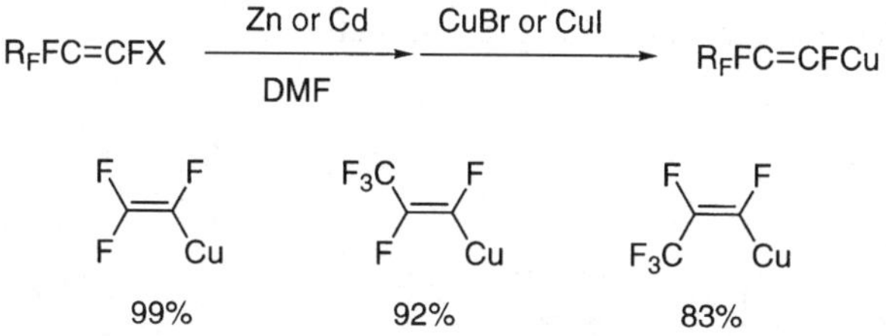

[Scheme showing (Z)-CF$_3$CF=CFCu reacting with:
- PhCOCl → F$_3$C/F C=C F/C(O)Ph, 80%
- Allyl-Br → F$_3$C/F C=C F/CH$_2$CH=CH$_2$, 94%
- PhI → F$_3$C/F C=C F/Ph, 56%
- MeI → F$_3$C/F C=C F/Me, 87%]

$$R_FFC=CFX \quad \xrightarrow[\text{DMF}]{\text{Zn or Cd}} \quad \xrightarrow{\text{CuBr or CuI}} \quad R_FFC=CFCu \tag{3.25}$$

F/F C=C F/Cu 99%   F$_3$C/F C=C F/Cu 92%   F/F C=C F$_3$C/Cu 83%

Fluorovinylsilanes and -stannanes are stable, readily available, and thus extremely versatile. Trifluorovinylsilanes are prepared by silylation of a trifluorovinyllithium or -magnesium reagent [33]. In particular, a procedure involving addition of butyllithium to a mixture of chlorosilanes and chlorotrifluoroethylene is highly effective [34].

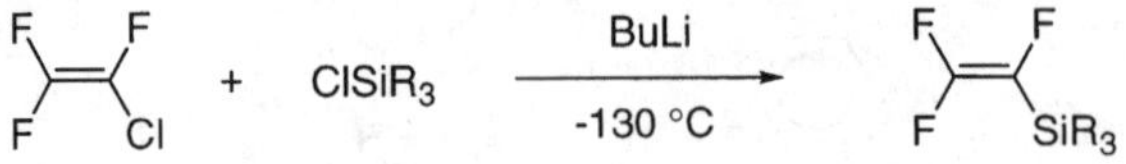

$$\underset{F}{\overset{F}{\diagdown}}C=C\underset{Cl}{\overset{F}{\diagup}} + ClSiR_3 \xrightarrow[-130\ °C]{BuLi} \underset{F}{\overset{F}{\diagdown}}C=C\underset{SiR_3}{\overset{F}{\diagup}} \qquad (3.26)$$

R$_3$Si : Et$_3$Si (85%); Pr$_3$Si (93%); PhMe$_2$Si (88%)

Trialkyl(trifluorovinyl)silanes serve as precursors of poly(difluoroacetylene). Thus, heating triethyl(trifluorovinyl)silane in the presence of a fluoride catalyst gives a black metallic polymer with exhibits conductivity (Scheme 3.18).

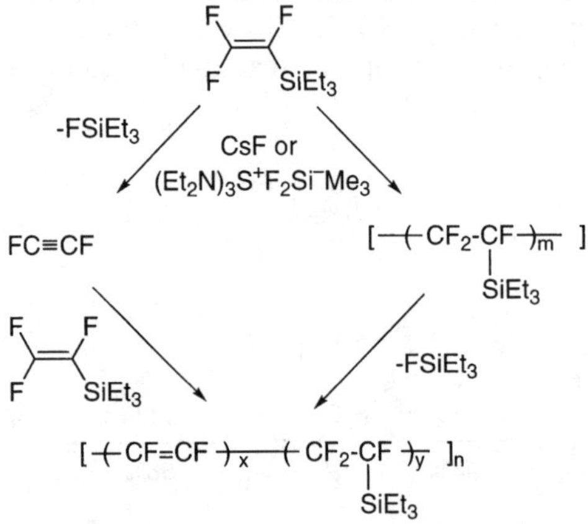

Scheme 3.18

Treatment of a trialkyl(trifluorovinyl)silane with butyllithium results in formation of 2-butyl-1,2-difluorovinyl(triethyl)silane via an addition-elimination process, as shown in Scheme 3.19 [35]. The resulting 1,2-difluoro-1-hexenylsilane is easily protodesilylated with KF or acetylated to give 3,4-difluoro-3-hexen-2-one with an AlCl$_3$ catalyst. Starting from various organolithium reagents and acid chlorides, a variety of α,β-difluoro-α,β-enones are readily available according to the present method.

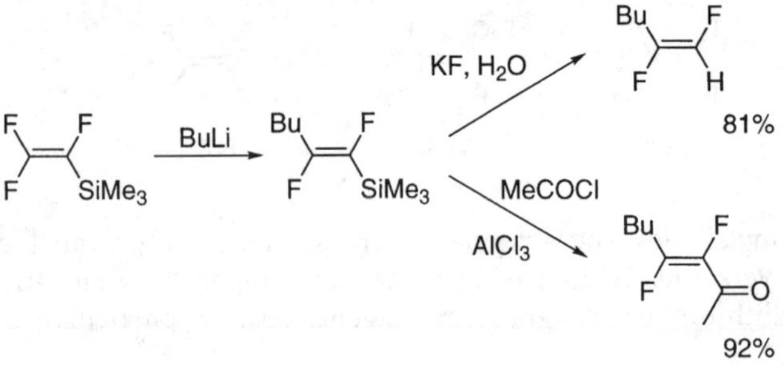

Scheme 3.19

3.1 Fluorine-Substituted Nucleophilic Reagents

Whereas trifluorovinylstannanes are available by treatment of trifluorovinyllithium or -magnesium with trialkylstannyl halides [36], 1,2-difluoro-1-alkenylstannanes are prepared from the corresponding vinylsilanes by transmetalation using KF [37]. The transformation takes place with retention of configuration.

$$R'CF=CFSiR_3 + ClSnBu_3 \xrightarrow[DMF]{KF} R'CF=CFSnBu_3 \quad (3.27)$$

R = Me, Et; R' = F, I, H, Bu, Ph   72 - 92%

The resulting tin reagent couples with aryl or alkenyl halides in the presence of a Pd catalyst to provide fluorinated styrene or conjugated dienes (Scheme 3.20) [38].

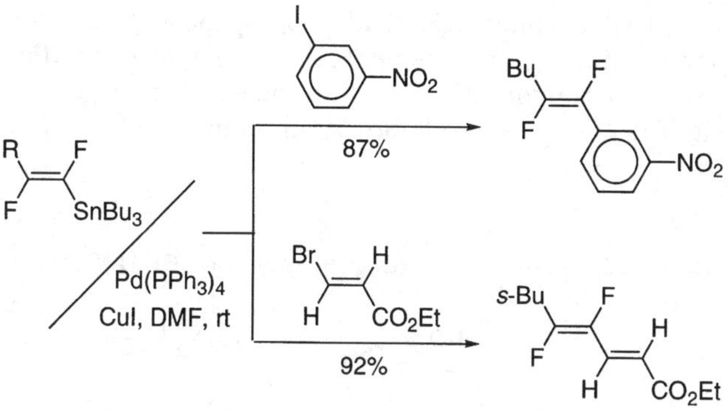

Scheme 3.20

3.1.3
Alkynylmetals

Although perfluoroalkynylmetals are assumed to be more stable than perfluoroalkyl- or -alkenylmetals because of the lack of the β-fluorine responsible for a β-elimination, synthetic studies on perfluoroalkynylmetals have been limited [39].

Due to the strong electron-withdrawing effect of a perfluoroalkyl group, metalation of perfluoroalkylethynes is easily achieved by treatment with an organolithium or -magnesium reagent. For example, reaction of 1,1,1-trifluoropropyne with butyllithium at −78 °C produces 3,3,3-trifluoropropynyllithium that is stable up to 0 °C and undergoes silylation and a carbonyl addition (Scheme 3.21) [40].

Trifluoropropynylmagnesium bromide is prepared by treatment of trifluoropropyne with ethylmagnesium bromide and reacts with acetone, giving rise to a trifluoromethylated propargyl alcohol [41].

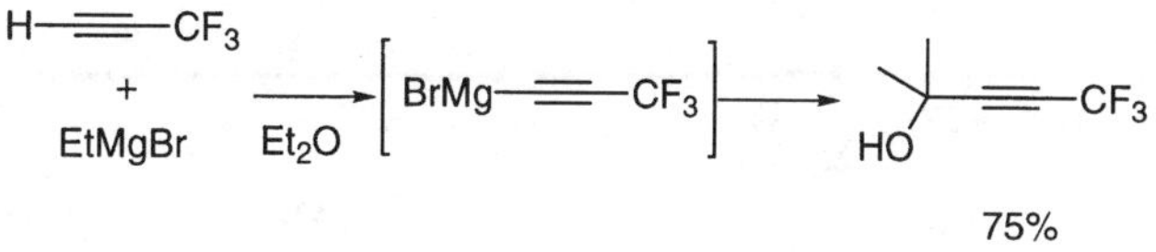

$$\underset{EtMgBr}{H\text{—}\!\!\equiv\!\!\text{—}CF_3} \xrightarrow{Et_2O} \left[BrMg\text{—}\!\!\equiv\!\!\text{—}CF_3 \right] \longrightarrow \underset{HO}{\underset{|}{\searrow}}\!\!\equiv\!\!\text{—}CF_3 \quad (3.28)$$

75%

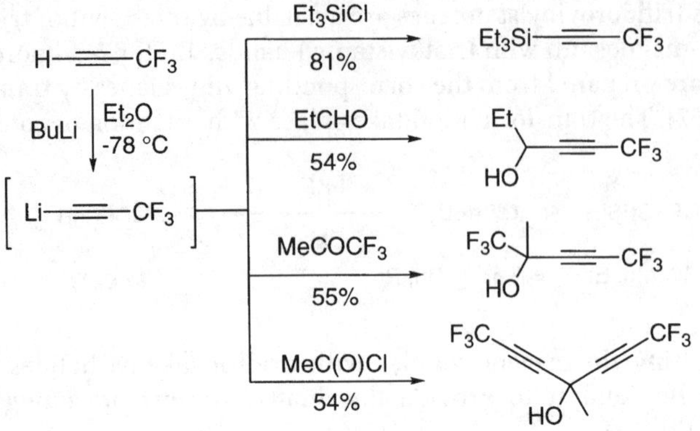

Scheme 3.21

$$H-\!\!\equiv\!\!-CF_3 \xrightarrow[-78\ °C]{BuLi,\ Et_2O} [Li-\!\!\equiv\!\!-CF_3]$$

Reactions:
- Et$_3$SiCl → Et$_3$Si−≡−CF$_3$ (81%)
- EtCHO → Et−CH(OH)−≡−CF$_3$ (54%)
- MeCOCF$_3$ → F$_3$C−C(OH)(Me)−≡−CF$_3$ (55%)
- MeC(O)Cl → F$_3$C−≡−C(OH)(CF$_3$)−≡−CF$_3$ (54%)

Reduction of 1,1,2-trichloro-3,3,3-trifluoropropene with zinc dust provides a trifluoropropynylzinc reagent (Scheme 3.22), which upon quenching with water gives 3,3,3-trifluoropropyne [42]. The fluorinated zinc reagent dimerizes by oxidation with CuCl$_2$ to give hexafluoro-2,4-hexadiyne [43].

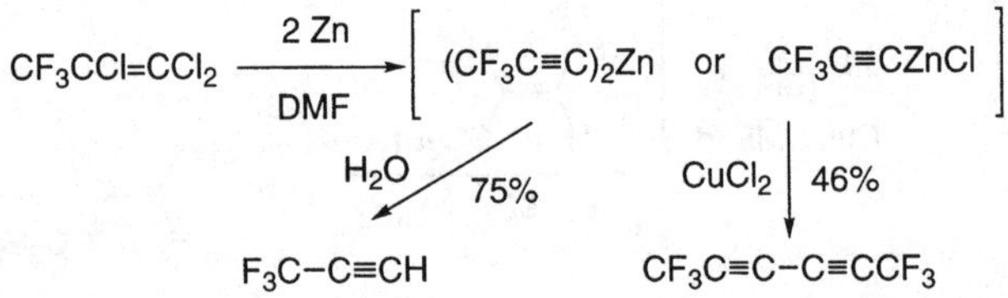

Scheme 3.22

$$CF_3CCl=CCl_2 \xrightarrow[DMF]{2\ Zn} [(CF_3C\equiv C)_2Zn\ \text{or}\ CF_3C\equiv CZnCl]$$

- H$_2$O → F$_3$C−C≡CH (75%)
- CuCl$_2$ → CF$_3$C≡C−C≡CCF$_3$ (46%)

Perfluoroalkynylzinc reagents are alternatively prepared by transmetalation of perfluoroalkynyllithiums with zinc chloride. Subsequent cross coupling with an aryl or alkenyl halide proceeds efficiently in the presence of Pd(PPh$_3$)$_4$. Scheme 3.23 illustrates some examples [44].

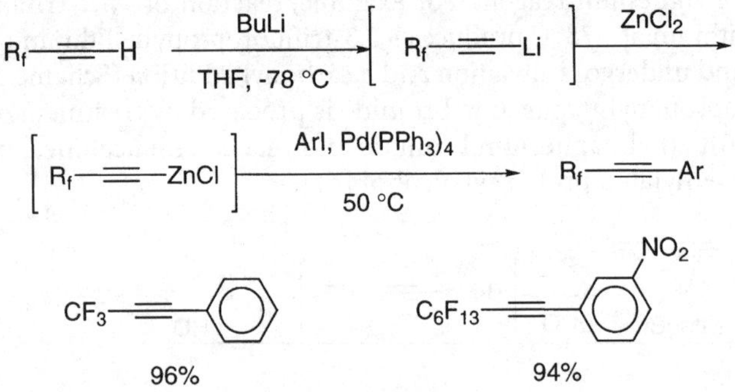

Scheme 3.23

$$R_f-\!\!\equiv\!\!-H \xrightarrow[THF,\ -78\ °C]{BuLi} R_f-\!\!\equiv\!\!-Li \xrightarrow{ZnCl_2} [R_f-\!\!\equiv\!\!-ZnCl] \xrightarrow[50\ °C]{ArI,\ Pd(PPh_3)_4} R_f-\!\!\equiv\!\!-Ar$$

Examples:
- CF$_3$−≡−C$_6$H$_5$ (96%)
- C$_6$F$_{13}$−≡−C$_6$H$_4$(NO$_2$) (94%)

3.1.4
Metal Enolates

Because metal enolates are the most basic species for carbon–carbon bond forming reactions in organic synthesis, metal enolates of fluorinated carbonyl compounds play a significant role in the synthesis of organofluorine compounds. This section describes such metal enolates in the order fluoro-, difluoro-, trifluoromethyl-substituted enolates, and perfluoroalkyl- and fluoro-substituted enolates, as summarized in Fig. 3.2.

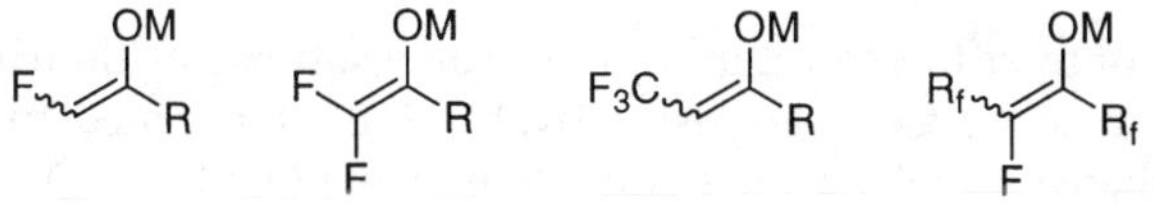

Fig. 3.2. Metal enolates containing fluorine

Precursors of α-fluorinated enolates are naturally α-fluorinated ketones, esters, amides, and thioesters. For example, 1-fluoro-3,3-dimethylbutan-2-one is deprotonated by treatment with lithium hexamethyldisilazide (LHMDS) in the presence of HMPA to generate a (Z)-enolate predominantly, which undergoes an aldol reaction with aldehydes, giving rise to α-fluoro β-hydroxy ketones stereoselectively [45].

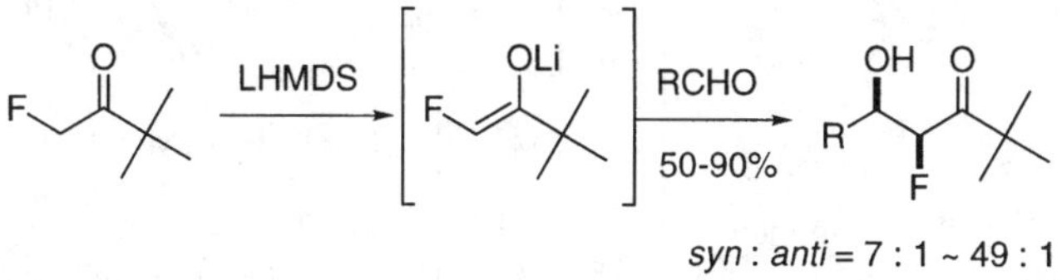

$$syn : anti = 7 : 1 \sim 49 : 1 \quad (3.29)$$

Esters and amides of fluoroacetic acid are similarly deprotonated with LHMDS or LDA to give the corresponding mixture of enolates (a 1:1 mixture of E- and Z-isomers) [46]. Thus, the diastereoselectivity of the subsequent aldol reaction is not necessarily high.

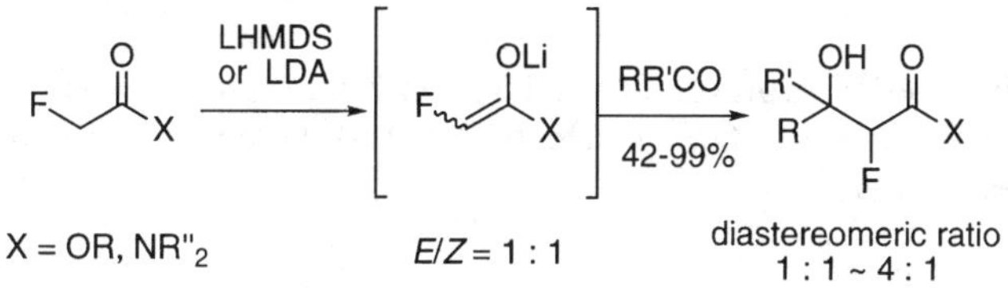

X = OR, NR''$_2$ E/Z = 1 : 1   diastereomeric ratio 1 : 1 ~ 4 : 1   (3.30)

Stereoselective synthesis of α-fluoro β-substituted α,β-unsaturated esters is achieved using an enolate generated from 2,4,6-trimethylphenyl fluoro(trimethylsilyl)acetate, via an aldol reaction and subsequent Peterson olefination [47]. An open transition state is proposed to account for the highly selective aldol reaction, because both (E)- and (Z)-enolates are confirmed to be generated.

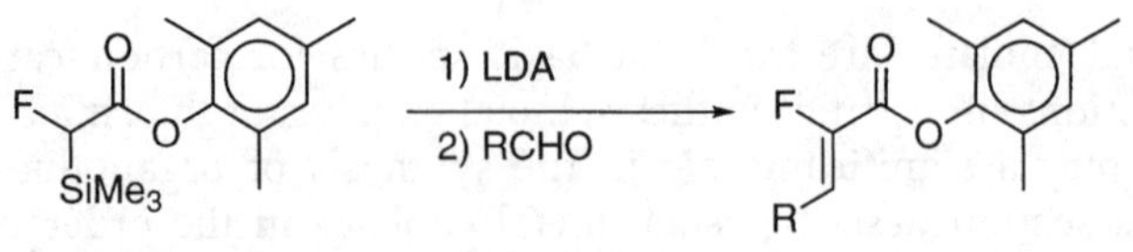

$$\underset{\underset{SiMe_3}{}}{F} \underset{O}{\overset{O}{\bigtriangleup}} \underset{}{\text{Ar}} \xrightarrow[\text{2) RCHO}]{\text{1) LDA}} F \underset{R}{\overset{O}{\bigtriangleup}} \text{OAr} \qquad (3.31)$$

R = Ph (49%, Z/E = 29 : 1); R = 4-MeOC$_6$H$_4$ (90%, Z/E = 32 : 1)
R = t-Bu (55%, Z/E = 24 : 1); R = 2-furyl (40%, Z/E = 50 : 1)

In sharp contrast to ester and amide enolates, those of fluorine-containing thiol esters are generated stereoselectively, as demonstrated in Scheme 3.24 [48]. Thus, phenyl 2-fluoropropanethioate is treated with to LDA to afford predominantly a (Z)-enolate, which reacts with aldehydes or imines to give the corresponding aldols or lactams, respectively, in moderate to good yields with high stereoselectivity. The reaction with aromatic aldehydes prefers *anti*-isomers, whereas *trans*-isomers of lactams are predominantly produced. In addition, a titanium enolate, generated by treatment of the lithium enolate with (*i*-PrO)$_3$TiCl, undergoes the *anti*-selective aldol reaction with both aromatic and aliphatic aldehydes.

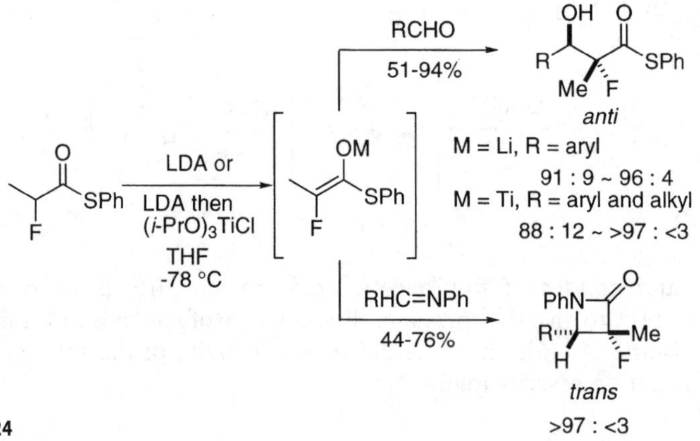

RCHO 51–94%

anti
M = Li, R = aryl
91 : 9 ~ 96 : 4
M = Ti, R = aryl and alkyl
88 : 12 ~ >97 : <3

LDA or
LDA then
(*i*-PrO)$_3$TiCl
THF
−78 °C

RHC=NPh 44–76%

trans
>97 : <3

Scheme 3.24

Reduction of ethyl bromofluoroacetates with zinc gives the corresponding zinc enolates which react with aldehydes and ketones to yield α-fluoro β-hydroxy esters [49].

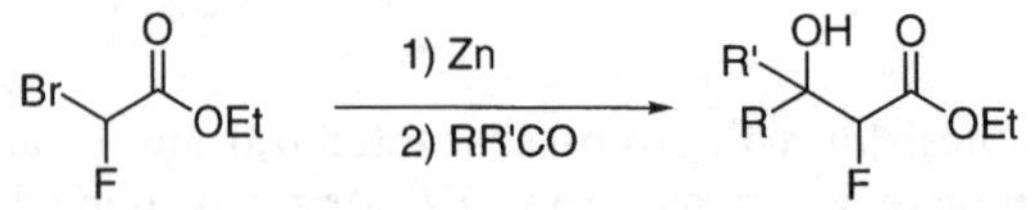

$$\underset{F}{\overset{Br}{\bigtriangleup}} \underset{}{\overset{O}{\bigtriangleup}} \text{OEt} \xrightarrow[\text{2) RR'CO}]{\text{1) Zn}} \underset{R}{\overset{R'}{\underset{F}{\bigtriangleup}}} \underset{}{\overset{OH}{\bigtriangleup}} \underset{}{\overset{O}{\bigtriangleup}} \text{OEt} \qquad (3.32)$$

PHCHO (68%), n-C$_6$H$_{13}$CHO (50%), CH$_3$CH=CHCHO (58%)
BuCOEt (56%), Me$_2$C=CHCOMe (44%)

3.1 Fluorine-Substituted Nucleophilic Reagents

The reaction of ethyl dibromofluoroacetate with two molar amounts of zinc and benzaldehyde in the presence of Et_2AlCl results in a double coupling reaction, giving rise to 2-fluoro-2-ethoxycarbonyl-1,3-diphenylpropane-1,3-diol [50].

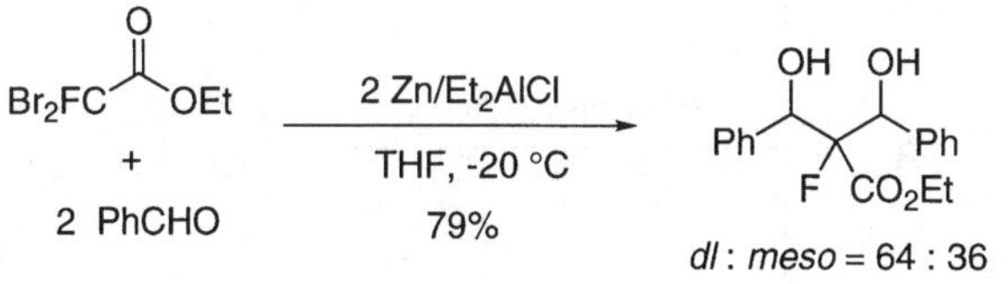

$$Br_2FC-C(O)-OEt + 2\ PhCHO \xrightarrow[\text{THF, -20 °C}]{2\ Zn/Et_2AlCl} \underset{F\ CO_2Et}{Ph\underset{|}{\overset{OH\ OH}{-}}Ph} \quad (3.33)$$

79%, dl : meso = 64 : 36

Enolates of difluorohalomethyl ketones and difluorohaloacetates are commonly prepared by a Reformatsky-type reaction [51]. Reduction of chlorodifluoromethyl ketones with zinc in the presence of aldehydes and ketones gives α,α-difluoro β-hydroxy carbonyl compounds in good yields. Addition of a metal salt is essential to assist the addition reaction depending on the kind of substituent R and the electrophile, as summarized in Scheme 3.25.

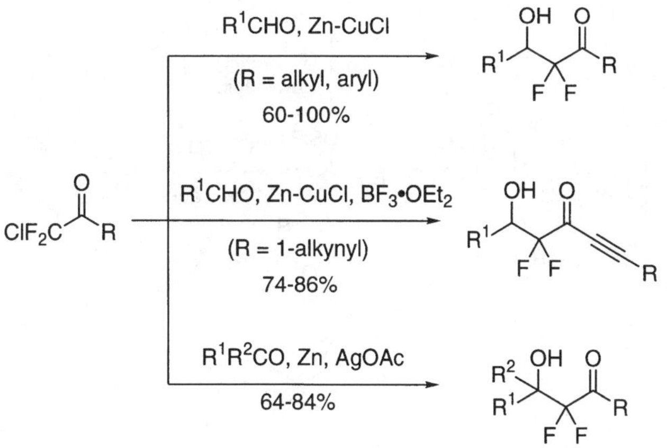

Scheme 3.25

$ClF_2C-C(O)-R$

- R^1CHO, Zn-CuCl (R = alkyl, aryl) 60–100% → R^1–CH(OH)–CF_2–C(O)–R
- R^1CHO, Zn-CuCl, $BF_3\cdot OEt_2$ (R = 1-alkynyl) 74–86% → R^1–CH(OH)–CF_2–C(O)–R
- R^1R^2CO, Zn, AgOAc 64–84% → R^1R^2C(OH)–CF_2–C(O)–R

Difluoro(halo)acetates are also reduced with zinc to give the corresponding enolates that react with aldehydes and imines to produce α,α-difluorinated β-hydroxy esters and β-lactams, as shown in Scheme 3.26 [52]. On the other hand, difluoroiodoacetate can be metalated with copper to give a copper enolate which is alkylated by various organic halides [53].

The zinc enolates of difluoromethyl ketones and difluoroacetates are easily trapped with chlorosilanes to give the corresponding difluorovinyl silyl ethers and ketene silyl acetals, respectively [54]. These silyl ethers undergo Lewis acid mediated 1,2-addition to aldehydes, ketones, and imines, and 1,4-addition to α,β-unsaturated carbonyl compounds effectively (Scheme 3.27).

Enolates of α-trifluoromethyl-substituted carbonyl compounds serve as versatile building blocks for a variety of trifluoromethyl-containing compounds. These compounds have received much interest because of their potent biological activities.

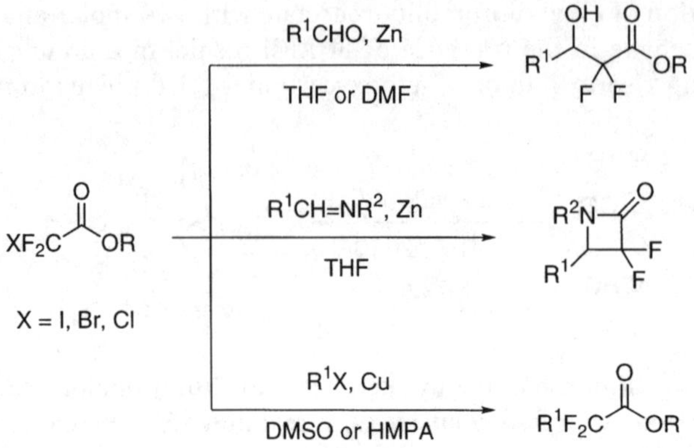

Scheme 3.26

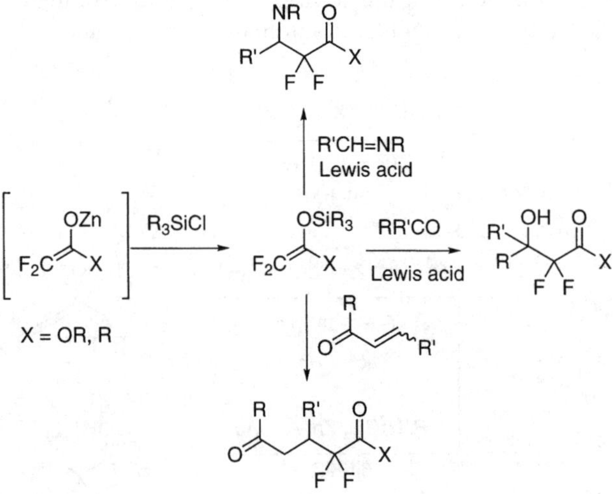

Scheme 3.27

A procedure for the generation of lithium enolates by deprotonation with LDA is not applicable to trifluoromethyl-substituted carbonyl compounds, because such lithium enolates, once generated, easily lose β-fluorine. To suppress such a side reaction, the enolate formation is carried out in the presence of an electrophile [55]. Thus, treatment of ethyl tetrafluoropropanoate with LDA in the presence of a carbonyl electrophile gives the corresponding adduct. The fact that no homo-aldol products are formed even in the reaction with cyclohexanone implies that ethyl tetrafluoropropanoate is deprotonated faster than cyclohexanone.

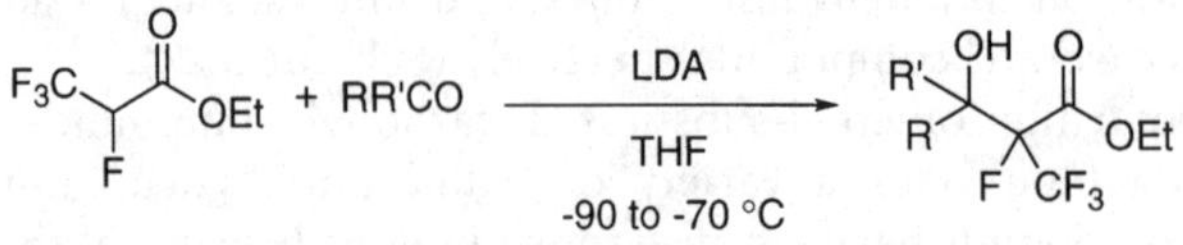

(3.34)

PhCHO (80%), EtCHO (73%), cyclohexanone (69%)

3.1 Fluorine-Substituted Nucleophilic Reagents

In addition to lithium enolates, boron and aluminum enolates of the type $CF_3CF=CX(OM)$ have been reported. Treatment of $CF_3CHFCONEt_2$ with Bu_2BOTf and then with i-Pr_2NEt produces the corresponding boron enolate solely an (E)-isomer. This enolate undergoes an aldol reaction with aldehydes in good yields with high *anti*-selectivities [56].

$$F_3C-CHF-C(O)NEt_2 \xrightarrow[\text{3) RCHO}]{\text{1) Bu}_2\text{BOTf, 2) (}i\text{-Pr)}_2\text{NEt}} R-CH(OH)-CF(CF_3)-C(O)NEt_2 \quad (3.35)$$

74–88%

anti : syn = 85 : 15 ~ 100 : 0

On the other hand, aluminum enolates of perfluoroalkyl ketones are generated by DIBAL-H reduction of perfluorinated 1-alkenyl phosphates, prepared easily from perfluoroalkyl ketones and sodium diethyl phosphite. Although the aluminum enolates react with aldehydes [57], the stereoselectivity is not generally high because of the modest E/Z selectivity of the aluminum enolate formation.

$$R_fFC=CR(OP(O)(OEt)_2) \xrightarrow[\text{2) R'CHO}]{\text{1) DIBAL}} R'-CH(OH)-CF(R_f)-C(O)R \quad (3.36)$$

56–88%

diastereomeric ratio
77 : 23 ~ 51 : 49

A Reformatsky-type reaction of methyl 2-bromo-3,3,3-trifluoropropanoate also provides a facile method for an α-trifluoromethyl-substituted ester enolate [58]. Use of 1,4-dioxane as the solvent and iodine as an activator is essential for the reaction. The resulting zinc enolate reacts with aldehydes in good yields but with low diastereoselectivity.

$$F_3C-CHBr-CO_2Me \xrightarrow[\text{2) RCHO}]{\text{1) Zn (I}_2\text{), dioxane}} R-CH(OH)-CH(CF_3)-CO_2Me \quad (3.37)$$

55–79%

diastereomeric ratio
57 : 43 ~ 77 : 23

Trifluoromethyl-substituted ketene silyl acetals are also useful building blocks for various trifluoromethyl-substituted targets. Such a ketene silyl acetal is prepared, for example, by treating methyl 3,3,3-trifluoropropanoate with trimethylsilyl triflate and triethylamine. The resulting ketene silyl acetal undergoes a trimethylsilyl triflate catalyzed aldol-type reaction with dimethyl acetals to give α-trifluoromethyl β-methoxy esters [59]. This procedure, when applied to 2-alkenyl trifluoropropionate, induces an ester enolate Claisen rearrangement to produce 2-trifluoromethyl-4-hexenoic acid conveniently.

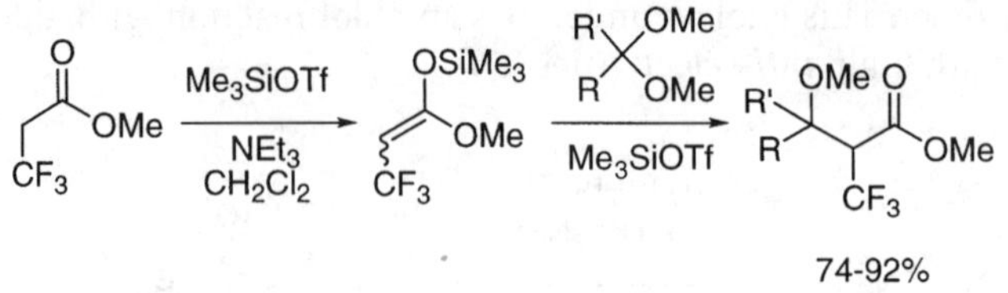

$$(3.38)$$

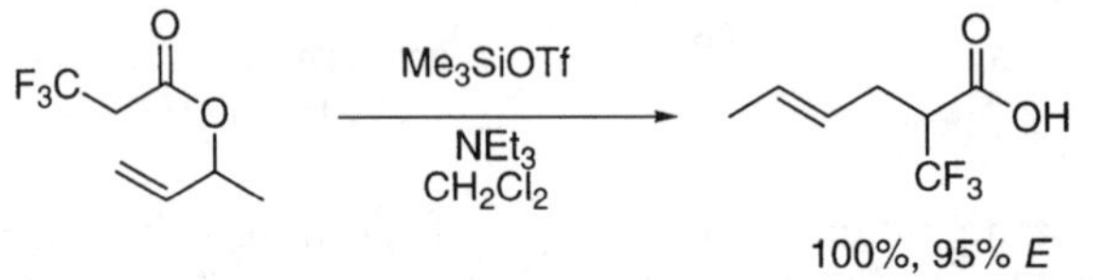

$$(3.39)$$

A facile method has been developed for the generation of perfluoro ketone enolates that react with both electrophiles and nucleophiles [60]. Treatment of $CF_3CH(OH)CF_3$ with two moles of butyllithium generates $CF_3C(OLi)=CF_2$ quantitatively that is stable even at room temperature (Scheme 3.28). The lithium enolate reacts as an ambient nucleophile with chlorotrimethylsilane at oxygen and with benzaldehyde at carbon, respectively. Noteworthy is that the enolate reacts also with a nucleophilic organometallic reagent: one of the fluorines on the enolate carbon is replaced by an alkyl group via an addition-elimination process. For example, treatment of the enolate with butyllithium followed by acetylation with acetyl chloride gives the corresponding butyl-substituted enol acetate.

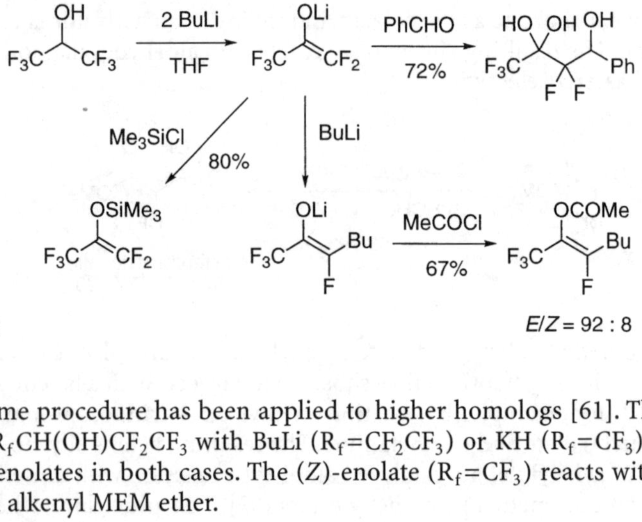

Scheme 3.28

The same procedure has been applied to higher homologs [61]. Thus, treatment of $R_fCH(OH)CF_2CF_3$ with BuLi ($R_f=CF_2CF_3$) or KH ($R_f=CF_3$) produces only (Z)-enolates in both cases. The (Z)-enolate ($R_f=CF_3$) reacts with MEMCl to give an alkenyl MEM ether.

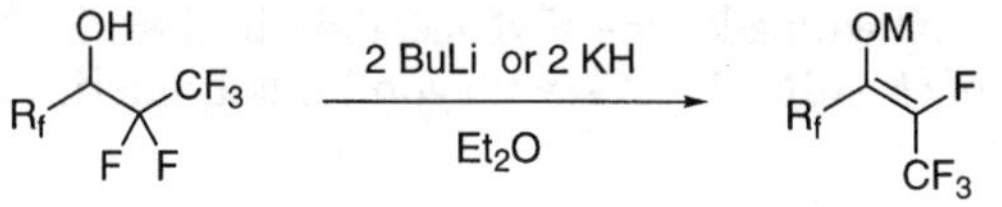

3.2 Fluorine-Substituted Electrophilic Reagents

$$\underset{\substack{F\ F}}{R_f}\!\!\!\!\!\!\!\!\!\!\!\!\!\!\overset{OH}{\underset{}{\diagup}}\!\!CF_3 \xrightarrow[Et_2O]{2\ BuLi\ or\ 2\ KH} R_f\!\!\diagup\!\!\!\!\overset{OM}{\underset{CF_3}{=}}\!\!\!F \quad (3.40)$$

$R_f = CF_3CF_2$, M = Li, 100% Z
$R_f = CF_3$, M = K, 100% Z

Reaction with BuLi and then with PhCOCl affords a butyl-substituted enol ester. In contrast, the reaction with aldehydes in lieu of PhCOCl does not take place even with CF_3CHO (Scheme 3.29).

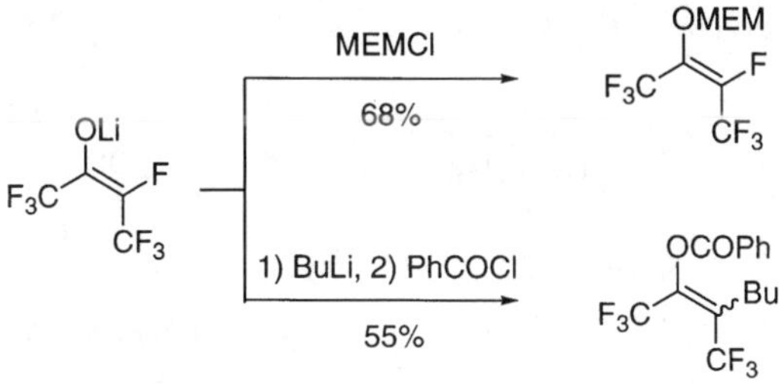

Scheme 3.29

3.2 Fluorine-Substituted Electrophilic Reagents

Fluorinated aldehydes, ketones, esters, α,β-unsaturated carbonyl compounds, and oxiranes behave as electrophiles in a manner similar to their non-fluorinated counterparts. This section focuses on characteristic examples of their reactions.

Due to the electron-withdrawing nature of its three fluorines, trifluoroacetaldehyde is a highly reactive electrophile which serves as a building block providing a $CF_3CH(OH)$ moiety. For example, CF_3CHO reacts with a chiral imide enolate to give (S,R,R)- and (S,R,S)-aldols only [62]. It is noteworthy that the carbonyl addition occurs from an *Si* face of the enolate carbon in contrast to a *Re* face attack by acetaldehyde.

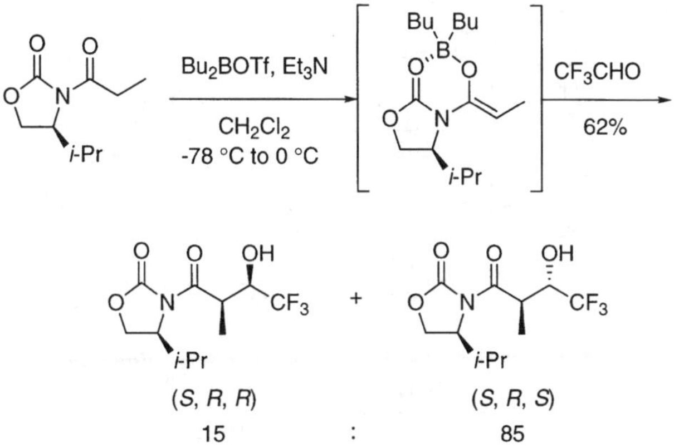

(3.41)

(S, R, R)   :   (S, R, S)
   15   :   85

Aldol reaction of CF_3CHO with a ketene silyl acetal derived from a thiol ester proceeds in the presence of (R)-BINOL-$TiCl_2$ with good enantiomeric excess [63].

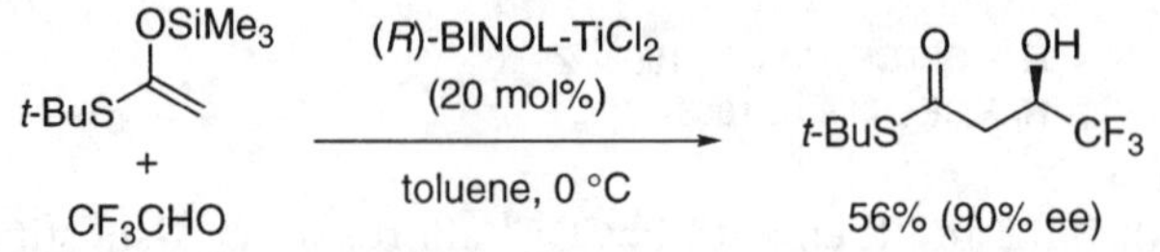

$$\underset{CF_3CHO}{\underset{+}{t\text{-BuS}\overset{OSiMe_3}{\diagup\!\!\!\diagdown}}} \xrightarrow[\text{toluene, 0 °C}]{(R)\text{-BINOL-TiCl}_2 \ (20 \text{ mol\%})} \underset{56\% \ (90\% \ ee)}{t\text{-BuS}\overset{O}{\underset{}{\diagdown}}\overset{OH}{\underset{}{\diagup}}CF_3} \qquad (3.42)$$

Using the same catalyst, CF_3CHO undergoes an ene reaction with 2-ethylidenecyclohexane to give a CF_3-containing homoallylic alcohol with high enantio- and diastereoselectivity.

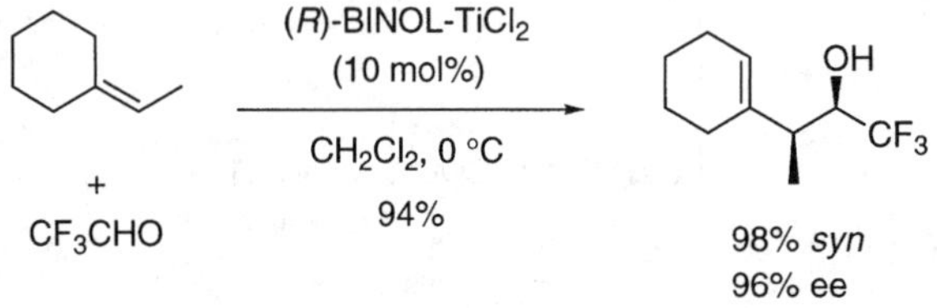

$$(3.43)$$

98% syn
96% ee

Optically active hemiacetal are produced in high yields and enantiomeric excess by the reaction of CF_3CHO with an alcohol in the presence of (R)-BINOL-Ti(O-i-Pr)$_2$ [64]. It is disclosed that the configuration of the hemiacetal is stable only at low temperatures.

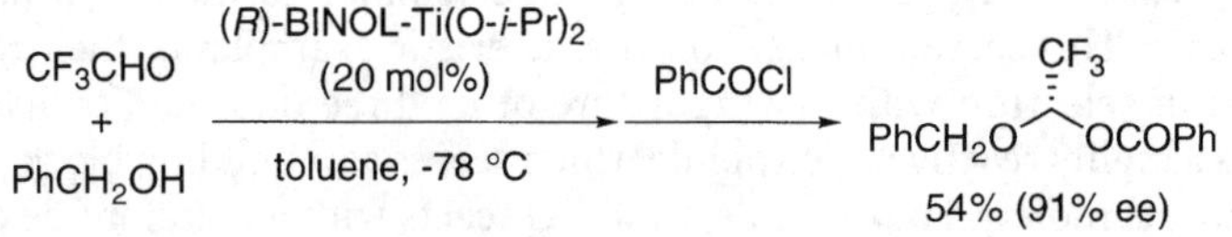

$$\underset{PhCH_2OH}{\underset{+}{CF_3CHO}} \xrightarrow[\text{toluene, -78 °C}]{(R)\text{-BINOL-Ti(O-}i\text{-Pr)}_2 \ (20 \text{ mol\%})} \xrightarrow{PhCOCl} \underset{54\% \ (91\% \ ee)}{PhCH_2O\overset{CF_3}{\underset{}{\diagup}}OCOPh} \qquad (3.44)$$

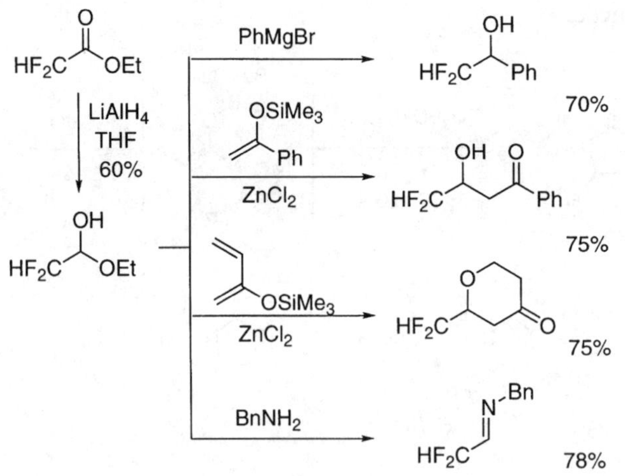

Scheme 3.30

3.2 Fluorine-Substituted Electrophilic Reagents

Difluoroacetaldedyde ethyl hemiacetal is prepared from ethyl difluoroacetate and $LiAlH_4$ and is used for the synthesis of difluoromethylated compounds. Upon treatment with Grignard reagents, enol silyl ethers, dienes, or amines, the hemiacetal undergoes carbonyl additions, a hetero-Diels–Alder reaction, or a condensation to form an imine, respectively (Scheme 3.30) [65].

Reduction of perfluoro alkanoates with DIBAL-H also gives hemiacetals, which react with allylstannanes in the presence of $ZnBr_2$ to give perfluorinated homoallylic alcohols [66].

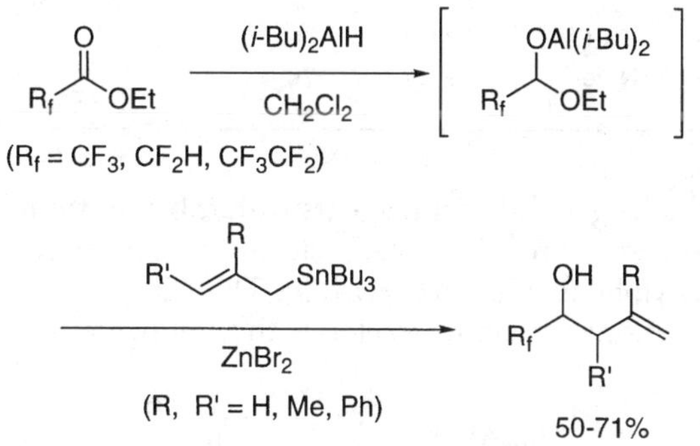

$$R_f\text{C(O)OEt} \xrightarrow[CH_2Cl_2]{(i\text{-}Bu)_2AlH} \left[R_f\text{CH(OAl(}i\text{-Bu)}_2\text{)OEt} \right] \quad (3.45)$$

($R_f = CF_3, CF_2H, CF_3CF_2$)

$$\xrightarrow[ZnBr_2]{R'CH=C(R)CH_2SnBu_3} R_f\text{CH(OH)CH(R)C(=CH_2)R'}$$

(R, R' = H, Me, Ph)   50–71%

Reaction of ethyl trifluoroacetate with an organolithium or -magnesium reagent gives trifluoromethyl ketones (Scheme 3.31) [67]. Remarkably, no tertiary alcohols are formed, because the tetrahedral intermediates, $CF_3(R)C(OEt)(OM)$, are stable enough to be trapped by acetyl chloride under the reaction conditions.

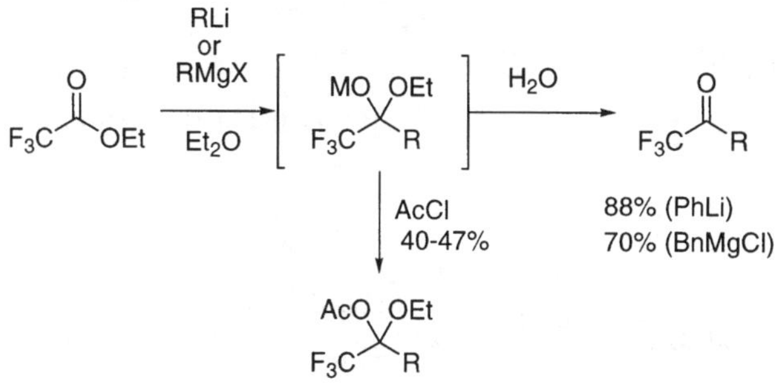

Scheme 3.31

$$F_3C\text{C(O)OEt} \xrightarrow[Et_2O]{RLi\text{ or }RMgX} \left[F_3C\text{C(OM)(OEt)R} \right] \xrightarrow{H_2O} F_3C\text{C(O)R}$$

88% (PhLi)
70% (BnMgCl)

↓ AcCl, 40–47%

$F_3C\text{C(OAc)(OEt)R}$

Michael addition of lithium enolates to ethyl (E)-3-(trifluoromethyl)acrylate is facilitated by a trifluoromethyl group and gives β-trifluoromethyl-substituted δ-keto esters [68]. The smooth reaction even with ketone enolates is ascribed to an interaction between a fluorine atom and a lithium ion in the initial Michael adducts. Such a fluorine-metal interaction is discussed in Chap. 4.3.

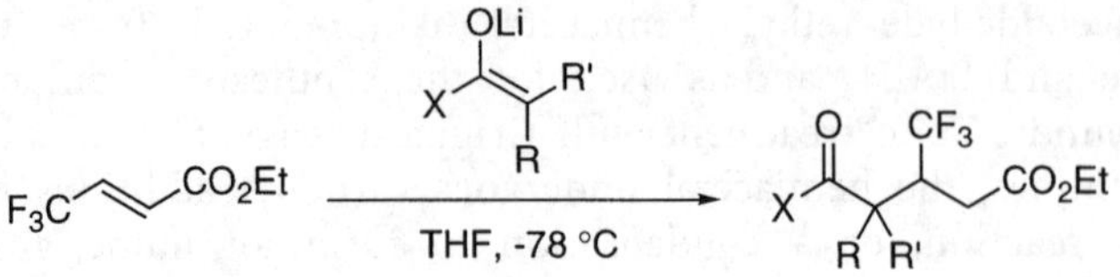

$$F_3C\diagup\!\!\!\diagdown CO_2Et \xrightarrow[\text{THF, -78 °C}]{\underset{R}{\overset{OLi}{X}}\diagdown\!\!\!\diagup R'} \underset{R\ R'}{X}\overset{O}{\diagdown}\overset{CF_3}{\diagup}CO_2Et \qquad (3.46)$$

R	R'	X	yield (%)	syn	:	anti
Me	H	Ph	98	<2	:	>98
H	Me	Et	97	<2	:	>98
Me	Me	OEt	98	–	:	–
H	Me	NMe$_2$	89	70	:	30

Conjugate addition of organolithium reagents with α,β-unsaturated ketones containing a trifluoromethyl group proceeds cleanly in the presence of aluminum tris(2,6-diphenylphenoxide) (ATPH) [69]. The CF$_3$ groups reduce the electron density of the enone system to accelerate the conjugate addition.

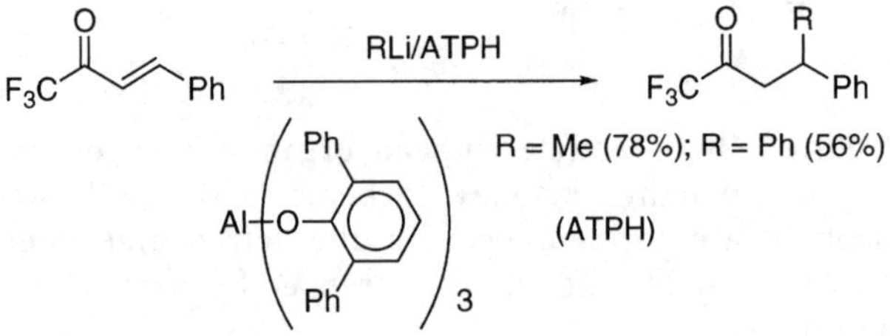

$$F_3C\overset{O}{\diagdown}\diagup\!\!\!\diagdown Ph \xrightarrow{RLi/ATPH} F_3C\overset{O}{\diagdown}\diagup\!\!\!\overset{R}{\diagdown}Ph \qquad (3.47)$$

R = Me (78%); R = Ph (56%)

(ATPH) = Al(O–C$_6$H$_3$(Ph)$_2$)$_3$

$$Ph\overset{O}{\diagdown}\diagup\!\!\!\diagdown CF_3 \xrightarrow{RLi/ATPH} Ph\overset{O}{\diagdown}\diagup\!\!\!\overset{R}{\diagdown}CF_3 \qquad (3.48)$$

R = Me (82%); R = Ph (97%)

$$BrF_2C\diagup\!\!\!\diagdown CO_2R \xrightarrow[\text{THF, -78 °C}]{\underset{}{t\text{-BuO}}\diagdown\!\!\!\overset{OLi}{\diagup}} \left[\underset{t\text{-BuO}_2C}{}\diagdown\!\!\!\overset{BrF_2C}{\diagup}\diagdown\!\!\!\overset{OR}{\diagup}\diagdown OLi\right] \xrightarrow[\text{THF-DMI}]{Et_3B\text{-}O_2} t\text{-BuO}_2C\diagdown\!\!\triangle\!\!\diagup CO_2R$$

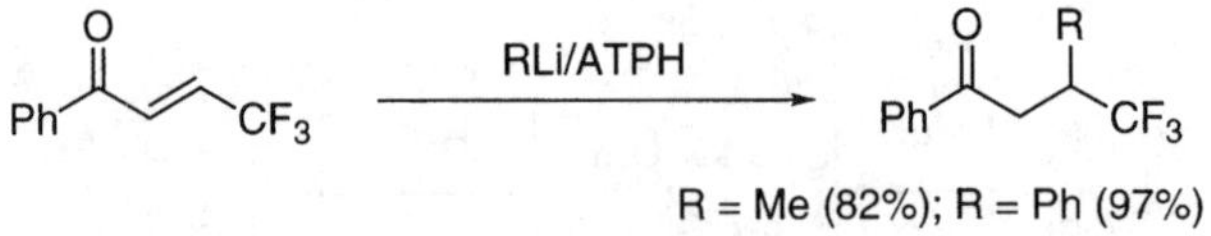

71%

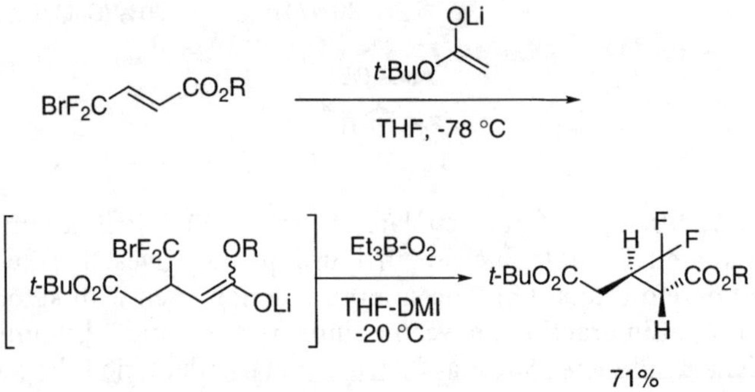

Scheme 3.32

gem-Difluorocyclopropanes are prepared by Michael addition of lithium enolates with 4-bromo-4,4-difluorocrotonate followed by a radical initiated intramolecular substitution reaction (Scheme 3.32) [70].

Optically active 3,3,3-trifluoropropene oxide is readily available by microbial oxidation of trifluoropropene or asymmetric reduction of bromomethyl trifluoromethyl ketone with B-chlorodiisopinocamphenylborane followed by cyclization. The chiral epoxide also serves as a versatile building block for the asymmetric synthesis of trifluoromethylated compounds [71]. Some transformations are summarized in Scheme 3.33.

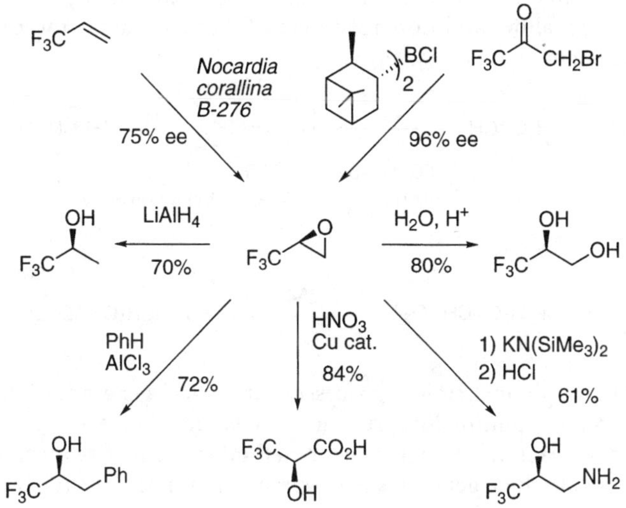

Scheme 3.33

3.3
Fluorine-Substituted Radicals

Perfluoroalkyl radicals show quite different behavior from their hydrocarbon counterparts [72]. For example, the 7-fluoro-7-norcaryl radical is configurationally stable and its bromine abstraction proceeds with retention of configuration in sharp contrast to the corresponding non-fluorinated cyclopropyl radical which undergoes rapid configurational interconversion under the conditions [73].

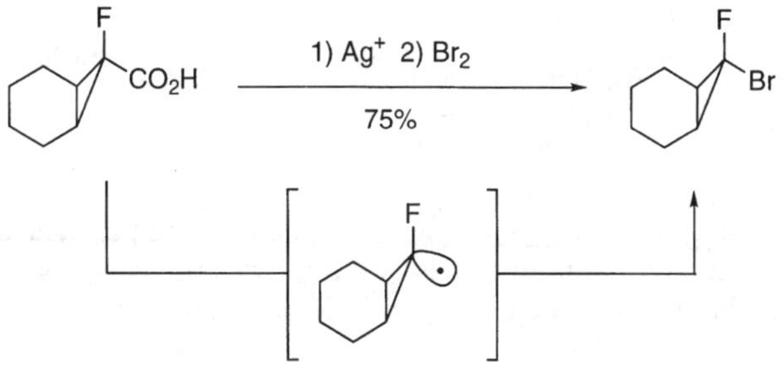

(3.49)

A number of methods for generating perfluoroalkyl radicals and their reactions have been studied. Common precursors are perfluoroalkyl iodides, perfluoroalkanoic acids, and perfluorodiacyl peroxides. Perfluoroalkyl iodides are the one most frequently used. Homolytic cleavage of a carbon–iodine bond is induced by heat, light, or a radical initiator and has been widely applied to addition reactions of perfluoroalkyl iodides into olefins since the discovery by Haszeldine, who treated trifluoromethyl iodide with ethylene with heating or irradiation to obtain 3,3,3-trifluoro-1-iodopropane in good yields along with telomers [74]. Because perfluoroalkyl radicals are electrophilic, olefins are limited mainly to electron-rich ones. Recently, low intensity light of 254 nm has been shown to allow addition reactions of perfluoroalkyl radicals even to electron-deficient olefins [75].

$$CF_3I + H_2C=CH_2 \xrightarrow[h\nu,\ 108\ h]{250\ °C,\ 48\ h} CF_3CH_2CH_2I + CF_3(CH_2CH_2)_nI \quad (3.50)$$
$$75\%$$
$$82\%\ (\text{based on conversion})$$

$$n\text{-}C_3F_7I + H_2C=CHCO_2Et \xrightarrow[24\ h,\ 83\%]{h\nu\ (254\ nm)} n\text{-}C_3F_7CH_2\overset{I}{C}HCO_2Et \quad (3.51)$$

Thermal and photochemical processes for radical generation often require high temperatures and/or long reaction times and thus are accompanied by undesired side reactions. An alternative method is a reductive initiation using a low valent metal, which generates perfluoroalkyl radicals effectively under mild conditions and effects their clean reaction with olefins. Various reductants are used such as Fe, Zn, Cu, Raney-Ni, Sn, Ti, Pd(PPh$_3$)$_4$, RhCl(PPh$_3$), Fe$_3$(CO)$_{12}$, Ru$_3$(CO)$_{12}$, Et$_3$B, and Me$_3$Al [72]. For example, Fe$_3$(CO)$_{12}$ catalyzes an addition reaction of heptafluoropropyl iodide to trimethylsilylacetylene are trimethylvinylsilane to give the respective adducts (Scheme 3.34) [76].

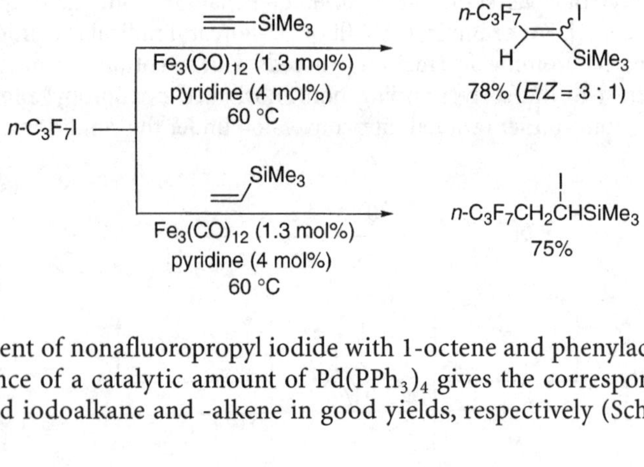

Scheme 3.34

Treatment of nonafluoropropyl iodide with 1-octene and phenylacetylene in the presence of a catalytic amount of Pd(PPh$_3$)$_4$ gives the corresponding perfluorinated iodoalkane and -alkene in good yields, respectively (Scheme 3.35) [77].

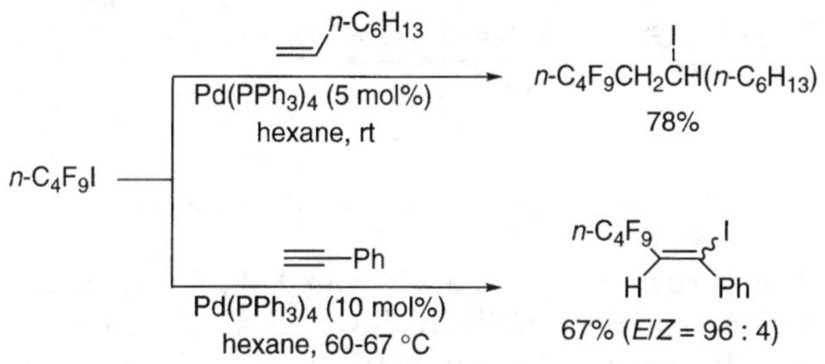

3.3 Fluorine-Substituted Radicals

Scheme 3.35

$n\text{-}C_4F_9I$ reacts with:
- $n\text{-}C_6H_{13}CH=CH_2$, Pd(PPh$_3$)$_4$ (5 mol%), hexane, rt → $n\text{-}C_4F_9CH_2CHI(n\text{-}C_6H_{13})$, 78%
- $PhC\equiv CH$, Pd(PPh$_3$)$_4$ (10 mol%), hexane, 60–67 °C → $n\text{-}C_4F_9C(H)=C(I)Ph$, 67% ($E/Z = 96:4$)

A three-component coupling reaction of $C_8H_{17}I$, 1-hexene, and carbon monoxide is achieved in the presence of a palladium catalyst and ethanol as solvent to give ethyl 2-butylheptadecafluoroundecanoate [78].

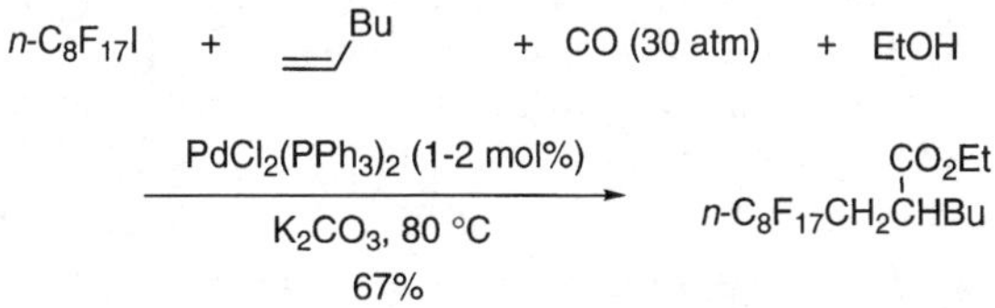

$$n\text{-}C_8F_{17}I + CH_2=CHBu + CO\,(30\,\text{atm}) + EtOH \xrightarrow[\text{K}_2\text{CO}_3,\ 80\,°\text{C}]{\text{PdCl}_2(\text{PPh}_3)_2\,(1\text{–}2\,\text{mol\%})} n\text{-}C_8F_{17}CH_2CH(CO_2Et)Bu \quad (3.52)$$
67%

For a radical addition of perfluoroalkyl iodide to carbon–carbon multiple bonds, BEt$_3$ is an effective catalyst [79]. The reaction proceeds at ambient temperatures or below with high regio- and stereoselectivity even when internal alkynes and alkenes are employed. Trifluoromethylation of terminal olefins with CF$_3$I and Me$_3$Al as initiator is performed at low temperature [80].

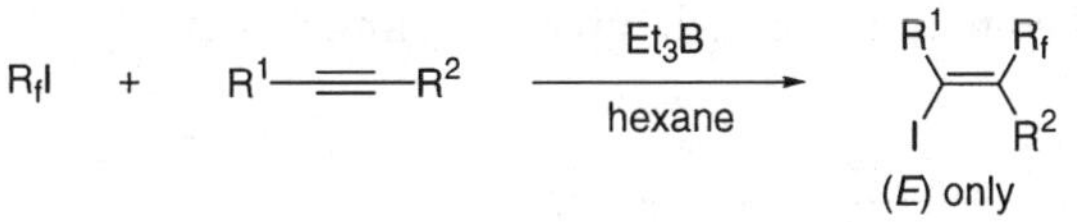

$$R_fI + R^1{\equiv}R^2 \xrightarrow[\text{hexane}]{\text{Et}_3B} \underset{(E)\ \text{only}}{R^1\!\!\diagdown\!\!C{=}C\!\!\diagup\!\!R^2\ (R_f,\ I)} \quad (3.53)$$

Rf	R^1	R^2	Temp (°C)	Yield (%)
C$_6$F$_{13}$	n-C$_{10}$H$_{21}$	H	25	94
	n-C$_5$H$_{11}$	n-C$_5$H$_{11}$	25	60
CF$_3$	n-C$_{10}$H$_{21}$	H	−24	76

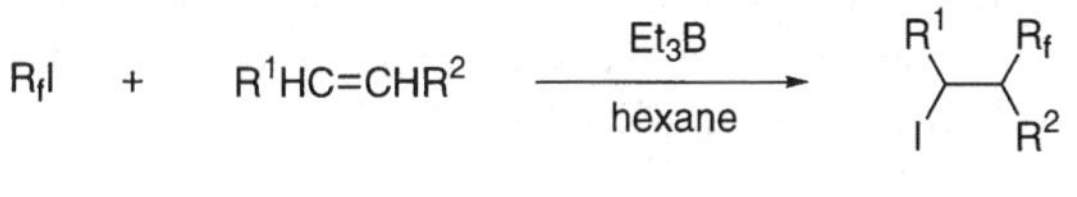

$$R_fI + R^1HC{=}CHR^2 \xrightarrow[\text{hexane}]{\text{Et}_3B} R^1\text{CH}(R_f)\text{CH}(I)R^2 \quad (3.54)$$

Rf	R^1	R^2	Temp (°C)	Yield (%)
C$_6$F$_{13}$	n-C$_{10}$H$_{21}$	H	25	93
	n-C$_5$H$_{11}$	n-C$_5$H$_{11}$	25	67[a]
CF$_3$	n-C$_5$H$_{11}$	n-C$_5$H$_{11}$	−24	61[a]

a) erythro/threo = 1 : 1

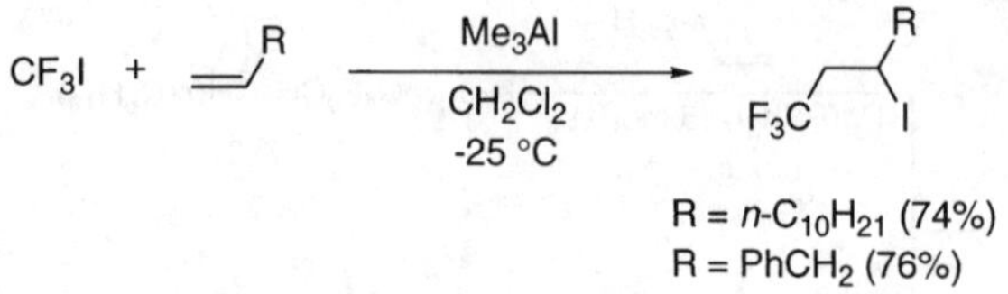

$$CF_3I + \underset{}{=\!\!\!/}^R \xrightarrow[-25\,°C]{Me_3Al \atop CH_2Cl_2} F_3C\!\!\underset{I}{\overset{R}{\wedge}}\!\! \qquad (3.55)$$

R = n-C$_{10}$H$_{21}$ (74%)
R = PhCH$_2$ (76%)

Treatment of perfluoroalkanoic acids with XeF$_2$ gives xenon carboxylates which decompose to produce perfluoroalkyl radicals (Scheme 3.36) [81]. A trifluoromethyl radical generated in this manner undergoes trifluoromethylation of aromatic compounds substituted by electron-withdrawing groups [82].

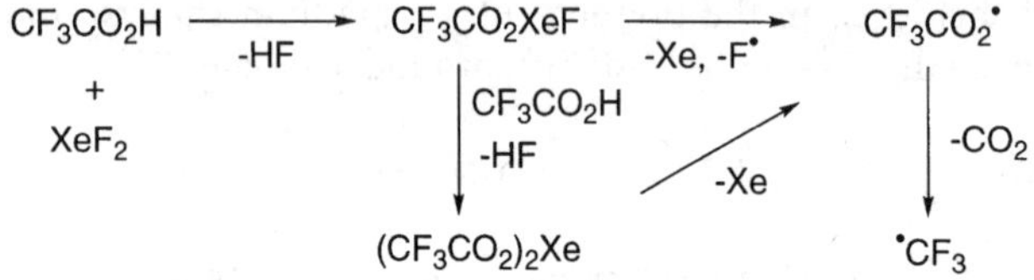

Scheme 3.36

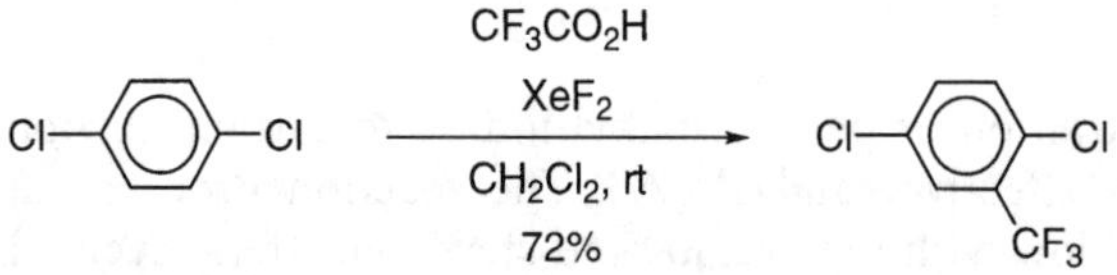

$$CF_3CO_2H + XeF_2 \xrightarrow{-HF} CF_3CO_2XeF \xrightarrow{-Xe,\,-F^\bullet} CF_3CO_2^\bullet \xrightarrow{-CO_2} {}^\bullet CF_3$$

with intermediate $(CF_3CO_2)_2Xe$ formed via CF_3CO_2H, $-HF$; and $-Xe$ pathway.

$$Cl\text{–}C_6H_4\text{–}Cl \xrightarrow[CH_2Cl_2,\,rt \atop 72\%]{CF_3CO_2H \atop XeF_2} Cl\text{–}C_6H_3(CF_3)\text{–}Cl \qquad (3.56)$$

An alternative method for a trifluoromethyl radical generation is performed by electrochemical oxidation of trifluoroacetic acid (Eq. 3.57) [83]. For instance, this method, as applied to fumaronitrile, gives 2-(trifluoromethyl)succinonitrile (Eq. 3.58).

$$CF_3CO_2H \xrightarrow[-CO_2]{-e} {}^\bullet CF_3 \qquad (3.57)$$

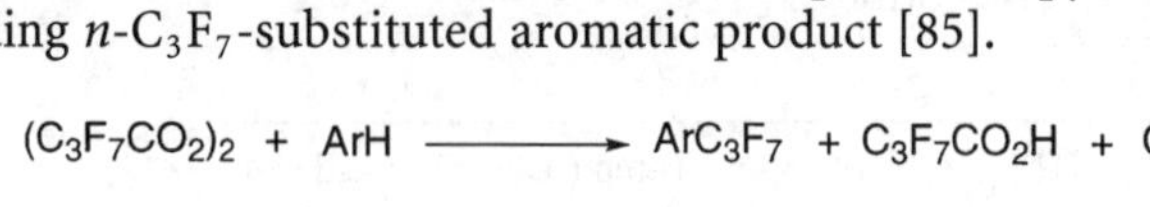

$$NC\!\!\diagup\!\!\diagdown\!\!CN \xrightarrow[-e \atop 65\%]{CF_3CO_2H\text{-}MeCN\text{-}H_2O} NC\!\!\underset{}{\overset{CF_3}{\wedge}}\!\!CN \qquad (3.58)$$

Perfluorodiacyl peroxides decompose at 20–40 °C to produce perfluoroalkyl radicals [84]. A thermal decomposition of $(n\text{-}C_3F_7CO_2)_2$ in an electron-rich aromatic solvent such as anisole, furan, thiophene, or pyrrole gives the corresponding $n\text{-}C_3F_7$-substituted aromatic product [85].

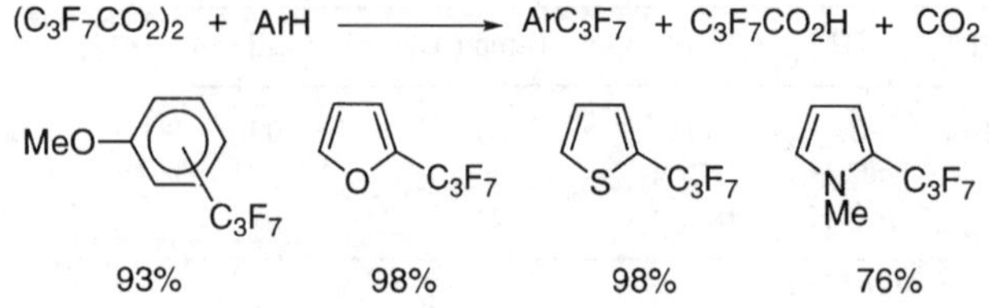

$$(C_3F_7CO_2)_2 + ArH \longrightarrow ArC_3F_7 + C_3F_7CO_2H + CO_2 \qquad (3.59)$$

MeO–C$_6$H$_4$–C$_3$F$_7$ (93%); furan-C$_3$F$_7$ (98%); thiophene-C$_3$F$_7$ (98%); N-Me-pyrrole-C$_3$F$_7$ (76%)

3.4
Fluorine-Substituted Carbenes

Since fluorine on a carbene carbon makes the labile divalent species stable and highly electrophilic, the reaction of fluorinated carbenes with alkenes and alkynes constitutes a facile method for the synthesis of fluorinated cyclopropanes and cyclopropenes [86].

A halogen-lithium exchange reaction of dibromofluoromethane with an organolithium reagent gives a lithium carbenoid, which undergoes α-elimination to give a fluorocarbene. The carbene reacts with alkenes to produce fluorocyclopropanes, but generally in low yields [87].

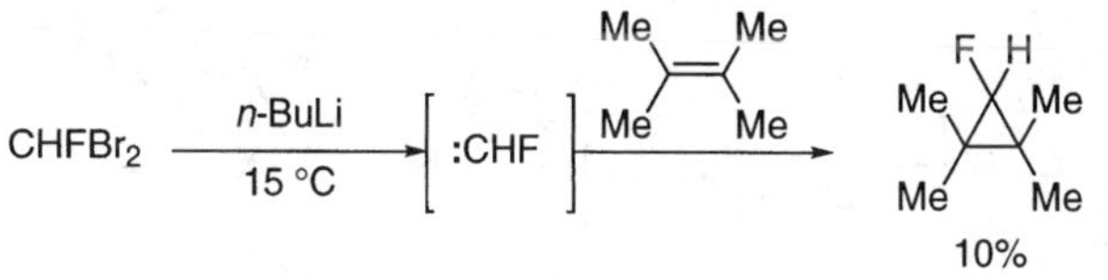

$$CHFBr_2 \xrightarrow[15\,°C]{n\text{-BuLi}} [:CHF] \xrightarrow{Me_2C=CMe_2} \underset{10\%}{Me_2C(F)(H)CMe_2} \quad (3.60)$$

On the other hand, reduction of fluorodiiodomethane with diethylzinc in the presence of cyclohexene affords 7-fluoronorcarane in a high yield as a diastereomeric mixture [88].

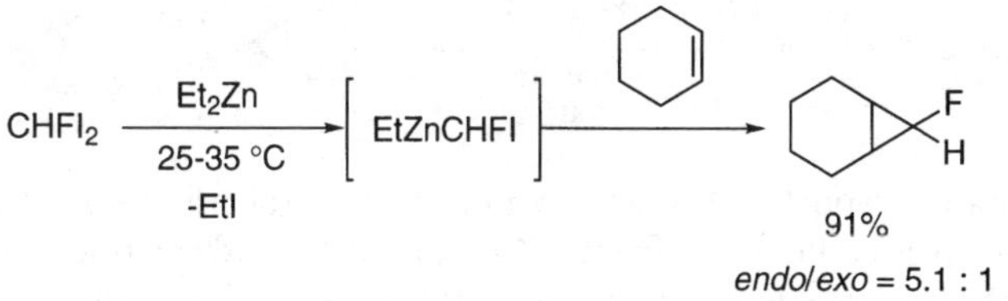

$$CHFI_2 \xrightarrow[\substack{25-35\,°C \\ -EtI}]{Et_2Zn} [EtZnCHFI] \xrightarrow{\text{cyclohexene}} \underset{\substack{91\% \\ endo/exo = 5.1:1}}{\text{7-fluoronorcarane}} \quad (3.61)$$

A diastereoselective fluorocyclopropanation of *N*-vinylcarbamate has been successfully applied to the synthesis of a key component of a new generation of quinolonecarboxylic acid, DU-6859 [89].

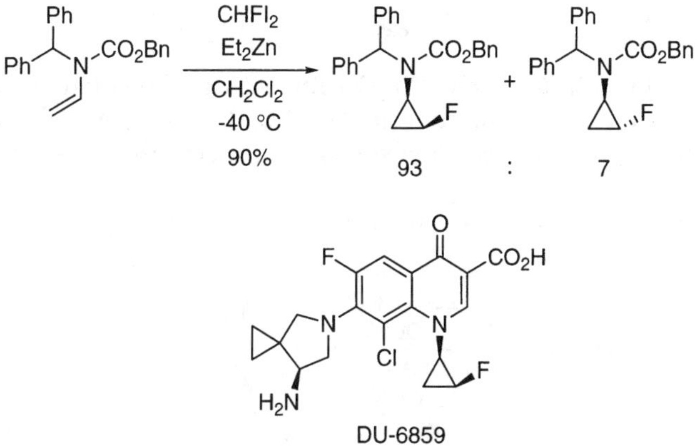

$$\underset{}{Ph_2C(H)N(CO_2Bn)CH=CH_2} \xrightarrow[\substack{CH_2Cl_2 \\ -40\,°C \\ 90\%}]{CHFI_2, Et_2Zn} \underset{93}{\text{cyclopropane-F (syn)}} + \underset{7}{\text{cyclopropane-F (anti)}} \quad (3.62)$$

DU-6859

Dehydrohalogenation of chlorodifluoromethane or bromodifluoromethane with an alkoxide or alkyllithium also produces difluorocarbene via an α-elimination [90] but in low yields due to a competitive reaction with the base. To prevent the side reaction, the concentration of the base is kept low [91] or generated on demand. For example, heating a mixture of chlorodifluoromethane, ethylene oxide, and tetraethylammonium bromide generates an alkoxide base and then difluorocarbene which further reacts with a co-existing acceptor 2,3-dimethyl-2-butene to quantitatively give a difluorocyclopropane product (Scheme 3.37).

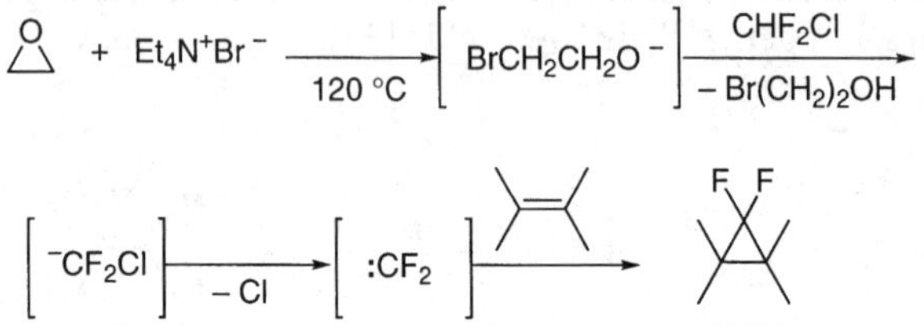

Scheme 3.37

Reduction of CBr_2F_2 with zinc metal also generates difluorocarbene that reacts with electron-rich olefins [92].

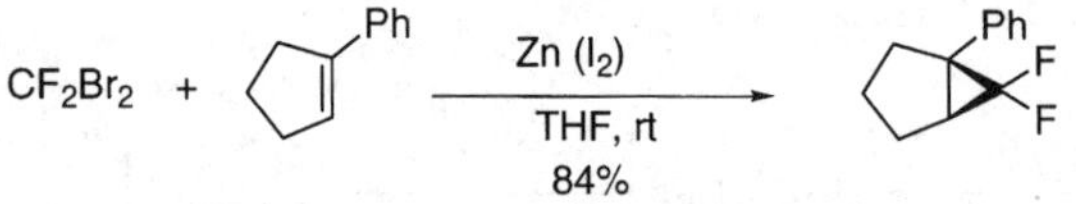

$$CF_2Br_2 + \text{(cyclopentadiene-Ph)} \xrightarrow[\text{THF, rt} \atop 84\%]{Zn (I_2)} \text{product} \quad (3.63)$$

An alternative report of the generation of difluorocarbene involves in situ formation of $[BrF_2CPPh_3]^+Br^-$ from CBr_2F_2 and PPh_3 and treatment with KF. In this way, difluorocarbene is produced and stereospecifically reacts with electron-rich olefins [93].

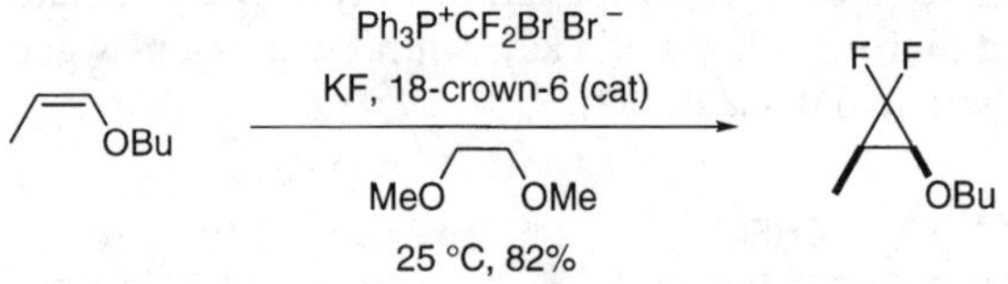

$$\text{(3.64)}$$

Pyrolysis of $CClF_2CO_2Na$ or $CClF_2CO_2K$ is a classical method for generating difluorocarbene [94]. The reaction proceeds by loss of carbon dioxide and a chloride ion to give, upon reaction with an olefin, a cyclized product.

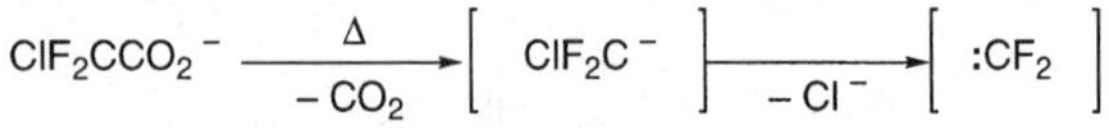

$$ClF_2CCO_2^- \xrightarrow{\Delta}_{-CO_2} [ClF_2C^-] \xrightarrow{}_{-Cl^-} [:CF_2] \quad (3.65)$$

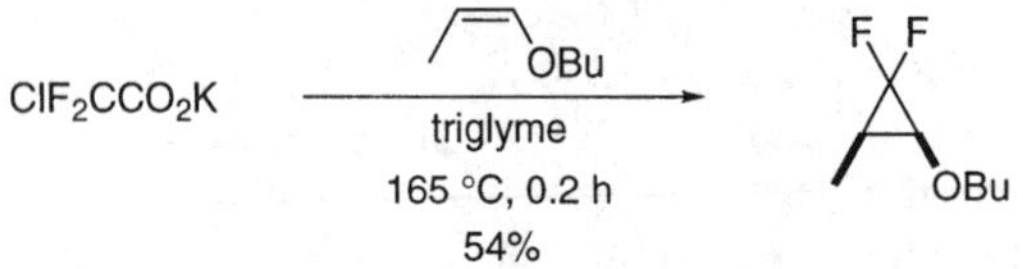

$$ClF_2CCO_2K \xrightarrow[\text{triglyme} \atop {165\ °C,\ 0.2\ h \atop 54\%}]{\text{OBu}} \text{product} \quad (3.66)$$

3.4 Fluorine-Substituted Carbenes

Another thermal procedure effective for the generation of difluorocarbene starts with commercially available hexafluoropropene oxide [95] which decomposes easily above 150 °C [96]. For example, heating a mixture of hexafluoropropene oxide and 1,1,2-trifluoroethene produces pentafluorocyclopropane conveniently.

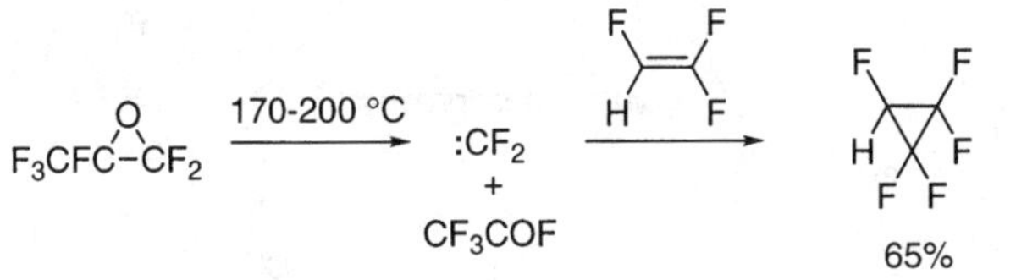

$$F_3CFC-CF_2 \xrightarrow{170-200 \,°C} :CF_2 + CF_3COF \longrightarrow \text{pentafluorocyclopropane} \quad 65\% \tag{3.67}$$

A CF_3-substituted organometallic reagent of Hg, Sn, or Si is a versatile source of difluorocarbene. Heating CF_3HgPh [97] or CF_3SnMe_3 [98] in the presence of NaI generates difluorocarbene that reacts with cyclohexene to give 7,7-difluoronorcarane.

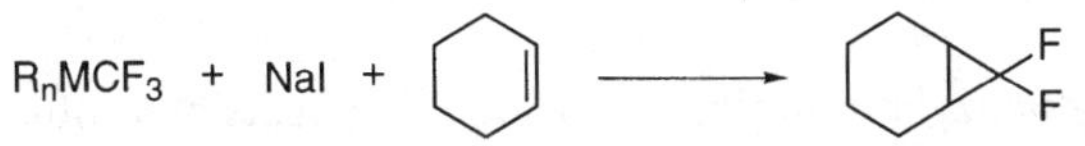

$$R_nMCF_3 + NaI + \text{cyclohexene} \longrightarrow \text{7,7-difluoronorcarane} \tag{3.68}$$

R_nMCF_3	conditions	yield (%)
$PhHgCF_3$	benzene, reflux	83
Me_3SnCF_3	DME, reflux	89

Trifluoro(trifluoromethyl)silane liberates difluorocarbene at 100 °C whose half-life time is estimated to be 7 min. This procedure has been applied to the synthesis of difluoromethylsilanes which otherwise are difficult to be prepared [99].

$$F_3SiCF_3 + ClSiH_3 \xrightarrow[60\%]{100\,°C,\ 2\,h} ClH_2Si(CF_2H) \tag{3.69}$$

In addition to cyclopropanation, difluorocarbene undergoes an apparent alkylation of a carbanion species. For example, a lithium acetylide, upon reaction with CBr_2ClF, gives rise to a bromodifluoromethylation product.

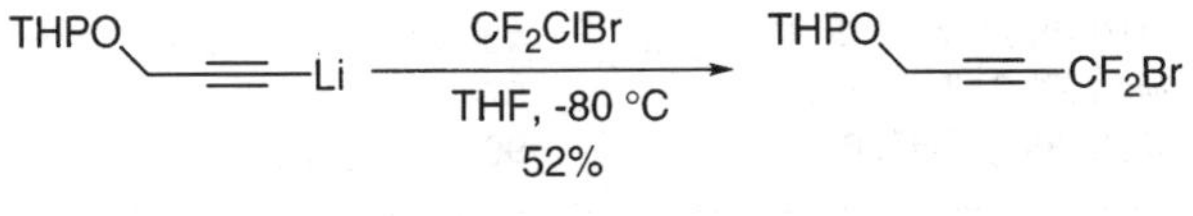

$$\text{THPO-C≡C-Li} \xrightarrow[\substack{THF,\ -80\,°C \\ 52\%}]{CF_2ClBr} \text{THPO-C≡C-CF}_2Br \tag{3.70}$$

The alkylation is understood in terms of an anion chain mechanism (Scheme 3.38) [100].

Chlorofluorocarbene is conveniently produced by dehydrochlorination of $CHFCl_2$ with NaOH or KOH in the presence of a phase-transfer catalyst such as

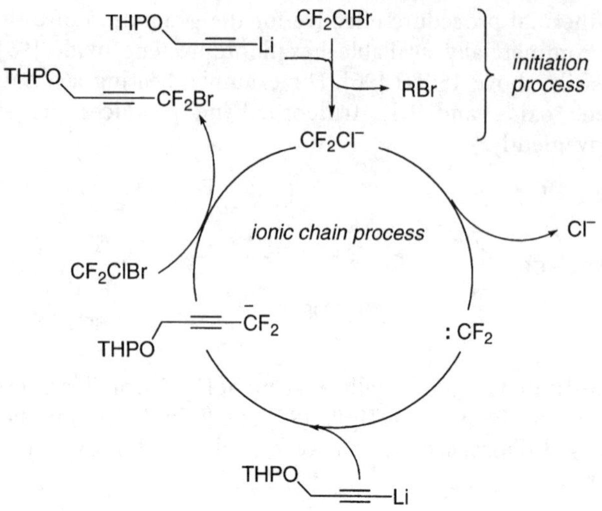

Scheme 3.38

trimethyl- or triethylbenzylammonium chloride and effects chlorofluorocyclopropanation of electron-rich olefins [101].

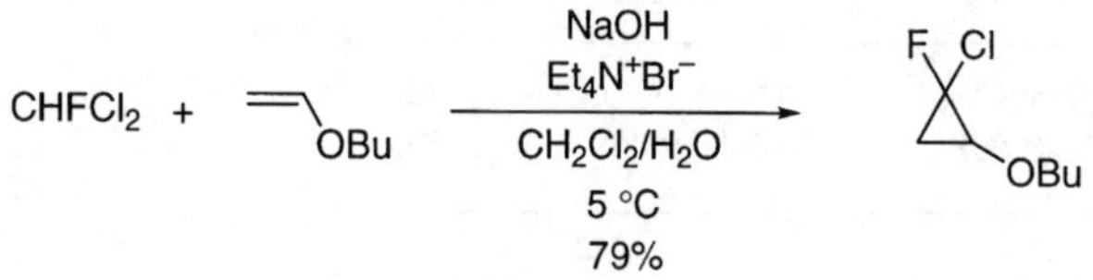

$$CHFCl_2 + \underset{OBu}{=} \xrightarrow[\substack{CH_2Cl_2/H_2O \\ 5\ °C \\ 79\%}]{\substack{NaOH \\ Et_4N^+Br^-}} \underset{OBu}{\overset{F\ Cl}{\triangle}} \quad (3.71)$$

Generation of chlorofluorocarbene from $CFCl_3$ is an alternative method and is carried out with various organometallic reagents. Treatment of $CFCl_3$ with BuLi in the presence of 2,3-dimethyl-2-butene affords an adduct in 49% yield [102]. The yield is improved with a reagent system consisting of Mg and LiCl [103] or of $TiCl_4$ and $LiAlH_4$ [104].

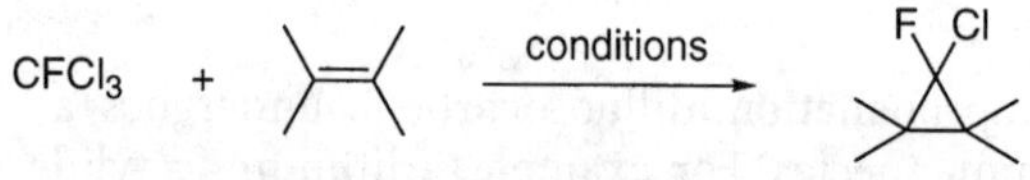

$$CFCl_3 + \underset{}{\bowtie} \xrightarrow{\text{conditions}} \overset{F\ Cl}{\underset{}{\triangle\!\!\!\bowtie}} \quad (3.72)$$

conditions	Yield (%)
BuLi, THF-hexane, -116 °C	49
Mg/LiCl, THF, -15 °C	85
$TiCl_4/LiAlH_4$, THF, 0 °C	90

Thermolysis of $Cl_2FCHgPh$ also gives chlorofluorocarbene, which reacts with various olefins including moderately nucleophilic cyclohexene and electron-deficient acrylonitrile (Scheme 3.39) [105].

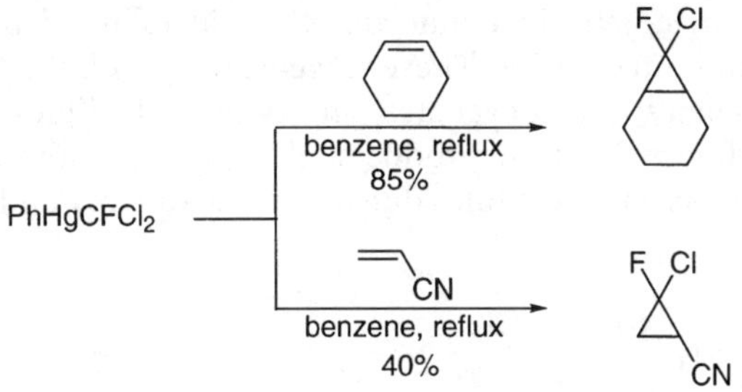

Scheme 3.39

Bromofluorocarbene is generated from $CHFBr_2$, $CFBr_3$, and Br_3CHgPh by a dehydrobromination, bromine–lithium exchange, or thermolysis, respectively, in exactly the same way as for chlorofluorocarbene [106].

3.5
Electrophilic Perfluoroalkylating Reagents

In sharp contrast to normal alkyl halides, perfluoroalkyl halides cannot undergo nucleophilic alkylation, because the electronegativities of perfluoroalkyl groups are higher than those of halogen atoms and thus nucleophiles attack at halogen instead of at the halogen-substituted carbon. To overcome this problem, perfluoroalkyl-group-containing hypervalent reagents involving (perfluoroalkyl)aryliodonium, (polyfluoroalkyl)aryliodonium, and (perfluoroalkyl)chalcogen salts, have been developed, some of which are now commercially available [107].

3.5.1
(Perfluoroalkyl)aryliodonium Salts

(Perfluoroalkyl)aryliodonium salts are generally prepared starting from perfluoroalkyl iodides (R_fI) via $R_fI(OCOCF_3)_2$, as summarized in Scheme 3.40. For example, (perfluoroalkyl)-*p*-tolyliodonium chlorides are generated by treat-

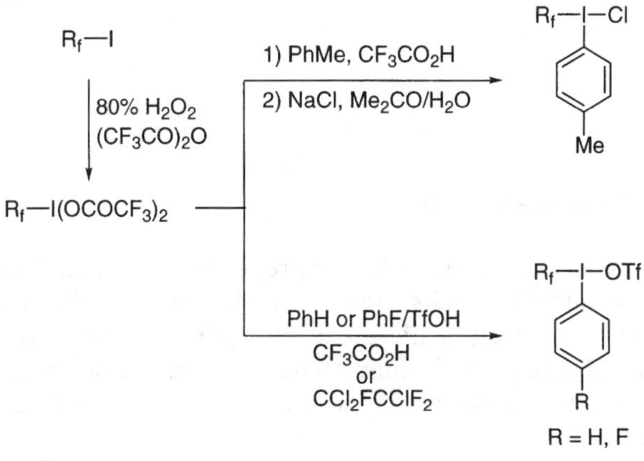

Scheme 3.40

ment of R$_f$I with toluene in trifluoroacetic acid followed by counter anion exchange with sodium chloride. These reagents react with thiolates, selenolates, nitrates, thiocyanates, selenocyanates, and N-methylanilines to give the corresponding perfluoroalkylated products, although available nucleophiles are limited to such hetero nucleophiles due to the low stability and reactivity of the reagents [108].

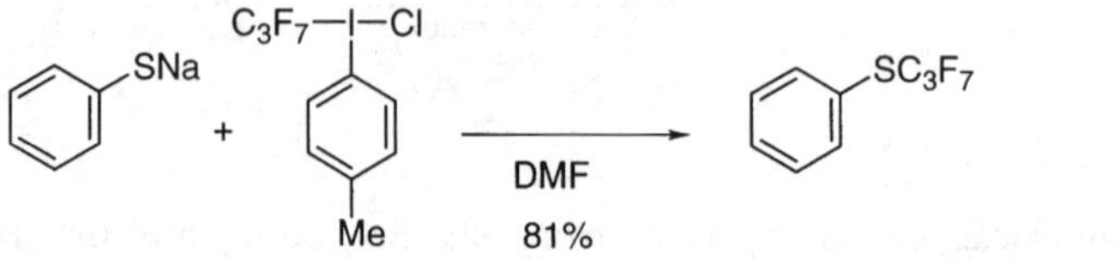

$$\text{PhSNa} + \text{C}_3\text{F}_7\text{-I-Cl-C}_6\text{H}_4\text{-Me} \xrightarrow{\text{DMF}} \text{PhSC}_3\text{F}_7 \quad 81\% \quad (3.73)$$

On the other hand, (perfluoroalkyl)phenyl(or p-fluorophenyl)iodonium triflates are more stable and more reactive. Various types of such (perfluoroalkyl)aryliodonium triflates can be prepared according to Scheme 3.40, except for the (trifluoromethyl)aryliodonium salt, which is not accessible due to the low stability of the synthetic intermediate, $CF_3I(OCOCF_3)_2$.

(Perfluoroalkyl)aryliodonium triflates react with thiols, carbanions, alkenes, alkynes, silyl enol ethers, and aromatics to afford various types of perfluoroalkylated compounds [107]. Examples are summarized in Scheme 3.41.

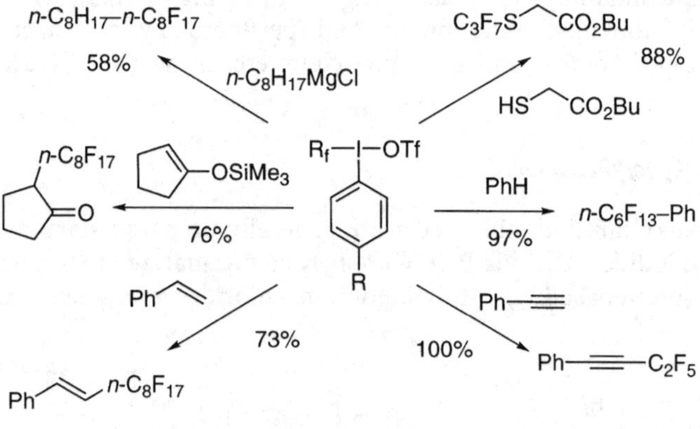

Scheme 3.41

3.5.2
(Polyfluoroalkyl)aryliodonium Salts

(Polyfluoroalkyl)aryliodonium salts are prepared in a manner similar to (perfluoroalkyl)aryliodonium salts and undergo electrophilic perfluoroalkylmethylation that is hardly achieved using R_fCH_2I as an electrophile [109]. Nucleophiles such as amines, lithium alkoxides, metal carboxylates, carbanions, enol silyl ethers, and electron-rich aromatics can be applied to this reaction (Scheme 3.42).

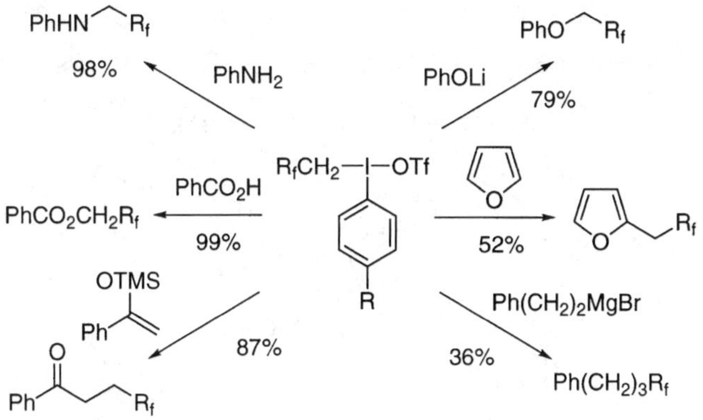

3.5 Electrophilic Perfluoroalkylating Reagents

Scheme 3.42

3.5.3
(Trifluoromethyl)chalcogenium Salts

Electrophilic trifluoromethylation is efficiently effected by the use of S-, Se-, or Te-(trifluoromethyl)dibenzochalcogenophenium salts that are prepared from 2-[(trifluoromethyl)thio-, seleno-, or telluro]biphenyls, respectively, as illustrated in Eq. (3.74).

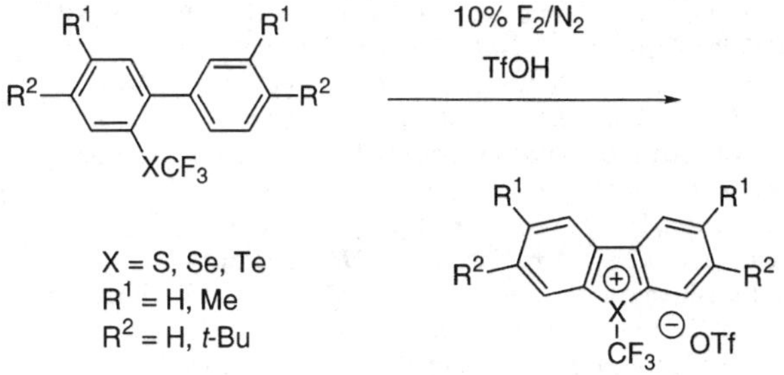

$$X = S, Se, Te$$
$$R^1 = H, Me$$
$$R^2 = H, t\text{-Bu}$$

(3.74)

The reactivity of electrophilic trifluoromethylation with these reagents depends on the combination of both chalcogen and substituents on the biphenyl ring and, hence, a variety of nucleophiles such as carbanions, electron-rich aromatics, silyl enol ethers, phosphines, and thiolate anions are trifluoromethylated by judicious choice of the reagents (Scheme 3.43) [110].

As described above, the building block approach allows the attachment of a variety of fluorine-containing structural units to diverse kinds of substrates in a regio-, stereo-, and enantioselective manner. Therefore, exploration of novel versatile organofluorine building blocks is still an important subject for facile, efficient, and environmentally benign syntheses of fluorine-containing target molecules.

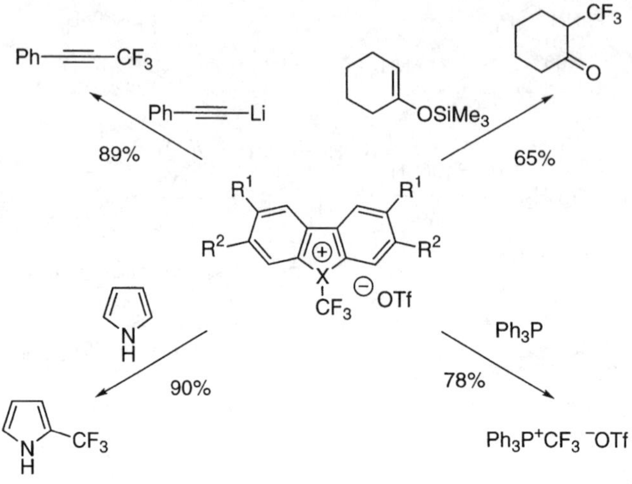

Scheme 3.43

References

1. (a) Gassman, P. G.; O.'Reilly, N. J. *Tetrahedron Lett.* 1985, *26*, 5243; (b) Gassman, P. G.; O'Reilly, N. J. *J. Org. Chem.* 1987, *52*, 2481; (c) Uno, H.; Shiraishi, Y.; Suzuki, H. *Bull. Chem. Soc. Jpn.* 1989, *62*, 2636; (d) Uno, H.; Okada, S.-i.; Ono, T.; Shiraishi, Y.; Suzuki, H. *J. Org. Chem.* 1992, *57*, 1504
2. Haszeldine, R. N. *J. Chem. Soc.* 1954, 1273
3. McBee, E. T.; Roberts, C. W.; Meiners, A. F. *J. Am. Chem. Soc.* 1957, *79*, 335
4. McLoughlin, V. C. R.; Thrower, J. *Tetrahedron* 1969, *25*, 5921
5. (a) Kobayashi, Y.; Kumadaki, I. *Tetrahedon Lett.* 1969, 4095; (b) Kobayashi, Y.; Yamamoto, K.; Kumadaki, I. *Tetrahedron Lett.* 1979, 4071
6. Kondratenko, N. V.; Vechirko, E. P.; Yagupolskii, L. M. *Synthesis* 1980, 932
7. Umemoto, T.; Ando, A. *Bull. Chem. Soc. Jpn.* 1986, *59*, 447
8. Wiemers, D. M.; Burton, D. J. *J. Am. Chem. Soc.* 1986, *108*, 832
9. Burton, D. J.; Wiemers, D. M. *J. Am. Chem. Soc.* 1985, *107*, 5014
10. (a) Kitazume, T.; Ishikawa, N. *Chem. Lett.* 1981, 1679; (b) Kitazume, T.; Ishikawa, N. *Chem. Lett.* 1982, 1453; (c) Kitazume, T.; Ishikawa, N. *Chem. Lett.* 1982, 137; (d) Kitazume, T.; Ishikawa, N. *J. Am. Chem. Soc.* 1985, *107*, 5186
11. (a) Fujita, M.; Hiyama, T. *Tetrahedron Lett.* 1986, *27*, 3659; (b) Fujita, M.; Morita, T.; Hiyama, T. *Tetrahedron Lett.* 1986, *27*, 2135; (c) Fujita, M.; Hiyama, T. *Tetrahedron Lett.* 1986, *27*, 3655; (d) Fujita, M.; Hiyama, T. *Bull. Chem. Soc. Jpn.* 1987, *60*, 4377
12. (a) Fujita, M.; Hiyama, T.; Kondo, K. *Tetrahedron Lett.* 1986, *27*, 2139; (b) Fujita, M.; Kondo, K.; Hiyama, T. *Bull. Chem. Soc. Jpn.* 1987, *60*, 4385
13. (a) Prakash, G. K. S.; Krishnamurti, R.; Olah, G. A. *J. Am. Chem. Soc.* 1989, *111*, 393; (b) Krishnamurti, R.; Bellew, D. R.; Prakash, G. K. S. *J. Org. Chem.* 1991, *56*, 984; (c) Wiedemann, J.; Heiner, T.; Mloston, G.; Prakash, G. K. S.; Olah, G. A. *Angew. Chem. Int. Ed.* 1998, *37*, 820
14. Urata, H.; Fuchikami, T. *Tetrahedron Lett.* 1991, *3 2*, 91
15. (a) Kuroboshi, M.; Yamada, N.; Takebe, Y.; Hiyama, T. *Synlett* 1995, 987; (b) Shimizu, M.; Yamada, N.; Takebe, Y.; Hata, T.; Kuroboshi, M.; Hiyama, T. *Bull. Chem. Soc. Jpn.* **71**, 2903 (1998)
16. (a) Kuroboshi, M.; Yamada, N.; Takebe, Y.; Hiyama, T. *Tetrahedron Lett.* 1995, *36*, 6271; (b) Shimizu, M.; Takebe, Y.; Kuroboshi, M.; Hiyama, T. *Tetrahedron Lett.* 1996, *37*, 7387

17. Hata, T.; Shimizu, M.; Hiyama, T. *Synlett* 1996, 831
18. (a) Burton, D. J.; Yang, Z.-Y.; Qiu, W. *Chem. Rev.* 1996, *96*, 1641; (b) Shimizu, M.; Hata, T.; Hiyama, T. *Tetrahedron Lett.* 1997, *38*, 4591; (c) Coutrot, P.; Grison, C.; Sauvetre, R. *J. Organomet. Chem.* 1987, *332*, 1; (d) Fujii, A.; Usuki, Y.; Iio, H.; Tokoroyama, T. *Synlett* 1994, 725
19. (a) Obayashi, M.; Kondo, K. *Tetrahedron Lett.* 1982, *23*, 2327; (b) Obayashi, M.; Ito, E.; Matsui, K.; Kondo, K. *Tetrahedron Lett.* 1982, *23*, 2323
20. (a) Lequeux, T. P.; Percy, J. M. *Synlett* 1995, 361; (b) Blades, K.; Lequeux, T. P.; Percy, J. M. *Tetrahedron* 1997, *53*, 10623
21. (a) Burton, D. J.; Ishihara, T.; Maruta, M. *Chem. Lett.* 1982, 755; (b) Burton, D. J.; Sprague, L. G. *J. Org. Chem.* 1988, *53*, 1523; (c) Burton, D. J.; Sprague, L. G. *J. Org. Chem.* 1989, *54*, 613; (d) Yokomatsu, T.; Suemune, K.; Murano, T.; Shibuya, S. *J. Org. Chem.* 1996, *61*, 7202; (e) Yokomatsu, T.; Murano, T.; Suemune, K.; Shibuya, S. *Tetrahedron* 1997, *53*, 815
22. (a) Hagiwara, T.; Fuchikami, T. *Synlett* 1995, 717; (b) Yudin, A. K.; Prakash, G. K. S.; Deffieux, D.; Bradley, M.; Bau, R.; Olah, G. A. *J. Am. Chem. Soc.* 1997, *119*, 1572
23. (a) Tarrant, P.; Johncock, P.; Savory, J. *J. Org. Chem.* 1963, *28*, 839; (b) Drakesmith, F. G.; Richardson, R. D.; Stewart, O. J.; Tarrant, P. *J. Org. Chem.* 1968, *33*, 286; (c) Normant, J.-F.; Foulon, J. P.; Masure, D.; Sauvetre, R.; Villieras, J. *Synthesis* 1975, 122; (d) Hahnfeld, J. L.; Burton, D. J. *Tetrahedron Lett.* 1975, 773; (e) Sauvetre, R.; Normant, J.-F. *Tetrahedron Lett.* 1981, *22*, 957; (f) Gillet, J.-P.; Sauvetre, R.; Normant, J.-F. *Synthesis* 1986, 355; (g) Normant, J.-F. *J. Organomet. Chem.* 1990, *400*, 19
24. Dubuffet, T.; Sauvetre, R.; Normant, J.-F. *J. Organomet. Chem.* 1988, *341*, 11
25. Watanabe, H.; Yamashita, F.; Uneyama, K. *Tetrahedron Lett.* 1993, *34*, 1941
26. (a) Nakai, T.; Tanaka, K.; Ishikawa, N. *Chem. Lett.* 1976, 1263; (b) Tanaka, K.; Nakai, T.; Ishikawa, N. *Tetrahedron Lett.* 1978, 4809
27. (a) Ichikawa, J.; Sonoda, T.; Kobayashi, H. *Tetrahedron Lett.* 1989, *30*, 1641; (b) Ichikawa, J.; Sonoda, T.; Kobayashi, H. *Tetrahedron Lett.* 1989, *30*, 5437; (c) Ichikawa, J.; Sonoda, T.; Kobayashi, H. *Tetrahedron Lett.* 1989, *30*, 6379; (d) Ichikawa, J.; Moriya, T.; Sonoda, T.; Kobayashi, H. *Chem. Lett.* 1991, 961; (e) Ichikawa, J.; Hamada, S.; Sonoda, T.; Kobayashi, H. *Tetrahedron Lett.* 1992, *33*, 337; (f) Ichikawa, J.; Minami, T.; Sonoda, T.; Kobayashi, H. *Tetrahedron Lett.* 1992, *33*, 3779; (g) Ichikawa, J.; Yonemaru, S.; Minami, T. *Synlett* 1992, 833
28. (a) Lee, J.; Tsukazaki, M.; Snieckus, V. *Tetrahedron Lett.* 1993, *34*, 415; (b) Howarth, J. A.; Owton, W. M.; Percy, J. M. *J. Chem. Soc., Chem. Commun.* 1995, 757; (c) Patel, S. T.; Percy, J. M.; Wilkes, R. D. *Tetrahedron* 1995, *51*, 9201; (d) Howarth, J. A.; Owton, W. M.; Percy, J. M.; Rock, M. H. *Tetrahedron* 1995, *51*, 10289
29. (a) Gillet, J.-P.; Sauvetre, R.; Normant, J.-F. *Tetrahedron Lett.* 1985, *26*, 3999; (b) Gillet, J.-P.; Sauvetre, R.; Normant, J.-F. *Synthesis* 1986, 538
30. (a) Tellier, F.; Sauvetre, R.; Normant, J.-F. *J. Organomet. Chem.* 1986, *303*, 309; (b) Tellier, F.; Sauvetre, R.; Normant, J.-F. *J. Organomet. Chem.* 1989, *364*, 17
31. (a) Hansen, S. W.; Spawn, T. D.; Burton, D. J. *J. Fluorine Chem.* 1987, *35*, 415; (b) Morken, P. A.; Burton, D. J. *J. Org. Chem.* 1993, *58*, 1167; (c) Nguyen, B. V.; Burton, D. J. *J. Org. Chem.* 1997, *62*, 7758; (d) Davis, C. R.; Burton, D. J. *J. Org. Chem.* 1997, *62*, 9217; (e) Heinze, P. L.; Burton, D. J. *J. Org. Chem.* 1988, *53*, 2714; (f) Jiang, B.; Xu, Y. *J. Org. Chem.* 1991, *56*, 7336
32. Burton, D. J.; Hansen, S. W. *J. Am. Chem. Soc.* 1986, *108*, 4229
33. (a) Seyferth, D.; Wada, T. *Inorg. Chem.* 1962, *1*, 78; (b) Tarrant, P.; Oliver, W. H. *J. Org. Chem.* 1966, *31*, 1143; (c) Drakesmith, F. G.; Stewart, O. J.; Tarrant, P. *J. Org Chem* 1968, *33*, 472
34. Hiyama, T.; Nishide, K.; Obayashi, M. *Chem. Lett.* 1984, 1765
35. Martin, S.; Sauvetre, R.; Normant, J.-F. *J. Organomet. Chem.* 1984, *264*, 155
36. (a) Kaesz, H. D.; Stafford, S. L.; Stone, F. G. A. *J. Am. Chem. Soc.* 1960, *82*, 6232; (b) Seyferth, D.; Brandle, K. A.; Raab, G. *Angew. Chem.* 1960, *72*, 77; (c) Seyferth, D.; Welch, D. E.; Raab, G. *J. Am. Chem. Soc.* 1962, *84*, 4266
37. Xue, L.; Lu, L.; Pedersen, S. D.; Liu, Q.; Narske, R. M.; Burton, D. J. *J. Org. Chem.* 1997, *62*, 1064

38. Lu, L.; Burton, D. J. *Tetrahedron Lett.* 1997, *38*, 7673
39. Burton, D. J.; Yang, Z.-Y.; Morken, P. A. *Tetrahedron* 1994, *50*, 2993
40. Drakesmith, F. G.; Stewart, O. J.; Tarrant, P. *J. Org. Chem.* 1968, *33*, 280
41. (a) Henne, A. L.; Nager, M. *J. Am. Chem. Soc.* 1952, *74*, 650; (b) Hanzawa, Y.; Kawagoe, K.-i.; Tanahashi, N.; Kobayashi, Y. *Tetrahedron Lett.* 1984, *25*, 4749
42. Finnegan, W. G.; Norris, W. P. *J. Org. Chem.* 1963, *28*, 1139
43. Norris, W. P.; Finnegan, W. G. *J. Org. Chem.* 1966, *31*, 3292
44. Yoneda, N.; Matsuoka, S.; Miyaura, N.; Fukuhara, T.; Suzuki, A. *Bull. Chem. Soc. Jpn.* 1990, *63*, 2124
45. Welch, J. T.; Seper, K. W. *Tetrahedron Lett.* 1984, *25*, 5247
46. (a) Welch, J. T.; Seper, K.; Eswarakrishnan, S.; Samartino, J. *J. Org. Chem.* 1984, *49*, 4720; (b) Welch, J. T.; Eswarakrishnan, S. *J. Org. Chem.* 1985, *50*, 5403
47. Welch, J. T.; Herbert, R. W. *J. Org. Chem.* 1990, *55*, 4782
48. (a) Ishihara, T.; Ichihara, K.; Yamanaka, H. *Tetrahedron Lett.* 1995, *36*, 8267; (b) Ishihara, T.; Ichihara, K.; Yamanaka, H. *Tetrahedron* 1996, *52*, 255
49. McBee, E. T.; Pierce, O. R.; Christman, D. L. *J. Am. Chem. Soc.* 1955, *77*, 1581
50. Ishihara, T.; Matsuda, T.; Imura, K.; Matsui, H.; Yamanaka, H. *Chem. Lett.* 1994, 2167
51. Kuroboshi, M.; Ishihara, T. *Bull. Chem. Soc. Jpn.* 1990, *63*, 428
52. (a) Hallinan, E. A.; Fried, J. *Tetrahedron Lett.* 1984, *25*, 2301; (b) Kitagawa, O.; Taguchi, T.; Kobayashi, Y. *Tetrahedron Lett.* 1988, *29*, 1803; (c) Lang, R. W.; Schaub, B. *Tetrahedron Lett.* 1988, *29*, 2943; (d) Taguchi, T.; Kitagawa, O.; Suda, Y.; Ohkawa, S.; Hashimoto, A.; Iitaka, Y.; Kobayashi, Y. *Tetrahedron Lett.* 1988, *29*, 5291
53. (a) Taguchi, T.; Kitagawa, O.; Morikawa, T.; Nishiwaki, T.; Uehara, H.; Endo, H.; Kobayashi, Y. *Tetrahedron Lett.* 1986, *27*, 6103; (b) Kitagawa, O.; Taguchi, T.; Kobayashi, Y. *Chem. Lett.* 1989, 389
54. (a) Yamana, M.; Ishihara, T.; Ando, T. *Tetrahedron Lett.* 1983, *24*, 507; (b) Kitagawa, O.; Taguchi, T.; Kobayashi, Y. *Tetrahedron Lett.* 1988, *29*, 1803; (c) Taguchi, T.; Kitagawa, O.; Suda, Y.; Ohkawa, S.; Hashimoto, A.; Iitaka, Y.; Kobayashi, Y. *Tetrahedron Lett.* 1988, *29*, 5291; (d) Kitagawa, O.; Hashimoto, A.; Kobayashi, Y.; Taguchi, T. *Chem. Lett.* 1990, 1307
55. Qian, C.-P.; Nakai, T. *Tetrahedron Lett.* 1990, *31*, 7043
56. (a) Ishihara, T.; Kuroboshi, M.; Yamaguchi, K. *Chem. Lett.* 1990, 211; (b) Kuroboshi, M.; Ishihara, T. *Bull. Chem. Soc. Jpn.* 1990, *63*, 1191
57. (a) Ishihara, T.; Yamaguchi, K.; Kuroboshi, M. *Chem. Lett.* 1989, 1191; (b) Ishihara, T.; Kuroboshi, M.; Yamaguchi, K.; Okada, Y. *J. Org. Chem.* 1990, *55*, 3107
58. Yokozawa, T.; Ishikawa, N.; Nakai, T. *Chem. Lett.* 1987, 1971
59. (a) Yokozawa, T.; Nakai, T.; Ishikawa, N. *Tetrahedron Lett.* 1984, *25*, 3991; (b) Yokozawa, T.; Nakai, T.; Ishikawa, N. *Tetrahedron Lett.* 1984, *25*, 3987
60. Qian, C.-P.; Nakai, T. *Tetrahedron Lett.* 1988, *29*, 4119
61. Qian, C.-P.; Nakai, T.; Dixon, D. A.; Smart, B. E. *J. Am. Chem. Soc.* 1990, *112*, 4602
62. Iseki, K.; Oishi, S.; Kobayashi, Y. *Tetrahedron* 1996, *52*, 71
63. Mikami, K.; Yajima, T.; Takasaki, T.; Matsukawa, S.; Terada, M.; Uchimaru, T.; Maruta, M. *Tetrahedron* 1996, *52*, 85
64. Poras, H.; Matsutani, H.; Yaruva, J.; Kusumoto, T.; Hiyama, T. *Chem. Lett.* 1998, 665
65. Kaneko, S.; Yamazaki, T.; Kitazume, T. *J. Org. Chem.* 1993, *58*, 2302
66. Ishihara, T.; Hayashi, H.; Yamanaka, H. *Tetrahedron Lett.* 1993, *34*, 5777
67. Creary, X. *J. Org. Chem.* 1987, *52*, 5026
68. Shinohara, N.; Haga, J.; Yamazaki, T.; Kitazume, T.; Nakamura, S. *J. Org. Chem.* 1995, *60*, 4363
69. Maruoka, K.; Shimada, I.; Imoto, H.; Yamamoto, H. *Synlett* 1994, 519
70. Taguchi, T.; Sasaki, H.; Shibuya, A.; Morikawa, T. *Tetrahedron Lett.* 1994, *35*, 913
71. (a) Takahashi, O.; Furuhashi, K.; Fukumasa, M.; Hirai, T. *Tetrahedron Lett.* 1990, *31*, 7031; (b) Bussche-Hunnefeld, C. v. d.; Cescato, C.; Seebach, D. *Chem. Ber.* 1992, *125*, 2795; (c) Katagiri, T.; Obara, F.; Toda, S.; Furuhashi, K. *Synlett* 1994, 507; (d) Ramachandran, P. V.; Gong, B.; Brown, H. C. *J. Org. Chem.* 1995, *60*, 41
72. Reviews of fluorinated radicals: (a) Dolbier, W. R. Jr. *Chem. Rev.* 1996, *96*, 1557; (b) Dolbier, W. R. Jr. *Top. Curr. Chem.* 1997, *192*, 97

73. Ishihara, T; Hayashi, K.; Ando, T.; Yamanaka, H. *J. Org. Chem.* 1975, *40*, 3264
74. Haszeldine, R. N. *J. Chem. Soc.* 1949, 2856
75. Qiu, Z.-M.; Burton, D. J. *J. Org. Chem.* 1995, *60*, 3465
76. Fuchikami, T.; Ojima, I. *Tetrahedron Lett.* 1984, *25*, 303
77. Ishihara, T.; Kuroboshi, M.; Okada, Y. *Chem. Lett.* 1986, 1895. For a related study using organotin compounds, see Matsubara, S.; Mitani, M.; Utimoto, K. *Tetrahedron Lett.* 1987, *28*, 5857
78. (a) Urata, H.; Yugari, H.; Fuchikami, T. *Chem. Lett.* 1987, 833; (b) Fuchikami, T.; Shibata, Y.; Urata, H. *Chem. Lett.* 1987, 521
79. (a) Takeyama, Y.; Ichinose, Y.; Oshima, K.; Utimoto, K. *Tetrahedron Lett.* 1989, *30*, 3159; (b) Miura, K.; Takeyama, Y.; Oshima, K.; Utimoto, K. *Bull. Chem. Soc. Jpn.* 1991, *64*, 1542
80. Maruoka, K.; Sano, H.; Fukutani, Y.; Yamamoto, H. *Chem. Lett.* 1985, 1689
81. Gregorcic, A.; Zupan, M. *J. Org. Chem.* 1979, *44*, 4120
82. Tanabe, Y.; Matsuo, N.; Ohno, N. *J. Org. Chem.* 1988, *53*, 4582
83. (a) Brookes, C. J.; Coe, P. L.; Pedler, A. E.; Tatlow, J. C. *J. Chem. Soc., Perkin Trans. 1* 1978, 202; (b) Renaud, R. N.; Champagne, P. J.; Savard, M. *Can. J. Chem.* 1979, *57*, 2617; (c) Muller, N. *J. Org. Chem.* 1983, *48*, 1370; (d) Uneyama, K.; Nanbu, H. *J. Org. Chem.* 1988, *53*, 4598; (e) Uneyama, K.; Morimoto, O.; Nanbu, H. *Tetrahedron Lett.* 1989, *30*, 109; (f) Uneyama, K.; Watanabe, S. *J. Org. Chem.* 1990, *55*, 3909
84. Reviews of fluorinated peroxides: (a) Sawada, H. *Chem. Rev.* 1996, *96*, 1779; (b) Chengxue, Z.; Renmo, Z.; Heqi, P.; Xiangshan, J.; Yangling, Q.; Chengjiu, W.; Xikui, J. *J. Org. Chem.* 1982, *47*, 2009
85. (a) Yoshida, M.; Amemiya, H.; Kobayashi, M.; Sawada, H.; Hagii, H.; Aoshima, K. *J. Chem. Soc., Chem. Commun.* 1985, 234; (b) Sawada, H.; Yoshida, M.; Hagii, H.; Aoshima, K.; Kobayashi, M. *Bull. Chem. Soc. Jpn.* 1986, *59*, 215; (c) Yoshida, M.; Yoshida, T.; Kobayashi, M.; Kamigata, N. *J. Chem. Soc., Perkin Trans. 1* 1989, 909
86. Reviews of fluorinated carbenes: (a) Burton, D. J.; Hahnfeld, J. L. *Fluorine Chem. Rev.* 1977, *8*, 119; (b) *Houben-Weyl Methoden der Organishen Chemie*; Regitz, M.; Giese, B., Eds.; Georg Thieme Verlag: Stuttgart, 1989; Vol. E19, Part B; (c) Smart, B. E. In: *Chemistry of Organic Fluorine Compounds II*; Hudlicky, M.; Pavlath, A. E., Eds.; American Chemical Society: Washington, DC, 1995; pp 767–796; (d) Brahms, D. L. S.; Dailey, W. P. *Chem. Rev.* 1996, *96*, 1585
87. (a) Schlosser, M.; Heinz, G. *Angew. Chem. Int. Ed. Engl.* 1968, *7*, 820; (b) Schlosser, M.; Heinz, G. *Chem. Ber.* 1971, *104*, 1934
88. Nishimura, J.; Furukawa, J. *J. Chem. Soc., Chem. Commun.* 1971, 1375
89. (a) Tamura, O.; Hashimoto, M.; Kobayashi, Y.; Katoh, T.; Nakatani, K.; Kamada, M.; Hayakawa, I.; Akiba, T.; Terashima, S. *Tetrahedron* 1994, *50*, 3889; (b) Akiba, T.; Tamura, O.; Hashimoto, M.; Kobayashi, Y.; Katoh, T.; Nakatani, K.; Kumada, M.; Hayakawa, I.; Terashima, S. *Tetrahedron* 1994, *50*, 3905
90. (a) Hine, J. H.; Langford, P. B. *J. Am. Chem. Soc.* 1957, *79*, 5497; (b) Hine, J.; Porter, J. J. *J. Am. Chem. Soc.* 1957, *79*, 5493
91. Weyerstahl, P.; Schwartzkopff, U.; Nerdel, F. *Liebigs Ann. Chem.* 1973, 2100
92. Dolbier, W. R. Jr.; Wojtowicz, H.; Burkholder, C. R. *J. Org. Chem.* 1990, *55*, 5420
93. (a) Burton, D. J.; Naae, D. G. *J. Am. Chem. Soc.* 1973, *95*, 8467; (b) Bessard, Y.; Muller, U.; Schlosser, M. *Tetrahedron* 1990, *46*, 5213; (c) Bessard, Y.; Schlosser, M. *Tetrahedron* 1991, *47*, 7323
94. (a) Birchall, J. M.; Cross, G. W.; Haszeldine, R. N. *Proc. Chem. Soc.* 1960, 81; (b) Taguchi, T.; Takigawa, T.; Tawara, Y.; Morikawa, T.; Kobayashi, Y. *Tetrahedron Lett.* 1984, *25*, 5689; (c) Bessard, Y.; Muller, U.; Schlosser, M. *Tetrahedron* 1990, *46*, 5213
95. Review of the chemistry of hexafluoropropylene oxide: Millauer, H.; Schwertfeger, W.; Siegemund, G. *Angew. Chem. Int. Ed. Engl.* 1985, *24*, 161
96. Sargeant, P. B. *J. Org. Chem.* 1970, *35*, 678
97. Seyferth, D.; Hopper, S. P.; Darragh, K. V. *J. Am. Chem. Soc.* 1969, *91*, 6536
98. (a) Clark, H. C.; Willis, C. J. *J. Am. Chem. Soc.* 1960, *82*, 1888; (b) Seyferth, D.; Dertouzos, H.; Suzuki, R.; Mui, J. Y.-P. *J. Org. Chem.* 1967, *32*, 2980

99. Burger, H.; Eujen, R.; Moritz, P. *J. Organomet. Chem.* 1991, *401*, 249
100. (a) Rico, I.; Cantacuzene, D.; Wakselman, C. *J. Chem. Soc., Perkin Trans. 1* 1982, 1063; (b) Kwok, P.-Y.; Muellner, F. W.; Chen, C.-K.; Fried, J. *J. Am. Chem. Soc.* 1987, *109*, 3684
101. (a) Weyerstahl, P.; Blume, G.; Muller, C. *Tetrahedron Lett.* 1971, 3869; (b) Chau, L. V.; Schlosser, M. *Synthesis* 1973, 112; (c) Molines, H.; Nguyen, T.; Wakselman, C. *Synthesis* 1985, 754
102. Burton, D. J.; Hahnfeld, J. L. *J. Org. Chem.* 1977, *42*, 828
103. Mu, C. H.; Tu, M. H. *J. Fluorine Chem.* 1994, *67*, 9
104. Dolbier, W. R. Jr.; Burkholder, C. R. *J. Org. Chem.* 1990, *55*, 589
105. Seyferth, D.; Darragh, K. V. *J. Org. Chem.* 1970, *35*, 1297
106. Dehydrobromination: Weyerstahl, P.; Blume, G.; Muller, C. *Tetrahedron Lett.* 1971, 3869; Bromine-metal exchange: Burton, D. J.; Hahnfeld, J. L. *J. Org. Chem.* 1977, *42*, 828; Thermolysis: Seyferth, D.; Hopper, S. P. *J. Organomet. Chem.* 1973, *51*, 77
107. Review: Umemoto, T. *Chem. Rev.* 1996, *96*, 1757
108. Yagupolskii, L. M.; Maletina, I. I.; Kondratenko, N. V.; Orda, V. V. *Synthesis* 1978, 835
109. Umemoto, T.; Gotoh, Y. *Bull. Chem. Soc. Jpn.* 1987, *60*, 3307
110. Umemoto, T.; Ishihara, S. *J. Am. Chem. Soc.* 1993, *115*, 2156

CHAPTER 4

Reactions of C–F Bonds

Due to the fact that it has the highest electronegativity, fluorine forms a bond to carbon with an energy higher than the other halides, and the reactions of C–F bonds exhibit a unique chemistry that is totally different from the chemistry of C–Cl, C–Br, and C–I bonds. This chapter describes the reactions of C–F bonds from the following viewpoints; reactions where fluorine acts as a leaving group, C–F bond activation by metal complexes, and interaction of fluorine with a proton or metal.

4.1
Fluorine Leaving Group

4.1.1
1-Fluoro Sugars

Among halides, fluorine has been recognized as the poorest leaving group and thus is rarely used for nucleophilic substitution of aliphatic substrates. In 1981, however, Mukaiyama showed that O-glycosidation [1] of a glycopyranosyl fluoride with an alcohol in the presence of $SnCl_2$ and $AgClO_4$ proceeded in good yields with high stereoselectivity (Eq. 4.1) [2].

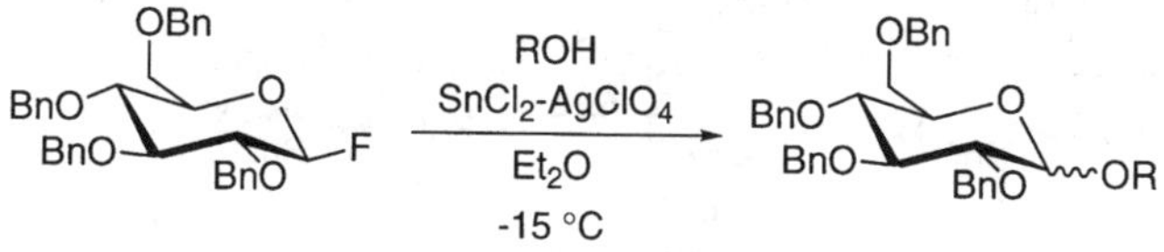

$$\text{BnO-sugar-F} \xrightarrow[\text{Et}_2\text{O}, \, -15\,°\text{C}]{\text{ROH, SnCl}_2\text{-AgClO}_4} \text{BnO-sugar-OR} \quad (4.1)$$

ROH	Yield (%)	α	:	β
MeOH	82	86	:	14
cyclohexanol	88	83	:	17
t-butyl alcohol	87	81	:	19
3β-cholestanol	96	92	:	8

Since then, glycosyl fluorides have been widely utilized as glycosyl donors for glycosidation because of their enhanced stability, ease of handling, and higher stereoselectivity compared with other glycosyl halides. Due to the poor reactivity of fluorine as a leaving group, the presence of a Lewis acid as an activator is generally essential, and various Lewis acids, such as $SnCl_2$-$AgClO_4$ [3], SiF_4 [4], Me_3SiOTf [4], $BF_3 \cdot OEt_2$ [5], TiF_4 [6], Cp_2MCl_2-AgX (M=Ti, Zr, Hf; X=ClO_4, OTf) [7], Me_2GaX (X=Cl, OTf) [8], Tf_2O [9], LnX_3 (Ln=Y, La, Yb; X=Cl, OTf, ClO_4) [10], and $LiClO_4$ [11], have been employed to promote the glycosidation of glycosyl fluorides.

A representative example is the N-glycosidation of a ribofuranosyl fluoride catalyzed by SiF_4 to give the β-isomer exclusively. Trimethylsilyl triflate promotes the condensation of methyl trimethylsilyl ether with a glycosyl fluoride. The diastereoselectivity of the reaction depends on the solvent irrespective of the anomeric configuration of the starting fluorides. Stereoselective allylation with allylsilane is accomplished using $BF_3 \cdot OEt_2$ as activator. The reagent system involving a Group 4 metallocene and $AgClO_4$ is effective for the O-glycosidation even with *t*-butyl alcohol. Gallium and rare earth metal salts can be employed for the O-glycosidation. A ytterbium counterion reverses the stereoselectivity.

With organomagnesium and -aluminum reagents, glycosyl fluorides undergo C-glycosidation without a Lewis acid catalyst. Dimethylaluminum cyanide

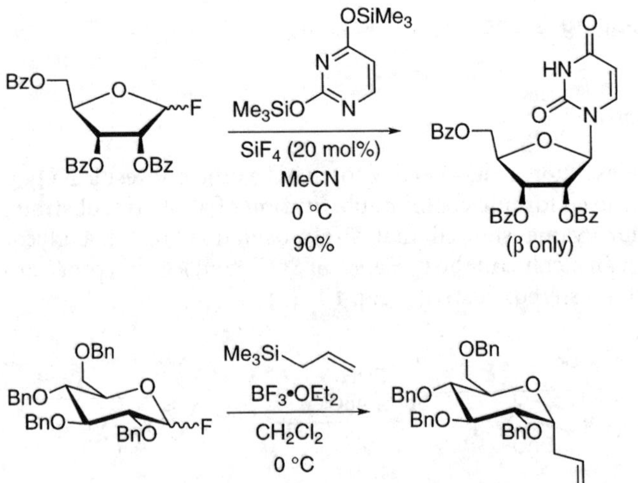

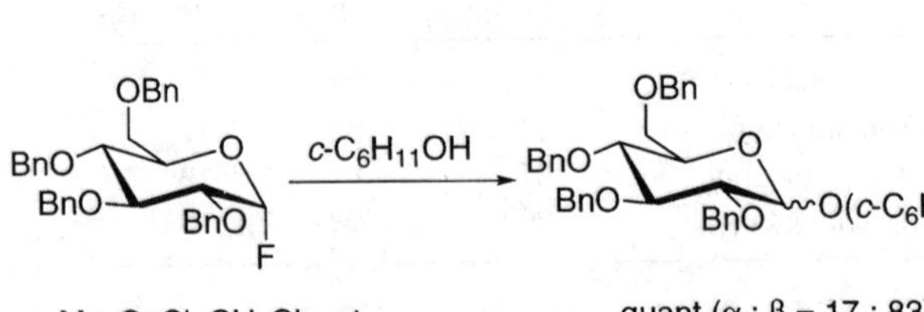

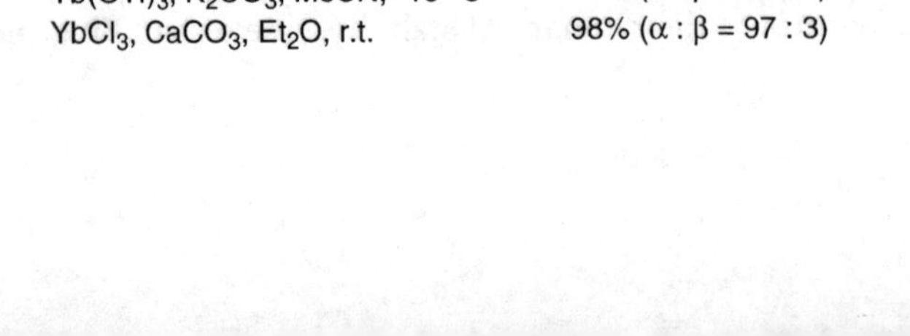

[12] and 2-thienylmagnesium bromide [13] give the corresponding C-glycosidation products with good stereoselectivities.

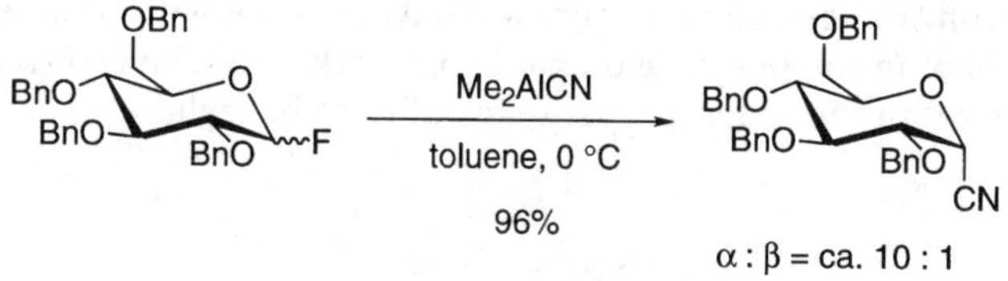

$$\text{(BnO, BnO, BnO, OBn)-F} \xrightarrow[\text{toluene, 0 °C}]{\text{Me}_2\text{AlCN}} \text{(BnO, BnO, BnO, OBn)-CN} \quad (4.5)$$

96% α : β = ca. 10 : 1

4.1.2
Aromatic Nucleophilic Substitution

Nucleophilic substitution is not common in aromatic compounds [14]. However, aromatic compounds substituted by electron-withdrawing groups (EWGs) at positions *ortho* and *para* to a leaving group undergo nucleophilic substitution ($S_N Ar$) smoothly (Eq. 4.6) [15].

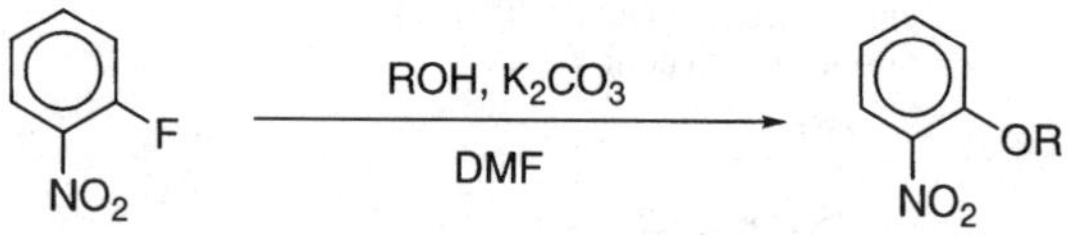

$$\text{o-F-C}_6\text{H}_4\text{-NO}_2 \xrightarrow[\text{DMF}]{\text{ROH, K}_2\text{CO}_3} \text{o-RO-C}_6\text{H}_4\text{-NO}_2 \quad (4.6)$$

ROH	Temp. (°C)	Yield (%)
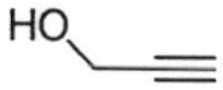 HO–CH₂–C≡CH	r.t.	85
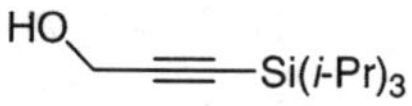 HO–CH₂–C≡C–Si(i-Pr)₃	r.t.	48
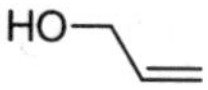 HO–CH₂–CH=CH₂	60	67
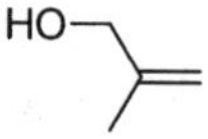 HO–CH(CH₃)=CH₂	80	91
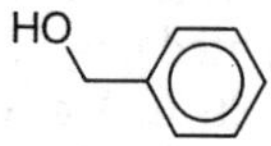 HO–CH₂–C₆H₅	110	97
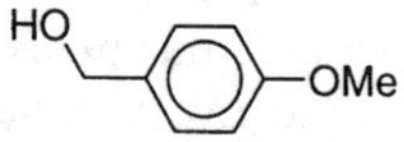 HO–CH₂–C₆H₄–OMe	105	72

A leaving group for $S_N Ar$ is NO_2, OR, or SR, in addition to halide, sulfate, sulfonate, and NR_3 that are common in aliphatic nucleophilic substitution. The most remarkable feature is that fluorine is the best leaving group. An approximate order of the leaving ability is $F > NO_2 > OTs > Cl, Br, I > N_3 > R_3N^+ > OR$. Providing that the $S_N Ar$ reaction occurs via an addition-elimination mechanism, it is most likely that the addition of a nucleophile is a rate-determining step and is accelerated by the strong inductive effect of an electron-withdrawing group as well as a fluorine substituent.

An alkoxide ion readily replaces a fluorine atom activated by an EWG. For example, 2-fluoronitrobenzene undergoes an S_NAr reaction with propargyl, allyl, or benzyl alcohol in the presence of K_2CO_3 in DMF [16]. The reactivity of the alcohol decreases in the order: propargylic > allylic > benzylic.

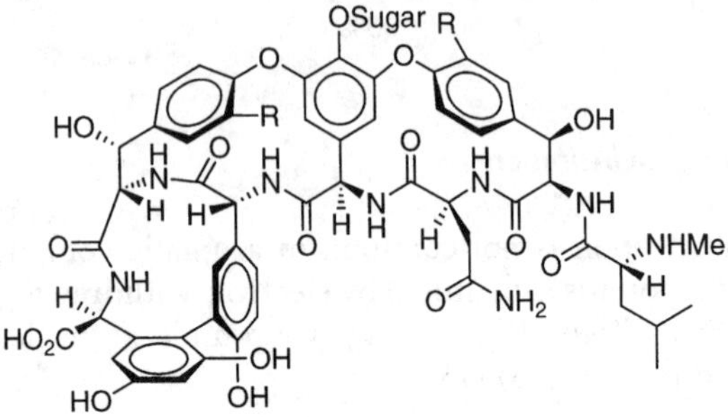

vancomycin aglycone (R = Cl)
orienticin aglycone (R = H)

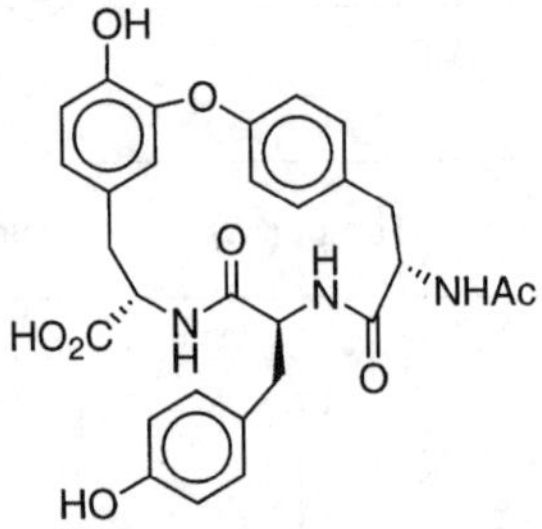

K-13

Fig. 4.1. Examples of macrocyclic biaryl ether compounds

Displacement of fluorine by a phenoxide ion constitutes a convenient way for the synthesis of diaryl ethers. An intramolecular S_NAr reaction of *o*-nitro-(fluoro)benzene with a phenoxide nucleophile is highlighted from the synthetic viewpoint of macrocyclic biaryl ether type natural products such as vancomycin, orienticin, and K-13 (Fig. 4.1) [17]. For example, treatment of a phenol derivative having an *ortho*-nitro(fluoro)benzene moiety with anhydrous potassium carbonate in DMF at 45–50 °C forms a 14-membered ring to give a diphenyl ether as the sole product [18]. It is worth noting that a high dilution is not necessary.

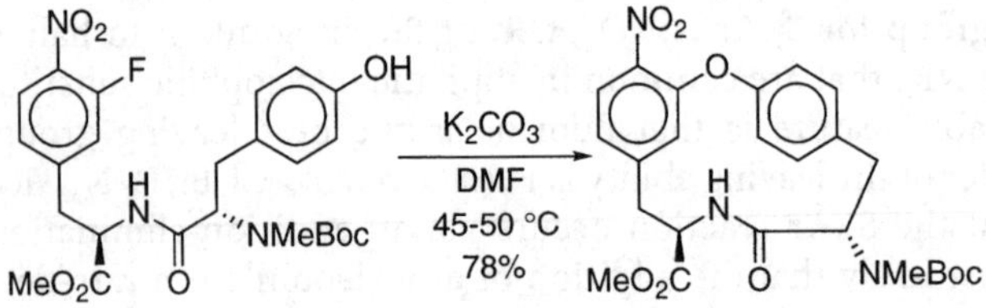

(4.7)

4.1 Fluorine Leaving Group

Both aliphatic and aromatic amines replace a fluorine atom in an aromatic ring when activated by *ortho-* and *para-*EWGs. For example, 2-cyanoaniline reacts with 4-nitro(fluoro)benzene with the aid of *t*-BuOK to give a diaryl amine [19], while 1-bromo-5-fluoro-2,3-dinitrobenzene gives an aniline derivative upon reaction with an allylic amine in the presence of $NaHCO_3$ at ambient temperatures [20].

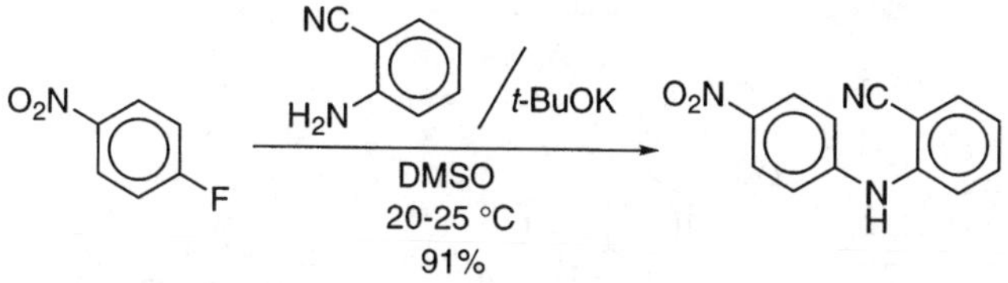

$$O_2N{-}C_6H_4{-}F + H_2N{-}C_6H_4{-}CN \xrightarrow[\substack{DMSO \\ 20\text{-}25\ °C \\ 91\%}]{t\text{-BuOK}} O_2N{-}C_6H_4{-}NH{-}C_6H_4{-}CN \quad (4.8)$$

$$\text{(O}_2\text{N, Br, NO}_2\text{-substituted fluorobenzene)} \xrightarrow[\substack{NaHCO_3 \\ dioxane\text{-}H_2O \\ 60\%}]{RNH_2} \text{(NHR product)} \quad (4.9)$$

$$RNH_2 \ :\ OHC{-}\underset{Me}{C}({=}CH){-}CH{=}CH{-}CH_2{-}NH_2 \quad (OMe)$$

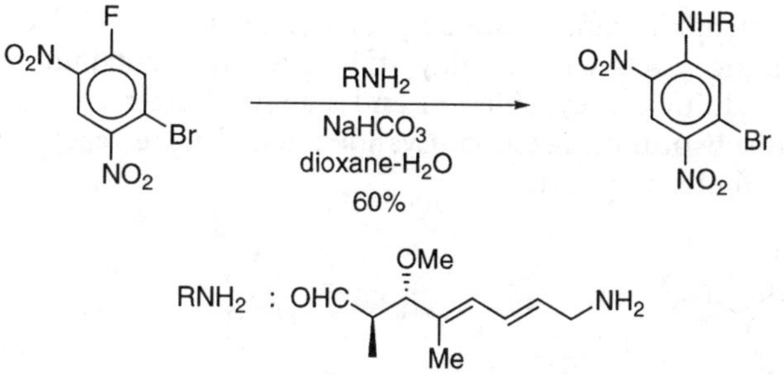

The substitution reaction is often accelerated by high pressure, as evidenced by the reactions shown in Scheme 4.1 [21].

An S_NAr reaction of 2,4-dinitrofluorobenzene is very useful for determining an N-terminal amino acid residue of polypeptides; the so-called Sanger method

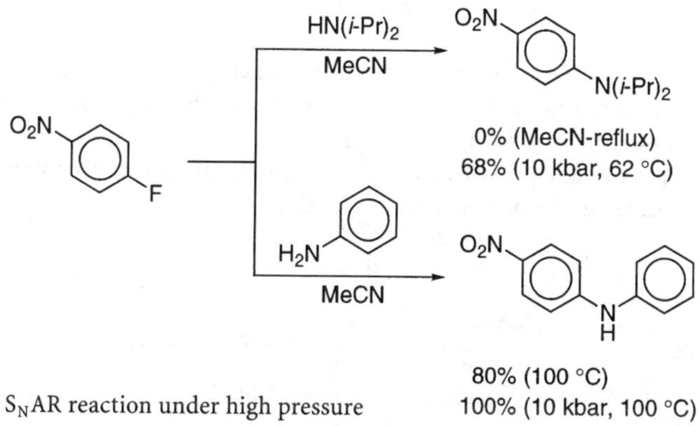

$$O_2N{-}C_6H_4{-}F \begin{cases} \xrightarrow[MeCN]{HN(i\text{-}Pr)_2} O_2N{-}C_6H_4{-}N(i\text{-}Pr)_2 \quad \begin{array}{l} 0\%\ (MeCN\text{-reflux}) \\ 68\%\ (10\ kbar,\ 62\ °C) \end{array} \\ \xrightarrow[MeCN]{H_2N{-}Ph} O_2N{-}C_6H_4{-}NH{-}Ph \quad \begin{array}{l} 80\%\ (100\ °C) \\ 100\%\ (10\ kbar,\ 100\ °C) \end{array} \end{cases}$$

Scheme 4.1. S_NAR reaction under high pressure

[22]. Thus, substitution with a free amino group of a polypeptide followed by hydrolysis gives a 2,4-dinitrophenyl-substituted amino acid which can be easily discriminated from other free amino acids.

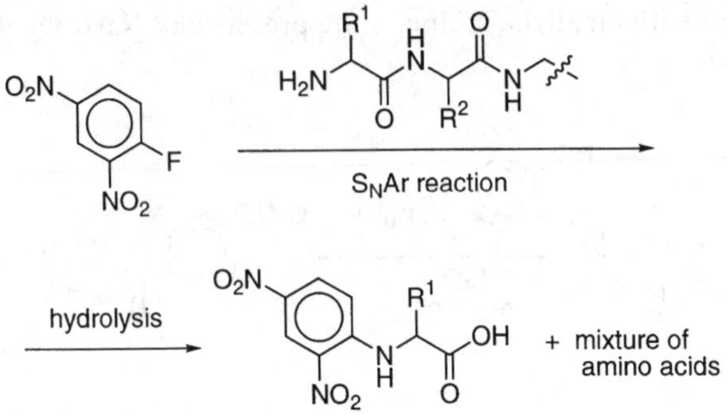

$$\text{(4.10)}$$

A carbon nucleophile can also undergo an $S_N Ar$ reaction [23]. For example, hexafluorobenzene reacts with an alkyllithium to give pentafluorotoluene in good yield. In addition, a cyanide ion and stabilized ester enolates are also capable of the substitution reaction. Cyanotrimethylsilane reacts with picryl fluoride to afford picryl cyanide.

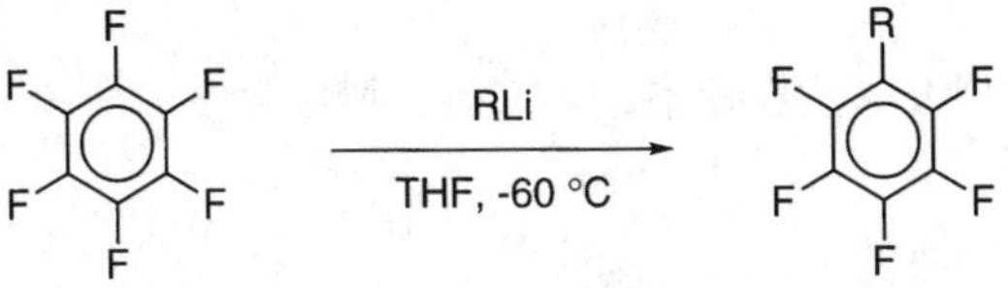

$$\text{(4.11)}$$

R = Me (70-75%); R = Bu (56%)

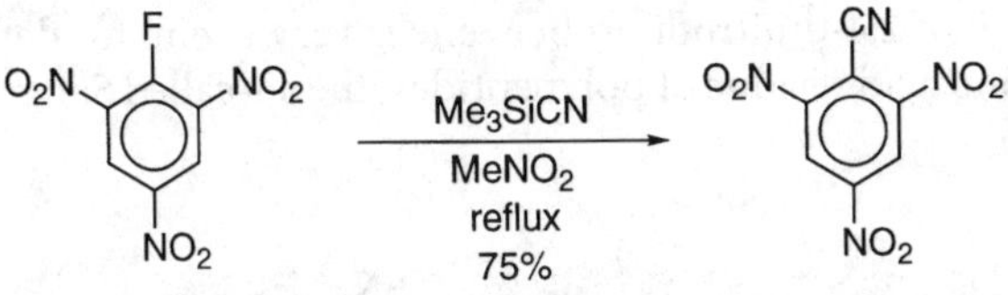

$$\text{(4.12)}$$

Fluorine-substituted aromatic substrates lacking an EWG can be activated towards an $S_N Ar$ reaction by complexation with chromium tricarbonyl. (Fluorobenzene)chromium tricarbonyl reacts with diethyl sodiomalonate smoothly to produce a substitution product [24]. An intramolecular version of this reaction gives a chromane structure [25].

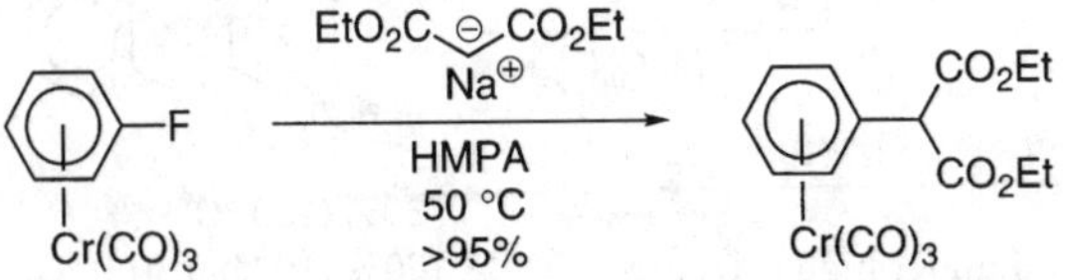

$$\text{(4.13)}$$

4.2
C–F Bond Activation by Metal Complexes

Due to the large bond energy of a carbon–fluorine bond, and to the high electronegativity of fluorine, perfluorinated hydrocarbons or fluorocarbons are inert, thermally stable, and useful in materials science. Very recently, growing interest has been focused on C–F bond activation by means of a metal complex, because the modification and functionalization of fluorocarbons via C–F bond activation should provide an efficient route to variously functionalized compounds useful for chemicals of industrial use [26]. In addition, activation of C–F bonds may lead to an effective destruction of chlorofluorocarbons that are responsible for global warming and ozone depletion. Thus, C–F bond activation is now studied extensively and is described here, excluding the reactions introduced in Chap. 4.1.

4.2.1
Activation of an Aliphatic C–F Bond

Activation of saturated fluorocarbons is relatively rare; in the past, the reaction was carried out under drastic conditions with little selectivity. Alkali metals react with fluorocarbons at 400–500 °C to give alkali metal fluorides and elemental carbon; perfluorodecalin upon treatment with iron at 450–500 °C gives perfluoronaphthalene [27]. Above 600 °C, the reaction of C_2F_6 with silica results in C–F bond cleavage to afford CO, CO_2, and SiF_4 [28].

Since a C–F bond possesses a low-lying σ*-orbital, a rational approach for the activation of a C–F bond is an electron-transfer process from a reducing reagent to fluorocarbons. In the past decade, therefore, several methods for C–F bond activation based on an electron-transfer reaction have been developed: the reactions are carried out under milder conditions with higher selectivity.

Treatment of polytetrafluoroethylene (PTFE) with lithium in liquid ammonia produces polyethylene (PE) in good yields [29]. Aromatization of a perfluorocarbon is seen in the reduction of perfluorodecalin with sodium benzophenone radical anion in THF to give perfluoronaphthalene [30]. The same transformation can be carried out with an excess amount of a reagent system consisting of cobaltocene and LiOTf (1:1) [31]. Use of cobaltocene alone yields cobaltocenium fluoride (Cp_2CoF), a very nucleophilic fluoride ion reagent. A catalytic version of the defluorination is a reaction carried out in THF solution at room temperature with a catalytic amount of Cp_2TiF_2 in the presence of $Al/HgCl_2$ as a terminal reductant [32]. The metallocene complex in this reaction serves as an *electron shuttle*, transporting electrons from the terminal reductant to the substrate.

$$\text{perfluorodecalin} \longrightarrow \text{perfluoronaphthalene} \tag{4.14}$$

Na[Ph$_2$CO], THF, -78 °C to rt, 62%
10 Cp$_2$Co/10 LiOTf, Et$_2$O, rt, 53%
Cp$_2$TiF$_2$/Al/HgCl$_2$, THF, rt, 52%

In a manner similar to the activation by a titanocene complex, bis(cyclopentadienyl)titanacyclobutane also activates a C(sp³)–F bond of tetrakis(trifluoromethyl)cyclopentadienone; the reagent and substrates are converted into 1,1-dimethylcyclopropane and a titanium fluoride complex [33].

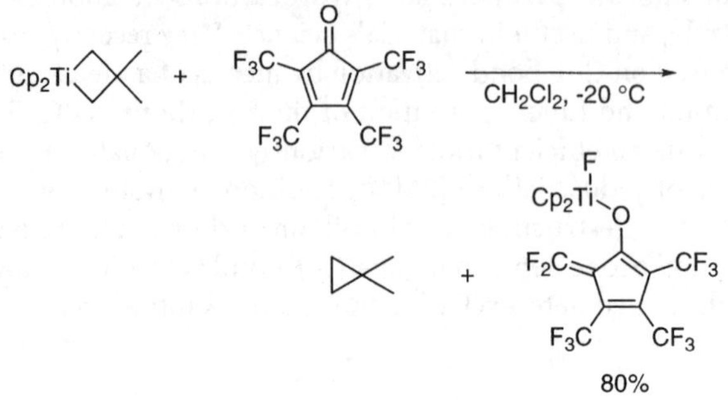

(4.15)

A convenient process for the dehalogenation of chlorofluorocarbons and perfluorocycloalkanes employs inexpensive and non-corrosive sodium oxalate as a reducing reagent [34]. By passing a stream of an inert gas saturated with a polyhalogenated substrate through a packed bed of powdered sodium oxalate at 270 °C, solid aromatic compounds are collected in an air-cooled condenser.

4.2.2
Activation of an Aromatic C–F Bond

Although aromatic C–F bonds are stronger in energy than aliphatic ones, more successful examples have been observed of C–F bond activation. The cross-coupling reaction of Grignard reagents with fluorobenzene is catalyzed by nickel-phosphine complexes. Thus, oxidative addition of a carbon–fluorine bond into nickel occurs. In this system, it is noteworthy that the reactivity of fluorobenzene is comparable with other halobenzenes [35a, b]. Hexafluorobenzene reacts with a Ni(PEt₃)₂(cod) complex to give in very low yield a thermally unstable square-planar $trans$-Ni(PEt$_3$)$_2$-(C$_6$F$_5$)(F) complex which gradually decomposes on standing at 30 °C [35c].

Intramolecular activation of a C–F bond through a chelation-assisted oxidative addition is observed in the reaction of the tungsten(0) complex W(CO₃)(NCEt)₃ with a (pentafluorophenyl)methylidene aniline to afford a stable tungsten(II) metallacycle with a fluorine atom bound to tungsten metal [36].

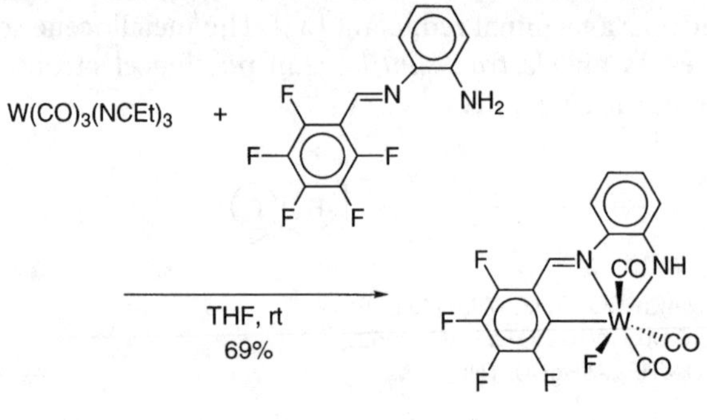

(4.16)

4.2 C–F Bond Activation by Metal Complexes

Intermolecular aromatic C–F bond activation is observed in the reaction of hexafluorobenzene with $(\eta^5\text{-CH}_3\text{C}_5\text{H}_4)_3\text{U}(t\text{-Bu})$ to give $(\eta^5\text{-CH}_3\text{C}_5\text{H}_4)_3\text{UF}$ along with C_6F_5 (t-Bu) (25%), isobutane (40%), isobutene (20%), and C_6F_5H (50%) [37]. Formation of such organic products suggests a radical mechanism. Similarly, hexafluorobenzene, upon reaction with a ruthenium dihydride complex, produces $L_2Ru(C_6F_5)H$ and HF, while a low-valent zirconium complex undergoes selective hydrogenolysis of aromatic fluorocarbons in good yields [38].

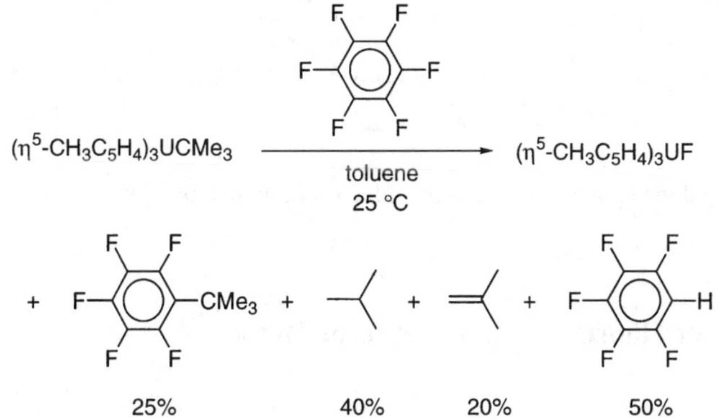

$$(\eta^5\text{-CH}_3\text{C}_5\text{H}_4)_3\text{UCMe}_3 \xrightarrow[25\ °C]{\text{toluene}} (\eta^5\text{-CH}_3\text{C}_5\text{H}_4)_3\text{UF} \quad (4.17)$$

+ C$_6$F$_5$–CMe$_3$ + isobutane + isobutene + C$_6$F$_5$H

25%   40%   20%   50%

Homogeneous *catalytic* C–F bond activation is achieved with a rhodium catalyst. For example, regioselective reduction of fluorine is attained by the reaction of C_6F_6 with $(\text{EtO})_3\text{SiH}$ in the presence of a $(\text{PMe}_3)_3\text{Rh}(C_6F_5)$ catalyst, giving pentafluorobenzene and $(\text{EtO})_3\text{SiF}$ [39].

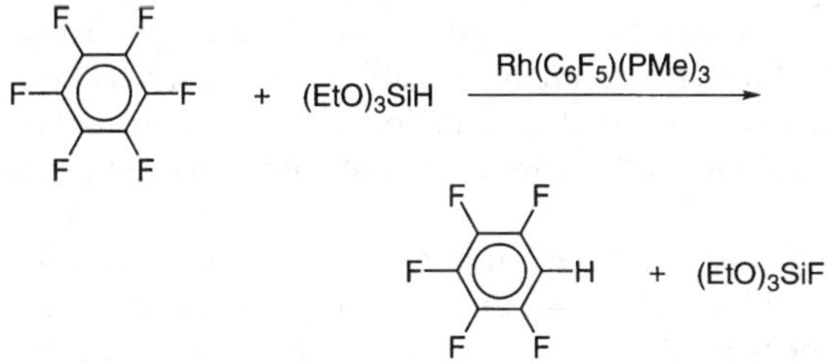

$$C_6F_6 + (\text{EtO})_3\text{SiH} \xrightarrow{\text{Rh}(C_6F_5)(\text{PMe}_3)} C_6F_5H + (\text{EtO})_3\text{SiF} \quad (4.18)$$

A mechanism proposed for this reaction is illustrated in Scheme 4.2. The hydrosilane may be replaced by a combination of molecular hydrogen and a base [40].

An silicon-fluorine exchange reaction of pentafluoroacetophenones is achieved with hexamethyldisilane and a $[\text{Rh}(\text{cod})_2]\text{BF}_4$ catalyst to induce selectively an *ortho* C–F bond cleavage [41]. Chelation assistance of a ketone carbonyl group is essential for the *ortho* C–F bond activation.

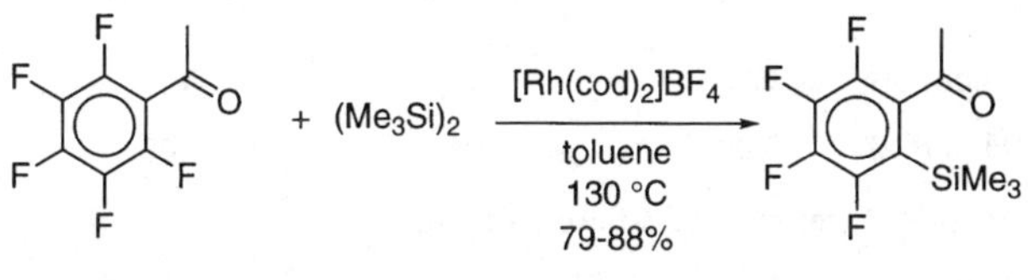

$$\text{C}_6\text{F}_5\text{C(O)CH}_3 + (\text{Me}_3\text{Si})_2 \xrightarrow[\substack{\text{toluene} \\ 130\ °C \\ 79-88\%}]{[\text{Rh}(\text{cod})_2]\text{BF}_4} \text{ortho-SiMe}_3\text{-C}_6\text{F}_4\text{C(O)CH}_3 \quad (4.19)$$

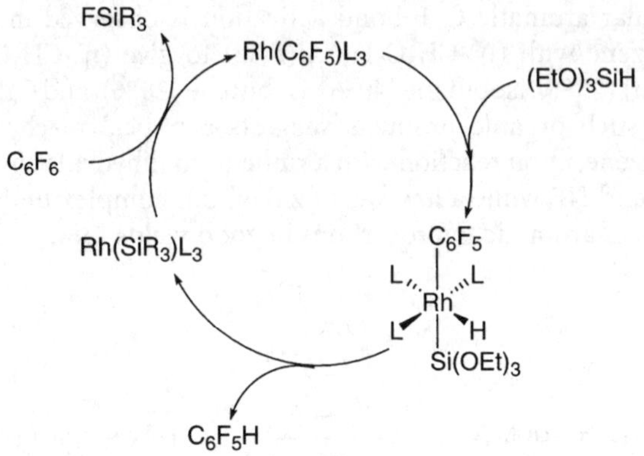

Scheme 4.2. Catalytic C–F bond activation with a rhodium complex

4.3
Interaction of Fluorine with a Proton or Metal

4.3.1
Fluorine-Hydrogen Interaction

Several efforts have been made to observe the possibility of a C–F⋯H–O bonding especially in 2-fluoroethanol [42, 43]. An electron diffraction study has shown that 2-fluoroethanol exists almost entirely (>90%) in a gauche conformation regarding fluorine and a hydroxyl group. In many cases, such intramolecular hydrogen bonding is proposed to explain the conformational preference of the gauche form. However, incompatible results are also available. For example, 2-fluoroethyl acetate that cannot form an intramolecular hydrogen bonding has also been shown to adapt a gauche conformation predominantly (>95%).

The relative energies of 2-fluoroethanol and an enol form of 2-fluoroacetaldehyde have been calculated (Figs. 4.2 and 4.3) [44]. The most stable structures for each are hydrogen-bonded conformers, GG and *cis-syn*, respectively. An energy difference of GG and GT conformers of 2-fluoroethanol is 1.93 kcal/mol, comparable to an average energy of a hydrogen bond. On the other hand, the energy gap of 0.12 kcal/mol between GT and TT conformers corresponds to a gauche effect. These results show that the predominance of a gauche form is due

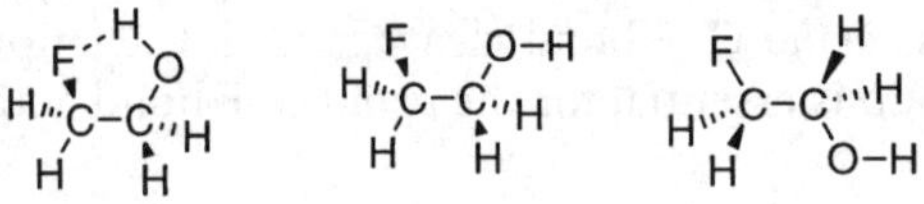

conformers	GG	GT	TT
ΔH (MP-2)	0.00	1.93	2.05

Fig. 4.2. Relative energies (kcal/mol) of 2-fluoroethanol

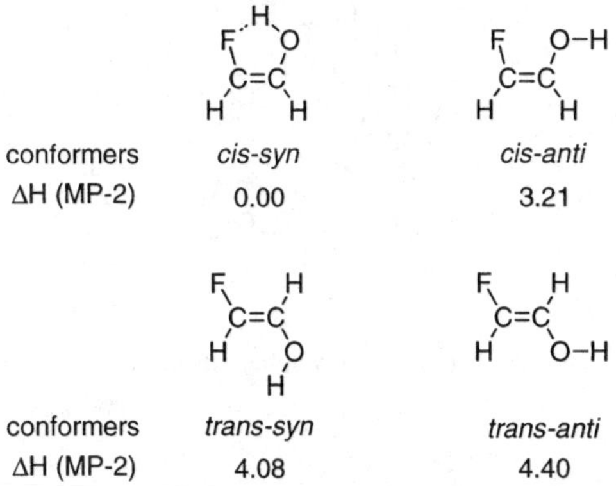

4.3 Interaction of Fluorine with a Proton or Metal

conformers	cis-syn	cis-anti
ΔH (MP-2)	0.00	3.21

conformers	trans-syn	trans-anti
ΔH (MP-2)	4.08	4.40

Fig. 4.3. Relative energies (kcal/mol) of 2-fluoroacetaldehyde enol

entirely to a hydrogen bonding. Of the enol forms of 2-fluoroacetaldehyde, a *cis-syn* conformer is 3.2 kcal/mol more stable than a *cis-anti* isomer. The energy difference between the *cis-syn* and *cis-anti* conformers is larger than that between GG and GT conformers. This may be attributed to the planar geometry of an enol of 2-fluoroacetaldehyde. Bond lengths between fluorine and hydrogen in the most stable conformers are 2.37 Å in the *cis-syn* conformers and 2.52 Å in the GG conformers; both are less than the sum of the van der Waals radii of H and F, indicating that a fluorine-hydrogen interaction clearly exists.

4.3.2
Fluorine-Metal Interaction

Interaction of fluorine with metal has also been assumed in order to understand the unique reactions and biological behavior of organofluorine compounds, and experimental and extensive theoretical studies on such an interaction have been made.

A survey of the X-ray database of the Cambridge Crystallographic Data Center has suggested possible interaction between fluorine and alkali metal [45]. There are seven compounds containing a C–F bond *and* alkali metal within a molecule; all are shown to have one or more contact(s) M(metal)···F with a length similar to an ionic $M^+ \cdots F^-$ interaction. Recently, 2,4,6-tris(trifluoromethyl)phenyllithium has been prepared and characterized by a low-temperature single-crystal X-ray analysis (Fig. 4.4) [46], which revealed that the lithium compound exists as a dimer. The two lithiums are coordinated by two fluorine atoms of *ortho*-CF_3 groups with the atomic distance being 2.227–2.293 Å.

Interaction between an alkali earth metal and fluorine is found in a fluorinated macrocyclic crown ether. X-ray analysis of the complex, 21-fluoro-2,5,8,11,14-pentaoxa[15]metacyclophane·$Ba(ClO_4)_2$, shows that a barium atom is coordinated to a fluorine atom with an atomic distance of 2.799(8) Å (Fig. 4.5) [47].

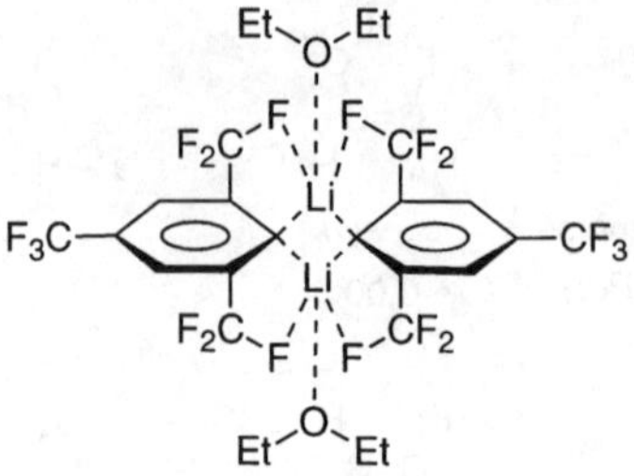

Fig. 4.4. Structure of 2,4,6-tris(trifluoromethyl)phenyllithium

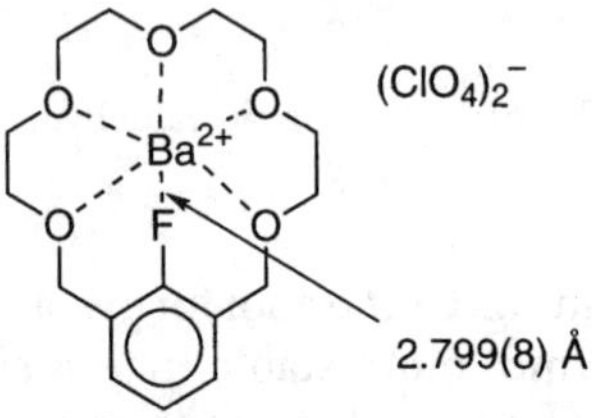

Fig. 4.5. Structure of the 21-fluoro-2,5,8,11,14-pentaoxa[15]metacyclophane·Ba(ClO$_4$)$_2$ complex

Interaction between fluorine and a transition metal is also observed in metal complexes such as Ru(η^2-SC$_6$F$_5$)(η^1-SC$_6$F$_5$)(PMe$_2$Ph) and IrH$_2$(Me$_2$CO)$_2$-(8-FC$_9$H$_6$N)$_2$. The respective atomic distances between the metal and fluorine are Ru–F = 2.489(6) Å and Ir–F = 2.514(8) Å, both shorter than the sum of the van der Waals radii [48].

Relevant to the calculation of the fluorine-substituted enol described in Chap. 4.3.1, the energies of the alkali metal enolates of 2-fluoroacetaldehyde have been computed (Table 4.1) [49]. At an MP-2 level, an alkali metal such as lithium, sodium, or potassium is suggested to stabilize the *cis*-enolate by

Table 4.1. Relative energies (kcal/mol) and bond lengths (Å) of 2-fluoroacetaldehyde enolate

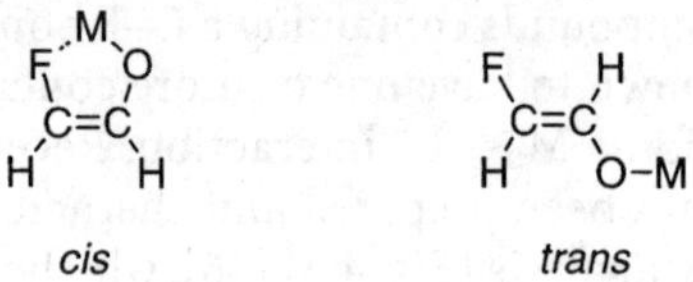

M	ΔE (*trans-cis*) MP-2	r(M–F) cis	r(M–O) cis
–	1.8	–	–
Li	13.2	1.86	1.75
Na	13.6	2.21	2.10
K	10.8	2.62	2.47

11–14 kcal/mol. This suggests a strong interaction between fluorine and an alkali metal. Such interaction is also demonstrated by the small difference in M–F and M–O bond lengths: 0.11 Å for M=Li and Na, and 0.15 Å for M=K.

Interaction of a lithium ion with a trifluoromethyl group in a substrate often induces stereoselective reactions [50]. For example, oxidation of an enolate of ethyl 3-methyl-4,4,4-trifluorobutyrate with an MoO_5-pyridine-HMPA complex gives an *anti*-α-hydroxyl ester with high stereoselectivity and is reasonably explained by chelation between a lithium ion and fluorine(s) of a trifluoromethyl group [50a].

Similar intramolecular fluorine-lithium interaction has been estimated by ab initio calculations of a lithium enolate model in order to explain the fact that ethyl 3-trifluoromethylacrylate is a good Michael acceptor even towards ketone enolates (Fig. 4.6) [51]. Four conformers of a (Z)-enolate are shown in Fig. 4.6; two of them involve an interaction between fluorine and lithium and are several kcal/mol more stable than the other two which lack such an interaction. In particular, the most stable one is a structure in which two fluorine atoms chelate a lithium atom connected to the enol oxygen.

Furthermore, ab initio molecular orbital (MO) calculations on a complete set of diastereomeric transition structures for the addition of lithium hydride to

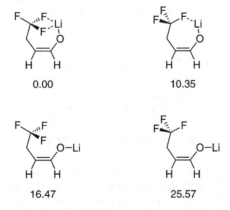

Fig. 4.6. Relative energies (kcal/mol) of a lithium enolate model (6–31G*/3–21G)

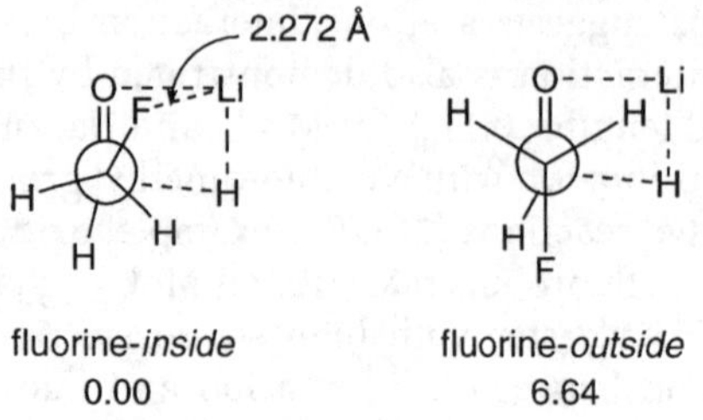

fluorine-*inside*
0.00

fluorine-*outside*
6.64

Fig. 4.7. Transition state structure and the relative energies (kJ/mol) for lithium hydride addition to 2-fluoroacetaldehyde

2-fluoroacetaldehyde demonstrate the possibility of an intermolecular fluorine-lithium interaction (Fig. 4.7) [52]. A transition state with fluorine inside is suggested to be the most stable one in which the distance between F and Li is only 2.27 Å. This transition state is 6.64 kJ/mol more stable than the one with fluorine outside. This conclusion contrasts sharply with the Anh–Eisenstein model that suggests that an antiperiplanar fluorine transition structure should be the most stable. These results demonstrate that the intermolecular electrostatic interaction between fluorine and lithium atoms dominates the Anh–Eisenstein electronic effect in determining the diastereoselectivity of the carbonyl addition reactions.

Dehydrobromination of 2-aryl-substituted 1,1-dibromo-1,2-difluoroethane with a base proceeds stereoselectively giving rise to difluoroolefins [53]. Thus, with a metal-containing base such as lithium 2,2,6,6-tetramethylpiperidide, (E)-olefins are produced exclusively, while (Z)-olefins are selectively formed with a metal-free base such as Bu_4NOH.

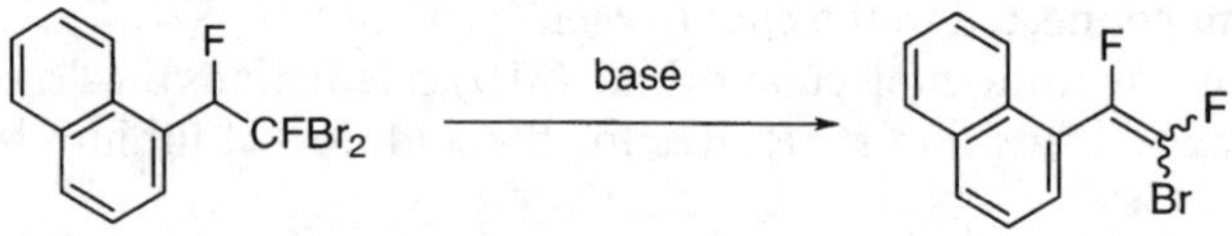

(4.21)

$LiN[(CMe_2CH_2)_2CH_2]$, THF, -98 °C, 82% (E : Z = >99 : <1)
Bu_4NOH, CH_2Cl_2/H_2O, -78 °C, 42%, (E : Z = 13 : 87)

These results are reasonably explained by an intermolecular lithium-fluorine interaction, as suggested in Scheme 4.3.

Deprotonation of fluorobenzene with an organolithium occurs regioselectively at an *ortho* position [54]. Examples are shown below. This phenomenon is explained in terms of a F-Li interaction, and ab initio calculations of a model reaction involving fluorobenzene and lithium hydride have been carried out (Fig. 4.8) [55].

A coordinated complex is computed to be 12.4 kcal/mol more stable than a separated species. The activation energy for *ortho* lithiation is estimated to be 18.9 kcal/mol relative to the coordination complex, while the corresponding value for *para* lithiation is 30.7 kcal/mol. It is noteworthy that the atomic

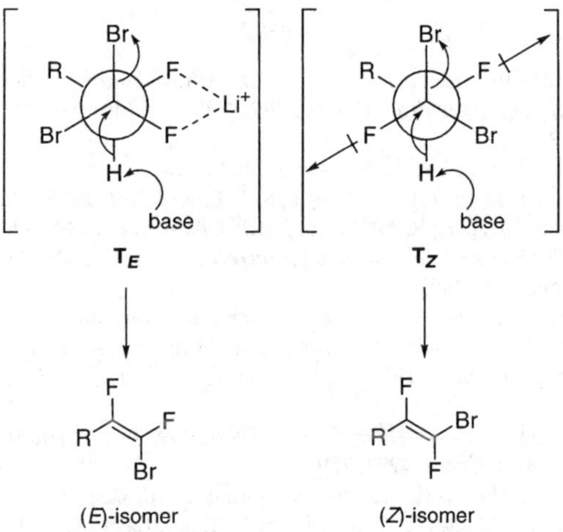

4.3 Interaction of Fluorine with a Proton or Metal

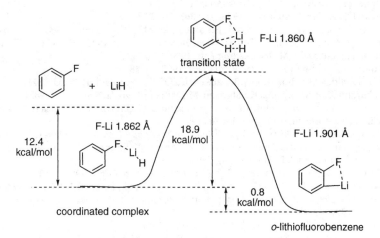

Scheme 4.3. Transition states of dehydrobromination

Fig. 4.8. Calculated structures for the lithiation of fluorobenzene with lithium hydride

distance of 1.860 Å between fluorine and lithium in a transition structure is shorter than those of the coordinated complex (1.862 Å) and *ortho*-lithiofluorobenzene (1.901 Å). These results suggest that the regioselectivity in the lithiation of fluoroaromatics is controlled by a fluorine-lithium interaction.

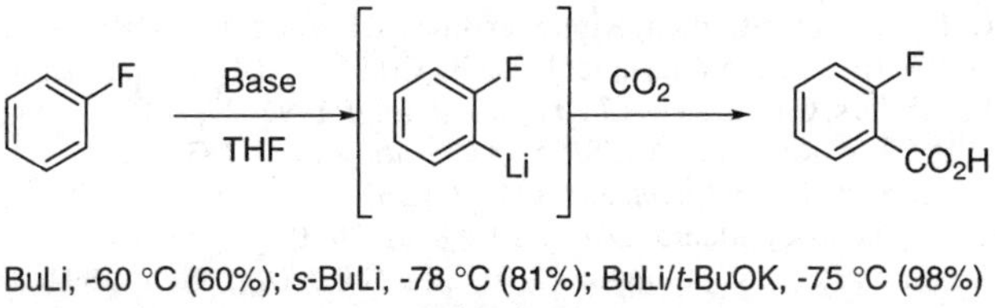

(4.22)

BuLi, -60 °C (60%); *s*-BuLi, -78 °C (81%); BuLi/*t*-BuOK, -75 °C (98%)

References

1. Reviews of glycosyl fluoride: (a) Penglis, A.A.E. *Adv. Carbohydr. Chem. Biochem.* **1981**, *38*, 195; (b) Card, P.J. *J. Carbohydrate Chem.* **1985**, *4*, 451; (c) Shimizu, M.; Togo, H.; Yokoyama, M. *Synthesis* **1998**, 799
2. Mukaiyama, T.; Murai, Y.; Shoda, S.-i. *Chem. Lett.* **1981**, 431
3. (a) Mukaiyama, T.; Hashimoto, Y.; Shoda, S.-I. *Chem. Lett.* **1983**, 935; (b) Nicolaou, K.C.; Dolle, R.E.; Papahatjis, D.P.; Randall, J.L. *J. Am. Chem. Soc.* **1984**, *106*, 4189
4. (a) Hashimoto, S.; Hayashi, M.; Noyori, R. *Tetrahedron Lett.* **1984**, *25*, 1379; (b) Noyori, R.; Hayashi, M. *Chem. Lett.* **1987**, 57
5. (a) Nicolaou, K.C.; Dolle, R.E.; Chucholowski, A.; Randall, J.L. *J. Chem. Soc., Chem. Commun.* **1984**, 1153; (b) Nicolaou, K.C.; Chucholowski, A.; Dolle, R.E.; Randall, J.L. *J. Chem. Soc., Chem. Commun.* **1984**, 1155; (c) Kunz, H.; Sager, W. *Helv. Chim. Acta* **1985**, *68*, 283
6. (a) Kreuzer, M.; Thiem, J. *Carbohydr. Res.* **1986**, *149*, 347; (b) Junnemann, J.; Lundt, I.; Thiem, J. *Liebigs Ann. Chem.* **1991**, 759
7. (a) Matsumoto, T.; Maeta, H.; Suzuki, K.; Tsuchihashi, G.-i. *Tetrahedron Lett.* **1988**, *29*, 3567; (b) Suzuki, K.; Maeta, H.; Matsumoto, T.; Tsuchihashi, G.-i. *Tetrahedron Lett.* **1988**, *29*, 3571; (c) Matsumoto, T.; Katsuki, M.; Suzuki, K. *Chem. Lett.* **1989**, 437
8. (a) Kobayashi, S.; Koide, K.; Ohno, M. *Tetrahedron Lett.* **1990**, *31*, 2435; (b) Koide, K.; Ohno, M.; Kobayashi, S. *Synthesis* **1996**, 1175
9. Wessel, H.P. *Tetrahedron Lett.* **1990**, *31*, 6863
10. (a) Hosono, S.; Kim, W.-S.; Sasai, H.; Shibasaki, M. *J. Org. Chem.* **1995**, *60*, 4; (b) Kim, W.-S.; Hosono, S.; Sasai, H.; Shibasaki, M. *Tetrahedron Lett.* **1995**, *36*, 4443; (c) Kim, W.-S.; Sasai, H.; Shibasaki, M. *Tetrahedron Lett.* **1996**, *37*, 7797
11. Bohm, G.; Waldmann, H. *Tetrahedron Lett.* **1995**, *36*, 3843
12. Macdonald, S.J.F.; Huizinga, W.B.; McKenzie, T.C. *J. Org. Chem.* **1988**, *53*, 3371
13. Yokoyama, M.; Toyoshima, H.; Shimizu, M.; Mito, J.; Togo, H. *Synthesis* **1998**, 409
14. March, J. *Advanced Organic Chemistry. Reactions Mechanisms, and Structure*; 4th edn.; Wiley, New York; chap 13, pp 641–676 (1992)
15. Patrick, T.B. In: *Chemistry of Organic Fluorine Compounds II. A Critical Review*; Hudlicky, M.; Pavlath, A.E., Eds.; American Chemical Society:Washington, D.C., pp 501–524 (1995)
16. Raeppel, S.; Raeppel, F.; Suffert, J. *Synlett* **1998**, 794
17. Burgess, K.; Lim, D.; Martinez, C.I. *Angew. Chem. Int. Ed. Engl.* **1996**, *35*, 1077
18. (a) Beugelmans, R.; Zhu, J.; Husson, N.; Bois-Choussy, M.; Singh, G.P. *J. Chem. Soc., Chem. Commun.* **1994**, 439; (b) Boger, D.L.; Zhou, J.; Borzilleri, R.M.; Nukui, S.; Castle, S.L. *J. Org. Chem.* **1997**, *62*, 2054
19. Gorvin, J.H. *J. Chem. Soc., Perkin Trans. 1* **1988**, 1331
20. Maehr, H.; Blount, J.F.; Leach, M.; Stempel, A. *Helv. Chim. Acta* **1974**, *57*, 936
21. Kotsuki, H.; Kobayashi, S.; Matsumoto, K.; Suenaga, H.; Nishizawa, H. *Synthesis* **1990**, 1147
22. Solomons, T.W.G. *Organic Chemistry*; 6th edn.; Wiley, New York; chap 24, pp 1156–1157 (1996)
23. (a) Birchall, J.M., Haszeldine, R.N. *J. Chem. Soc.* **1961**, 3719; (b) Chaykovsky, M.; Adolph, H.G. *Synth. Commun.* **1986**, *16*, 205; (c) Filler, R.; Fiebig, A.E.; Pelister, M.Y. *J. Org. Chem.* **1980**, *45*, 1290
24. Semmelhack, M.F.; Hall, H.T. *J. Am. Chem. Soc.* **1974**, *96*, 7091
25. (a) Houghton, R.P.; Voyle, M.; Price, R. *J. Chem. Soc., Chem. Commun.* **1980**, 884; (b) Houghton, R.P.; Voyle, M. *J. Organomet. Chem.* **1983**, *259*, 183
26. Reviews of C–F bond activation: (a) Kiplinger, J.L.; Richmond, T.G.; Osterberg, C.E. *Chem. Rev.* **1994**, *94*, 373; (b) Burdenuiuc, J.; Jedlicka, B.; Crabtree, R.H. *Chem. Ber./Recueil* **1997**, *130*, 145; (c) Saunders, G.C. *Angew. Chem. Int. Ed. Engl.* **1996**, *35*, 2615
27. Gething, B.; Patrick, C.R.; Stacey, M.; Tatlow, J.C. *Nature* **1959**, *183*, 588
28. White, L. Jr.; Rice, O.K. *J. Am. Chem. Soc.* **1947**, *69*, 267
29. Chakrabarti, N.; Jacobus, J. *Macromolecules* **1988**, *21*, 3011
30. Marsella, J.A.; Gilicinski, A. G.; Coughlin, A.M.; Pez, G.P. *J. Org. Chem.* **1992**, *57*, 2856

31. Bennett, B.K.; Harrison, R.G.; Richmond, T.G. *J. Am. Chem. Soc.* **1994**, *116*, 11165
32. Kiplinger, J.L.; Richmond, T.G. *J. Am. Chem. Soc.* **1996**, *118*, 1805
33. Burk, M.J.; Staley, D.L.; Tumas, W.*J. Chem. Soc., Chem. Commun.* **1990**, 809
34. Burdeniuc, J.; Crabtree, R.H. *Science* **1996**, *271*, 340
35. (a) Kiso, Y.; Tamao, K.; Kumada, M. *J. Organomet. Chem.* **1973**, *50*, C12; (b) Tamao, K.; Sumitani, K.; Kiso, Y.; Zembayashi, M.; Fujioka, A.; Kodama, S.-i.; Nakajima, I.; Minato, A.; M, K. *Bull. Chem. Soc. Jpn.* **1976**, *49*, 1958; (c) Fahey, D.R.; Mahan, J.E. *J. Am. Chem. Soc.* **1977**, *99*, 2501
36. (a) Richmond, T.G.; Osterberg, C.E.; Arif, A.M. *J. Am. Chem. Soc.* **1987**, *109*, 8091; (b) Lucht, B.L.; Poss, M.J.; King, M.A.; Richmond, T.G. *J. Chem. Soc., Chem. Commun.* **1991**, 400
37. Weydert, M.; Andersen, R.A.; Bergman, R.G. *J. Am. Chem. Soc.* **1993**, *115*, 8837
38. (a) Whittlesey, M.K.; Perutz, R.N.; Moore, M.H. *Chem. Commun.* **1996**, 787; (b) Kiplinger, J.L.; Richmond, T.G. *Chem. Commun.* **1996**, 1115
39. Aizenberg, M.; Milstein, D. *Science* **1994**, *265*, 359
40. (a) Aizenberg, M.; Milstein, D. *J. Am. Chem. Soc.* **1995**, *117*, 8674; (b) Young, B.J. Jr.; Grushin, V.V. *Organometallics* **1999**, *18*, 294
41. Ishii, Y.; Chatani, N.; Yorimitsu, S.; Murai, S. *Chem. Lett.* **1998**, 157
42. Review of interaction of fluorine with proton or metal: Yamazaki, T.; Kitazume, T. *Yuuki Gosei Kagaku Kyokaishi* **1996**, *54*, 665
43. (a) Krueger, P.J.; Mettee, H.D. *Can. J. Chem.* **1964**, *42*, 326; (b) Buckton, K. S.; Azrak, R.G. *J. Chem. Phys.* **1970**, *52*, 5652; (c) Hagen, K.; Hedberg, K. *J. Am. Chem. Soc.* **1973**, *95*, 8263; (d) Abraham, R.J.; Monasterios, J.R. *Org. Mag. Res.* **1973**, *5*, 305; (e) Griffith, R.C.; Roberts, J.D. *Tetrahedron Lett.* **1974**, 3499; (f) Hobza, P.; Mulder, F.; Sandorfy, C. *J. Am. Chem. Soc.* **1981**, *103*, 1360; (g) Curtiss, L.A.; Frurip, D.J.; Blander, M. *J. Am. Chem. Soc.* **1983**, *100*, 79; (h) Huang, J.; Hedberg, K. *J. Am. Chem. Soc.* **1989**, *111*, 6909
44. Dixon, D.A.; Smart, B.E. *J. Phys. Chem.* **1991**, *95*, 1609
45. Murray-Rust, P.; Stallings, W.C.; Monti, C.T.; Preston, R.K.; Glusker, J.P. *J. Am. Chem. Soc.* **1983**, *105*, 3206
46. Stalke, D.; Whitmire, K.H. *J. Chem. Soc., Chem. Commun.* **1990**, 833
47. Plenio, H.; Diodone, R. *Angew. Chem. Int. Ed. Engl.* **1994**, *33*, 2175
48. (a) Catala, R.M.; CruzGarritz, D.; Hills, A.; Hughes, D.L.; Richards, R.L.; Sosa, P.; Torrens, H. *J. Chem. Soc., Chem. Commun.* **1987**, 261; (b) Kulawiec, R.J.; Holt, E.M.; Lavin, M.; Crabtree, R.H. *Inorg. Chem.* **1987**, *26*, 2559
49. Dixon, D.A.; Smart, B.E. In *Selective Fluorination in Organic and Bioorganic Chemistry*; Welch, J.T., Ed.; ACS, Washington, D.C.; pp 18–35 (1991)
50. (a) Morizawa, Y.; Yasuda, A.; Uchida, K. *Tetrahedron Lett.* **1986**, *27*, 1833; (b) Hanamoto, T.; Fuchikami, T. *J. Org. Chem.* **1990**, *55*, 4969
51. Yamazaki, T.; Haga, J.; Kitazume, T.; Nakamura, S. *Chem. Lett.* **1991**, 2171
52. Wong, S.S.; Paddon-Row, M.N. *J. Chem. Soc., Chem. Commun.* **1991**, 327
53. (a) Kuroboshi, M.; Yamada, N.; Takebe, Y.; Hiyama, T. *Tetrahedron Lett.* **1995**, *36*, 6271; (b) Shimizu, M.; Yamada, N.; Takebe, Y.; Hata, T.; Kuroboshi, M.; Hiyama, T. *Bull. Chem. Soc., Jpn* **1998**, *71*, 2903
54. (a) Gilman, H.; Soddy, T.S. *J. Org. Chem.* **1957**, *22*, 1715; (b) Schlosser, M.; Katsoulos, G.; Takagishi, S. *Synlett* **1990**, 747; (c) Bridges, A.J.; Lee, A.; Maduakor, E.C.; Schwartz, C.E. *Tetrahedron Lett.* **1992**, *33*, 7495
55. Hommes, N.J.R. v. E.; Schleyer, P. v. R. *Angew. Chem. Int. Ed. Engl.* **1992**, *31*, 755

CHAPTER 5

Biologically Active Organofluorine Compounds

5.1
Fluorine Effect in Biological Activity

One of the important applications of fluorinated organic compounds is in medicinal chemistry. Recent progress in organofluorine chemistry has contributed significantly to the great advances in modern medical treatments. With the aid of the known influence of a fluorine atom on physical, chemical, and biological phenomena therapeutic efficacy has been increased and pharmacological properties improved [1–3].

Resources of naturally occurring fluorinated organic compounds are limited to a small number of tropical and subtropical plants and microorganisms. They are monofluoroacetic acid, ω-fluoro fatty acids, nucleocidin, and 4-fluorothreonine, etc. (Fig. 5.1). In 1944, Marais identified fluoroacetate in the leaves of the South African shrub *Dichapetalum cymosum*, and its biomedical mechanism of toxicity in mammals was elucidated by the research of Peters [4].

In 1953, Fried published a pioneering work on the preparation of 9α-fluorohydrocortisone acetate (Fig. 5.2) [5]. Utilization of fluorination as a tool to en-

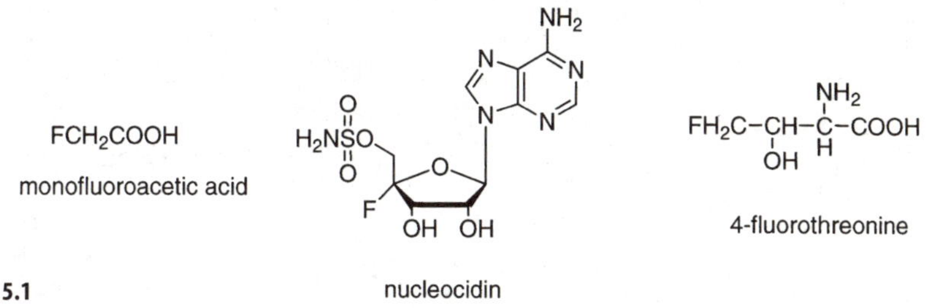

Fig. 5.1 FCH$_2$COOH monofluoroacetic acid H$_2$NSO (nucleocidin) FH$_2$C–CH–C–COOH with OH and H 4-fluorothreonine

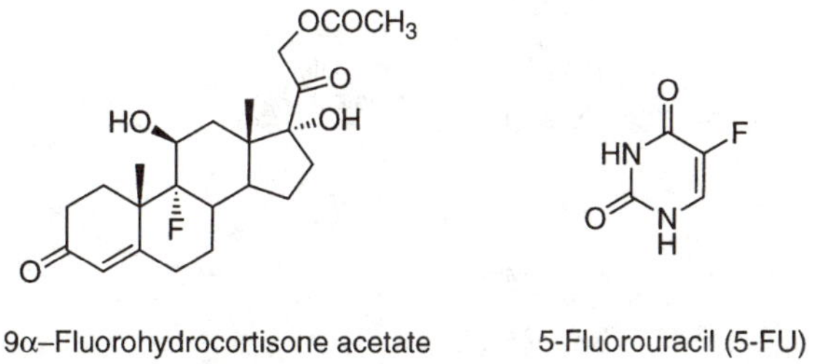

Fig. 5.2 9α–Fluorohydrocortisone acetate   5-Fluorouracil (5-FU)

hance biological activity and to improve the versatility of a biologically active substance was thus clearly demonstrated. The report by Fried stimulated pharmaceutical researchers to routinely introduce fluorine as a substituent to modify original biological activity.

Another significant finding in the late 1950s was that 5-fluorouracil (5-FU, Fig. 5.2) exhibits significant tumor-inhibiting activity [6], a discovery made through the synthesis of novel nucleic acids by substituting fluorine for hydrogen in naturally occurring pyrimidines and purines. 5-FU and its nucleoside, 5-fluoro-2′-deoxy-β-uridine (FdUR), are anabolized to 5-fluoro-2′-deoxyuridylate (FdUMP), and the FdUMP competitively inhibits thymidylate synthetase, an enzyme which normally converts 2′-deoxyuridylic acid into thymidylic acid (dTMP), an essential component of DNA (Scheme 5.1). 5-FU has been employed with success for the treatment of human breast cancer and several other types of malignancies [1e].

In the 1960s and 1970s, the development of novel fluorine-containing drugs and biomedical applications became a steady stream thanks to new reagents and techniques for site-selective introduction of fluorine into organic molecules. The development of new fluorinating agents and modifications to fluori-

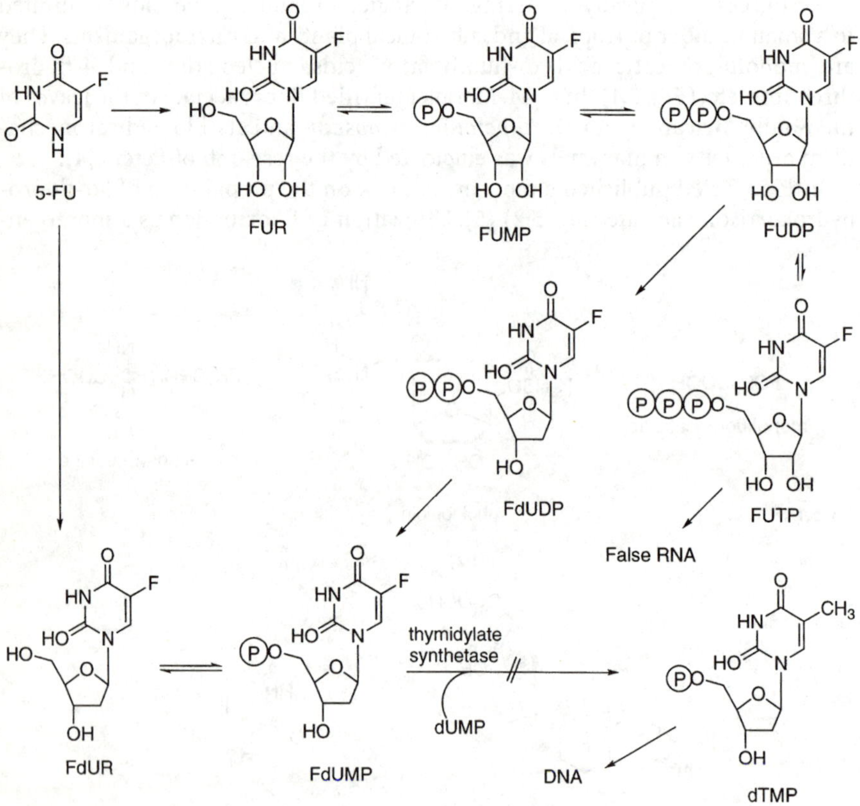

Scheme 5.1. Biological mechanism of 5-FU

nation procedures using conventional agents contributed greatly to the present rapid progress in this field.

An understanding of the fundamental biochemical mechanisms, coupled with knowledge of the physicochemical properties affected by fluorine substitution, has aided the rational design of many pharmaceutical drugs and pharmacological agents. Fluorinated analogs also serve as excellent probes for biochemical mechanisms. For example, many ^{19}F-NMR studies have demonstrated the utility of fluorine-labeled proteins as mechanistic tools [7]. A positron-emitting isotope of ^{18}F is also used to label a variety of organic molecules for studying biochemical transformations and distribution in the mammalian body.

A wide variety of effective fluoromedicines have been developed and put into the pharmaceutical marketplace, including steroidal and non-steroidal (NSAIDs) anti-inflammatory agents, anticancer and antiviral agents, antihypertensive agents, and central nervous system drugs for the management of mental illnesses such as depression and psychoses.

Fluorine has a small van der Waals radius of 1.35 Å (Pauling), close to that of hydrogen (1.20 Å), and thus mimics hydrogen at an enzyme receptor site with respect to steric requirement. On the other hand, the van der Waals radius estimated by Bondi is 1.47 Å, only 20% larger than that (1.20 Å) of hydrogen [8]. The C–F bond length, 1.38 Å, is comparable to that of the C–H, 1.10 Å, and C–O bond lengths, 1.43 Å. This means that such substitution will have little effect on the steric bulk of a molecule. However, the steric effect (Es) of a CF_3 group, according to an estimation based on the hydrolysis constant of α-substituted (R)-acetic acid ester, is –2.40, closer to –1.78 of an isopropyl group and larger than –1.24 of a methyl group (Table 5.1) [9, 10].

Table 5.1. Substituent constants of aliphatic and aromatic compounds

Substituent (R)	Steric effect (Es)[a]	Aliphatic compound hydrophobicity (π_R)[b]	Aromatic compound hydrophobicity (π_R)[b]
H	0.00	0.00	0.00
F	–0.46	–0.16	0.14
Cl	–0.97	–0.17	0.71
Br	–1.16	–0.03	0.86
OH	–0.55	–1.87	–0.49
NO_2	–1.01	–1.39	–0.28
CF_3	–2.40	0.06	0.88
CH_3	–1.24	0.54	0.56
C_2H_5	–1.31	1.08	1.02
$(CH_3)_2CH$	–1.78		
$(CH_3)_3C$	–2.78		1.98
CN	–0.51	–1.50	–0.57

[a] Es: Steric effect, estimated based on the hydrolysis constant of α-substituted (R)-acetic acid ester.
[b] π_R: Hydrophobicity; $\pi_R = \log P_R - \log P_H$, P = partition coefficient between 1-octanol and water.

Fluorine is the most electronegative element: the Pauling electronegativity is 4.0; cf. 3.5 for oxygen, 3.0 for chlorine and 2.8 for bromine. The high electronegativity of fluorine influences the electron distribution in a molecule, affecting the acidity or basicity of the neighboring group, the dipole moment within the molecule, and the overall reactivity and stability and, consequently, changes the chemical and physical properties drastically.

A C–F bond is the strongest among halogen–carbon bonds: heat of formation of a C–F bond is 456–486 kJ/mol; that of a C–Cl bond is roughly 350 kJ/mol, comparable to a C–H bond of 356–435 kJ/mol. The strong bond energy of C–F bonds contributes to the high thermal and oxidative stabilities of organofluorine compounds.

Introduction of a fluorine atom usually increases lipid solubility, enhancing rates of absorption and transport of drugs in vivo. As shown in Table 5.1, the substitution of fluorine for hydrogen slightly increases the lipophilicity. In contrast, a polyfluorinated substituent, such as a trifluoromethyl group, induces lipophilicity much more than a methyl group or a chlorine substituent [3b]. This fact often contributes to the improvement in pharmacological activity.

The electronic properties of a F–C bond and its biological effects are understood in the way discussed in Chap. 1.1.3. A fluorine atom on an sp^3 carbon causes a pronounced electron-withdrawing effect through a relay of induced dipoles along a chain of bonded atoms. This effect is called the sigma withdrawing effect, $-I_\sigma$. The electron-withdrawing effect may also be induced by a through space electrostatic interaction. This is known as a field effect. Thus, a fluorine atom in a monofluorinated molecule often acts as an acceptor of a hydrogen bond [3b] (see Fig. 1.1, Chap 1).

The electronic effect of fluorine attached directly to an sp^2 carbon in a π-system is especially complex, because unshared electrons on fluorine may be donated back to the π-system. This resonance effect is called a $+I_\pi$ repulsive interaction. Thus, the electron-withdrawing $-I_\sigma$ effect of a fluorine substituent on an aromatic ring is canceled by the electron-donating $+I_\pi$ effect particularly in an *ortho* or *para* position. Consequently, monofluorination of an aromatic compound brings little steric and electronic effect. However, since a C–F bond is stable and able to form preferentially an intramolecular hydrogen bond, fluorine substitution is considered a tool for drug design. In contrast, a trifluoromethyl group on an sp^3 or sp^2 carbon shows a strong electron-withdrawing effect.

It is well recognized that fluorine stabilizes an α-cationic center by the interaction of a vacant p-orbital of a carbocation with the filled orbitals of fluorine. However, a CF_3 group strongly destabilizes an α-cationic center in sharp contrast to a methyl group.

In general, a carbocation at a β-position of a fluorine atom is destabilized by the $-I$ effect. Thus, it is difficult to generate a β-fluoro carbocation. This distinct substitution effect has been successfully applied to the stabilization of biomolecules otherwise unstable under neutral or physiological pH circumstances. Some examples are seen in prostaglandins and thromboxanes (see Fig. 1.4, Chap. 1).

In contrast, a carbanionic center substituted by fluorine is generally destabilized through I_π repulsive interactions, but the one that has a fluorine atom

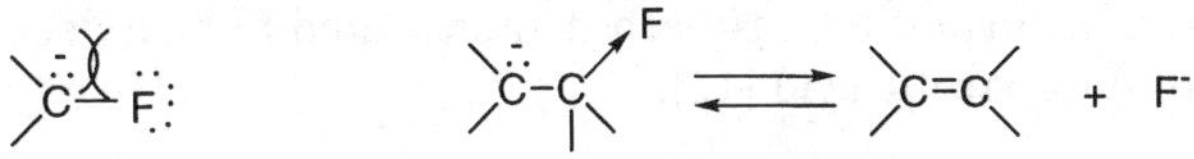

Fig. 5.3   α-Carbanion   β-Carbanion

at a β-position is stabilized by an electron-withdrawing effect (Fig. 5.3); sometimes an elimination reaction may be induced and vice versa. An example accounted for by these characteristics is a suicide inhibitor such as α-fluoromethylglycine.

5.2 Strategies for Design and Synthesis

5.2.1
Structure-Activity Relationship

Of all the elements, fluorine is the most electronegative and forms the strongest single bond with carbon, a bond of an extremely large dissociation energy, whereas the presence of fluorine in a molecule has minor steric but potentially drastic electronic consequences.

Fatty acids having fluorine at a ω-carbon, $FCH_2(CH_2CH_2)_nCOOH$, are metabolized and degraded to fluoroacetic acid by β-oxidation in a way similar to normal fatty acids [4] (see Scheme 1.2, Chap. 1). The resulting fluoroacetic acid is incorporated into the metabolic system of mammals in a similar way to acetic acid, because the bulkiness of a fluorine atom is not discriminated. Fluoroacetic acid is first activated to fluoroacetyl-CoA by citrate synthase into (2R,3R)-2-fluorocitric acid, a common co-metabolite of fluoroacetic acid, arising from the condensation of fluoroacetyl-CoA with oxaloacetic acid. The biosynthetic pathway is illustrated in Scheme 1.1 in Chap. 1.

The stereochemical course of the citrate synthase reaction with fluoroacetyl-CoA is stereospecific and the 2-*pro-S* hydrogen of fluoroacetyl-CoA is abstracted exclusively. This observation demonstrates restricted rotation around the C–C bond of the fluoroacetyl moiety when bound to the enzyme. Thus, stereoelectronic control is operating during the reaction. The condensation reaction proceeds with inversion of configuration at C-2 to generate (2R,3R)-2-fluorocitric acid as the only stereoisomer as shown below. The enzyme of mammals distinguishes between the hydrogen and fluorine atoms and, moreover, recognizes the *pro-S* and *pro-R* hydrogen atoms.

The toxicity of fluoroacetic acid against mammals is attributed to the *lethal synthesis* of (2R,3R)-2-fluorocitric acid by citrate synthase; the other three stereoisomers are nontoxic [11]. (2R,3R)-2-Fluorocitric acid is a competitive inhibitor of aconitase, an enzyme that follows the citrate synthase in the citric acid cycle and interconverts citric and isocitric acids. Consequently, fluoroacetic acid blocks the citric acid cycle as a poison. More recently, the inhibition of the citrate transformation has been demonstrated to be a decisive factor in the toxicity of fluoroacetic acid [12].

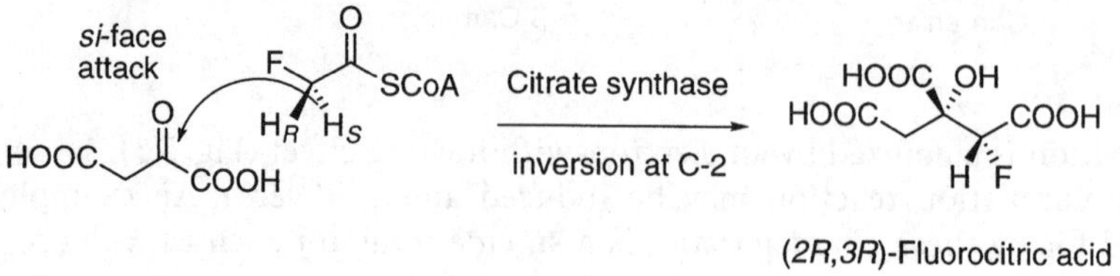

$$\text{(5.1)}$$

(2R,3R)-Fluorocitric acid

On the other hand, an enzyme that mediates the conversion of fluoroacetic acid to glycolic acid has been isolated from a number of microorganisms using fluoroacetic acid as the sole carbon source, e. g. *Pseudomonas* spp. and *Fusarium solani* [4]. A mechanism for this transformation is proposed as follows: a thiol group of the enzyme attacks fluoroacetic acid, displacing fluorine to form an α-thioacetic acid, which is subsequently attacked by a water molecule to release glycolic acid, as illustrated in Scheme 5.2.

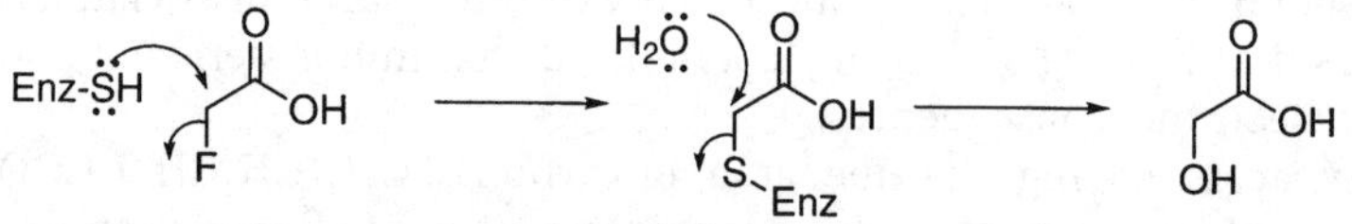

Scheme 5.2. Metabolic transformation of fluoroacetic acid

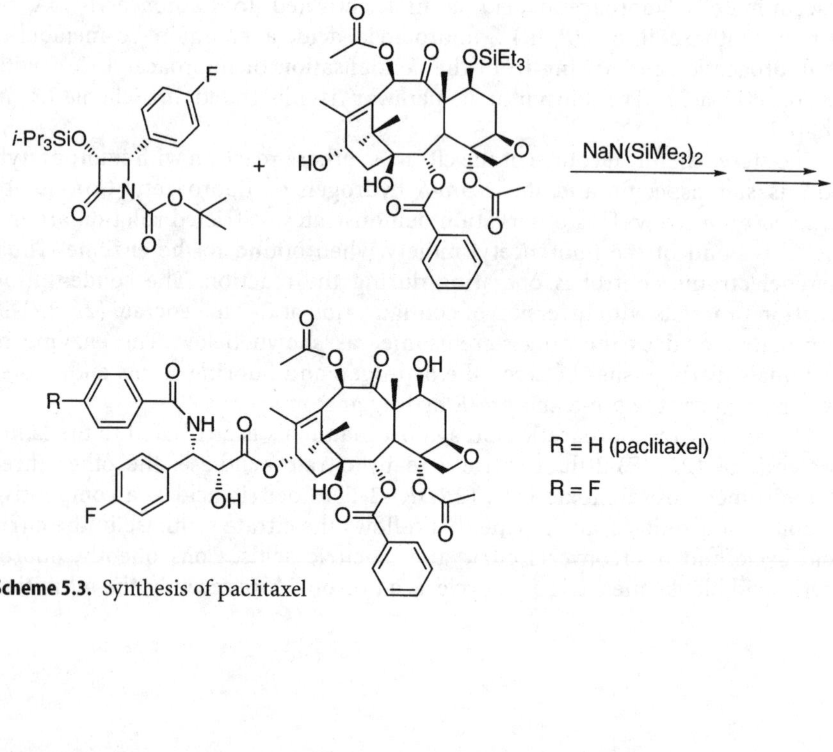

Scheme 5.3. Synthesis of paclitaxel

R = H (paclitaxel)
R = F

5.2 Strategies for Design and Synthesis

Replacement of hydrogen in an agent by fluorine can alter the reactivity and thus change the biological efficiency. Structural modifications in relation to biological activities, i.e. structure-activity relationships (SARs), have been studied extensively on the basis of molecular modeling and molecular mechanics calculations. There are numerous data on SARs of fluorinated agents such as quinolone carboxylic acid antibacterials [13], β-lactam antibiotics [14], vitamin D_3 analogs [15], HIV protease inhibitors [16], prostacyclins [17], angiotensin converting enzyme (ACE) inhibitors [18], and taxoids [19], etc. In particular, a series of fluorine-containing taxoids, extremely potent new therapeutic agents in the treatment of metastatic breast and ovarian cancers, have been synthesized, and the activities studied in relation to the conformation in solution (Scheme 5.3).

With respect to a central nervous system agent, a fluorine substituent in a drug may increase the lipophilicity to facilitate its transport across a blood-brain barrier, a principal diffusion barrier separating brain and blood [20] and, as a result, enhance the absorption rate.

5.2.2
Commercially Available Fluorinated Materials

Thanks to the remarkable growth in the fluorochemical industry, many fluorinated organic compounds of relatively low molecular weights are now available. Examples are HCFCs (hydrochlorofluorocarbons), HFCs (hydrofluorocarbons), alternatives of CFCs (chlorofluorocarbons), and key intermediates of fluoropolymers, which are used as building blocks for the construction of a variety of desired fluorine-containing molecules [21]. Chlorodifluoromethane, for example, is employed for the synthesis of difluoromethyl sulfides and difluoromethyl ethers [22].

$$\text{(5.2)}$$

Tetrafluoroethylene (TFE, $CF_2=CF_2$) and chlorotrifluoroethylene (CTFE, $CF_2=CFCl$) are transformed to tetrafluoroethyl ethers and chlorotrifluoroethyl ethers [23], respectively. An application to glycosidation is illustrated in Eq. (5.3). Hexafluoropropene and hexafluoropropene oxide (HFPO) are versatile building blocks for the synthesis of fluorinated nucleoside analogs.

$$\text{(5.3)}$$

Fluorosulfonyldifluoroacetates ($FSO_2CF_2CO_2R$) are some of the most important intermediates for the preparation of perfluorinated ion-exchange

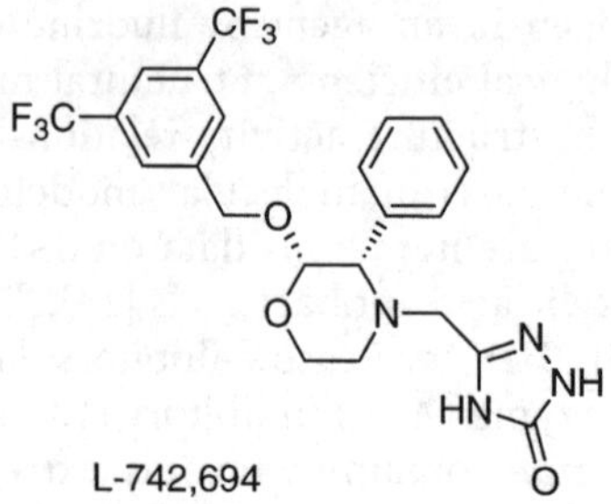

Fig. 5.4 L-742,694

membranes. The ester undergoes a trifluoromethylation reaction of aromatic halides in the presence of a copper catalyst at 80–120 °C (Chap. 3.1.1) [24, 25]. The reagent CF_3-Cu is considered to be responsible for the reaction. Difluorohaloacetates (XCF_2COOR; X=Cl, Br, I) are widely employed for the synthesis of difluorinated sugars [26], nucleosides [27], prostanoids [28], and peptide mimetics [29], etc.

Numerous fluorinated aromatic compounds are commercially available and can be converted into many biologically active compounds [30]. Recently, a 3,5-bis(trifluoromethyl)benzyl unit has been widely used for potent pharmaceutical drugs. A typical example is the morpholine-based human NK-1 antagonist exemplified in Fig. 5.4 [31].

5.3
Fluorinated Amino Acids and Carbohydrates

5.3.1
Amino Acids

Elucidation of the physiological roles of a specific enzyme has led to the design of highly selective enzyme inhibitors useful for practical applications. Irreversible inactivation of an enzyme is caused by the incorporation of an inactivator into an enzyme-active site and by a normal catalysis through proton abstraction, isomerization, elimination, or oxidation leading to a reactive species which is capable of reacting irreversibly with the active site. Since the enzyme becomes inactivated by its own action mechanism, such inhibitors are referred to as *suicide inhibitors* [1–3]. Many fluorinated amino acids are potent enzyme inactivators [32, 33].

Fluoromethyl-substituted amino acids are known to inhibit an enzymatic decarboxylation reaction. Loss of carbon dioxide and elimination of a fluoride ion from an intermediate Schiff base, formed from pyridoxal phosphate and a fluoromethyl amino acid, are thought to induce an enzymatic inactivation (Scheme 5.4). Elimination of a fluoride ion generates a reactive Michael-type acceptor, which is then attacked by a nucleophilic functional group in the enzyme. The covalently bound enzyme is no longer free to bind an additional substrate.

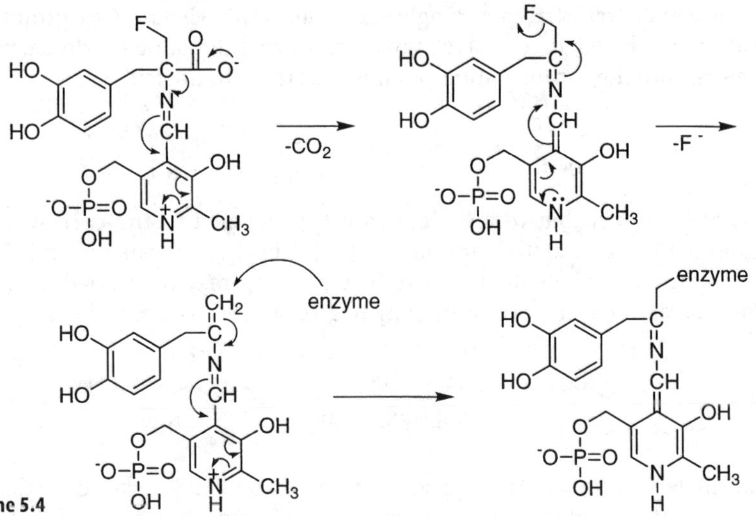

5.3 Fluorinated Amino Acids and Carbohydrates

Scheme 5.4

Amino acids substituted by a difluoromethyl group at an α-position are also effective inhibitors of decarboxylases. Difluoromethyl ornithine [34] was found to inhibit ornithine decarboxylase and intrude polyamine synthesis, a step needed for growth and development, and was implied to be an effective antigestinal, antitrypanosomal, antioccidal, and antitumor agent. In general, the introduction of a fluoromethyl group at an α-carbon enhances the potency and efficiency of α-amino acids. α-Monofluoromethyl-dopa, for example, has selective peripheral activity (Fig. 5.5) [2].

Syntheses of fluorinated amino acids are carried out mainly by two methods. One is based on an amino acid synthesis using fluorinated building blocks, the other is the fluorination of non-fluorinated precursors. A straightforward route is an amine substitution of an α-halo acid or ester with ammonia or an equivalent such as sodium azide or potassium phthalimide (Scheme 5.5).

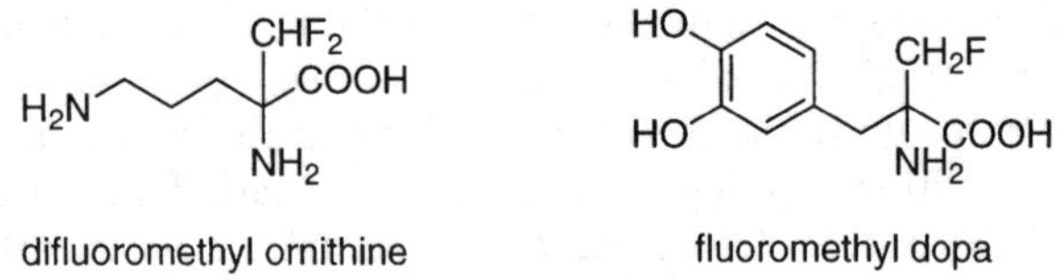

Fig. 5.5   difluoromethyl ornithine   fluoromethyl dopa

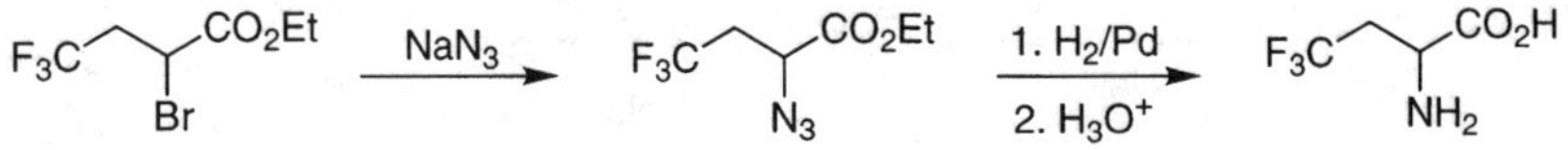

Scheme 5.5

When a substrate has a strongly electronegative CF_3 or CF_2 group at a β-position, the halogenated ester may be converted into an α-azido derivative without elimination or formation of an isomeric β-amino acid.

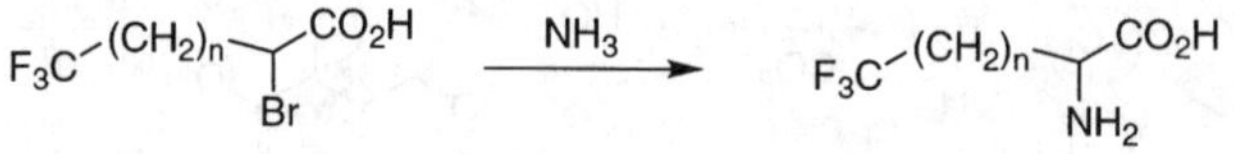

$$F_3C\text{-}(CH_2)_n\text{-}CHBr\text{-}CO_2H \xrightarrow{NH_3} F_3C\text{-}(CH_2)_n\text{-}CH(NH_2)\text{-}CO_2H \quad (5.4)$$

It is well known that α-keto acids are transformed to the corresponding α-amino acids by reductive amination [35]. The key intermediates [33] are imines, oximes, or phenylhydrazones. The reaction may be carried out by an enzymatic reaction using a transaminase to give optically active amino acids [36].

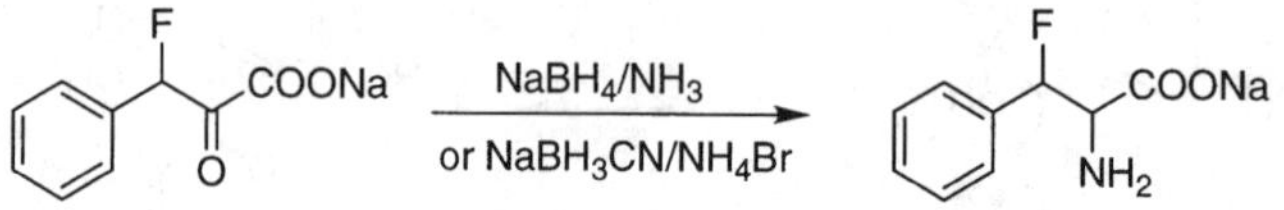

$$\text{Ph-CF(CO-COONa)} \xrightarrow[\text{or NaBH}_3\text{CN/NH}_4\text{Br}]{NaBH_4/NH_3} \text{Ph-CF(CH(NH}_2\text{)-COONa)} \quad (5.5)$$

Fluorinated amino acids can also be prepared by the Strecker or hydantoin synthesis from aldehydes. The Erlenmeyer azlactone synthesis can also be applied to the synthesis of aromatic and aliphatic fluorinated amino acids under two-carbon elongation (Scheme 5.6) [33]. Enantioselective hydrogenation of dehydro amino acids in the presence of an asymmetric catalyst is now a powerful tool for the asymmetric synthesis [32, 37].

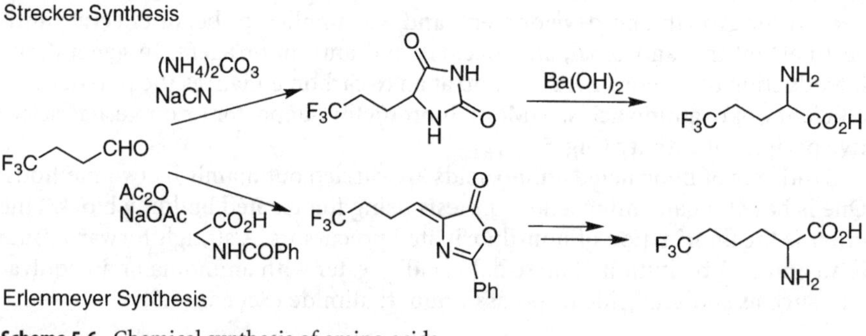

Strecker Synthesis

Erlenmeyer Synthesis

Scheme 5.6. Chemical synthesis of amino acids

Alkylation of acetamidomalonates with a fluorinated electrophile is convenient for the synthesis of α-amino acids. For example, ω-fluoroalkyl bromides are converted into the corresponding ω-fluoro amino acids under two-carbon elongation. Conjugate addition of acetamidomalonates to a trifluoromethyl-substituted Michael acceptor, phenyl 3,3,3-trifluoro-1-propenyl sulfone, gives, after reduction, 4,4,4-trifluorovaline (Scheme 5.7) [38].

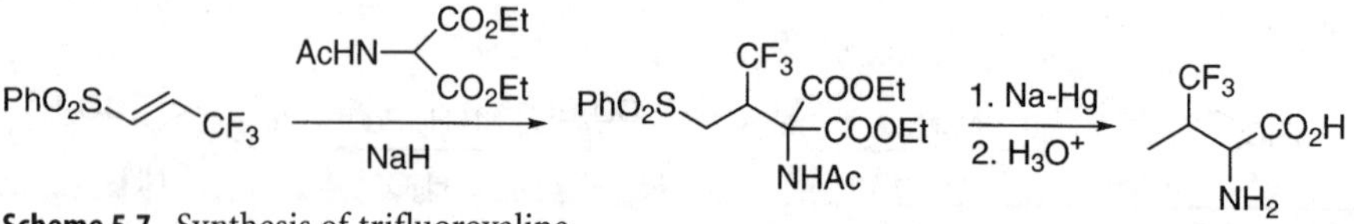

Scheme 5.7. Synthesis of trifluorovaline

5.3 Fluorinated Amino Acids and Carbohydrates

Introduction of fluorine or a fluorine-containing group into a synthetic intermediate and a natural amino acid is an alternative approach to a diversity of unnatural fluorinated amino acids. For example, alkylation of α-amino acids with fluorinated alkyl halides, such as CF_2CHCl, gives α-fluoroalkyl α-amino acids. Using an electrophilic fluorinating reagent, α-fluoro α-amino acids are easily prepared. Starting with this amino acid as a component, fluorinated oligopeptides are prepared [39].

Fluorination of tyrosine with F_2/N_2 gives 3′-fluorotyrosine exclusively [40, 41].

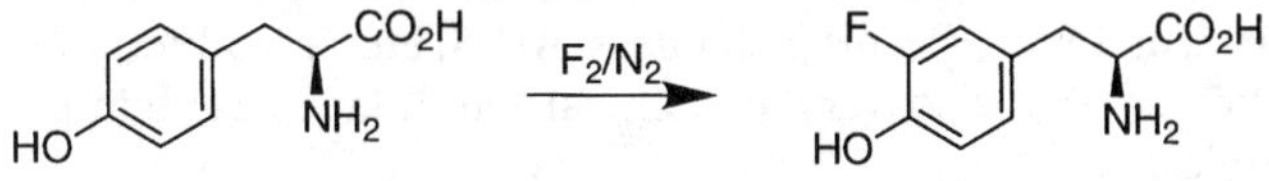

$$\text{HO-C}_6\text{H}_4\text{-CH}_2\text{-CH(NH}_2\text{)-CO}_2\text{H} \xrightarrow{F_2/N_2} \text{3'-F-4-HO-C}_6\text{H}_3\text{-CH}_2\text{-CH(NH}_2\text{)-CO}_2\text{H} \qquad (5.6)$$

Treatment of a 4-hydroxyproline derivative with DAST gives a 4-fluoro derivative with inversion of configuration. This process has been applied to the synthesis of all four stereoisomers of 4-fluoroglutamic acid (Scheme 5.8) [42].

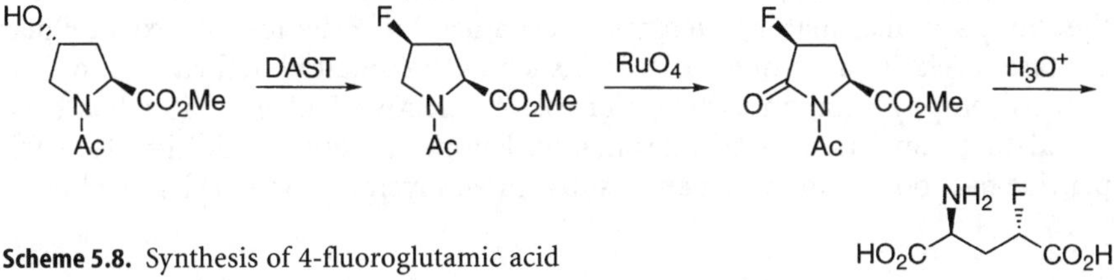

Scheme 5.8. Synthesis of 4-fluoroglutamic acid

The amidocarbonylation reaction (Wakamatsu reaction) of 2-(trifluoromethyl)propanal and 3-(trifluoromethyl)propanal gives 4,4,4-trifluorovaline and 5,5,5-trifluoronorvaline, respectively. The starting aldehydes are derived from regioselective hydroformylation of 3,3,3-trifluoropropene catalyzed by $Co_2(CO)_8$ or $Co_2(CO)_8$-$Rh_6(CO)_{16}$ (Scheme 5.9) [43]. Asymmetric syntheses of

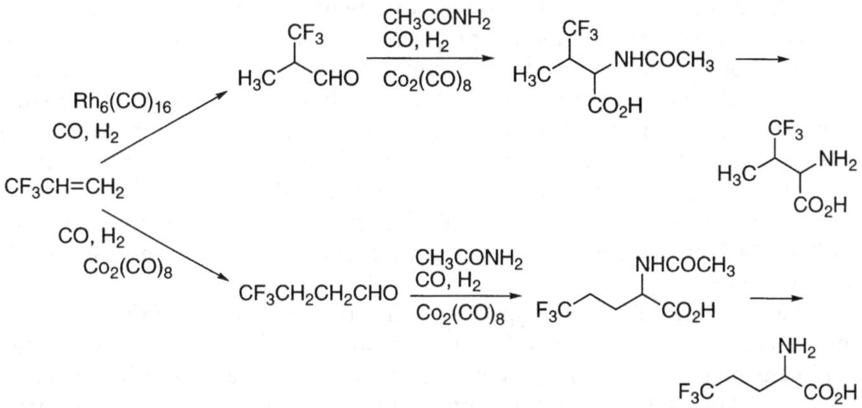

Scheme 5.9. Synthesis of trifluoromethyl-containing amino acids from trifluoropropene

fluorinated amino acid derivatives by an aldol reaction of isocyanoacetate esters and further transformation to fluorine-containing peptides have also been achieved [32, 37, 44].

5.3.2
Protease Inhibitors

The search for a specific, orally active protease inhibitor has become an important strategy for the effective treatment of such diseases as metastatic cancer, malaria, arthritis, sleeping sickness, AIDS, etc. Introduction of fluorine is now the key to the discovery of clinical candidates, as demonstrated by several examples.

A trifluoromethyl ketone is highly electrophilic when connected to an enzyme via a hydrate form, a tetrahedral structure that mimics the transition state of an enzymatic hydrolysis. The tetrahedral intermediate is a potent reversible inhibitor. A typical example is an inhibitor of human leukocyte elastase (HLE). Serine protease, produced by neutrophiles, is closely related to the body's inflammatory defense mechanism. Imbalance of extracellular elastase levels induces many diseases such as rheumatoid arthritis, smoking-induced emphysema, and cystic fibrosis. A selective HLE inhibitor has been found that consists of a trifluoromethyl ketone moiety in addition to a tripeptide backbone containing an N-substituted glycine residue (Fig. 5.6) [45].

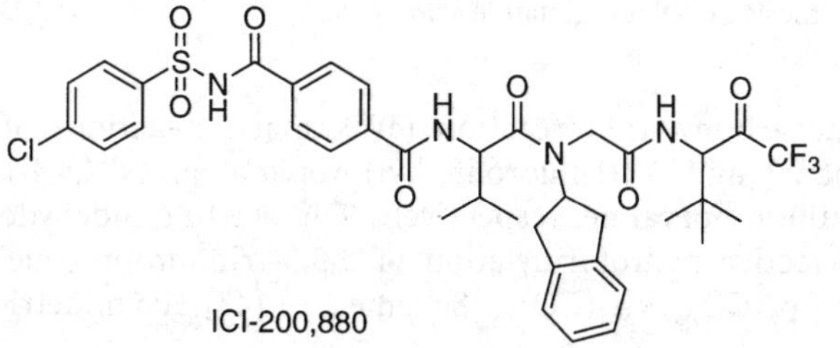

Fig. 5.6 ICI-200,880

Its key component, an α-amino trifluoromethyl ketone or its precursor trifluoromethyl α-amino alcohol, is prepared by condensation of a nitroalkane with trifluoroacetaldehyde [46], by trifluoromethylation of an α-amino aldehyde with CF_3SiMe_3 catalyzed by TBAF, or by trifluoroacetylation of oxazolidinone followed by hydrolysis (Scheme 5.10) [47, 48]. A nitro aldol condensation (Henry reaction) is also effective in the synthesis of α-amino monofluoromethyl carbonyl derivatives [49]. A hydrochloride salt of valine pentafluoroethyl ketone is prepared by alkylation of *N*-Boc valine methyl ester with pentafluoroethyllithium prepared in situ from ICF_2CF_3 and MeLi·LiBr at −78 C [50].

Aspartyl protease, called renin, cleaves the protein substrate angiotensinogen into decapeptide angiotensin I which, in turn, is cleaved into octapeptide angiotensinogen II. Highly potent inhibitors of renin are revealed to commonly contain statine, a novel amino acid present in a naturally occurring pepsin inhibitor called pepstatin. A difluorostatine-containing peptide is also

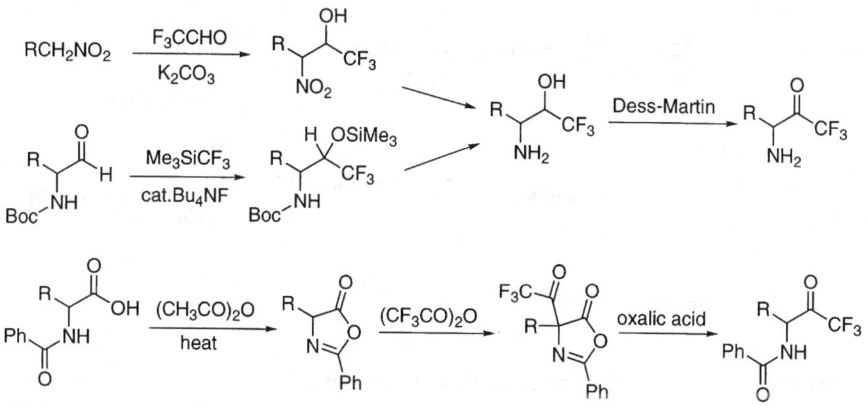

5.3 Fluorinated Amino Acids and Carbohydrates

Scheme 5.10

a potent renin inhibitor, more potent than the non-fluorinated analog. Due to fluorine functionality, the carbonyl group in difluorostatone becomes much more electrophilic than statine to facilitate addition of water to form a tetrahedral species, a transition state analog, akin to the one formed during the enzyme-catalyzed hydrolysis of a peptide bond. A typical example is shown in Fig. 5.7 [51].

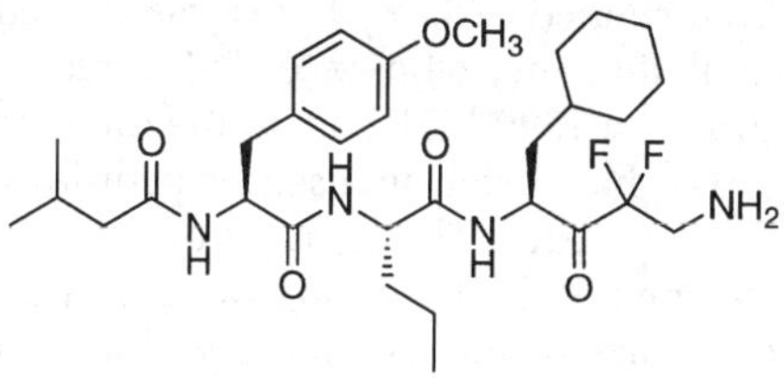

Fig. 5.7

A human immunosuppressive virus (HIV)-encoded protease, required for the post-translational processing of polyprotein gag and gag/pol gene products, has attracted enormous attention as a potential chemotherapeutic target for the treatment of HIV infection. Cellular penetration is required in the design of effective HIV protease inhibitors. New inhibitors, small and lipophilic molecules based on the difluorostatone-type transition state mimic, have been suggested [52].

A peptidomimetic having a difluoro ketone moiety has been demonstrated recently in the study of amyloid plaque formation and the pathogenesis of Alzheimer's disease. The difluoromethylene unit is prepared by a Reformatsky reaction of ethyl bromodifluoroacetate with an N-protected amino aldehyde followed by oxidation. The peptide analog of α,α-difluoro β-keto amide (see below) enhances the permeability and localization to the membrane in a cell-based assay through the retained N- and C-terminal protecting groups and selectively inhibits the production of a β-amyloid precursor protein (Scheme 5.11) [53].

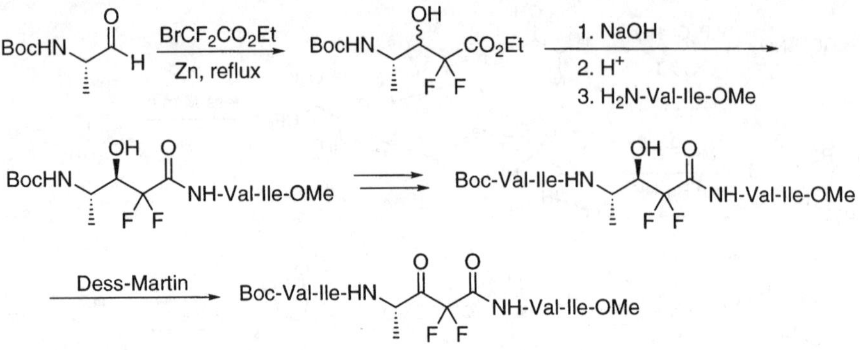

Scheme 5.11. Synthesis of fluorine-containing oligopeptides

5.3.3
Carbohydrates

Selectively fluorinated carbohydrates [54] serve as proof of biochemical mechanisms and for the modification of the glycoside activities and thus have many applications in biochemistry, medicinal chemistry, and pharmacology. The first synthesis of fluorinated carbohydrates was carried out by Moissan, the discoverer of elemental fluorine, who recorded a reaction between F_2 and glucose in his monograph in 1890 [55]. Due to the biological potency of fluoro sugars, growing interest has been focused expeditiously on fluoro sugar chemistry over the last 30 years. The activity of fluoro sugars is mostly attributed to the size (or the bond length plus van der Waals radius) 2.74 Å of a C–F bond, similar to the size of the C–O moiety of a C–OH group, 2.83 Å, and also to the capability of a C–F bond to participate in hydrogen bonding owing to the large electronegativity of fluorine. In other words, replacement of a hydroxyl group by fluorine in a carbohydrate residue causes a profound electronic effect to neighboring groups with a very minor steric perturbation of the original structure or conformation.

Fluorinated carbohydrates are synthesized by a sophisticated combination of modern chemical and enzymatic techniques, because requirements are strict for the stereochemical control at multiply continuing asymmetric centers. Numerous methods have been invented for the preparation of fluorinated neutral sugars, amino sugars, and branched sugars. Typical fluorination methods are a fluoride displacement reaction of sulfonate esters (reagent: KF, CsF, R_4NF), a fluorinating dehydroxylation with diethylaminosulfur trifluoride (DAST) or reagents of a similar type, epoxide or aziridine ring opening by a fluoride ion (reagent: HF, $(HF)_x \cdot$ pyridine, R_4NF, KHF_2), an addition reaction of fluorine to a C=C bond (reagent: F_2, XeF_2, CF_3OF, CH_3COOF), and a fluorination reaction of >C=O groups (reagent: DAST) [54, 55]. In particular, a profound effect of neighboring group participation at a cationic center and a 1,3-diaxial or 1,2-steric effect should be considered in

5.3 Fluorinated Amino Acids and Carbohydrates

these transformations. A typical example is the reaction of methyl α-D-*gluco*-pyranoside with neat DAST to give 4,6-dideoxy-4,6-difluoro-α-*galacto*-pyranoside [55].

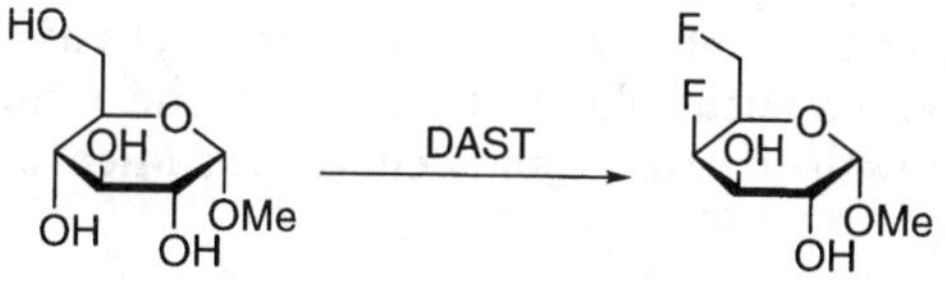

(5.7)

Recently, a stereoselective synthesis of 2-deoxy-2-fluoro sugars was accomplished by the reaction of the corresponding glycals with a commercially available electrophilic fluorination reagent, 1-chloromethyl-4-fluoro-1,4-diazoniabicyclo[2.2.2]octane bis(tetrafluoroborate) (Selectfluor) [56].

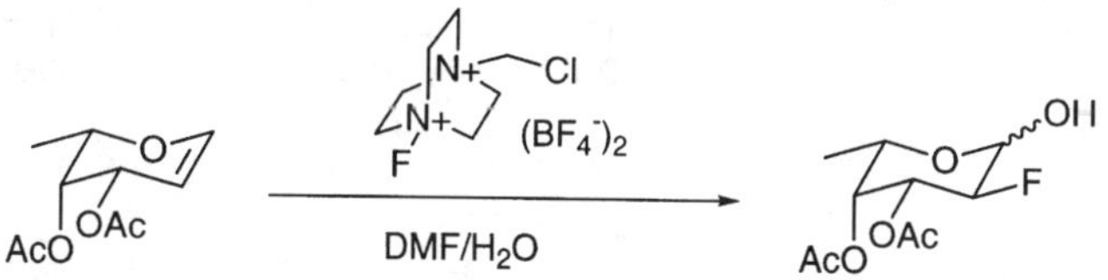

(5.8)

5.3.4
Nucleosides

The adenine-containing antibiotic nucleocidin (Fig. 5.1), the sole naturally occurring fluorinated nucleoside [57], displays a broad spectrum of antibiotic activities and is a particularly effective antitrypanosomal agent [58].

It is well known that substitution of a 3- or 5-hydroxyl group of a nucleoside with fluorine inhibits its transformation to a phosphonate ester, and fluorine at C-2, namely next to a glycosyl bond, generally enhances the stability of the glycosyl bond through an electron-withdrawing effect.

Anticancer Agents. For the treatment of leukemia and solid tumors, many fluorinated pyrimidine and purine nucleosides (and their nucleotides) have been synthesized that are less toxic and more effective than 5-fluorouracil. Presently, 5-fluorouracil is prepared by direct fluorination of uracil in acetic acid (Scheme 5.12). 2′-Deoxy-5-fluorouridine (FdUR) exhibits superior antitumor activities to its mother compound [1c].

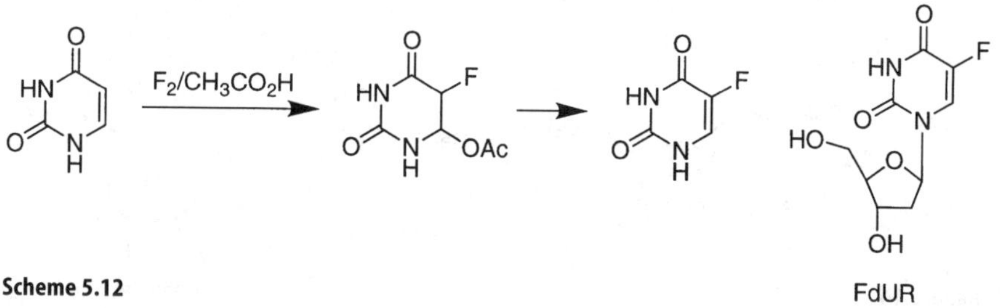

Scheme 5.12

FdUR

The reaction of uridine with $(CF_3)_2Hg$ is effective for the introduction of a trifluoromethyl group to give 5-trifluoromethyluridine [59]. This is prepared conveniently by the coupling reaction of 5-iodouridine with trifluoromethyl iodide in the presence of copper powder in hexamethylphosphoramide [60]. 5-Trifluoromethyluridine exhibits antitumor and antiviral activities by irreversible amide formation with a nucleophilic site of an enzyme. A fluoroolefin is generated by the reaction with an enzyme and behaves as a Michael acceptor, as shown in Scheme 5.13 [61].

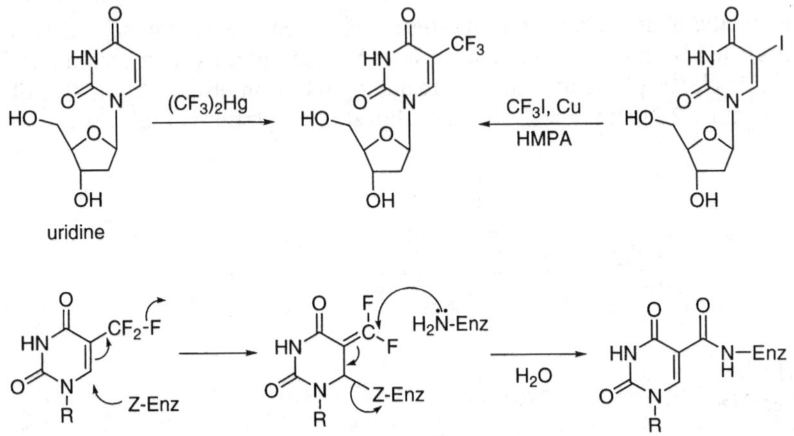

Scheme 5.13

Gemcitabine, a *gem*-difluorinated analog of deoxycytidine, was initially developed as an antiviral agent, but later was shown to exhibit a narrow therapeutic index as a drug for the treatment of human solid tumors, especially non-small cell lung and pancreatic cancer [27b]. A Reformatsky reaction of ethyl bromodifluoroacetate with (*R*)-2,3-*O*-isopropylideneglyceraldehyde affords a 3:1 mixture of α,α-difluoro β-hydroxy ester. Separation and ring closure to a lactone followed by synthetic manipulations give a mesylate which is condensed with trimethylsilylated cytosine. Subsequent deprotection and separation give the desired gemcitabine, as illustrated in Scheme 5.14.

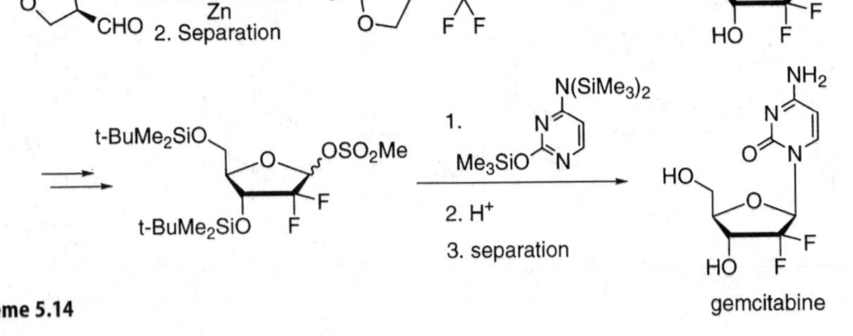

Scheme 5.14

5.3 Fluorinated Amino Acids and Carbohydrates

A fluoroolefin analog of a cytidine nucleoside inhibits a ribonucleotide diphosphate reductase targeted for the treatment of tumors. This is synthesized by the introduction of a fluoroolefin using a fluorinated Horner–Emmons reagent [62]. The resulting fluorovinyl sulfone was transformed to the final (E)-fluoroolefin by stannylation and deprotection (Scheme 5.15).

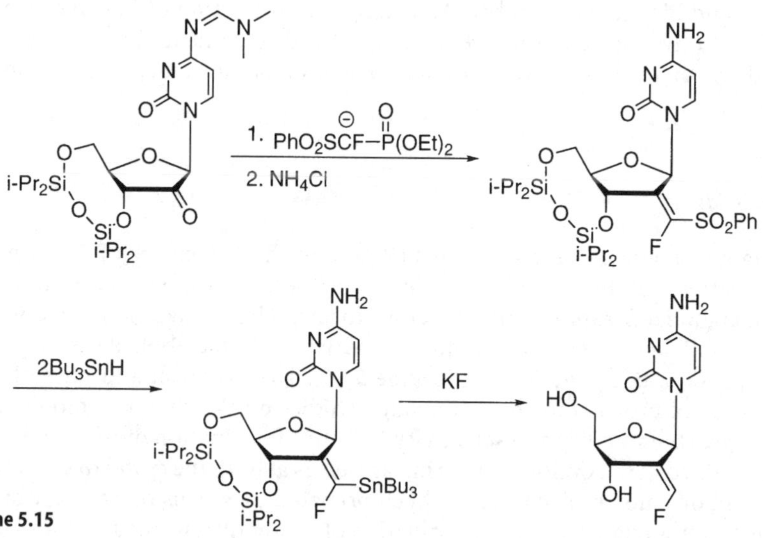

Scheme 5.15

Antiviral Agents. Many fluorinated pyrimidine and purine nucleosides have been prepared (Fig. 5.8) in order to find an alternative to 3'-azidothymidine (AZT), an inhibitor of HIV-1 that is the putative cause of acquired immune deficiency syndrome (AIDS) [63]. Of these, 2',3'-dideoxy-3'-fluorouridine and 3'-deoxy-3'-fluorothymidine (FLT) have been pursued extensively [64]. Another potent therapeutic agent against HIV is (−)-2',3'-dideoxy-5-fluoro-3'-thiacytidine (FTC) that serves as an extremely potent and selective inhibitor of HIV replication in vitro and in vivo [65].

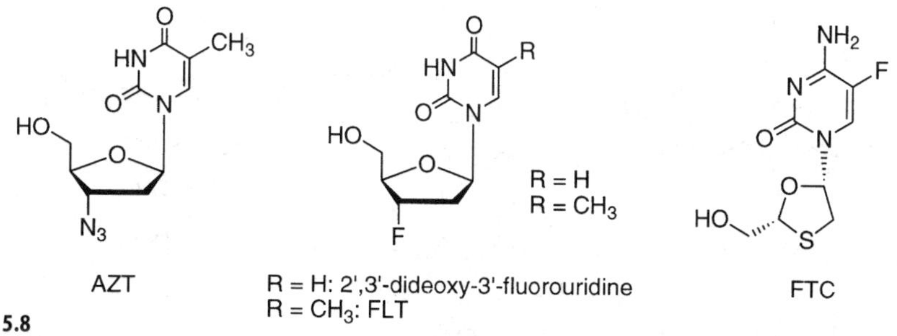

AZT

R = H: 2',3'-dideoxy-3'-fluorouridine
R = CH$_3$: FLT

FTC

Fig. 5.8

5.4
Fluorine-Containing Pharmaceuticals

Recently, it has been demonstrated that the selective substitution of fluorine for hydrogen or a hydroxyl group in a biologically active compound and the introduction of a trifluoromethyl, difluoromethyl, fluoromethyl, or fluorovinyl substituent are highly effective for the discovery of effective drugs, diagnostic agents, and biochemical probes. Accordingly, introduction of fluorine into biologically active molecules is a very powerful and versatile tool for the design of new drugs on the basis of the rational elucidation of molecular recognition processes.

5.4.1
Prostanoids

Prostanoids, namely prostaglandins (PGs) and thromboxanes (TXs), exhibit a wide variety of biological activities at extremely low concentrations and are metabolized rapidly into inactive forms. The biosynthetic pathway of prostanoids begins with arachidonic acid which is metabolized in organs to PGH_2, $PGF_{2\alpha}$, PGI_2 and TXA_2 (Scheme 5.16). The chemical and physiological instability of prostanoids has stimulated studies on the chemical modification to improve the stability and selectivity. Selective fluorination of prostanoids has been effected at the cyclopentane ring as well as at both the α and ω side chains. Target fluorinated molecules have been prepared by means of a variety of synthetic technologies, including fluorination of synthetic intermediates and use of fluorinated building blocks [1c, 2, 66].

Prostaglandin I_2 (PGI_2), a potent inhibitor of platelet aggregation, has a half-life time $t_{1/2}$ of less than 10 min under physiological conditions and its vinyl ether moiety is easily hydrolyzed to give inactive 6-oxo-prostaglandin

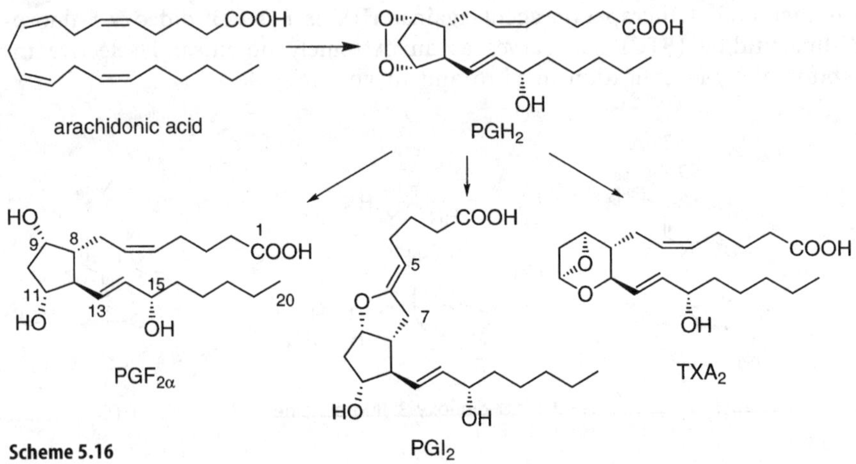

Scheme 5.16

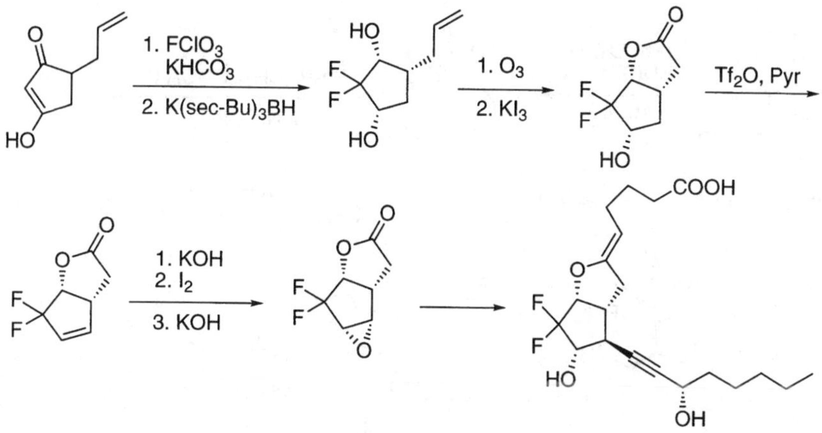

5.4 Fluorine-Containing Pharmaceuticals

Scheme 5.17

10,10-difluoro-13,14-dehydro PGI$_2$

$F_{1\alpha}$. Introduction of fluorine near an acid-labile vinyl ethereal moiety results in the stabilization of the structure of PGI$_2$. The antiplatelet aggregatory and hypotensive effects of 10,10-difluoro-13,14-dehydro-PGI$_2$ are preserved with a remarkably long half-life time ($t_{1/2}$ = ca. 24 h) [67]. A key intermediate difluoro lactone is synthesized by fluorination of 5-allylcyclopentane-1,3-dione with perchloryl fluoride. The difluorinated lactone is converted into an epoxy lactone by iodolactonization followed by treatment with a base (Scheme 5.17).

A remarkable stabilization effect by two fluorine atoms introduced at C-7 of PGI$_2$ is seen in 7,7-difluoro-PGI$_2$, which retains an extremely high inhibitory activity on platelet aggregation even in an oral administration [68]. Amazingly, its $t_{1/2}$ is more than 90 days! 7,7-Difluoro-18,19-dehydro-16,20-dimethyl-PGI$_2$ has been prepared by a manganese salt catalyzed electrophilic fluorination of the Corey lactone and a subsequent stereoselective Wittig reaction of the resulting difluoro lactone (Scheme 5.18).

In contrast to PGI$_2$, thromboxane A$_2$ (TXA$_2$) contracts the aorta and induces platelet aggregation with $t_{1/2}$ = 32 s at 37 °C under physiological conditions [69]. The short half-life time is attributed to its 2,6-dioxabicyclo-[3.1.1]heptane skeleton. However, 10,10-difluoro-TXA$_2$ is hydrolytically stable with four to five times more potency than natural TXA$_2$ with respect to stimulation of platelet aggregation. Moreover, difluorinated TXA$_2$ shows differential effects on platelet aggregation and aorta. The difluoro part is introduced by the reaction of a starting aldehyde with BrZnCF$_2$COOEt. Following transformations lead to a protected form of a dioxabicyclo[3.1.1]heptane intermediate (Scheme 5.19).

PGs are generally metabolized to form biologically inactive 15-oxo analogs by 15-dehydrogenase in the lung. Similarly, introduction of two fluorines at C-16 enhances the resistance to such metabolism and at the same time improves the biological selectivity. Indeed, the 15-hydroxyl group of 16,16-di-

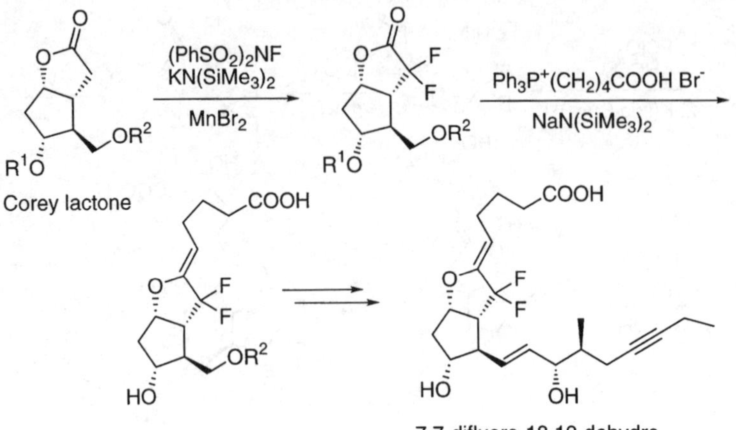

Scheme 5.18

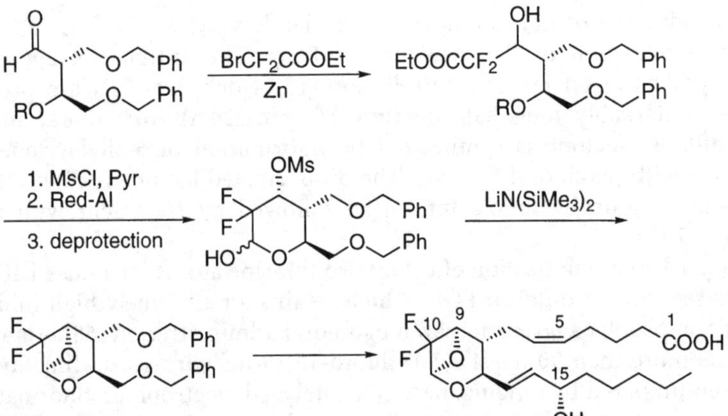

7,7-difluoro-18,19-dehydro-16,20-dimethyl PGI$_2$

Scheme 5.19

fluoro-PGE$_1$ (Fig. 5.9) is not metabolized by 15-dehydrogenase. Worthy of note is that the antifertility activity of 16,16-difluoro-PGF$_{2\alpha}$ is enhanced with a side effect being suppressed against smooth muscle contraction. The difluoro-PGF$_{2\alpha}$ is prepared by the reaction of a difluorinated Horner–Emmons reagent with the Corey lactone (Scheme 5.20) [70].

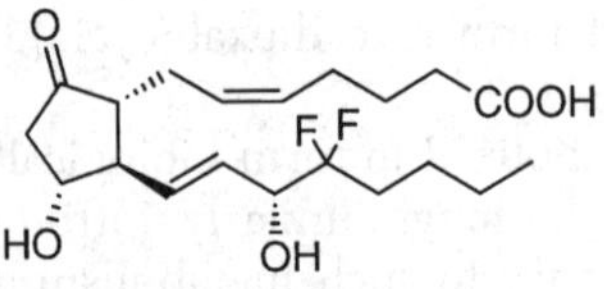

Fig. 5.9   16,16-difluoro PGE$_1$

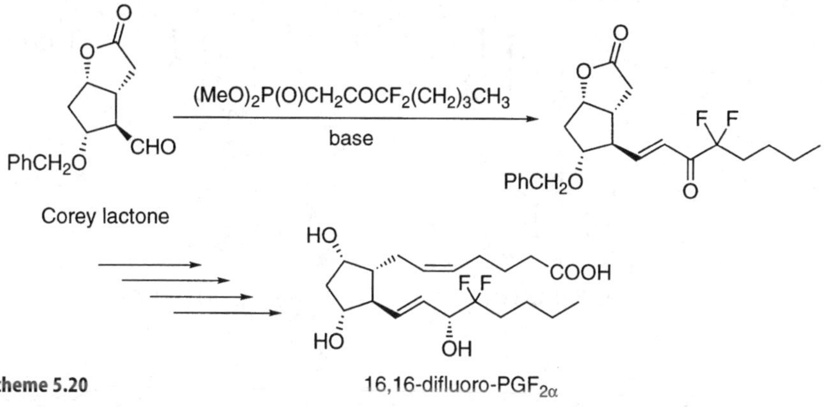

Scheme 5.20

16,16-difluoro-PGF$_{2\alpha}$

5.4.2
Vitamin D$_3$

Since the early work of Fried [5] it has been known that fluorinated steroids exhibit significant biological activities. Studies on vitamin D$_3$ have proved that a novel fluorination technique leads to improvement in the selectivity and utility of biologically active agents. Introduction of fluorine at the position where metabolic hydroxylation takes place is effective to prevent such metabolism owing to a minimal steric alternation (mimic effect) and the strong C–F bond. Vitamin D$_3$ is metabolized in the liver to give an activated form by hydroxylation at C-25 (25-OH D$_3$). Further hydroxylation at C-1 (1α) in the kidney gives active steroid hormone 1,25-(OH)$_2$ D$_3$. 25-OH D$_3$ is alternatively hydroxylated at C-23, C-24, or C-26. Among the polyhydroxy derivatives of vitamin D$_3$, 1,25-(OH)$_2$ D$_3$ is the most active; the hydroxylation at the side chain is a deactivation process leading to excretion.

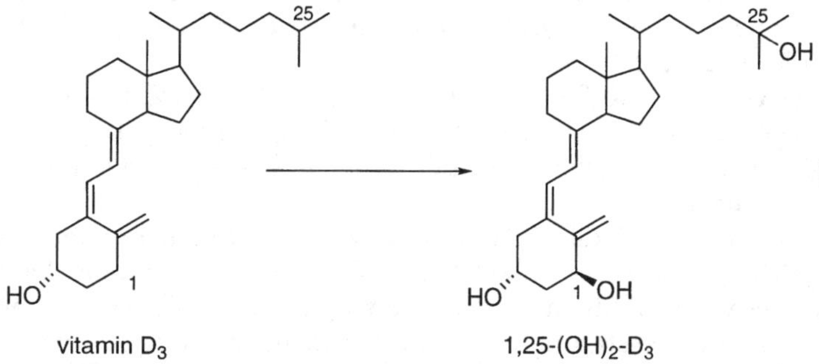

$$(5.9)$$

vitamin D$_3$ 1,25-(OH)$_2$-D$_3$

A fluorinated derivative of 1,25-(OH)$_2$ D$_3$ with fluorine at C-25 and C-1 tolerates the hydroxylation reaction. This fact manifests the role of these hydroxyl groups in various aspects of the activity of vitamin D$_3$. 1,25-(OH)$_2$ D$_3$ with fluorine at C-23, C-24, or C-26 blocks further hydroxylation reactions, a process of metabolic deactivation of vitamin D$_3$ (Fig. 5.10) [58].

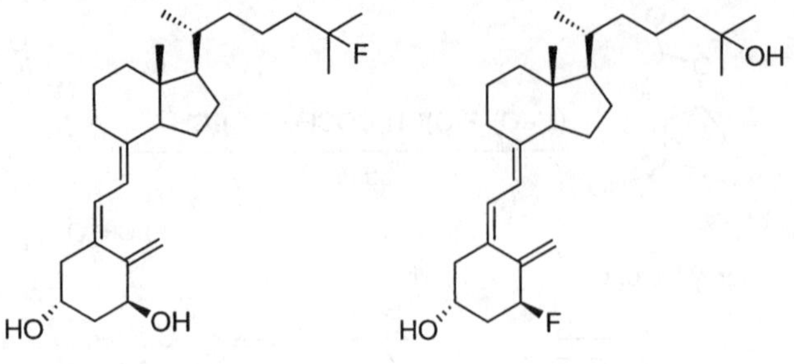

25-F, 1-OH D$_3$

1α-F, 25-OH D$_3$

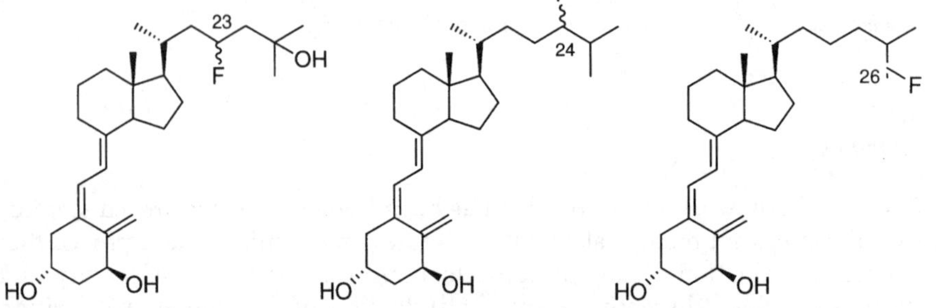

23-F, 1,25-(OH)$_2$ D$_3$

24-F, 1-OH D$_3$

26-F, 1-OH D$_3$

Fig. 5.10

A fluorinated D$_3$ derivative, 1α-fluoro-25-hydroxy-vitamin D$_3$, was prepared by a regio- and stereoselective ring-opening reaction of 6β-acetoxy-1β,2β-epoxy-5α-cholestan-3β-ol with potassium hydrogen fluoride [71] followed by protection to give an acetonide, which was further transformed to 1α-F,25-OH D$_3$ by synthetic manipulations including reductive removal of the 2β-hydroxyl group, a photochemical cyclohexadiene-hexatriene reaction, and a 1,7-hydride shift (Scheme 5.21).

1,25-(OH)$_2$ D$_3$ with fluorine at C-24, C-26, or C-2 adjacent to the 25- or 1-hydroxyl group (Fig. 5.11) is useful for determining the role of the hydroxyl group at C-25 or C-1 during the interaction with a hydrogen-bond donor or acceptor in a receptor [58].

The fact that 1α,25-dihydroxy-26,26,26,27,27,27-hexafluoro-vitamin D$_3$ (26-F$_3$, 27-F$_3$-1,15-(OH)$_2$ D$_3$) is more active than 1,25-(OH)$_2$ D$_3$ means that the hexafluoro derivative binds the receptor less tightly. A condensation reaction of a phenylsulfonylated starting material with hexafluoroacetone followed by desulfonylation gives a key intermediate, a 25-hydroxyhexafluorocholesterol derivative, which is deprotected and oxidized with 2,3-dichloro-5,6-dicyano-1,4-benzoquinone (DDQ) to afford a cross-conjugated triene. This is then transformed to the hexafluoro analog of 1,25-(OH)$_2$ D$_3$ (Scheme 5.22) [72].

Other fluorinated examples are a 16-ene-23-yne analog of 26-F$_3$, 27-F$_3$, 1,25-(OH)$_2$ D$_3$ that possesses potential pharmacological properties for the treatment

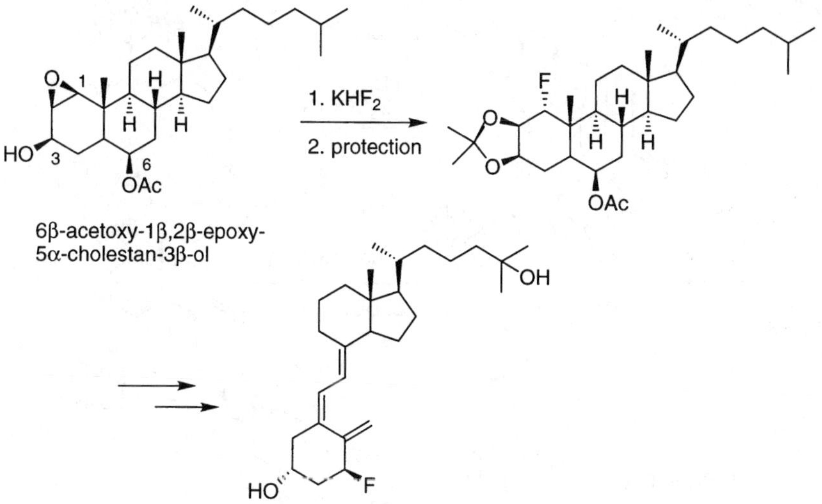

5.4 Fluorine-Containing Pharmaceuticals

Scheme 5.21

6β-acetoxy-1β,2β-epoxy-5α-cholestan-3β-ol

1. KHF$_2$
2. protection

→ → 1α–F, 25-OH D$_3$

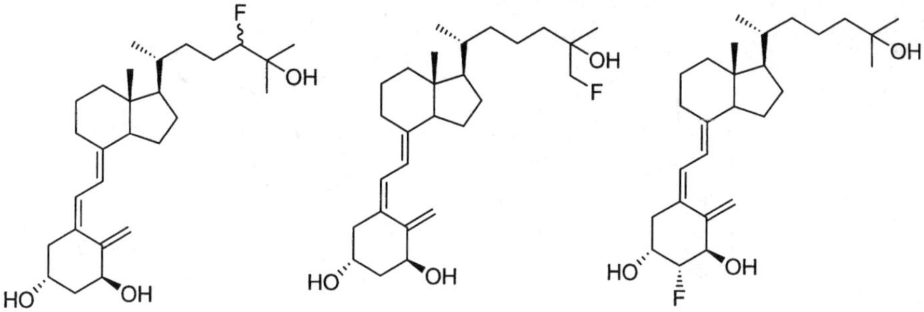

24-F, 1,25-(OH)$_2$ D$_3$ 26-F, 1,25-(OH)$_2$ D$_3$ 2-F, 1,25-(OH)$_2$ D$_3$

Fig. 5.11

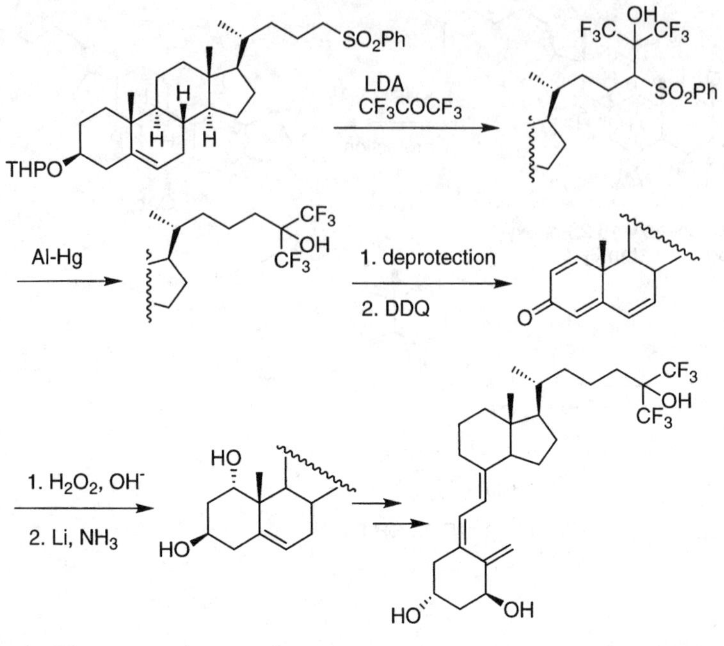

Scheme 5.22

26-F$_3$, 27-F$_3$, 1,25-(OH)$_2$ D$_3$

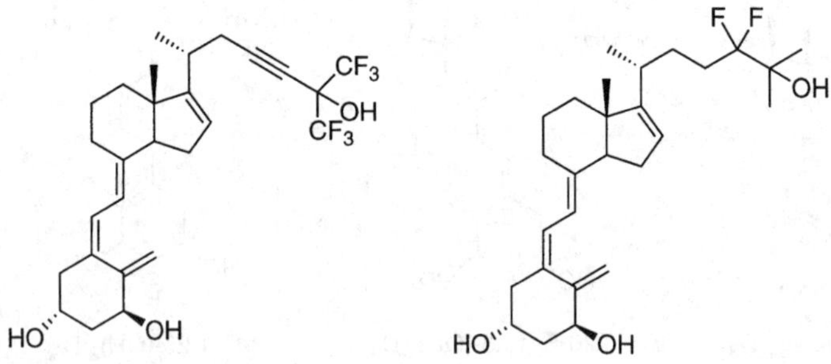

Fig. 5.12

of proliferative and immunological diseases and a 16-ene-24,24-difluoro analog of 1,25-(OH)$_2$ D$_3$ (Fig. 5.12) [73].

5.4.3
Central Nervous System Agents

A great success in the development of new fluorine-containing agents is highlighted in the treatment of disorders of the central nervous system (CNS). As the lipid solubility of fluorinated aromatic compounds is significantly improved, the rate of absorption and transport of a drug across the blood-brain barrier is greatly enhanced by a fluorine substituent [74].

5.4 Fluorine-Containing Pharmaceuticals

Fig. 5.13   diazepam

The most important popular prototype drug is diazepam (Fig. 5.13) which has a 1,4-benzodiazepine structure and is prescribed as a sedative and tranquilizer, controlling mild to moderate degrees of anxiety and tension. An *ortho*-fluorophenyl derivative is shown to increase the hypnotic activity of diazepam.

Fluorine-containing pharmaceuticals have played a key role in the treatment of a major debilitating illness. These agents selectively inhibit re-uptake of serotonin. Fluoxetine [75] was launched in 1986 as a very efficient and successful antidepressant. Paroxetine [76], introduced in 1991, exhibits activity similar to fluoxetine (Fig. 5.14) but with a shorter duration of action.

Fig. 5.14   fluoxetine   paroxetine

5.4.4
Antibacterials and Antifungals

New antibacterials, fluoroquinolone carboxylic acids [77, 78], are essential to combat infections caused by microorganisms resistant to the traditional antibacterials: penicillins, cephalosporins, and tetracyclines. In the early days, fluorine-containing agents were limited to flucloxacillin (vide infra), a penicillinase-stable penicillin [79], and β-fluoro-D-alanine [80]. Since the introduction of norfloxacin in the early 1980s, fluoroquinolone carboxylic acids have been competitively developed, and new more effective members have emerged (Fig. 5.15). Fluoroquinolone and naphthyridine carboxylic acids inhibit a DNA gyrase bacterial enzyme, exhibiting a broad spectrum of activity against various aerobic and anaerobic gram-positive and gram-negative bacteria. Norfloxacin, enoxacin, ciprofloxacin, sparfloxacin, and tosufloxacin contain fluorine at C-6 and a piperazyl group (except for tosufloxacin) at C-7 to increase the activity. Levofloxacin, an (S)-isomer of ofloxacin, is especially effective in lowering respiratory, urinary tract and prostate infections, and in treating sexually transmitted diseases.

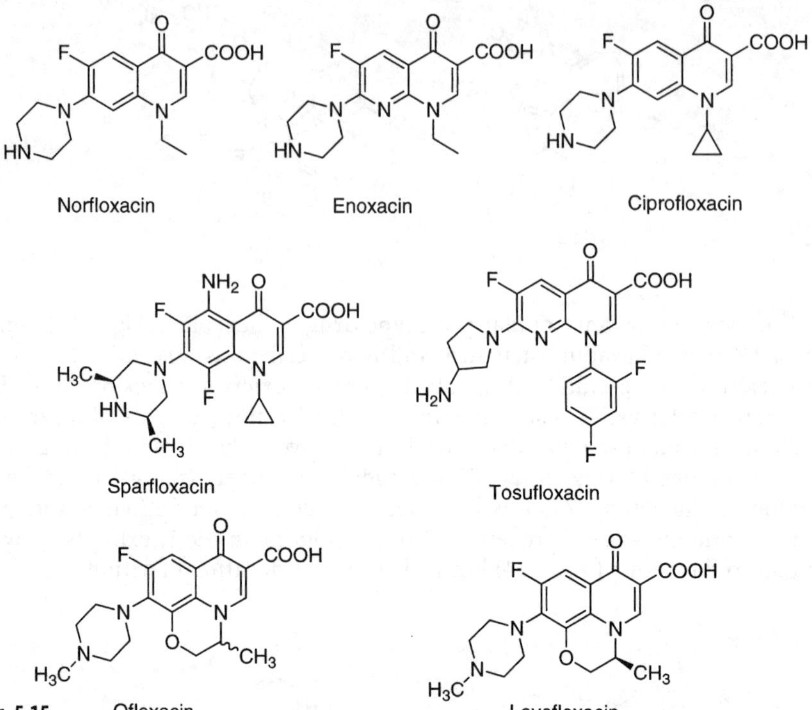

Fig. 5.15 Norfloxacin, Enoxacin, Ciprofloxacin, Sparfloxacin, Tosufloxacin, Ofloxacin, Levofloxacin

Norfloxacin is prepared as follows: reaction of 3-chloro-4-fluoroaniline with ethoxymethylenemalonate and cyclization (Gloud–Jacobs reaction) give hydroxyquinoline carboxylate. After N-ethylation and saponification, chlorine at C-7 is substituted by piperazine by activation with boron trifluoride (Scheme 5.23).

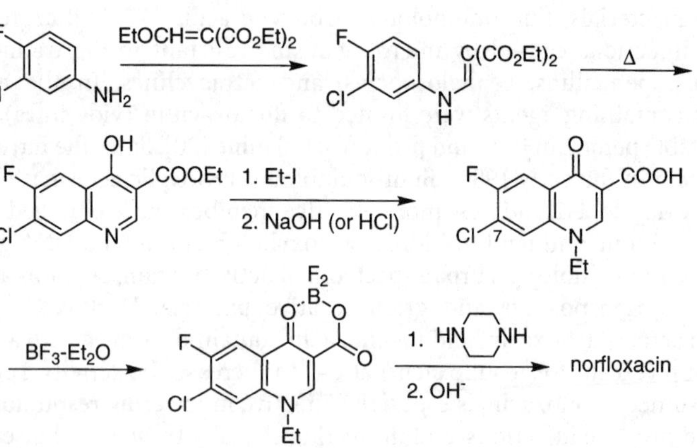

Scheme 5.23

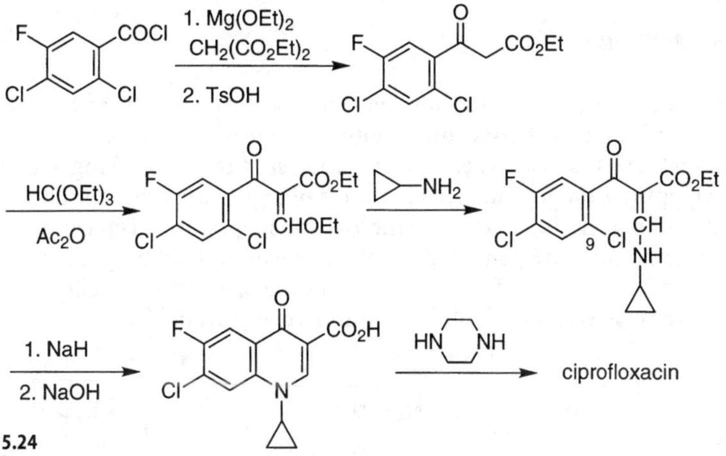

5.4 Fluorine-Containing Pharmaceuticals

Scheme 5.24

On the other hand, the quinoline ring of ciprofloxacin can be constructed in an alternative manner by a nucleophilic substitution of Cl at C-9 with an aminocyclopropyl group. A key intermediate enamino ester is prepared by the reaction of the ethoxymethylene derivative with cyclopropylamine (Scheme 5.24).

Malaria is caused by parasitic protozoa, primarily *Plasmodium falciparum*. The antiprotozoal drug mefloquine (Fig. 5.16) [81] is one of the main agents for the current treatment of malaria, used singly or in combination with chloroquine. Since no useful vaccines are yet available, the compound still plays a significant role in the treatment.

Stable, orally active, and topical antifungal agents have recently emerged which have a 2-fluorophenyl(triazolyl)methane moiety in common. In particular, bistriazole fluconazole, launched in 1988, is effective against dermal, vaginal, and other infections, inhibiting fungal ergosterol synthesis. A new antifungal, flutrimazole, exhibits significant activity, being used singly or in combination with fluconazole or flucytosine (Fig. 5.16).

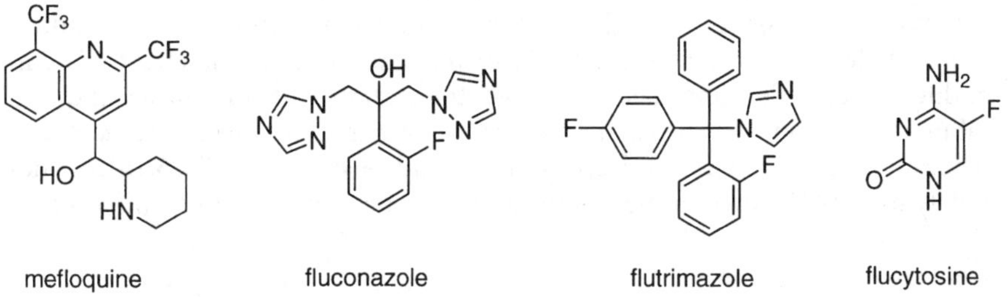

mefloquine   fluconazole   flutrimazole   flucytosine

Fig. 5.16

5.4.5
β-Lactam Antibiotics

A very large number of β-lactams have antibacterial activities by disrupting bacterial cell wall synthesis and inhibiting one or more penicillin-binding proteins (PBPs) that catalyze, as transprotease, the crosslinking reactions of D-alanylpeptides on peptidoglycan strands of a growing cell wall. The β-lactam ring acts as an acylating reagent to inhibit transpeptidases. Typical examples of β-lactam antibiotics are penicillins, cephalosporins, cephamycins, carbapenems, and monobactams. Some fluorinated compounds, e.g. flucloxacillin [79b] and flomoxef [82] (Fig. 5.17), are already commercially available.

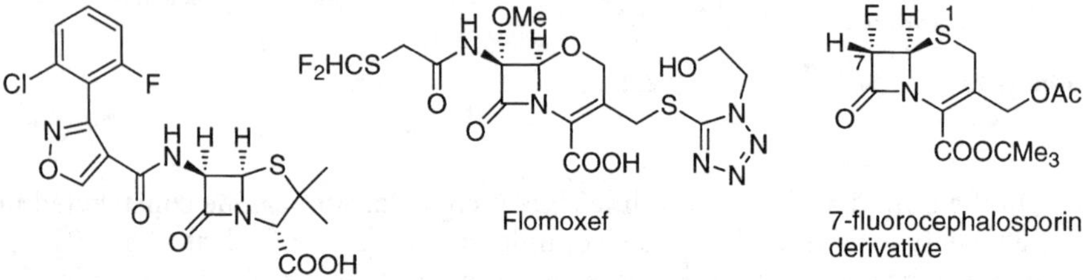

Fig. 5.17

Flucloxacillin, Flomoxef, 7-fluorocephalosporin derivative

Introduction of fluorine at C-6 of a penicillin structure is carried out on the basis that the replacement of hydrogen by fluorine does not induce any significant steric consequences at β-lactamase binding sites. However, fluorine raises the acidity of the geminal C-6 hydrogens and thus the vicinal β-lactam carbonyl group becomes readily acylated. Following a similar strategy, fluorine can be introduced at C-7 of the cephalosporin structure.

5.4.6
Anesthetics

In the 1950s, fluoroxene and halothane (Fig. 5.18) became available as anesthetics. Whereas fluoroxene is flammable at high concentrations and induces nausea, halothane is nonflammable with relatively minor side effects. The introduction of halothane has influenced the practice of anesthesia and contributed to major advances in medical treatment and health care. A new generation of fluoro anesthetics, desflurane and sevoflurane (Fig. 5.18), are now widely used [83]. Sevoflurane has properties of fast uptake and elimination.

$CF_3CH_2OCH=CH_2$ — Fluoroxene

$CF_3CHClBr$ — Halothane

$CF_3CHFOCHF_2$ — Desflurane

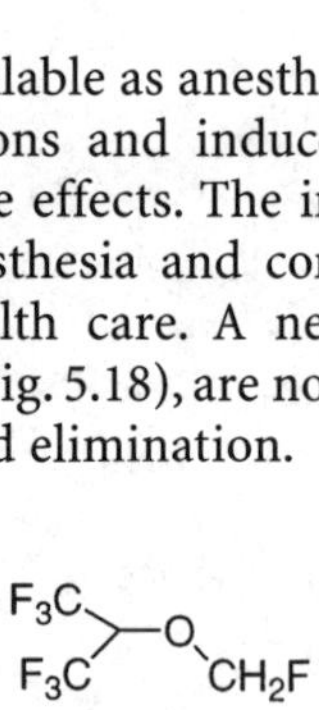

$(F_3C)_2CHOCH_2F$ — Sevoflurane

Fig. 5.18

5.4 Fluorine-Containing Pharmaceuticals

5.4.7
Artificial Blood Substitutes

Blood is a fluid which consists of many components and has a variety of functions, namely, transport of metabolic substrates and removal of metabolic products, maintenance of ion balance, and regulation of the immune system. The most important role is the transport of oxygen to tissues and the removal of carbon dioxide from the tissues. Gas exchange by blood, a function of blood corpuscles, can be carried out by artificial blood substitutes [84].

In 1966, Clark demonstrated that a mouse submerged in perfluoro 2-butyltetrahydrofuran could survive for up to ten minutes [85]. The animal could receive a sufficient amount of oxygen from the liquid and, after removal from the liquid, did not show any apparent ill effects, exhibiting the high oxygen carrying capacity and low toxicity of perfluorocarbons (PFCs). However, intravenous infusion of liquid PFCs causes the death of animals because of the immiscibility in water and blood. Therefore, the PFC must be dispersed as fine particles for use of an artificial blood substitute.

Examples of PFCs, consisting of carbon, fluorine and, in some cases, of such heteroatoms as oxygen or nitrogen, are illustrated in Fig. 5.19. PFCs are chemi-

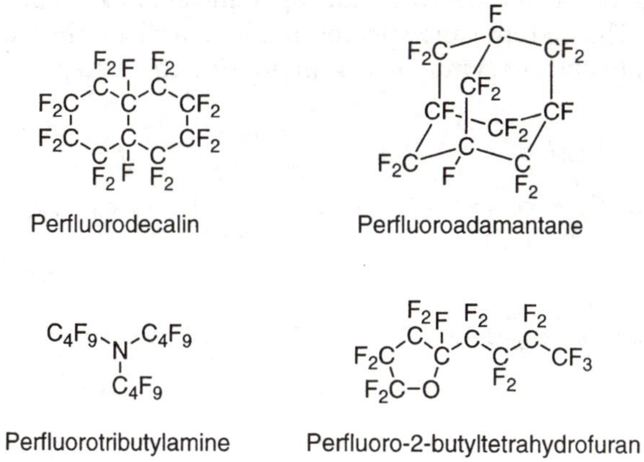

Fig. 5.19 Perfluorodecalin, Perfluoroadamantane, Perfluorotributylamine, Perfluoro-2-butyltetrahydrofuran

Table 5.2. Solubility of oxygen and carbon dioxide in CFCs

Compound	Temp. (°C)	Solubility (ml/100 ml)	
		Oxygen	Carbon dioxide
ethanol	25	24.2	247.9
water	25	2.9	75.9
	37	2.4	57.0
perfluorodecalin	37	45.0	134.0
perfluorotributylamine	37	38.9	142.0
perfluorotetrahydrofuran	37	58.0	160.0

cally extremely stable and are excreted by animals without being metabolized. Boiling points of PFCs are relatively low (perfluorodecalin: 142 °C) and they exhibit low surface tensions, 9–16 mN/m, compared with those of alkanes and alcohols, 25–35 mN/m. Generally, the surface tension of a liquid relates well to the solubility of a gas in the liquid (Table 5.2). The gas exchange functions of blood corpuscles can be carried out by artificial blood substitutes. An emulsion consisting of perfluorodecalin and perfluoroisopropylamine has already been launched for the oxygen-carrying agent.

5.4.8
^{18}F-Labeled Tracers for Positron Emission Tomography

Fluorine-18 is a positron-emitting isotope of fluorine and has a long half-life of 110 min. Positron emission tomography (PET) [86], in conjunction with appropriate radiotracers labeled with fluorine-18, has been used to study biochemical transformation, drug pharmacokinetics, and pharmacodynamics in the human and animal body. Typical agents are 2-deoxy-2-[^{18}F]fluoro-D-glucose (a PET tracer for 2-deoxy-D-glucose) and 6-[^{18}F]fluoro-L-DOPA (a PET tracer for L-DOPA), as shown in Fig. 5.20. Recently, PET has also been used to assess the functional and neurochemical parameters in a normal and diseased human brain. Thus, a rapid and effective method for the introduction of ^{18}F is a great concern to current synthetic organofluorine chemistry.

Fig. 5.20   a PET tracer for 2-deoxy-D-glucose   a PET tracer for L-DOPA

5.5
Fluorine-Containing Agrochemicals

Agrochemicals are indispensable for the large-scale production of high quality crops and for an improvement in the harsh labor conditions in the fields. Because use of large amounts of chemicals may cause problems associated with safety and the environment, use of small amounts of highly potent, selective, and negligibly nontoxic agents are desirable that are effective only for a certain period and then will be decomposed rapidly to totally nontoxic compounds. To this end, many novel agrochemicals have been developed that allow applications in much smaller amounts. Fluorine has played a key role in the design of novel highly potent agrochemicals [87–91].

The fluorine effect in agrochemicals is understood in terms of an increase in lipophilicity, mimic effect, electronic effect and block effect, as described in previous sections of this chapter.

5.5.1
Insecticides

A typical class of fluorine-containing insecticides is *N*-2,6-difluorobenzoyl-*N'*-arylurea and its derivatives. Some examples are listed in Fig. 5.21. The agents inhibit chitin biosynthesis and are often called insect growth regulators (IGRs). As chitin does not affect mammals, the insecticides are selective to specific kinds of insects and nontoxic to mammals and much less influential against their natural enemies and honey bees.

A typical example is diflubenzuron, developed as a fluorine analog of 2,6-dichlorobenzoyl urea TH-6038 (Scheme 5.25) [91]. Triflumuron was designed

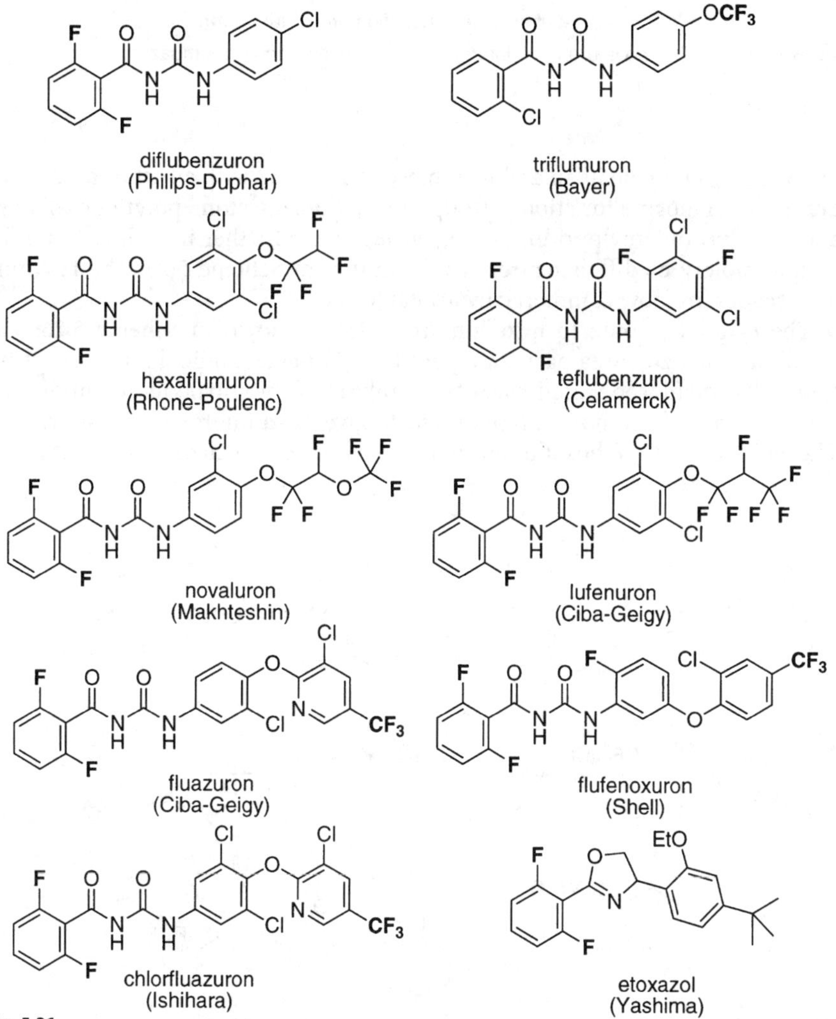

Fig. 5.21

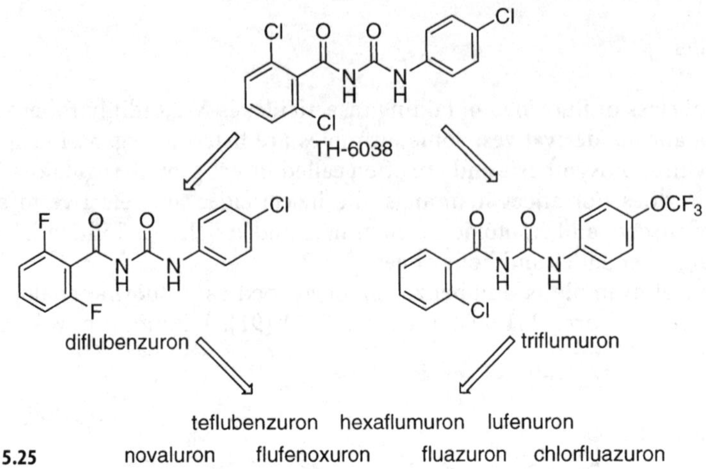

Scheme 5.25

[TH-6038 → diflubenzuron, triflumuron → teflubenzuron, novaluron, flufenoxuron, hexaflumuron, fluazuron, lufenuron, chlorfluazuron]

by replacing chlorine in an aniline moiety of TH-6038 with a trifluoromethoxyl group [92]. Fluorine functional groups, like a fluorine atom, polyfluoroalkoxyl, and 3-trifluoromethylpyridin-2-yloxy groups were further introduced into the aniline moiety of diflubenzuron, as illustrated in Scheme 5.25. The resulting insecticides are now commercially available.

The synthetic route to hexaflumuron [93] is shown in Scheme 5.26. The 2,6-difluorobenzoylurea part is prepared by (1) nucleophilic fluorine substitution of 2,6-dichlorobenzonitrile, (2) hydrolysis of the nitrile to an amide, and (3) reaction of the amide with phosgene to give 2,6-difluorobenzoylisocyanate. The aniline part of hexaflumuron is prepared by nucleophilic addition of

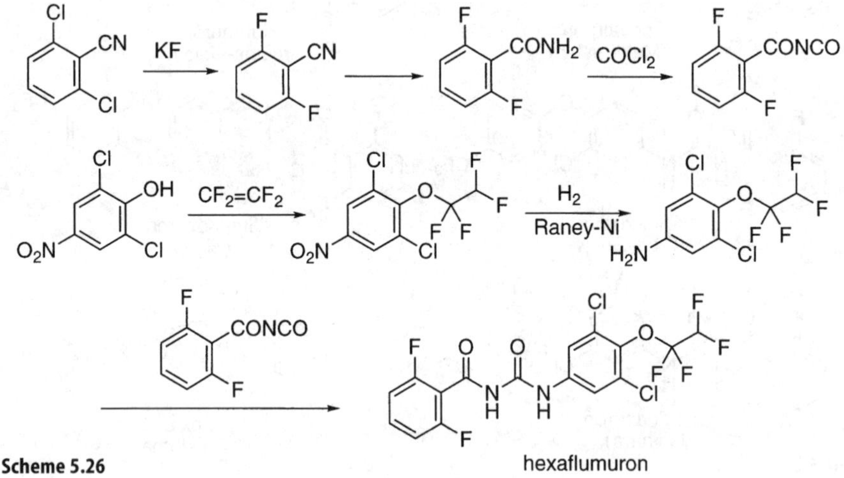

Scheme 5.26

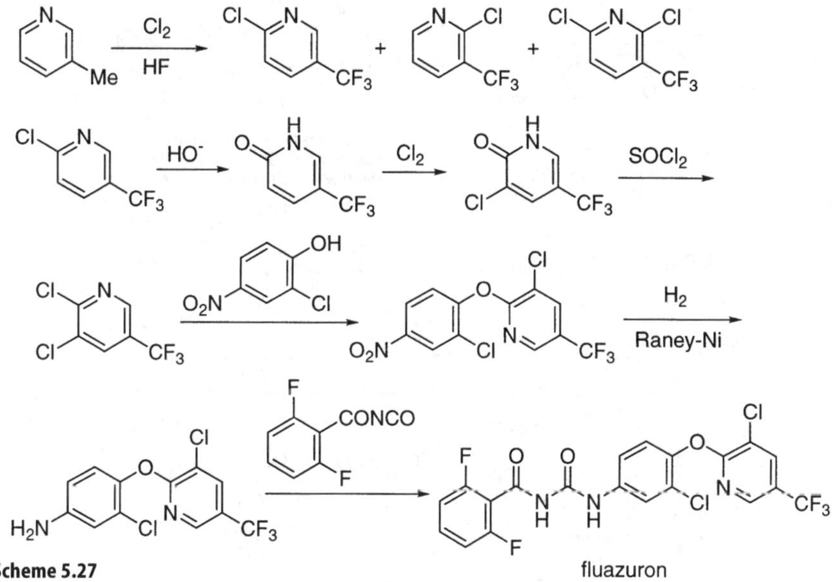

Scheme 5.27

2,6-dichloro-4-nitrophenol to tetrafluoroethene followed by reduction. The resulting aniline derivative is allowed to react with the benzoylisocyanate to give hexaflumuron. Use of $CF_2=CFCF_3$ or $CF_2=CFOCF_3$ in lieu of $CF_2=CF_2$ gives lufenuron or novaluron, respectively.

The synthesis of fluazuron starts with chlorinative fluorination of β-picoline, as summarized in Scheme 5.27 [94]. Separation of 2-chloro-5-trifluoromethylpyridine to form a mixture of regioisomers, hydrolysis to pyridone, stepwise dichlorination, nucleophilic substitution of the resulting dichloropyridine with 2-chloro-4-nitrophenol, and catalytic reduction of the nitro group followed by reaction with 2,6-difluorobenzoylisocyanate give fluazuron.

The second important class of insecticides are the pyrethroids that affect the nervous system of insects. It is known that the major components of a pyrethrum are pyrethrin I and pyrethrin II (Fig. 5.22), highly potent insecticides that are almost nontoxic to mammals. These are, however, so labile against sunlight that their activity is rapidly lost when used in the fields.

Since the use of polychlorinated insecticides is now prohibited for environmental reasons, the natural pyrethroids have been extensively modified, leading to the discovery of permethrin [95] and cypermethrin [96] that are extremely potent and stable for certain periods against sunlight but then decompose gradually. To enhance the insecticidal activity and, especially, the acaricidal activity, trifluoromethyl analogs were introduced. For example, bifenthrin [97], developed by FMC in 1982, has about 55-fold acaricidal activity in addition to an insecticidal activity comparable to permethrin [98]. Tefluthin is a tetrafluorobenzyl ester analog of biphenthrin and thus has a high vapor pressure which makes it more effective against vermin in soil [99].

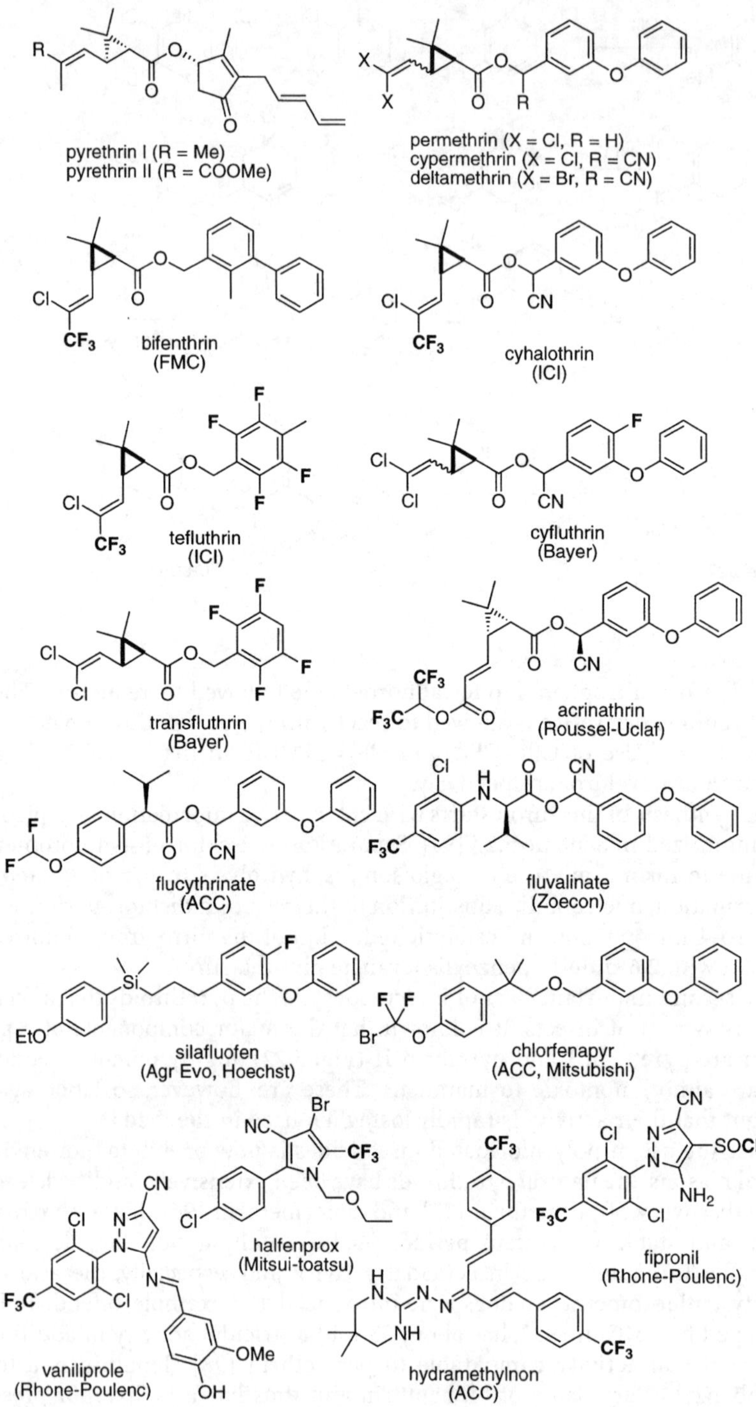

pyrethrin I (R = Me)
pyrethrin II (R = COOMe)

permethrin (X = Cl, R = H)
cypermethrin (X = Cl, R = CN)
deltamethrin (X = Br, R = CN)

bifenthrin
(FMC)

cyhalothrin
(ICI)

tefluthrin
(ICI)

cyfluthrin
(Bayer)

transfluthrin
(Bayer)

acrinathrin
(Roussel-Uclaf)

flucythrinate
(ACC)

fluvalinate
(Zoecon)

silafluofen
(Agr Evo, Hoechst)

chlorfenapyr
(ACC, Mitsubishi)

halfenprox
(Mitsui-toatsu)

fipronil
(Rhone-Poulenc)

vaniliprole
(Rhone-Poulenc)

hydramethylnon
(ACC)

Fig. 5.22

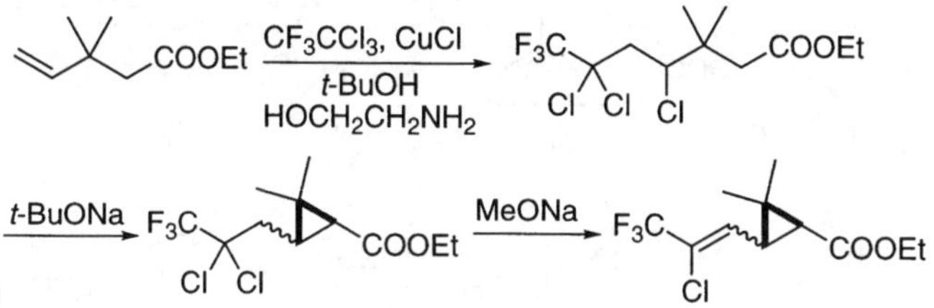

5.5 Fluorine-Containing Agrochemicals

Scheme 5.28

Synthesis of the trifluoromethyl-substituted pyrethroids is achieved by various methods using 1,1,1-trichlorotrifluoroethane (CFC-113a) as a CF_3 building block. The first method involves a copper(I) chloride catalyzed addition of CFC-113a to ethyl 3,3-dimethyl-4-butenoate. The resulting adduct is then treated with a base to induce ring closure and olefin formation, as summarized in Scheme 5.28 [100]. This method usually gives a mixture of *cis*- and *trans*-cyclopropanecarboxylates with ratios controllable to some extent by proper choice of the base.

The second approach involves addition of ethyl diazoacetate to 3,3,3-trifluoro-2,2-dichloropropyl-substituted isobutene or 1,1,1-trifluoro-1-chloro-5-methyl-2,4-hexadiene (Schemes 5.29 and 5.30) [101]. Although straightforward and stereospecific, the reaction is not necessarily effective due to the electron-deficient nature of the olefinic moiety.

Ring contraction is the third method. Cycloaddition of a ketene with isobutene gives a cyclobutanone which, upon treatment with a base, undergoes a Favorskii rearrangement to give preferentially a *cis*-cyclopropanecarboxylic acid. The sequence of the reactions is illustrated in Scheme 5.31 [102].

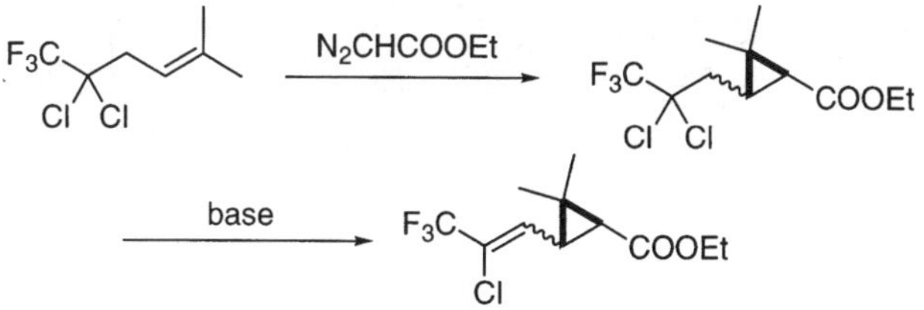

Scheme 5.29

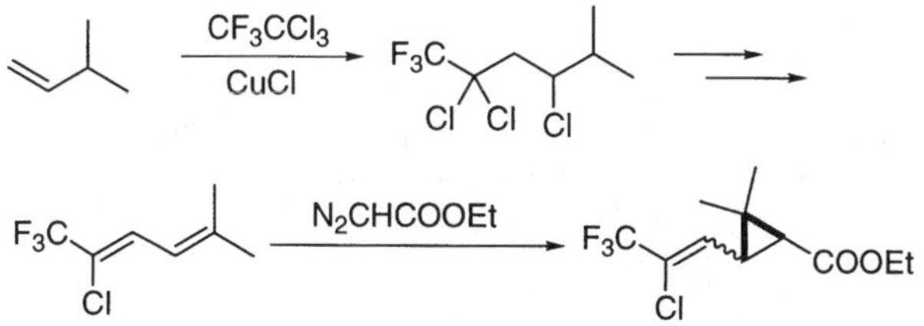

Scheme 5.30

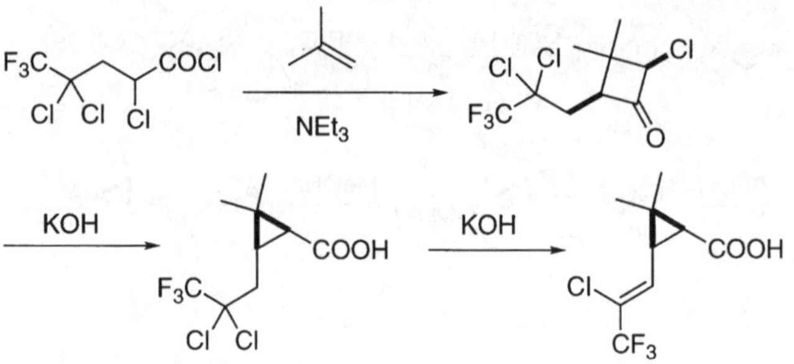

Scheme 5.31

The last approach is carbonyl addition of a zinc carbenoid derived from CFC-113a. The resulting adduct is acetylated and then reduced with zinc dust to give the target ester (Scheme 5.32). Since an adduct derived from *cis*-Caron aldehyde forms a lactone ring and is easily separated, only *trans*-cyclopropanecarboxylate results [103].

Selective synthesis of more potent *cis*-cyclopropanecarboxylates than the *trans*-isomers is stereospecifically achieved by addition of the same zinc carbenoid to 3-methyl-2-butenal, diazoacetylation, and intramolecular carbene addition, followed by zinc reduction, as shown in Scheme 5.33 [103].

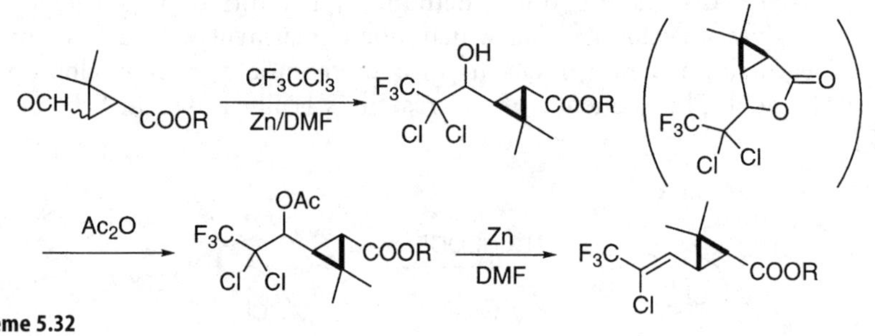

Scheme 5.32

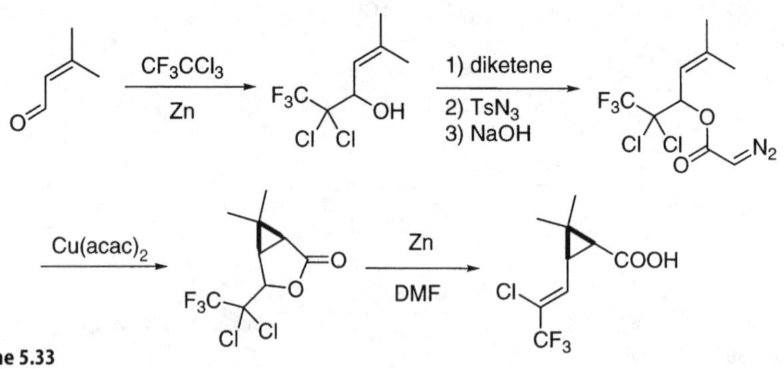

Scheme 5.33

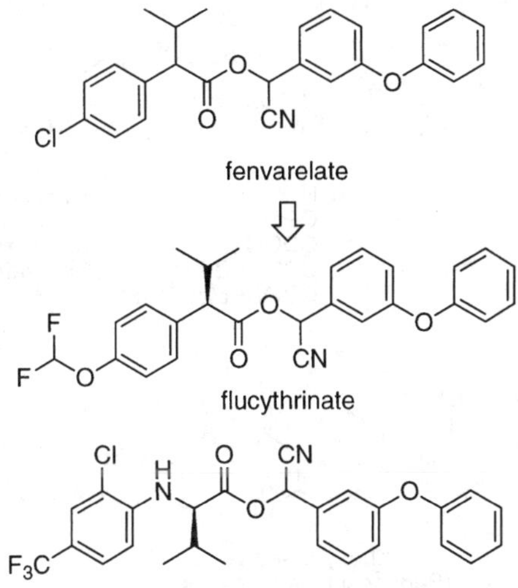

5.5 Fluorine-Containing Agrochemicals

Fig. 5.23

fenvarelate

⇩

flucythrinate

fluvalinate

Further design of synthetic pyrethroid insecticides has led to novel structures that lack a cyclopropane ring. A typical example is fenvarelate [104], illustrated in Fig. 5.23. The chlorine in fenvarelate is replaced by a difluoromethoxyl group to give flucythrinate (Fig. 5.22) [105]. Further modification of 4-chlorophenyl in fenvalerate to 2-chloro-4-trifluorophenylamino gives fluvalinate [106]. Whereas fenvarelate was patented as a mixture of 4 diastereomers, these new compounds have been patented as a diastereomeric mixture with the configurations of the carbon α to the ester carbonyl being both R. Fluvalinate is as active as fenvalerate against the tobacco moth and the house fly and has an acaricidal activity seven times higher.

Later, it was revealed that the insecticidal activities were retained in the absence of the ester carbonyl group of fenvalerate. Thus, silafluofen [101] and halfenprox (Fig. 5.22) were designed and patented. Silafluofen was designed by replacing a CMe_2 group in fenvalerate with $SiMe_2$. Although synthetic pyrethroids are generally toxic to fish, silafluofen is much less toxic and can be applied to rice fields.

5.5.2
Herbicides

A typical class of herbicides are the diphenyl ethers, including nitrofen, shown in Fig. 5.24. This type of compounds inhibits the protection mechanism for the biosynthesis of chlorophyll and allows singlet oxygen generated by sunlight irradiation to oxygenate lipids in membranes of weeds and/or forbs to wither them [87]. A chlorine substituent in nitrofen can be replaced by a CF_3 group to give nitrofluorfen, to which various substituents can be introduced at a position *ortho* to a nitro group. Of these, acifluorfen, fomesafen, and lactofen are post agents against forbs of soybeans.

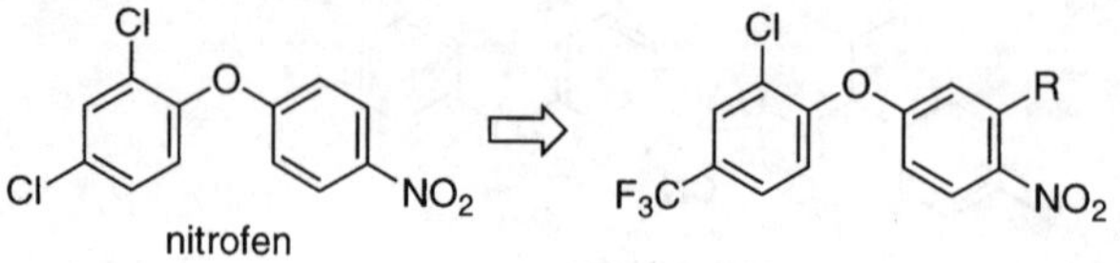

Fig. 5.24

nitrofen

R = H — nitrofluorfen
OEt — oxyfluorfen (Rohn & Hass)
COONa — acifluorfen-Na
CONHSO$_2$Me — fomesafen (ICI)
COOCHMeCOOEt — lactofen (PPG)

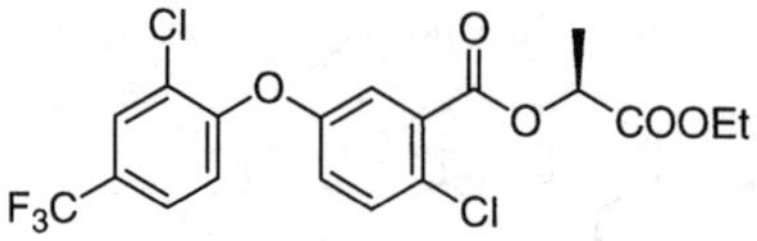

ethoxyfen-ethyl
(Budapest Chem. Works)

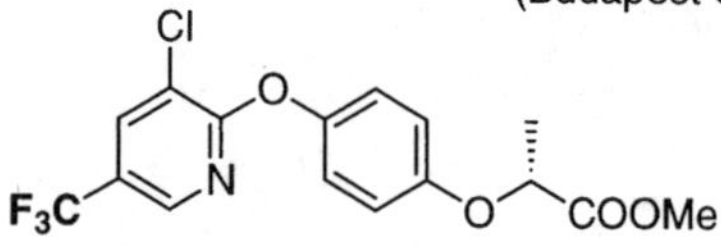

haloxyfop-*R*-methyl
(Dow Elanco)

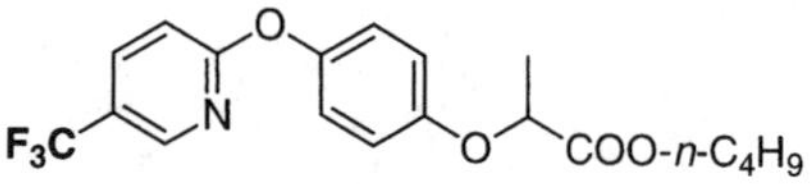

fluazifop-*p*-butyl
(Ishihara)

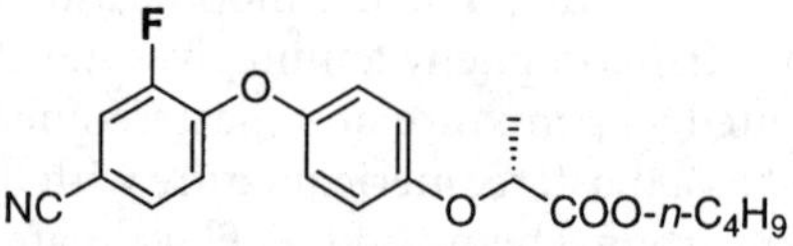

cyhalofop-butyl
(Dow Elanco)

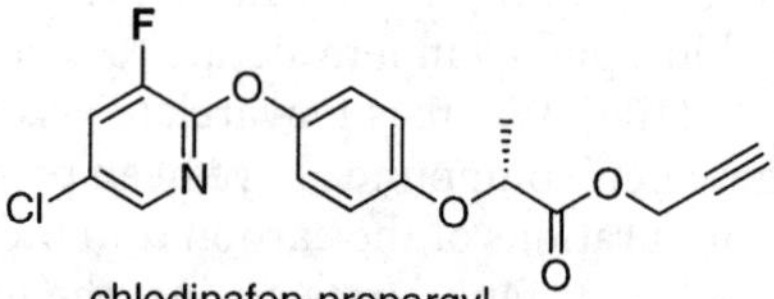

chlodinafop-propargyl
(Ciba-Geigy)

Fig. 5.25

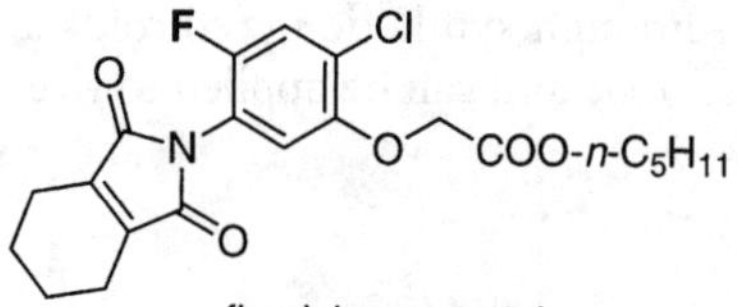

flumiclorac-pentyl
(Sumitomo)

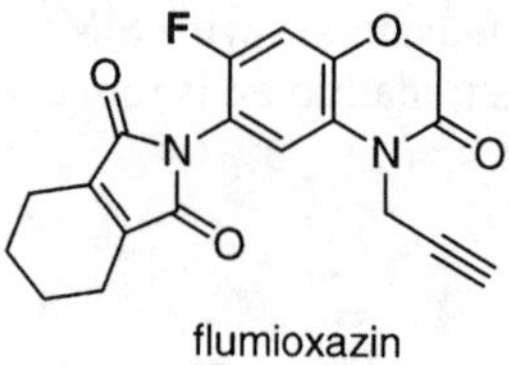

flumioxazin
(Sumitomo)

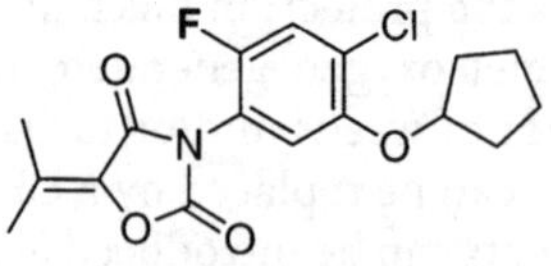

pentaoxazone
(Kaken)

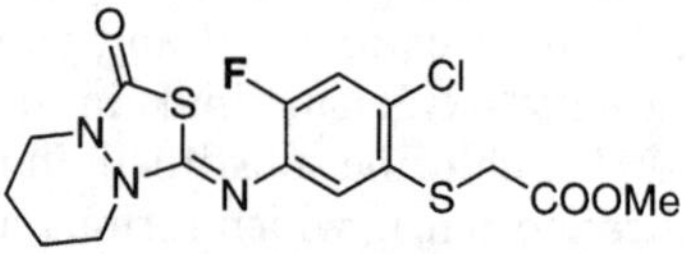

fluthiacet-methyl
(Kumiai)

Fig. 5.26

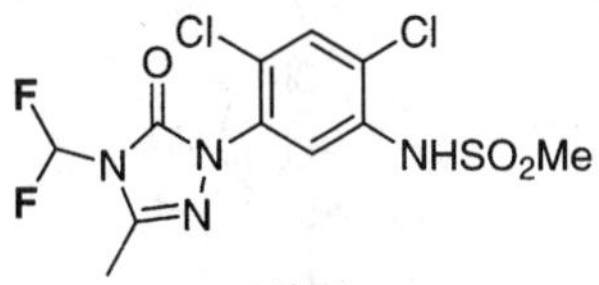

5.5 Fluorine-Containing Agrochemicals

sulfentrazon (FMC)

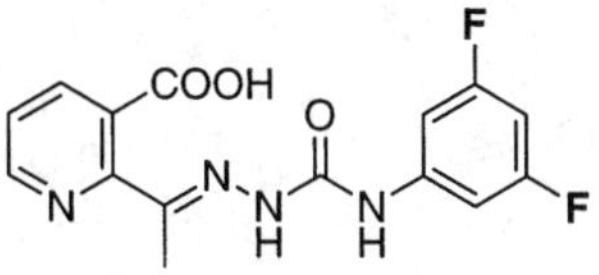

duflufrnzopyr (BASF)

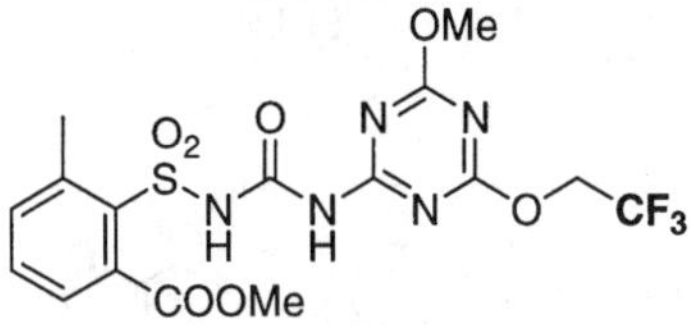

triflusulfuron-methyl (Du Pont)

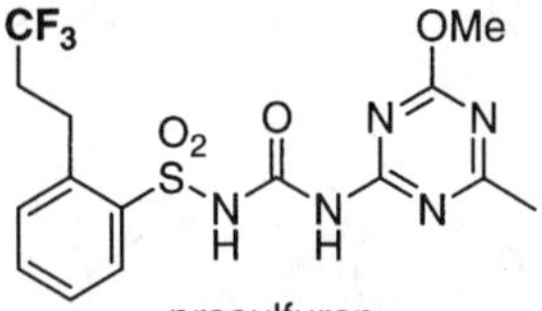

prosulfuron (Ciba-Geigy)

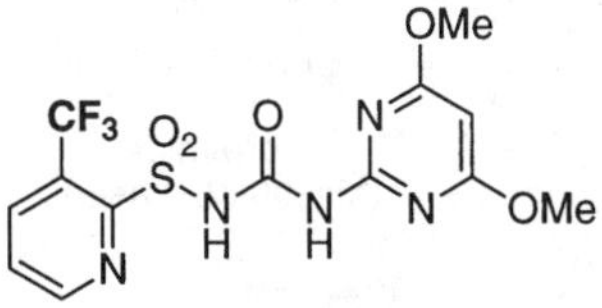

flazasulfuron (Ishihara)

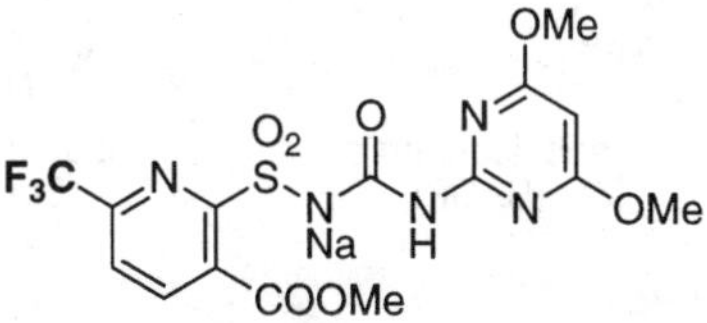

flupyrsulfuron-methyl-sodium (Du Pont)

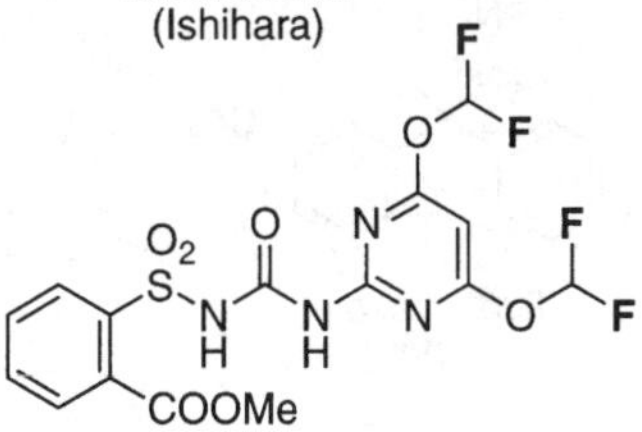

primisulfuron-methyl (Ciba-Geigy)

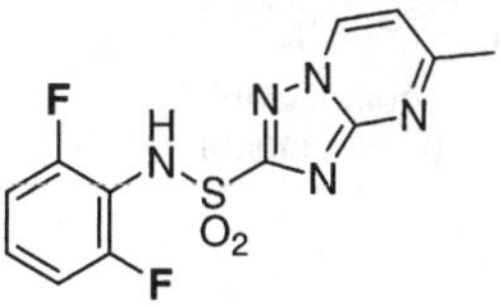

flumetsulam (Dow Elanco)

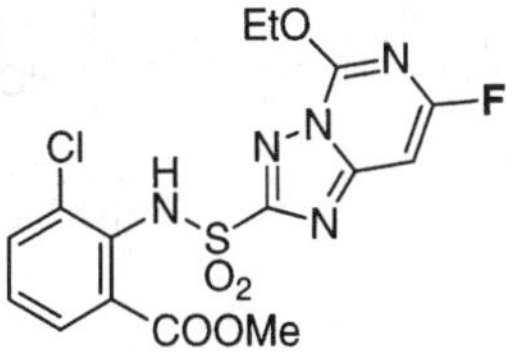

cloransulam-methyl (Dow Elanco)

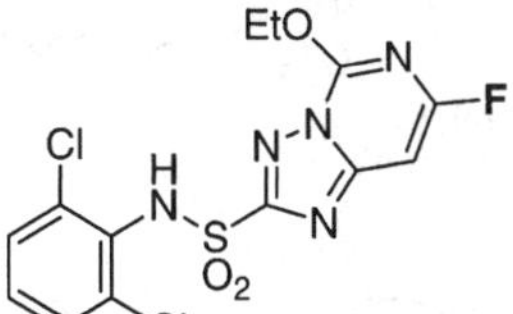

diclosulam (Dow Elanco)

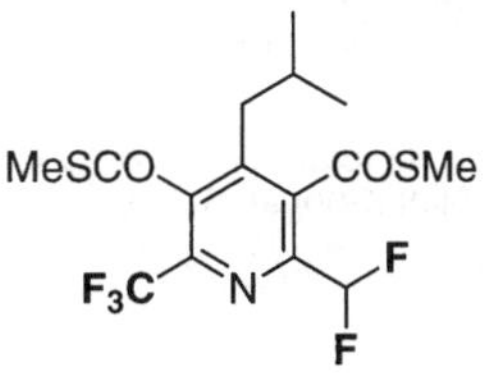

dithiopyr (Monsanto)

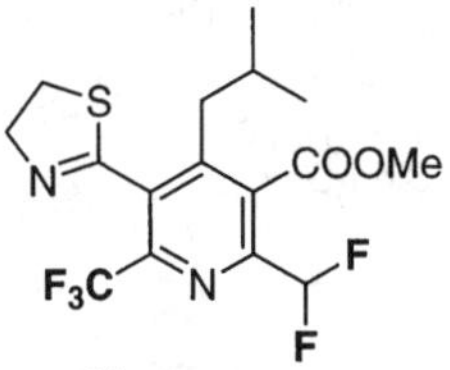

thiazopyr (Monsanto)

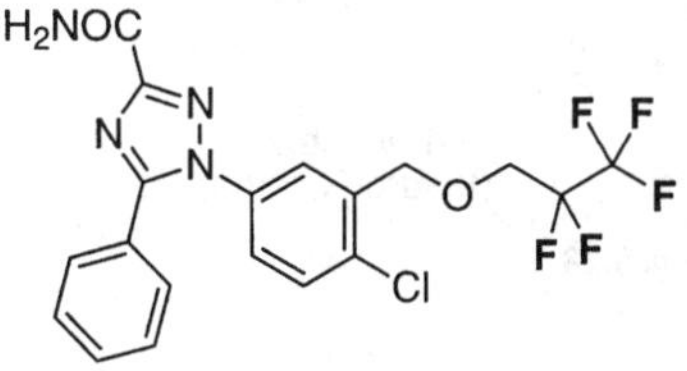

flupoxam (Kureha)

Fig. 5.27

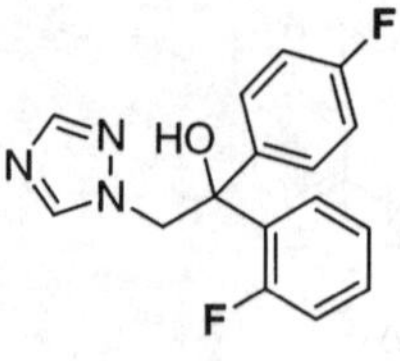

flutriafol
(ICI)

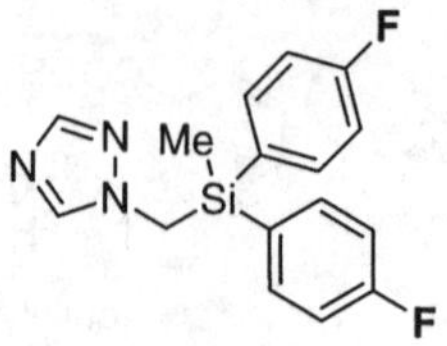

flusilazole
(Du Pont)

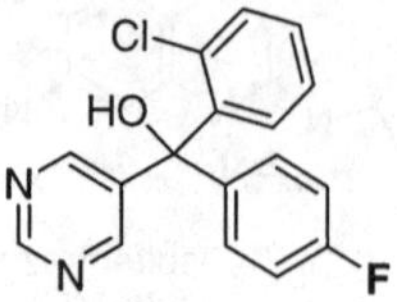

nuarimol
(Eli Lilly)

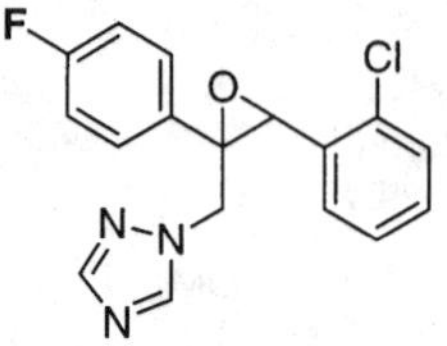

epoxiconazole
(BASF)

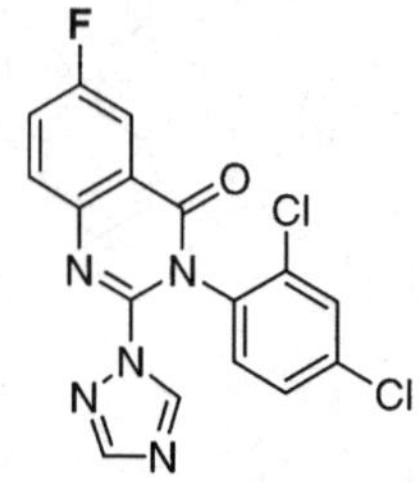

fluquinconazole
(Schering)

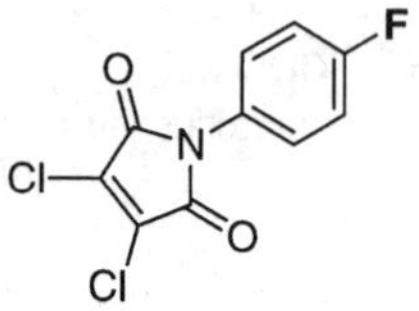

fluoroimide
(Mitsubishi-kasei)

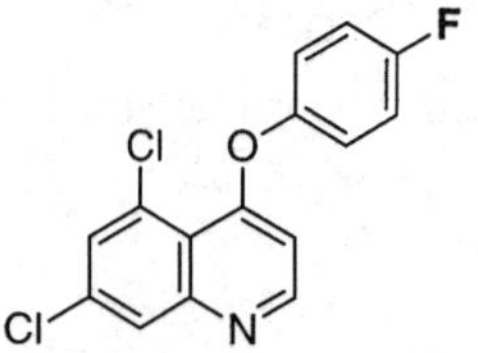

quinoxyfen
(Dow Elanco)

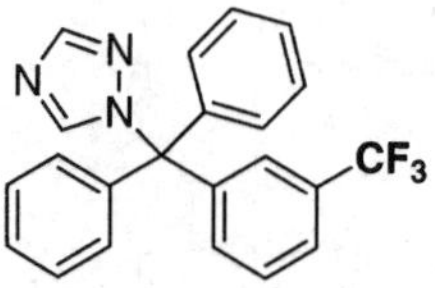

fluotrimazole
(Bayer)

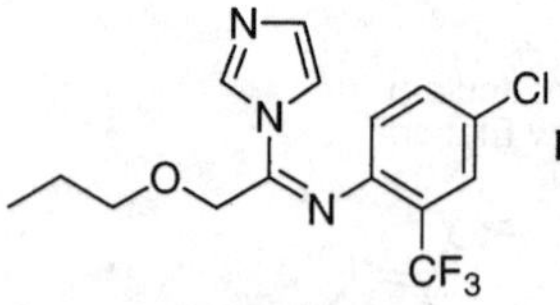

triflumizole
(Nihon-soda)

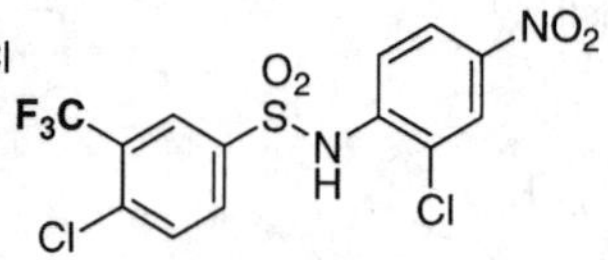

flusulfamide
(Mitsui-toatsu)

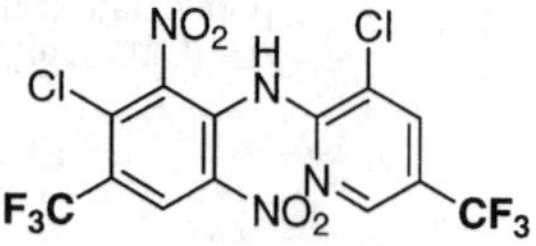

fluazinam
(Ishihara)

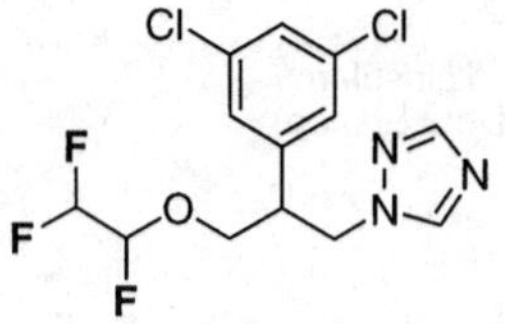

tetraconezole
(Isagro-Ricerca)

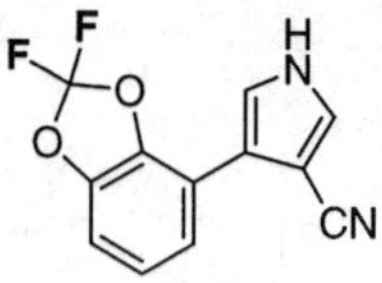

fludioxonil
(Ciba-Geigy)

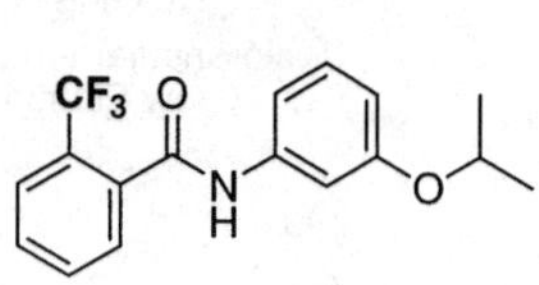

flutolanil
(Nihon-noyaku)

Fig. 5.28

5.5 Fluorine-Containing Agrochemicals

An analog of lactofen is ethoxyfen-ethyl (Fig. 5.25) which contains a lactic acid moiety [107]. This structural modification led to a general structural class of 2-(4-aryloxy)phenoxypropionates that show high and selective herbicidal activity against grasses. Of these, haloxyfop-*R*-methyl [108], cyhalofop-butyl [109] and chlodinafop-propargyl [110] have been launched.

Another type of photochemical herbicides are cyclic imides such as flumiclorac-pentyl [111], flumioxazin [112], pentaoxazone [113], and fluthiacet-methyl [114]. A common structural feature is a 4,5-disubstituted 2-fluorophenyl group substituted to an imide nitrogen or the isomers, as illustrated in Fig. 5.26.

The third important class of herbicides are sulfonylureas and triazopyrimidines that inihibit acetolactate synthetase. Examples [86] that contain a fluorine, difluoromethoxyl, or trifluoromethyl group are listed in Fig. 5.27.

5.5.3
Fungicides

Few fungicides are known that contain a fluorine functionality. Some examples are listed in Fig. 5.28. A synthetic route to flutriafol is shown in Scheme 5.34. The Friedel–Crafts reaction of fluorobenzene with 2-fluorobenzoyl chloride gives difluorobenzophenone, which is then allowed to react with a Grignard reagent to give flutriafol that is active against mildew of wheat.

Flusilazole is prepared by the route summarized in Scheme 5.35. 4-Bromofluorobenzene is lithiated and silylated with dichloro(chloromethyl)-methylsilane to give $Cl(CH_2)SiMe(4\text{-}F\text{-}C_6H_4)_2$ which is then reacted with sodio-

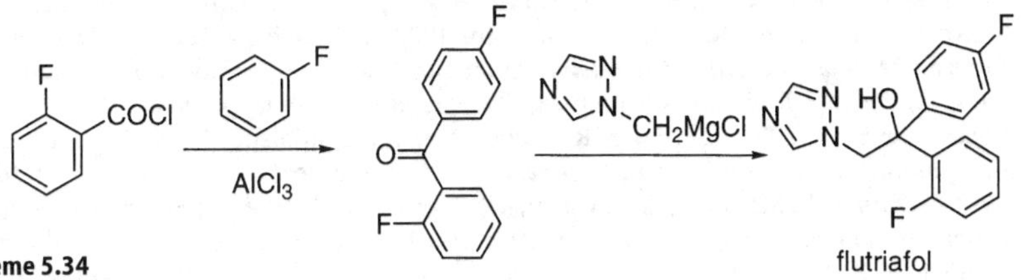

Scheme 5.34

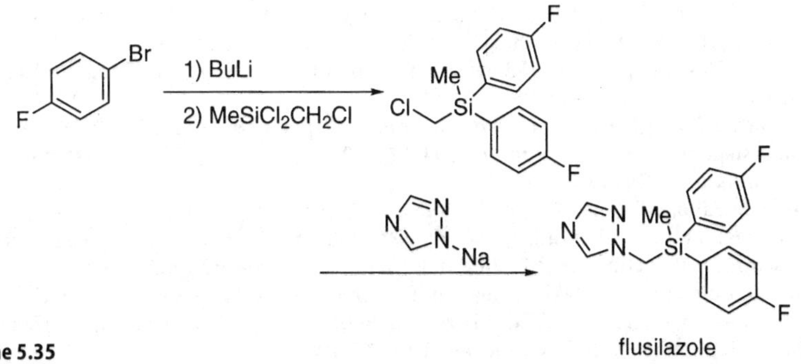

flutriafol

Scheme 5.35

flusilazole

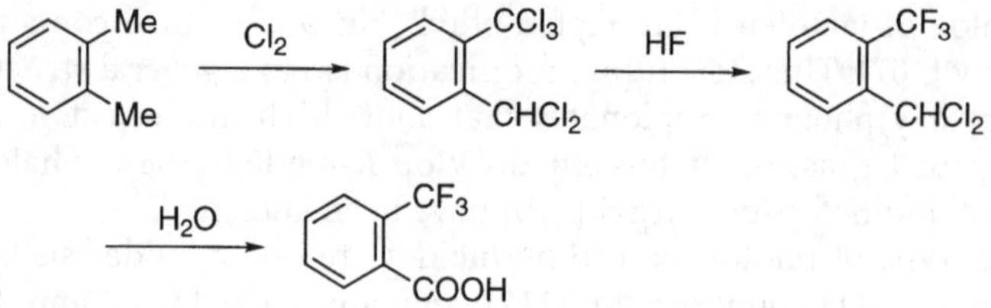

Scheme 5.36

triazole to give flusilazole. This compound can be applied to wheat, apples, grapes, peanuts, and sugar beet [87].

The *o*-trifluoromethylbenzoic acid moiety of flutolanil is prepared on a large scale by pentachlorination of *o*-xylene, fluorine substitution, and hydrolysis (Scheme 5.36).

The design and synthesis of agrochemicals have been carried out by replacing a substituent in a prototype lead compound with a fluorine functional group, e.g. CH_3 to CF_3, Cl to CF_3, CF_3O or $CHCF_2O$. The structure-activity relationship is now expressed in various numerical ways. Accordingly, exploitation of novel agrochemicals containing a variety of fluorine functionalities will be accelerated in the near future, and they will be marketed if the synthetic costs are reasonable.

References

1. (a) Filler, R.; Kobayashi, Y. *Biomedicinal Aspects of Fluorine Chemistry* ; Kodansha Ltd.: Tokyo, Elsevier Biomedical Press: Amsterdam, 1982; (b) Welch, J.T. *Selective Fluorination in Organic and Bioorganic Chemistry*, American Chemical Society: Washington, D.C., 1991; (c) Welch, J.T.; Eswarakrishnan, S. *Fluorine in Bioorganic Chemistry*, John Wiley & Sons: New York, 1991; (d) Filler, R.; Kobayashi, Y.; Yagupolskii, L. M. *Organofluorine Compounds in Medicinal Chemistry and Biomedical Applications*, Elsevier: Amsterdam, 1993; (e) Ojima, I.; McCarthy, J.R.; Welch, J.T. *Biomedical Frontiers of Fluorine Chemistry*, American Chemical Society: Washington, D.C., 1996; (f) Liebman, J.F.; Greenberg, A.; Dolbier, W.R. Jr. *Fluorine-containing Molecules, Structure, Reactivity, Synthesis, and Applications*, VCH Publishers, Inc.: New York, 1988
2. Welch, J.T. *Tetrahedron* 1987, *43*, 3123
3. (a) Ishikawa, N.; Kobayashi, Y. In: *Fusso no Kagobutsu*; Kodansha: Tokyo, 1979; Chapter V; (b) Kitazume, T.; Ishihara, T.; Taguchi, T. In: *Fusso no Kagaku*; Kodansha: Tokyo, 1993; Chap 5; (c) Yamabe, M.; Matsuo, M. In: *Fusso kei Zairyo no Saishin Doko*: Shi Emu Shi; Tokyo, 1994; Chap 3; (d) Kobayashi, Y.; Kumadaki, I.; Taguchi, T. *Fusso Yakugaku*; Hirokawa Shoten: Tokyo, 1993
4. (a) Harper, D.B.; O'Hagan, D. *Natural Product Reports* 1994, *11*. 123; (b) Marais, J.S.C. *Onderstepoort J. Vet. Sci. Anim. Int.* 1943, *18*, 203; (c) Marais, J.S.C. *Onderstepoort J. Vet. Sci. Anim. Int.* 1944, *20*, 67
5. Fried, J.; Sabo, E.F. *J. Am. Chem. Soc.* 1953, *75*, 2273; 1954, *76*, 1455
6. Heidelberger, C.; Chaudhuri, N. K.; Danneberg, P.; Mooren, D.; Griesbach, L.; Duschinsky, R.; Schnitzer, R.J.; Pleven, E.; Schneiner, J. *Nature* 1957, *179*, 663
7. For example: (a) Kumadaki, I.; Taguchi, T.; Shimokawa, K. In *Fusso Yakugaku*; Kobayashi, Y.; Kumadaki, I.; Taguchi, T. Eds.; Hirokawa Shoten: Tokyo, 1993; Chap 3; (b) Nakada, T.; Kwee, I.L.; Conboy, C.B. *J. Neurochem*, 1986, *48*, 198

8. Bondi, A. *J. Phys. Chem.* 1964, *68*, 441
9. Kumadaki, I. In: *Fusso Yakugaku*; Kobayashi, Y.; Kumadaki, I.; Taguchi, T. Eds.; Hirokawa Shoten: Tokyo, 1993; Chap 2
10. Welch, J.T. In: *Selective Fluorination in Organic and Bioorganic Chemistry*, Welch, J.T. Ed.; American Chemical Society: Washington, D.C., 1991; Chap 1
11. Card, R.J.; Hitz, W. D. *J. Am. Chem. Soc.* 1984, *106*, 5348
12. Walsh, C. *Adv. Enzymol* 1982, *55*, 197
13. Chu, D.T.W. In: *Organofluorine Compounds in Medicinal Chemistry and Biomedical Applications*, Filler, R.; Kobayashi, Y.; Yagupolskii, L. M. Eds.; Elsevier: Amsterdam, 1993; p 165
14. (a) Mascaretti, O. A.; Boschetti, C.E.; Danelon, G.O. In: *Organofluorine Compounds in Medicinal Chemistry and Biomedical Applications*, Filler, R.; Kobayashi, Y.; Yagupolskii, L. M. Eds.; Elsevier: Amsterdam, 1993; p 135; For an example of an antibacterial agent; (b) Li, Q.; Chu, T.W.; Claiborne, A.; Cooper, C.S.; Lee, C.M.; Raye, K.; Berst, K.B; Donner, P.; Wang, W.; Hasvold, L.; Fung, A.; Ma, Z.; Tufano, M.; Flamm, R.; Shen, L. L.; Baranowski, J.; Nilius, A.; Alder, J.; Meulbroek, J.; Marsh, K.; Crowell, D.; Hui, Y.; Seif, L.; Melcher, L. M.; Henry, R.; Spanton, S.; Faghih, R.; Klein, L. L.; Tanaka, K.; Plattner, J.J. *J. Med. Chem.* 1996, *39*, 3070
15. Yamada, S.; Yamamoto, K.; Masuno, H.; Ohta, M. *J. Med. Chem.* 1998, *41*, 1467
16. Schirlin, D.; Rondeau, J.M.; Podlogan, B.; Tardif, C.; Tarnus, C.; Van Dorsselaer, V.; Farr, R. In: *Biomedical Frontiers of Fluorine Chemistry*, Ojima, I.; McCarthy, J.R.; Welch, J.T. Eds.; American Chemical Society: Washington, D.C., 1996; Chap 13
17. Matsumura, Y.; Nakano, T.; Asai, T.; Morizawa, Y. In: *Biomedical Frontiers of Fluorine Chemistry*, Ojima, I.; McCarthy, J.R.; Welch, J.T. Eds.; American Chemical Society: Washington, D.C., 1996; Chap 6
18. Sham, H. L. In: *Biomedical Frontiers of Fluorine Chemistry*, Ojima, I.; McCarthy, J.R.; Welch, J.T. Eds.; American Chemical Society: Washington, D.C., 1996; Chap 14
19. Ojima, I.; Kuduk, S.D.; Chakravarty, S.; Ourevitch, M.; Begue, J.P. *J. Am. Chem. Soc.* 1997, *119*, 5519
20. Elliott, A. J. In: *Organofluorine Compounds in Medicinal Chemistry and Biomedical Applications*, Filler, R.; Kobayashi, Y.; Yagupolskii, L. M. Eds.; Elsevier: Amsterdam, 1993; p 209
21. Knunyants, I.L.; Yakobson, G.G. *Synthesis of Fluoroorganic Compounds*; Springer-Verlag: Berlin, 1985
22. Houston, M. E. Jr.; Honek, J.F. *J. Chem. Soc., Chem. Commun.* 1989, 761
23. Halazy, S.; Ehrhard, D.A.; Gerhart, F. *J. Am. Chem. Soc.* 1989, 111, 3484
24. (a) Chen, Q.-Y. *J. Fluorine Chem.* 1995, *72*, 241; (b) Chen, Q.-Y.; Wu, S.-W. *J. Chem. Soc., Chem. Commun.* 1989, 705
25. For reviews; (a) Uneyama, K. *J. Synth. Org. Chem. Jpn.* 1991, *49*, 612; (b) Yamazaki, T.; Kitazume, T. *J. Synth. Org. Chem. Jpn.* 1991, *49*, 721
26. (a) Hertel, L. W.; Kroin, J.S.; Misner, J.W.; Tustin, J.M. *J. Org. Chem.* 1988, *53*, 2406; (b) Kitagawa, O.; Taguchi, T.; Kobayashi, Y. *Tetrahedron Lett.* 1988, *29*, 1803; (c) Matsumura, Y.; Fujii, H.; Nakayama, T.; Morizawa, Y.; Yasuda, A. *J. Fluorine Chem.* 1992, *57*, 203
27. (a) Chou, T.S.; Heath, P.C.; Patterson, L. E.; Poteet, L. M.; Lakin, R.E.; Hunt, A. H. *Synthesis* 1992, 565; (b) Hertel, L. W.; Kroin, J.S.; Grossmann, C.S.; Grindey, G.B.; Dorr, A. F.; Storniolo, A. M. V.; Plunkett, W.; Gandhi, V.; Huang, P. In *Biomedical Frontiers of Fluorine Chemistry*, Ojima, I.; McCarthy, J.R.; Welch, J.T. Eds.; American Chemical Society: Washington, D.C., 1996; Chap 19
28. Fried, J.; John, V.; Szwedo, M. J. Jr.; Chen, C.-K.; Yang, C.-O.; Morinelli, T.A.; Okwu, A. K.; Halushka, P.V. *J. Am. Chem. Soc.* 1989, *111*, 4510
29. For example: (a) Doherty, A. M.; Sircar, I.; Kornberg, B.E.; Quin, J. III; Winters, R.T.; Kaltenbronn, J.S.; Taylor, M. D.; Batley, B.L.; Rapundalo, S.R.; Ryan, M. J.; Painchaud, C.A. *J. Med. Chem.* 1992, *35*, 2; (b) Yuan, W.; Munoz, B.; Wong, C.-H. *J. Med. Chem.* 1993, *36*, 211; (c) Robinson, R.P.; Donahue, K.M. *J. Org. Chem.* 1992, *57*, 7309

30. Schlosser, M. *Angew. Chem. Int. Ed.* 1998, *110*, 1496
31. Hale, J.J.; Mills, S.G.; MacCoss, M.; Shah, S.K.; Qi, H.; Marthre, D.; Cascieri, M. A.; Sadowski, S.; Strader, C.D.; MacIntyre, D.E.; Metzger, J.M. *J. Med. Chem.* 1996, *39*, 1760 and references cited therein
32. Kukhar, V.P; Soloshonok, V.A. *Fluorine-containing Amino Acids, Synthesis and Properties*, Kukhar, V.P; Soloshonok, V.A. Eds.; John Wiley & Sons: Chichester, 1994
33. Tolman, V. *Amino Acids* 1996, *11*, 15
34. Seidenfield, J.; Marton, L. J. *Biochem. Biophys. Res. Commun.* 1979, *86*, 1192
35. Tsushima, T.; Nishikawa, J.; Sato, T.; Tanida, H.; Tori, K.; Tsuji, T.; Misaki, S.; Suefuji, M. *Tetrahedron Lett.* 1980, *21*, 3593
36. Matsumura, Y.; Urushihara, M. *Fluorine-containing Amino Acids, Synthesis and Properties*, Kukhar, V.P; Soloshonok, V.A. Eds.; John Wiley & Sons: Chichester, 1994; Chap 6
37. For example: Soloshonok, V.A. In: *Biomedical Frontiers of Fluorine Chemistry*, Ojima, I.; McCarthy, J.R.; Welch, J.T. Eds.; American Chemical Society: Washington, D.C., 1996; Chap 2
38. Taguchi, T.; Tomizawa, G.; Nakajima, M.; Kobayashi, Y. *Chem. Pharm. Bull.* 1985, *33*, 4077
39. Takeuchi, Y.; Kamezaki, M.; Kirihara, K.; Haufe, G.; Laue, K.W.; Shibata, N. *Chem Pharm. Bull.* 1998, *46*, 1062
40. Chirakal, R.; Brown, K.L.; Firnau, G.; Garnett, E.S.; Hughes, D.W.; Sayer, B.G.; Smith, R.W. *J. Fluorine Chem.* 1987, *37*, 267
41. Kaneko, C.; Chiba, J.; Toyota, A.; Sato, M. *Chem. Pharm. Bull.* 1995, *43*, 760
42. Hudlicky, M. *J. Fluorine Chem.* 1993, *60*, 193
43. (a) Ojima, I.; Kato, K.; Nakahashi, K. *J. Org. Chem.* 1989, *54*, 4511; (b) Ojima, I. *Chem. Rev.* 1988, *88*, 1011
44. Soloshonok, V.A.; Soloshonok, I.V.; Kukhar, V.P.; Sxedas, V.K. *J. Org. Chem.* 1998, *63*, 1878
45. (a) Skiles, J.W.; Fuchs, V.; Miao, C.; Sorcek, R.; Grozinger, K.G.; Mauldin, S.C.; Vitous, J.; Mui, W.; Jacober, S.; Chow, G.; Matteo, M.; Skoog, M.; Weldon, S.M.; Possanza, G.; Keirns, J.; Letts, G.; Rosenthal, A. S. *J. Med. Chem.* 1992, *35*, 641; For an example of an inhibitor of human neutrophile elastase: (b) Edwards, P.D.; Andisik, D.W.; Bryant, C.A.; Ewing, B.; Gomes, B.; Lewis, J.J.; Rakiewicz, D.; Steelman, A.; Trainor, D.A.; Tuthill, P.A.; Mauger, R.C.; Veale, C.A.; Wildonger, R.A.; Williams, J.C.; Wolanin, D.J.; Zottola, M. *J. Med. Chem.* 1997, *40*, 1876; (c) Veals, C.A.; Bernstein, P.R.; Bohnert, C.M.; Brown, F. J.; Bryant, C.; Damewood J.R. Jr.; Earley, R.; Feeney, S.W.; Edwards, P.D.; Gomes, B.; Hulsizer, J.M.; Kosmider, B.J.; Krell, E.D.; Moore, G.; Salcedo, T.W.; Shaw, A.; Siberstein, D.S.; Steelman, G.B.; Stein, M.; Strimpler, A.; Thomas, R.M.; Vacek, E.P.; Williams, J.C.; Wolanin, D.J.; Wooson, S. *J. Med. Chem.* 1997, *40*, 3173
46. Walter, M.W.; Adlington, R.M.; Baldwin, J.E.; Schofield, C.J. *Tetrahedron* 1997, *53*, 7275
47. Walter, M.W.; Adlington, R.M.; Baldwin, J.E.; Chan, J.; Schofield, C.J. *Tetrahedron Lett.* 1995, *36*, 7761
48. Peet, N. P.; Burkhart, J.P.; Angelastro, M.R.; Giroux, E.L.; Mehdi, S.; Bey, P.; Kolb, M.; Neises, B.; Schirlin, D. *J. Med. Chem.* 1990, *33*, 394
49. Chatterjee, S.; Ator, M.A.; B.-Coyne, D.; Josef, K.; Wells, G.; Trpathy, R.; Iqbal, M.; Bihovsky, R.; Senadhi, S.E.; Mallya, S.; O'Kane, T.M.; McKenna, B.A.; Siman, R.; Mallamo, J.P. *J. Med. Chem.* 1997, *40*, 3820
50. Cregge, R.J.; Durham, S.L.; Farr, R.A.; Gallion, S.L.; Hare, C.M.; Hoffman, R.V.; Janusz, M.J.; Kim, H.-O.; Koehl, J.R.; Mehdi, S.; Metz, W. A.; Peet, N. P.; Pelton, J.T.; Schreuder, H. A.; Sunder, S.; Tardif, C. *J. Med. Chem.* 1998, *41*, 2461
51. (a) Shirlin, D.; Tarnus, C.; Galtzer, S.; Remy, J.M. *Bioorg. Med. Chem. Lett.* 1992, *2*, 651; (b) For an example of an inhibitor of human renin: Patel, D.V.; R.-Gauvin, K.; Ryono, D.E.; Free, C.A.; Smith, S.A.; Petrillo, E.W. Jr. *J. Med. Chem.* 1993, *36*, 2431
52. Kirk, K.L.; Filler, R. In: *Biomedical Frontiers of Fluorine Chemistry*, Ojima, I.; McCarthy, J.R.; Welch, J.T. Eds.; American Chemical Society: Washington, D.C., 1996; Chap 1
53. Wolfe, M.S.; Citron, M.; Diehl, T.S.; Xia, W.; Donkor, I.O.; Selkoe, D.J. *J. Med. Chem.* 1998, *41*, 6

54. (a) Taylor, N. F. *Fluorinated Carbohydrates, Chemical and Biochemical Aspects*; American Chemical Society: Washington, D.C., 1988; (b) Penglis, A. A. E. *Adv Carbohydr Chem. Biochem.* 1981, *38*, 195
55. Iida, T. In: *Fusso Yakugaku*; Kobayashi, Y.; Kumadaki, I.; Taguchi, T. Eds.; Hirokawa Shoten: Tokyo, 1993; Chap 3 (Moissan, H. L. *Fluor et Composes*, Steinheil, G.: Paris, p 244.)
56. Burkart, M.D.; Zhang, Z.; Hung, S.-C.; Wong, C.-H. *J. Am. Chem. Soc.* 1997, *119*, 11743
57. (a) Herdewijn, P.; van Aerschot, A.; Kerremans, L. *Nucleosides Nucleotides* 1989, *8*, 65; (b) Pankiewicz, K.W.; Watanabe, K.A. *J. Fluorine Chem.* 1993, *64*, 15; (c) Bergstrom, D.E.; Swartling, D.J. In: *Fluorine-containing Molecules, Structure, Reactivity, Synthesis, and Applications*, Liebman, J.F.; Greenberg, A.; Dolbier, W. R. Jr. Eds.; VCH Publishers, Inc.: New York, 1988; Chap 11
58. Welch, J.T. In: *Selective Fluorination in Organic and Bioorganic Chemistry*, Welch, J.T. Ed.; American Chemical Society: Washington, D.C., 1991; Chap 1
59. Schwartz, B.; Cech, D.; Reefschlager, J. *J. Prakt. Chem.* 1984, *326*, 985
60. Kobayashi, Y.; Kumadaki, I.; Yamamoto, K. *J. Chem. Soc., Chem. Commun.* 1977, 536
61. Santi, D.; Pogolotti, A. L. Jr.; Newman, E.M.; Wataya, Y. In: *Biomedicinal Aspects of Fluorine Chemistry*; Filler, R.; Kobayashi, Y. Eds.; Kodansha Ltd.: Tokyo, Elsevier Biomedical Press: Amsterdam, 1982; p 123
62. (a) McCarthy, J.R.; Sunkara, P.S.; Matthews, D.P.; Bitonti, A. J.; Jarvi, E.T.; Sabol, J.S.; Resvick, R.J.; Huber, E.W.; van der Donk, W.; Yu, G.; Stubbe, J. In *Biomedical Frontiers of Fluorine Chemistry*, Ojima, I.; McCarthy, J.R.; Welch, J.T. Eds.; American Chemical Society: Washington, D.C., 1996; Chap 18; (b) Matsuda, A.; Itoh, H.; Takenuki, K.; Susaki, T.; Ueda, T. *Chem. Pharm. Bull*. 1988, *36*, 945; (c) Matthews, D.P.; Persichetti, R.A.; Sabol, R.A.; Stewart, N. T.; McCarthy, J.R. *Nucleosides Nucleotides* 1993, *12*, 115
63. De Clercq, E. *Clinical Microbiology Reviews* 1997, *10*, 674
64. Prous, J.R. *Drugs Fut.* 1994, *19*, 221
65. Prous, J.R. *Drugs Fut.* 1995, *20*, 761
66. Yasuda, A, In: *Organofluorine Compounds in Medicinal Chemistry and Biomedical Applications*, Filler, R.; Kobayashi, Y.; Yagupolskii, L. M. Eds.; Elsevier: Amsterdam, 1993; p 275
67. Fried, J.; Mitra, M.; Nagarajan, M.; Mehrotra, M.M. *J. Med. Chem.* 1980, *23*, 234.
68. (a) Nakano, T.; Makino, M.; Morizawa, Y.; Matsumura, Y. *Angew. Chem. Int. Ed. Engl.* 1996, *35*, 1019; (b) Chang, C.-S.; Negishi, M.; Nakano, T.; Morizawa, Y.; Matsumura, Y.; Ichikawa, A. *Prostaglandins* 1997, *53*, 83
69. Fried, J.; Hallinan, E.A.; Szewedo, M. Jr. *J. Am. Chem. Soc.* 1984, *106*, 3871
70. Magerlein, B.J.; Miller, W. L. *Prostaglandins* 1975, *9*, 527
71. Ohshima, E.; Sai, H.; Takatsuto, S.; Ikekawa, N.; Kobayashi, Y. Tanaka, Y. DeLuca, H. F. *Chem. Pharm. Bull.* 1984, *32*, 3525
72. (a) Kobayashi, Y.; Taguchi, T.; Mitsuhashi, S.; Eguchi, T.; Ohshima, E.; Ikekawa, N. *Chem. Pharm. Bull.* 1982, *30*, 4297; (b) Taguchi, T. In: *Fusso Yakugaku*; Kobayashi, Y.; Kumadaki, I.; Taguchi, T. Eds.; Hirokawa Shoten: Tokyo, 1993; Chap 3
73. Posner, G.H.; Lee, J.K.; Wang, Q.; Peleg, S.; Burke, M.; Brem, H.; Dolan, P.; Kensler, T.W. *J. Med. Chem.* 1998, *41*, 3008
74. Elliott, A. J. In: *Organofluorine Compounds in Medicinal Chemistry and Biomedical Applications*, Filler, R.; Kobayashi, Y.; Yagupolskii, L. M. Eds.; Elsevier: Amsterdam, 1993; p 209
75. Molloy, B.; Schmiegel, K.K. *German Patent* 2,500,110 (1975)
76. Christensen, J.A.; Squires, R.F. *German Patent* 2,404, 113 (1974)
77. (a) Chu, D.T.W. In: *Organofluorine Compounds in Medicinal Chemistry and Biomedical Applications*, Filler, R.; Kobayashi, Y.; Yagupolskii, L. M. Eds.; Elsevier: Amsterdam, 1993; p 165; (b) Hayakawa, I.; Atarashi, S. In: *Fusso Yakugaku*; Kobayashi, Y.; Kumadaki, I.; Taguchi, T. Eds.; Hirokawa Shoten: Tokyo, 1993; Chap 3
78. Fernandes, P.B. *International Telesymposium on Quinolones, Volume for Discussion*, J.R. Prous Science Publishers: Barcelona, 1989

79. (a) Nagata, W.; Sendo, Y. In: *Fusso Yakugaku*; Kobayashi, Y.; Kumadaki, I.; Taguchi, T. Eds.; Hirokawa Shoten: Tokyo, 1993; Chap 3; (b) Sutherland, R.; Croydon, E.A.P.; Rolinson, G.N. *Brit. Med. J.* 1970, *4*, 455
80. (a) Tsushima, T.; Kawada, K. In: *Fusso Yakugaku*; Kobayashi, Y.; Kumadaki, I.; Taguchi, T. Eds.; Hirokawa Shoten: Tokyo, 1993; Chap 3; (b) Walsh, C. *Tetrahedron* 1982, *38*, 871
81. Filler, R. In: *Organofluorine Compounds in Medicinal Chemistry and Biomedical Applications*, Filler, R.; Kobayashi, Y.; Yagupolskii, L. M. Eds.; Elsevier: Amsterdam, 1993; p 1
82. Tsuji, T.; Satoh, H.; Narisada, M.; Hamashima, Y.; Yoshida, T. *J. Antibiot.* 1965, *38*, 466
83. Halpern, D.F. In: *Organofluorine Compounds in Medicinal Chemistry and Biomedical Applications*, Filler, R.; Kobayashi, Y.; Yagupolskii, L. M. Eds.; Elsevier: Amsterdam, 1993; p 101
84. (a) Yokoyama, K.; Suyama, T.; Naito, R. In: *Biomedicinal Aspects of Fluorine Chemistry*; Filler, R.; Kobayashi, Y. Eds.; Kodansha Ltd.: Tokyo, Elsevier Biomedical Press: Amsterdam, 1982; p 191; (b) Yokoyama, K. *Fusso Yakugaku*; Kobayashi, Y.; Kumadaki, I.; Taguchi, T. Eds.; Hirokawa Shoten: Tokyo, 1993; Chap 3
85. Clark, L. C. Jr.; Gollan, F. *Science*, 1966, *152*, 1755
86. Fowler, J.S. In: *Organofluorine Compounds in Medicinal Chemistry and Biomedical Applications*, Filler, R.; Kobayashi, Y.; Yagupolskii, L. M. Eds.; Elsevier: Amsterdam, 1993; p 309
87. K. Ikura, *Development of New Agrochemicals*, CMC: Tokyo, 1997
88. *Pesticide Data Book*, 3rd Edn, Soft Science Publications: Tokyo, 1997
89. Yoshioka, H.; Takayama, C.; Matsuo, N. *Yuki Gosei Kagaku Kyokai Shi*, 1984, *42*, 809
90. Kumai, S. *Newest Aspects of Fluoro Functional Material*, CMC: Tokyo, 1994; p 209
91. Van Daalen, J.J.; Maltzer, J.; Mulder, R.; Wellinga, K. *Naturwissenshaften*, 1972, *59*, 312
92. Zoebelein, G.; Hammann, I.; Sirrenberg, W. *Z. Ang. Ent.*, 1980, *89*, 289
93. EP 71,279; *Chem. Abstr.*, 1983, *99*, 53395
94. EP 79,311; *Chem. Abstr.*, 1983, *99*, 139780
95. Elliott, M.; Farnham, A.; Janes, N.; Needham, R.; Pulman, D. Stevenson, J. *Nature*, 1973, *246*, 169
96. Elliott, M.; Farnham, A.; Janes, N.; Needham, R.; Pulman, D. *Pestic. Sci.*, 1975, *6*, 537
97. USP 4,341,796; *Chem. Abstr.*, 1982, *97*, 194606
98. Engel, J.F.; Plummer, E.L.; Stewart, R.R.; VanSaun, W. A.; Montgomery, R.E.; Cruickshank, P.A.; Harnish, W. N.; Nethery, A. A.; Crosby, G.A. *Pesticide Chemistry*, Pergamon Press, Vol. 1, 1982; p 101
99. McDonald, E.; Pnja, N.; Justsum, A.R. *Proc. Brit. Crop Protect. Conf.-Pests & Diseases*, 1986, 199
100. Arlt, D.; Jautelat, M.; Lantzsch, R. *Angew. Chem. Int. Ed. Engl.*, 1981, *20*, 703
101. Laider, D.A.; Milner, D.J. *J. Organomet. Chem.*, 1984, *270*, 121
102. Bellus, D. *Pure Appl. Chem.*, 1985, *57*, 1827
103. Fujita, M.; Hiyama, T. *Yuki Gosei Kagaku Kyokai Shi*, 1987, *45*, 664; Fujita, M.; Kondo, K.; Hiyama, T. *Bull. Chem. Soc. Jpn.*, 1987, *60*, 4385; Fujita, M.; Hiyama, T. *Tetrahedron Lett.*, 1986, *27*, 3659; Fujita, M.; Hiyama, T.; Kondo, K. *Tetrahedron Lett.*, 1986, *27*, 2139
104. Ohno, N.; Fujimoto, K.; Okuno, Y.; Mizutani, T.; Hirano, M.; Itaya, N.; Honda, T.; Yoshioka, H. *Pestic. Sci.*, 1976, *7*, 241
105. Whitney, W.K.; Wettstein, K. *Proc. Brit. Crop Protect. Conf. -Pests & Diseases*, 1979, 387
106. Henrik, C.A.; Garsia, B.A.; Staal, G.B.; Cerf, D.C.; Anderson, R.J.; Gill, K.; Chinn, H.R.; Labobitz, J.N.; Leippe, M.M.; Woo, S.L.; Carney, R.L.; Gordon, D.C.; Kohn, G.K. *Pestic. Sci.*, 1980, *11*, 224
107. DE 3731609, JP 01–113396; *Chem. Abstr.*, 1989, *111*, 58018
108. EP 3,890; *Chem. Abstr.*, 1980, *92*, 76305
109. EP 302,203; *Chem. Abstr.*, 1989, *111*, 96861
110. EP 191,736; *Chem. Abstr.*, 1986, *105*, 221012
111. EP 83,055, JP 58–110566; *Chem. Abstr.*, 1983, *99*, 175591
112. JP 61–30586; *Chem. Abstr.*, 1986, *105*, 56350
113. JP 60–226270, WO 87/2357; *Chem. Abstr.*, 1987, *107*, 198300
114. EP 273417; *Chem. Abstr.*, 1988, *109*, 211069

CHAPTER 6
Fluorine-Containing Materials

6.1
Fluorine Effect in Materials

In general, the elements of Group 17 are electronegative and make a strong bond to carbon. As Table 1.1 in Chap. 1 shows, fluorine is the most electronegative of all elements. As fluorine forms a covalent bond using $2s$- and $2p$-orbitals, its van der Waals and covalent bond radii are the smallest next to those of hydrogen. The C–F bond energy is the largest among carbon–halogen bonds. Accordingly, various kinds of perfluorocarbons are stable compounds in sharp contrast to perchlorocarbons: only carbon tetrachloride is stable. The structures of perfluorodecane and its hydrocarbon counterpart are compared in Fig. 6.1. Calculation by MOPAC gives an in-plane zigzag conformation for decane and a helical cylinder-like structure for perfluorodecane due to the 1,3 steric repulsion caused by the large size of fluorine and also to the electronic repulsion between unshared electron pairs of fluorine atoms. Thus, a perfluoropolymethylene chain resembles a stiff rod covered in unshared electrons. The characteristic features correspond to the stability of PFCs against biological, chemical and physical stimulus. In contrast, partial fluorination induces a strong local dipole moment that causes various electrical effects useful for

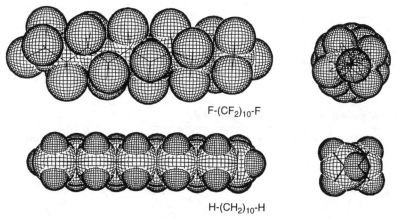

F-(CF$_2$)$_{10}$-F

H-(CH$_2$)$_{10}$-H

Fig. 6.1. Conformations of F-(CF$_2$)$_{10}$-F and H-(CH$_2$)$_{10}$-H calculated by MOPAC

functional materials. In the following sections, the salient physical features of organofluorine compounds will be briefly reviewed [1–5].

6.1.1
Boiling Points and Melting Points

Because fluorine has a much larger atomic weight (19) and only a 10% larger van der Waals radius than hydrogen, PFCs usually have specific gravities larger than hydrocarbons. For example, the specific gravity of hexane is 0.659, while that of perfluorohexane is 1.680, roughly 2.50 times larger. On the basis of the high specific gravities and molecular weights, PFCs should have higher boiling points than hydrocarbons; however, the differences are extremely small. As Table 6.1 shows, PFCs of less than 3 carbons indeed have higher boiling points, but those of more than 4 carbons show lower boiling points. This fact is attributed to the low polarity and low intermolecular attractive forces of PFCs.

Table 6.2 shows the boiling points of fluoro- and chloromethanes. It is easy to see that the boiling points of chloromethanes rise according to the number of chlorine atoms. In contrast, the boiling points of CH_3F and CH_2F_2 indeed rise in proportion to the number of fluorines, but those of CHF_3 and CF_4 become lower. The boiling point of CF_4 is the lowest among fluoromethanes owing to the properties discussed above. The fact that CH_2F_2 has a boiling point higher than that of CHF_3 is attributed to the polar effect and hydrogen bonding of $H \cdots F$.

Table 6.1. Boiling points of $F\text{-}(CF_2)_n\text{-}F$ and $H\text{-}(CH_2)_n\text{-}H$

$F\text{-}(CF_2)_n\text{-}F$	Boiling Point (°C)	$H\text{-}(CH_2)_n\text{-}F$	Boiling Point (°C)
CF_4	−128	CH_4	−161
C_2F_6	−78	C_2H_6	−88
C_3F_8	−37	C_3H_8	−42
$n\text{-}C_4F_{10}$	−2.2	$n\text{-}C_4H_{10}$	−0.5
$n\text{-}C_5F_{12}$	59	$n\text{-}C_5H_{12}$	36
$n\text{-}C_6F_{14}$	58	$n\text{-}C_6H_{14}$	69
$n\text{-}C_7F_{16}$	83	$n\text{-}C_7H_{16}$	98
$n\text{-}C_8F_{18}$	103	$n\text{-}C_8H_{18}$	126
$n\text{-}C_9F_{20}$	125	$n\text{-}C_9H_{20}$	151

Table 6.2. Boiling points of $CH_{4-n}F_n$ and $CH_{4-n}Cl_n$

$CH_{4-n}F_n$	Boiling Point (°C)	$CH_{4-n}Cl_n$	Boiling Point (°C)
CH_4	−161	CH_4	−161
CH_3F	−78	CH_3Cl	−24
CH_2F_2	−52	CH_2Cl_2	40
CHF_3	−82	$CHCl_3$	61
CF_4	−128	CCl_4	77

Table 6.3. Boiling points of $C_6H_{6-n}F_n$

Structure	Benzene	Fluorobenzene	1,2-difluorobenzene	1,3-difluorobenzene	1,4-difluorobenzene
BP	80 °C	85 °C	92 °C	83 °C	89 °C

Structure	1,2,3-triF	1,2,4-triF	1,3,5-triF	1,2,3,4-tetraF
BP	95 °C	89 °C	76 °C	95 °C

Structure	1,2,3,5-tetraF	1,2,4,5-tetraF	pentaF	hexaF
BP	88 °C	89 °C	89 °C	82 °C

The boiling points of fluorobenzenes, listed in Table 6.3, fall in the range 76 to 95 °C, regardless of the number of fluorines and are slightly dependent on the symmetry of substitution. Those which induce large polarization, e.g. o-difluorobenzene, 1,2,3-trifluorobenzene and 1,2,3,4-tetrafluorobenzene, boil at relatively higher temperatures of 92 to 95 °C, whereas the symmetric and less polarized derivatives 1,3,5-trifluorobenzene and hexafluorobenzene boil at 76 and 82 °C, respectively.

For high molecular weight compounds, PFCs have low boiling points, weak intermolecular interactions and thus high vapor pressures. Therefore, PFCs are generally volatile and readily sublime even below their melting points and, for this reason, handling and storage should be carried out with care.

Owing to the salient features described above, fluorine-containing compounds have found many industrial applications. In particular, chlorofluorocarbons (CFCs) have been used for refrigerants due to their stability and low boiling points. Nowadays CFCs are being replaced by hydrofluorocarbons due to their high ozone-depletion potential.

The melting points of linear hydrocarbons and their perfluorinated derivatives are shown in Fig. 6.2. Melting points of hydrocarbons with even numbers of carbons are, in general, higher than those with an odd number of carbons. The odd-even rule is attributed to the symmetry of the zigzag conformation of hydrocarbons (vide supra). In contrast, the odd-even rule is only applicable to PFCs with less than 4 carbons, probably because in PFCs with more than 5 carbons, the molecule becomes stiff enough to lose the symmetry

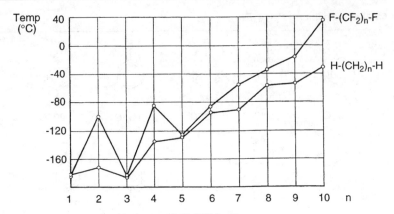

Fig. 6.2. Melting points of $F\text{-}(CF_2)_n\text{-}F$ and $H\text{-}(CH_2)_n\text{-}H$

difference depending on the number of carbons. Partially fluorinated compounds melt at temperatures only slightly higher than those of the parent compounds.

6.1.2
Solubility

PFCs are insoluble both in water and hydrocarbons. Since PFCs have higher specific gravities, they separate from hydrocarbons to lie below the organic phase. Organic solvents that dissolve PFCs are limited to aromatics, acetone, and chlorohydrocarbons in addition to polyfluorinated compounds. The fact that PFCs are capable of dissolving oxygen and carbon dioxide is attributed to the stiffness of PFC molecules which generate more vacant space where the gaseous molecules can slip in, rather than to attractive interaction of fluorine with the gaseous molecules. The unique features of PFCs have found applications in artificial blood.

In contrast, partially fluorinated compounds are highly polar and thus dissolve polar materials. For example, 2,2,2-trifluoroethanol and hexafluoro-2-propanol dissolve polar resins like nylons. In addition, trifluoroacetaldehyde and hexafluoroacetone give stable hydrate forms, hemiacetals, in addition to acetals.

6.1.3
Surface Tension

The surface tensions (γ) of some PFCs are listed in Table 6.4. These fall in the range 10 to 15 dyne/cm, smaller than those of the corresponding hydrocarbons (15–25 dyne/cm) and much smaller than water (72 dyne/cm). The surfaces of solid PFCs also have small critical surface tensions (γ_c): 18 dyne/cm for poly(tetrafluoroethene); cf. polyethylene, 31 dyne/cm; poly(vinyl chloride), 39 dyne/cm. Thus, poly(tetrafluoroethene) is water- and oil-repellent. Such properties

Table 6.4. Surface tensions of fluorocarbons and hydrocarbons at 20 °C

Fluorocarbons	γ (dyne/cm)	Hydrocarbons	γ (dyne/cm)
$n\text{-}C_5F_{12}$	9.9	$n\text{-}C_5H_{12}$	13.0
$n\text{-}C_7F_{16}$	7.1	$n\text{-}C_7H_{16}$	20.3
$n\text{-}C_8F_{18}$	13.6	$n\text{-}C_8H_{18}$	21.8
$n\text{-}C_4F_9\text{-}O\text{-}n\text{-}C_4F_9$	13.4	$n\text{-}C_4H_9\text{-}O\text{-}n\text{-}C_4H_9$	22.9
$n\text{-}C_3F_7COOH$	16.3	$n\text{-}C_3H_7COOH$	26.8

are also found in polymers which have a perfluoroalkyl side chain and perfluoroalkanoic acids of an appropriate length. Accordingly, these are used for water- and oil-repellents and surface modifiers [6].

6.1.4
Refractive Index

In general, fluorine compounds have small refractive indices due to the small size of fluorine and the stiffness of the molecules. For example, the refractive index of perfluoropentane is 1.245, smaller than that of pentane (1.358). The refractive indices of partially fluorinated compounds are also smaller than the parent compounds: 2,2,2-trifluoroethanol (1.291) and ethanol (1.362). Such properties apply to polymers with a perfluoroalkyl group also; these polymers have found application as optical fibers [6]. Since PFCs are transparent in the near infrared region, in contrast to hydrocarbons, PFC polymers have the advantage of a small loss of optical transmission in addition to heat resistance. Accordingly, PFC materials have potential wide application for high quality optical fibers.

6.1.5
Viscosity

Figure 6.3 shows the absolute viscosity and kinetic viscosity of heptane and perfluoroheptane in relation to temperature. As is easily seen, perfluoroheptane is more viscous than heptane. However, the kinetic viscosities, namely the viscosity per specific gravity, are comparable. Moreover, as the temperature dependency of the viscosity of perfluoroheptane is larger than that of heptane, perfluoroheptane becomes less viscous at higher temperatures. This phenomenon is attributed to the mobility of stiffness and the larger molecular weight of perfluoroheptane.

The viscosity of partially fluorinated compounds may increase due to increased polarity. However, compared with other halogens, such an effect is much smaller and thus fluorine is often used for liquid crystals of low viscosity and high polarity.

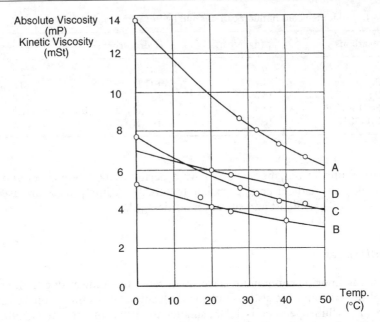

Fig. 6.3. Viscosities of perfluoroheptane and heptane: *A* absolute viscosity of perfluoroheptane; *B* absolute viscosity of heptane; *C* kinetic viscosity of perfluoroheptane; *D* kinetic viscosity of heptane

6.2 Chlorofluorocarbons, Hydrochlorofluorocarbons, Hydrofluorocarbons, and Alternatives

Chlorofluorocarbons (CFCs), halogenated volatile compounds with chlorine and fluorine atoms, have typical characteristics such as an easy phase change between liquid and vapor, owing to a small heat of vaporization, appropriate boiling points, thermal and chemical stability, low flammability, and low toxicity. These compounds have wide applications for refrigeration, air conditioning, cleaning, degreasing of metals, solder flux removal, dewatering of printed circuit boards, degreasing semiconductors, foaming (blowing) agents to create rigid plastic foams, and paint and aerosol formulations [7–10].

6.2.1
Brief History

In 1974, it was conjectured by Rowland and Molina that the high stability of chlorofluorocarbons enabled the compounds to pass unchanged through the troposphere to the stratosphere, where the intense UV radiation from the sun could cause the C–Cl bonds to homolytically break down, releasing chlorine atoms, which were known from laboratory experiments to catalyze the destruction of ozone [11]. The mechanism suggested for the reaction is illustrated in Scheme 6.1.

6.2 Chlorofluoro-, Hydrochlorofluoro-, Hydrofluorocarbons, and Alternatives

Scheme 6.1
$$CF_2Cl_2 \xrightarrow{h\nu} Cl\bullet + \bullet CF_2Cl$$
$$Cl\bullet + O_3 \longrightarrow ClO\bullet + O_2$$
$$ClO\bullet + O \longrightarrow Cl\bullet + O_2$$

The trace gas ozone, present throughout the Earth's atmosphere at peak concentrations of 10 ppm, plays a critical role for life on Earth by absorbing biologically harmful UV radiation in the short-wavelength regions below 320 nanometers (nm) (Scheme 6.2). The chlorine atom (Cl•) in the stratosphere inhibits the photochemical generation of ozone from oxygen. On the other hand, bromine-containing molecules (Halons), including CF_3Br, are suspected to be more harmful to the ozone layer. Because of their long atmospheric lifetime and large infrared absorption, CFCs are also cited as possible contributors to global warming.

Scheme 6.2
$$O_2 \xrightarrow{h\nu (<240 \text{ nm})} 2 O$$
$$O + O_2 \longrightarrow O_3$$
$$O_3 \xrightarrow{h\nu (210-320 \text{ nm})} O + O_2$$

In September 1987, the first international agreement for the protection of the global environment worldwide concerning the depletion of the ozone layer, known as the Montreal Protocol, was signed. It was subsequently revised in June 1990. The 1990 Montreal Protocol mandated a total phase-out of production and consumption of CFCs, Halons, and carbon tetrachloride to be achieved by the year 2000. In 1992, the Montreal Protocol was extensively revised at an intergovernmental conference in Copenhagen, bringing forward the date of the deadline for the phase-out of CFCs and, for the first time, introducing a specific timetable for the phase-out of hydrochlorofluorocarbons (HCFCs). The regulation of HCFCs was tightened even more at the Vienna Conference in 1995, as shown in Table 6.5.

Many chemical manufacturers have focused on the development of hydrochlorofluorocarbons (HCFCs) and hydrofluorocarbons (HFCs) to provide substitutes which possess performance and properties similar to those of CFCs. HCFCs have much reduced ozone-depletion potentials (ODPs), and HFCs have zero ODP compared with CFCs. Typical CFCs, HCFCs, and HFCs, including chemical formulae and physical properties, are listed in Table 6.6. A fundamental nomenclature by code number follows: (1) the first digit on the right is the number of fluorine atoms in a CFC compound; (2) the second digit from the right is the number of hydrogen atoms *plus one*; (3) the third digit is the number of carbons *minus one*, and omitted if zero; (4) any discrepancy with a valency of four for carbon is made up with chlorine(s); (5) if bromine is present, the letter B is put with its number; (6) a cyclic isomer is denoted as starting with

Table 6.5. Regulatory reduction in the consumption of CFCs and HCFCs in % (Montreal Protocol agreed and in force at present)

Cap level[c]	CFCs[a]	Other CFCs[b]	HCFCs 3.1%
Base year	1986	1989	1989
1993	–	– 20	–
1994	– 75	– 75	–
1995	– 75	– 75	–
1996	–100	–100	freeze
2000			freeze
2004			– 35
2007			– 35
2008			– 35
2010			– 65
2012			– 65
2013			– 65
2015			– 90
2020			– 99.5
2030			–100

[a] CFCs include 11, 12, 113, 113a, 114, 114a, and 115.
[b] Other CFCs are 13, 112, 112a, 111, and the fully halogenated C_3 CFCs.
[c] The cap is the allowable maximum consumption of HCFCs defined as follows: consumption of HCFCs in 1989 + (consumption of CFCs in 1989 × 0.028). The amount of each substance is adjusted by its ozone-depletion potential (ODP), a measure of its potential to deplete stratospheric ozone relative to that of CFC-11.

Table 6.6. Chemical formulae and physical properties of some CFCs, HCFCs and HFCs

	Chemical formula	Boiling point at 1 atm (°C)	Critical temp (°C)	Critical pressure (atm)
CFC-11	CCl_3F	23.8	198.0	43.2
CFC-12	CCl_2F_2	– 29.8	112.0	40.6
CFC-113	CCl_2FCClF_2	47.6	214.1	33.7
HFC-134a	CF_3CH_2F	– 26.5	100.6	40.0
HFC-125	CF_3CHF_2	– 48.8	66.0	35.7
HFC-32	CH_2F_2	– 51.7	78.1	57.0
HFC-143a	CF_3CH_3	– 47.6	72.8	37.2
HCFC-22	$CHClF_2$	– 40.8	96.0	48.7
HCFC-141b	CH_3CCl_2F	32.0	210.3	45.8
HCFC-123	$CHCl_2CF_3$	27.9	185.0	37.4
HCFC-225ca	$CF_3CF_2CHCl_2$	51.1	203.6	30.1
HCFC-225cb	$CClF_2CF_2CHClF$	56.1	211.7	30.9

C; (7) if many isomers are available, lower-case letters, a, b, c, ..., are put at the end with a being designated to the most symmetrical one.

6.2.2
Production of Chlorofluorocarbons and Hydrochlorofluorocarbons

Chlorofluorocarbons are generally synthesized by partial fluorination of polychlorocarbons. A prevailing industrial method is fluorination with hydrogen fluoride (HF) in the presence of antimony pentachloride ($SbCl_5$) [12], the process developed by Swarts. Synthetic methods for HCFC-22, CFC-11, CFC-12 and CFC-113 are shown in Fig. 6.4.

HCFCs and CFCs are produced in the liquid phase: polychlorinated compounds such as chloroform ($CHCl_3$) or carbon tetrachloride (CCl_4) are heated with anhydrous HF and $SbCl_5$. To prevent corrosion of the reaction vessel, the water content is kept below 0.1%. The reactions are carried out in the temperature range 60–150 °C depending on the raw materials and desired products. A continuous gas-phase reaction system is also applied to the production using HF gas in the presence of a catalyst such as Al_2O_3 or Cr_2O_3.

Solubility of CFCs in water is generally low. However, HCFC-22 ($CHClF_2$) has relatively high solubility in water: 0.060 g/100g-water (0 °C, 1 atm) as compared with CFC-11: 0.0036 and CFC-12: 0.0026. On the other hand, CFCs are commonly soluble in nonpolar or slightly polar organic solvents, such as hydrocarbons, chlorinated hydrocarbons, alcohols and ketones, and insoluble in polar solvents such as aniline, phenol, benzyl alcohol, benzophenone, formamides and nitromethane. In contrast, HCFCs are miscible even with polar solvents probably due to hydrogen-bond formation to an oxygen or nitrogen atom of the solvent.

The most important feature of CFCs is their chemical stability: CFC-12 is recovered unchanged after storage for one month at 175 °C and is also stable to concentrated sulfuric acid or alkali solutions [13]. CFCs do not react with metals such as stainless steel, cast iron, steel, copper, tin, zinc, lead, and aluminum at ambient temperatures, but react with fused aluminum and magnesium and their alloys.

The toxicity of CFCs to mammals is extremely low. CFC-12 exhibits particularly low toxicity: harmless for 2 h at a concentration of 20 vol% [14].

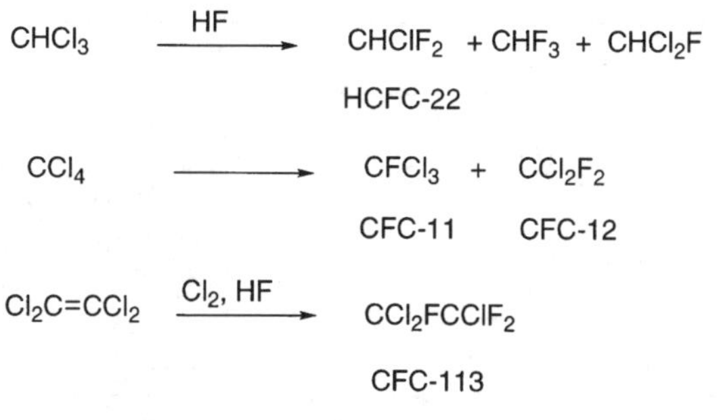

$$CHCl_3 \xrightarrow{HF} \underset{HCFC-22}{CHClF_2} + CHF_3 + CHCl_2F$$

$$CCl_4 \longrightarrow \underset{CFC-11}{CFCl_3} + \underset{CFC-12}{CCl_2F_2}$$

$$Cl_2C=CCl_2 \xrightarrow{Cl_2,\ HF} \underset{CFC-113}{CCl_2FCClF_2}$$

Fig. 6.4

6.2.3
Syntheses of CFC Alternatives

As Molina and Rowland suggested, CFCs emitted are so stable in the tropospheric atmosphere that they finally reach the stratosphere through an eddy diffusion within several years. Consequently, alternatives are desirable that will be duly decomposed in the troposphere before reaching the stratosphere.

An example is HCFC-22 which contains hydrogen atom(s) in the molecule. HCFC-22 is expected to react with a hydroxyl radical (·OH), an active species present in the troposphere, to be degraded with a short lifetime. On the other hand, HFCs lacking a chlorine atom do not provide any species reactive with ozone and thus will not destroy the ozone layer, even if some of them do reach the stratosphere. Thus, the guiding principle for the development of CFC alternatives based on industrial production is the construction of molecules which conserve the original properties but which do not have the toxicity and the adverse effect on the global environment.

HCFCs and HFCs are now widely recognized as potential substitutes for CFCs. Representative CFCs and their substitute HFCs and HCFCs, as well as industrial applications, are summarized in Table 6.7 [15].

CFCs such as CFC-11 (CCl_3F) and CFC-12 (CCl_2F_2) are synthesized by a simple liquid-phase fluorination process catalyzed by an antimony catalyst, the process developed by Swarts in the 1890s. On the other hand, the synthesis of alternative CFCs requires multi-step reactions: (1) fluorination, (2) hydrogenation, (3) halogen exchange, (4) chlorofluorination, (5) isomerization, (6) disproportionation, (7) chlorination, and (8) C–C bond formation via the Friedel–Crafts reaction. One of the most important factors in the process development of these reactions in the liquid or vapor phase is discovery of an effective catalyst. Typical syntheses of HFCs for refrigerants are described in Figs. 6.5 and 6.6.

There are many reports for the synthesis of HFC-134a [16]. A straightforward route to HFC-134a includes the reaction of trichloroethene (TCE) with HF to furnish CF_3CH_2Cl (HCFC-133a), followed by replacement of the remaining chlorine with fluorine in a liquid phase catalyzed by $SbCl_5$ (or SbF_5) [17]. Alternatively, the reaction can be carried out in the vapor phase catalyzed by a chromium-based system [18]. Because of a thermodynamically controlled

Table 6.7. Applications of some HCFCs and HFCs

Application	Current CFC	Alternative
Refrigerants	CFC-12	HFC-134a
	HCFC-22	HFC-32/HFC-125/HFC-134a
		HFC-32/HFC-125
Blowing agents	CFC-11	HCFC-141b
		HFC-245fa
Cleaning agents	CFC-113	HCFC-225ca
		HCFC-225cb

6.2 Chlorofluoro-, Hydrochlorofluoro-, Hydrofluorocarbons, and Alternatives

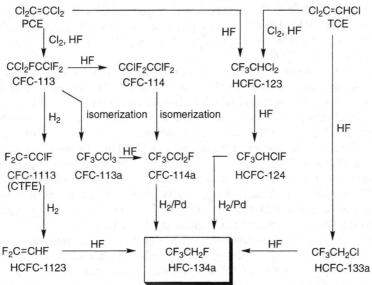

Fig. 6.5

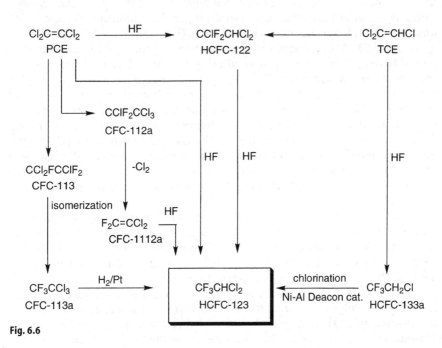

Fig. 6.6

endothermic reaction, HF in excess is required to achieve the conversion of HCFC-133a to HFC-134a. For example, a single-pass conversion is achieved using 6–10 moles of HF per mole of HCFC-133a at 350–400 °C. High selectivity is observed with a Cr, Co, or Ni catalyst.

The second method for the preparation of CFC-134a is hydrogenolysis of CFC-114a prepared from tetrachloroethylene (PCE). PCE is chlorinated in situ to produce hexachloroethane, which is then reacted with HF in the liquid phase using a conventional antimony catalyst, or in the vapor phase with a chromium-based catalyst system [19]. Halogen scrambling occurs to some extent in a vapor phase system, producing an isomeric product, CFC114a. In one approach, CFC-113 is first isomerized in the liquid phase to CFC-113a by a Friedel–Crafts catalyst such as anhydrous aluminum chloride [20]. Then CFC-113a is treated with HF to give CFC-114a. Hydrogenolysis of CFC-114a is attained in a vapor phase. A palladium catalyst is very effective for the reduction [21] in combination with a Group 11 or lanthanide metal salt [22, 23].

Fluorination of trichloroethene (TCE) or tetrachloroethene (PCE) followed by hydrogenation is the third method. CFC-124 is prepared directly from PCE and HF in the presence of a Cr, Co, or Ni catalyst on alumina [24]. An attractive commercial approach is dehydrochlorination of CFC-113 to give CFC-1113 (CTFE) and subsequent replacement of the vinylic chlorine in CTFE to produce trifluoroethene (HCFC-1123) [25]. A combined system of a palladium catalyst with a bismuth or thallium salt is useful [26], and an addition reaction of HF toward HCFC-1123 proceeds quantitatively [27].

As with HFC-134a, several promising synthetic processes for the manufacture of HCFC-123 have been proposed using TCE or PCE as the starting material. These are summarized in Fig. 6.6. A straightforward method is the reaction of PCE with HF. An HF addition reaction to the carbon–carbon double bond of PCE, followed by a halogen-exchange reaction, gives predominantly a thermodynamic product, 1,1-dichloro-2,2,2-trifluoroethane [28, 29]. A conventional liquid-phase process starting with PCE, Cl_2, and HF to produce CFC-113, followed by isomerization, gives CFC-113a, whose hydrogenolysis with a platinum or rhodium catalyst on carbon affords HCFC-123 [30, 31]. Reduction of CFC-113a is also carried out with aqueous Na_2SO_3 solution or zinc in a protic solvent [32–34].

The process starting with TCE involves HF addition and chlorination of the resulting HCFC-133a. A critical step in this process is control of overchlorination of the desired product HCFC-123. Photochemical chlorination of HCFC-133a is carried out by co-feeding O_2, HCl, and HCFC-133a over a nickel-on-alumina Deacon catalyst at high temperatures [35].

Alternatives to CFC-11 as a blowing agent for rigid plastic foams, e.g. polyurethane and polyisocyanurate foams, are listed in Table 6.7. Because a part of the blowing agent is trapped in the foams, the low thermal conductivity of the blowing agent affects the insulating properties of the final products.

HCFC-141b is now used as an alternative to the blowing agent CFC-11. HCFC-225ca and HCFC-225cb are employed as substitutes for CFC-113. These two isomers can be prepared in a ratio of ca. 3:2 via the Prins reaction of tetrafluoroethylene (TFE) with $CHFCl_2$ (HCFC-21) at 15 °C cata-

Fig. 6.7

$$CF_2=CF_2 \xrightarrow[\text{cat.}]{CHCl_2F} CF_3CF_2CHCl_2 + CClF_2CF_2CHClF$$

$$\phantom{CF_2=CF_2 \xrightarrow[\text{cat.}]{CHCl_2F}} \text{HCFC-225ca} \text{HCFC-225cb}$$

lyzed by aluminum chloride [36]. A halogen-exchange reaction of the resulting halogenated propanes is carried out in the gas phase with a metal oxide catalyst such as Al_2O_3 or Cr_2O_3 to afford two isomers as the major products (Fig. 6.7) [37].

6.2.4
Evaluation of Safety and Environmental Effects

In 1988 refrigerant manufacturers jointly funded and set up the "Programme for Alternative Fluorocarbon Toxicity Testing" (PAFT) in order to not only manage such toxicity tests as acute toxicity, sub-acute toxicity, mutagenicity, chronic toxicity, and metabolism testing, etc., but also to publish the acquired data of such candidates as HCFC-123, HFC-134a, HCFC-141b, HCFC-124, HFC-125, HCFC-225ca, HCFC-225cb and HFC-32. The refrigerant manufacturers have also extended the research to the atmospheric aspects of fluorocarbon fluids and funded the "Alternative Fluorocarbons Environmental Acceptability Study" (AFEAS). AFEAS intends to make clear the influence of HCFC and HFC on the environment by understanding the complex interactions of chemical species in the atmosphere and to publish the obtained scientific information [38].

An important aspect of these projects is the determination of ozone-depletion potentials (ODPs) of chlorine-containing compounds [39]. CFC-11 is assigned an ODP of unity; the ODPs of other compounds are normalized to that of CFC-11 on a mass-for-mass basis. The lifetime and chlorine content of a compound are considered in determining its ODP. As shown in Table 6.8, HCFCs have much lower ODPs than CFCs, due mainly to the shorter atmospheric lifetimes of HCFCs. HFCs lacking chlorine show zero ODPs [40].

The relative contribution to global warming by alternative CFCs and other greenhouse gases are defined as Halocarbon Global Warming Potentials (HGWPs) and Global Warming Potentials (GWPs). CFC-11 and carbon dioxide are assigned an HGWP and a GWP of unity, and the HGWPs and GWPs of other compounds are normalized to those of CFC-11 and carbon dioxide on a mass-for-mass basis, respectively. Recently, GWP has been the most frequently used criterion because of the easy comparison between CFC alternatives and other greenhouse gases, for example, methane and nitrous oxide. Examples are also shown in Table 6.8 [41].

Table 6.8. Influence of the alternatives on the environment

Compound	Formula	ODP	HGWP	GWP
CFC-11	CCl_3F	1.0	1.0	4000
CFC-12	CCl_2F_2	1.0	3.4	8500
CFC-113	CCl_2FCClF_2	0.8	1.4	5000
CFC-114	$CClF_2CClF_2$	1.0	4.1	9300
CFC-115	CF_3CClF_2	0.6	7.5	9300
trichloroethane	CCl_3CH_3	0.12	0.02	110
HCFC-123	CF_3CHCl_2	0.02	0.02	93
HCFC-124	CF_3CHClF	0.022	0.10	480
HCFC-141b	CH_3CCl_2F	0.11	0.12	630
HCFC-142b	CH_3CClF_2	0.065	0.42	2000
HCFC-225ca	$CF_3CF_2CHCl_2$	0.025	0.04	170
HCFC-225cb	$CClF_2CF_2CHClF$	0.033	0.15	530
HCFC-22	$CHClF_2$	0.055	0.36	1700
HFC-134a	CF_3CH_2F	0	0.25	1300
HFC-125	CF_3CHF_2	0	0.84	2800
HFC-143a	CF_3CH_3	0	1.1	3800
HFC-152a	CHF_2CH_3	0	0.03	140
HFC-32	CH_2F_2	0	0.13	650

HGWP: AFEAS report in 1990 (CFC-11 = 1.0). GWP: IPCC report, Radiative Forcing of Climate Change in 1994. (CO_2 = 1.0, ITH (Integration Time Horizon) = 100 years) [39–41].

6.2.5
Alternatives to the Third Generation

Recently it has been reported that fluorinated ethers have useful properties as candidates for CFC and HCFC alternatives, as potential refrigerants and as potential blowing agents [42]. Typical examples are *hydrofluoro ethers* HFE-347mcc ($CF_3CF_2CF_2OCH_3$), HFE-347mmy [$(CF_3)_2CFOCH_3$], HFE-245mc ($CF_3CF_2OCH_3$) and HFE-227me ($CF_3CHFOCF_3$). The code number nomenclature is extended to four or more carbon atom compounds. In the abbreviation, the letters c, e and m indicate a CF_2, a CHF and a CF_3 group, respectively. At present, the physical properties, thermal stability, thermodynamic properties, toxicity, and environmental effects of these HFEs are being studied. Currently, extensive efforts are being directed to research aimed at the development of new compounds with structures totally different from those of CFCs, HCFCs, or HFCs, which exhibit no contribution to the ozone-layer depletion, are degraded in the atmosphere soon after their useful life has ended, which exhibit little greenhouse effect, and which have properties similar to those of CFCs and HCFCs, i.e. non-flammability and low toxicity.

6.3
Fluorine-Containing Liquid Crystals

The invention of a display device using liquid crystals in 1963 [43] stimulated extensive research on the various electrical applications of liquid crystals. Since

6.3 Fluorine-Containing Liquid Crystals

then a variety of driving modes have been proposed. Of these, a display device using twisted nematic (TN) liquid crystals appeared in 1970 [44] and was applied to displays on watches, calculators, and game machines. Later, super twisted nematic (STN) and thin film transistor (TFT) modes were invented which made it possible to commercialize larger size and colored displays. In recent years ferroelectric and antiferroelectric liquid crystals have been shown to respond much faster than those with nematic liquid crystals. In this section, the role of fluorine in fluorine-containing liquid crystals will be discussed with possible applications of these materials.

6.3.1
Nematic Liquid Crystals

Liquid crystals used for a TN display device are rod-like molecules comprising a six-membered ring unit (mesogen) like benzene and cyclohexane, a connecting group like -COO-, an alkyl substituent, and a polar group like CN [45, 46]. Typical examples are shown in Fig. 6.8. The cyano group induces a dipole moment directed to the long axis of each molecule.

When these molecules are spread on a director film (orientation film) such as polyimide, all the molecules align in a direction parallel to the direction of the film induced by rubbing. When the compounds are placed in a glass cell whose inner surfaces have been coated with such director films, and each glass is twisted through 90°, the molecules of each liquid crystal align in a twisted fashion to form a 90° helix, each terminal face directed along the director film. When a polarizer is placed outside of each cell glass and twisted as above, the incident light to the top glass is twisted through 90° to come out through the bottom glass as illustrated in Fig. 6.9 A. Thus, the cell looks bright. When an electric field is applied to the cell, the inner liquid crystalline molecules align along the electric field. Thus, the incident light can no longer pass through the bottom glass. Thus the cell appears dark (Fig. 6.9 B). The principle discussed here is the one for a TN mode display device. The STN mode employs a 270° twist.

Most nematic liquid crystalline molecules have a dipole moment along their long axis to have a positive dielectric anisotropy ($+\Delta\varepsilon$), abbreviated as the Np type, whereas those which have a dipole moment perpendicular to the long axis of nematic liquid crystalline molecules are denoted as Nn-type materials and show a negative dielectric anisotropy ($-\Delta\varepsilon$) [47–49]. When an Np-type com-

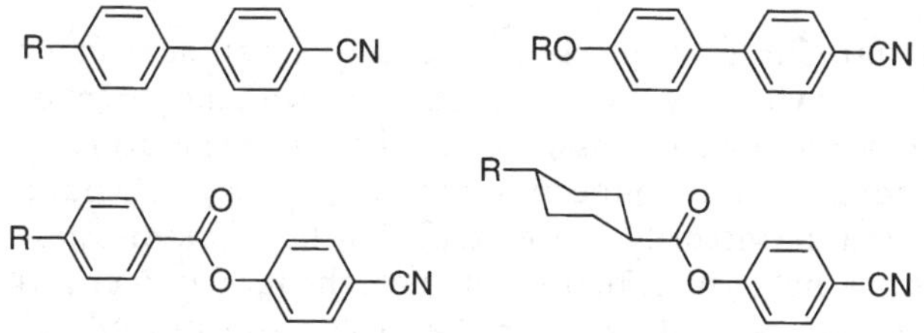

Fig. 6.8. Typical nematic liquid crystals

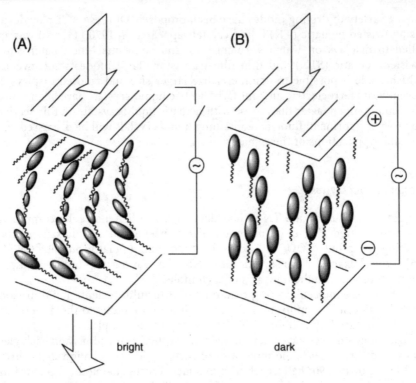

Fig. 6.9. Switching principle of twisted nematic liquid crystals

pound, e.g. 4-butyloxy-2,3-dicyanophenyl 4-pentylcyclohexanecarboxylate (see Fig. 6.10), is added to nematic liquid crystals, it often improves the quality of the display. However, the miscible amount of the compounds is limited mainly due to the steric perturbation by a cyano group. Replacement of cyano by fluorine is shown to improve this limitation [47-49]. Since fluorine is similar to hydrogen in size and induces a large dipole, the fluorine-containing liquid crystals retain liquid crystallinity with much reduced viscosity. Although the Δε-values of the difluoro derivatives in Fig. 6.10 are not as large in absolute value as that of dicyano compounds, the difluoro compounds show a stable nematic phase [50]. In particular, 4-(4-propylcyclohexyl)cyclohexylmethyl 4-ethoxy-2,3-difluorophenyl ether has a nematic phase over a wide temperature range of 62-154 °C [51].

Fluorine introduced at an alkyl or alkoxy substituent of nematic liquid crystals raises the viscosity slightly as compared with the parent compound and tends to induce a smectic phase rather than a nematic phase. In contrast, fluorine introduced at a connecting group appears to stabilize the nematic phase and lower the viscosity as exemplified by the difluoro-*trans*-stilbene type compounds [52] in Fig. 6.11. To prevent photochemical isomerization, materials of the type 1-(4-substituted cyclohexyl)-2-(4-substituted phenyl)-1,2-difluoroethene are now under development.

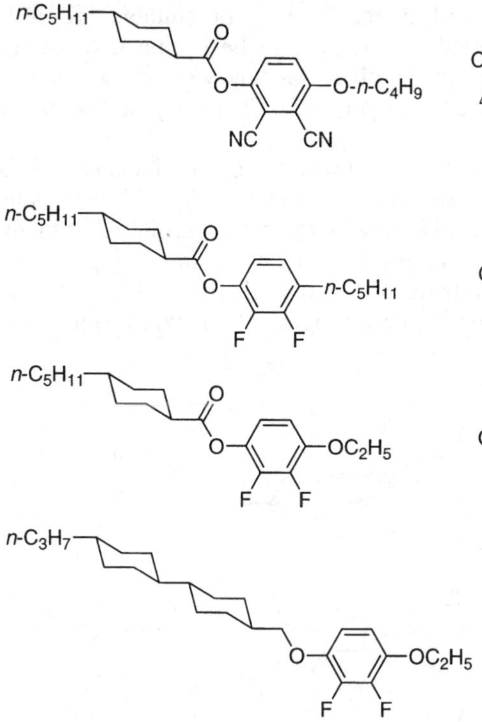

Cr 93 (N 64) Iso
Δε -22

Cr 13 N 29 Iso
Δε -2.1

Cr 51 N 63 Iso
Δε -4.6

Cr 62 N 154 Iso
Δε -4.6

Fig. 6.10. Liquid crystals with negative Δε

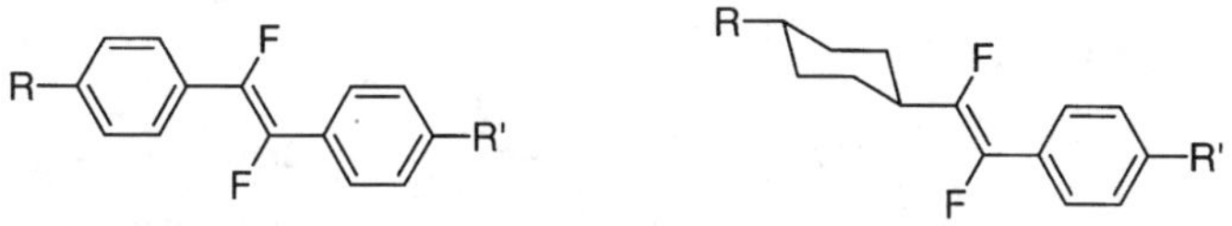

Fig. 6.11. Liquid crystals with fluorines at a connecting group

For Np-type liquid crystals, materials containing fluorine in lieu of a cyano group have been prepared but have found little application in TN and STN mode display devices. This is due to the poor solubility in the prevailing cyano-substituted materials. However, when a TFT mode liquid crystal display based on an active matrix was developed, fluorine-substituted materials were re-evaluated. The TFT mode drive allows fast switching, high quality, and full-colored displays. Materials needed for TFT displays should have a nematic phase over a wide temperature range, be driven at low voltage, and have high resistance (voltage holding ratio). The cyano-substituted materials are highly viscous and driven only at high voltage, although they have a large positive Δε value. Moreover, the voltage holding ratio of the materials tends to decrease upon a long period of UV irradiation, and ionic impurities in the director film are found to be incorporated into the materials. Accordingly, the cyano-substituted materials are not suitable for TFT displays. Instead, fluorine-substituted materials have been shown to be appropriate, since they have a high voltage holding ratio, low viscosity, and appropriate Δε. Some typical materials used currently for TFT displays are shown in Fig. 6.12 [53–57].

Monofluorobenzene with a 4-(*trans*-4-propylcyclohexyl)cyclohexyl substituent has a nematic phase between 88 and 159 °C with a Δε value of +4.8, whereas its difluoro derivative lowers the upper limits of both the nematic phase to 118 °C and the melting point to 44 °C, increasing Δε to +5.3 with similar viscosity and birefringence. Although the trifluoro derivative has a larger Δε of +8.3, its nematic phase ranges only between 65 and 94 °C. Loss of

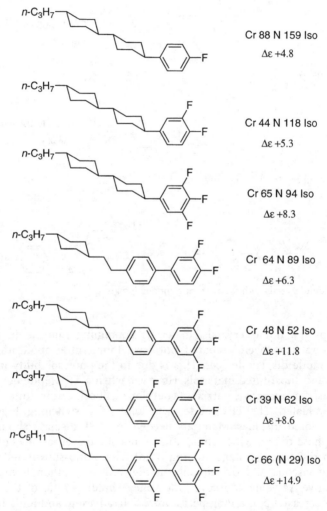

Fig. 6.12. Liquid crystals for TFT displays

liquid crystallinity is observed also with 3,4,5-trifluoro-4'-[2-(*trans*-4-propylcyclohexyl)ethyl]biphenyl. Introduction of fluorine at C-2 and/or C-6 also raises Δε, but the liquid crystallinity is again reduced.

Fluorine functionalities such as a CF_3, CF_3O or a CHF_2O group can also be introduced into liquid crystalline materials (Fig. 6.13). Although a CF_3 group has a large dipole moment of 2.56 Debye and induces a large Δε, it induces also high viscosity and tends to cause a loss of liquid crystallinity. Thus, liquid crystals containing a CF_3 group are not suitable materials. In contrast, those having a CF_3O (0.36 Debye) or CHF_2O group (2.46 Debye) have excellent liquid crystallinity and low viscosity. Accordingly, they, in combination with fluorine, are often employed in liquid crystalline materials for TFT displays.

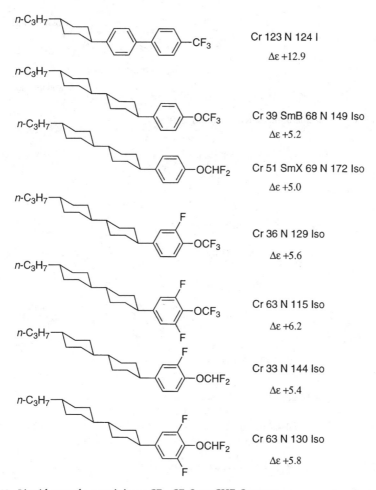

Fig. 6.13. Liquid crystals containing a CF_3, CF_3O, or CHF_2O group

6.3.2
Ferroelectric Liquid Crystals

In 1975, Meyer and co-workers were the first to prepare a chiral liquid crystalline compound, 2-methylbutyl 4-{4-decyloxybenzylideneamino}cinnamate (Fig. 6.14), and disclose that this compound exhibits ferroelectricity [58]. The ferroelectric compound has both a polar functional group and a chiral center to form a smectic C (SmC) phase, a tilted structure of a smectic A phase (SmA).

Figure 6.15 shows the classification of liquid crystalline phases. The nematic phase has an order of one direction, namely rod-like molecules align so that all molecules face mostly in a single direction. If liquid crystals form a layer (two-dimensional order), a smectic phase is produced. If molecules align perpendicular to a layer, the phase is called SmA. When molecules are tilted from the normal layers, the resulting phase is called SmC. Further ordering such as a

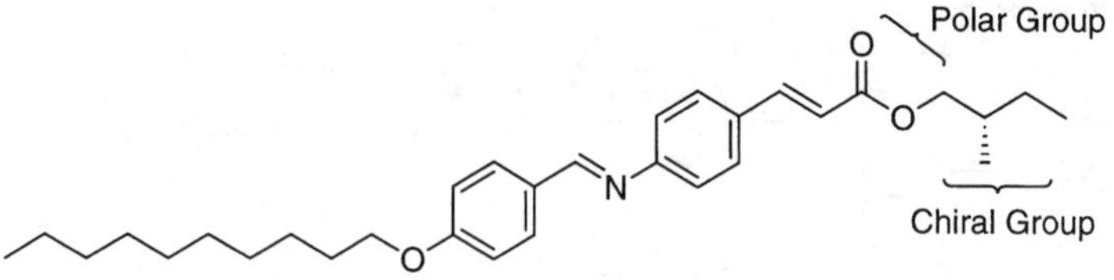

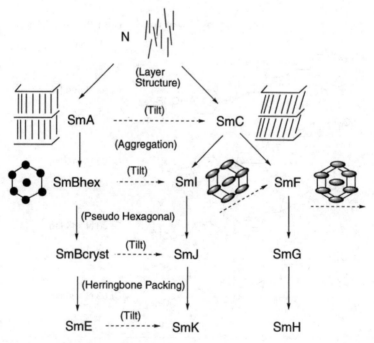

Fig. 6.14. First example of ferroelectric liquid crystals

Fig. 6.15. Classification of smectic phases

hexagonal structure gives smectic I, F, J, K, and H, etc. These higher ordered phases are not yet well understood. When liquid crystalline molecules are optically active, an asterisk is put after the name of the phase, e.g. N* and SmC*. These chiral phases form helical structures.

In 1980, Clark and Lagerwall [59] discovered that ferroelectric liquid crystals have a memory effect and could be applied to fast switching devices. Since then, many attempts have been made at practical applications. Except for the SmC* phase, liquid crystals showing ferroelectricity are generally too viscous to be applied to display materials. Accordingly, ferroelectric liquid crystals are mostly used that have an SmC* phase.

The switching model of ferroelectric liquid crystals is schematically illustrated in Fig. 6.16. Liquid crystalline molecules form a layer structure with a tilt. The tilt direction changes slightly between layers and thus helical structures along a normal of layers are produced. When these molecules are inserted in a cell of 2 μm thickness with a director film inside the surface, the molecules on each cell surface direct uniformly along the director film with a tilt. When an electric field E is applied to the cell, all the molecules tilt in one direction owing to the chirality and the dipole moment of each molecule. The total dipole moment perpendicular to the tilt is called spontaneous polarization (Ps). When the applied electric field is inversed, the tilt is also inversed. The time required for this change is several tens of milliseconds, about one hundredth of the loosing rate of the helical structure of SmC* liquid crystals.

The response time (τ) of ferroelectric liquid crystals is roughly proportional to the viscosity (η) of the material, the spontaneous polarization (Ps), and the electric field (E). Namely, $\tau = \eta/(Ps \cdot E)$. Therefore, chiral liquid crystalline materials having a large Ps and small η have been exploited extensively. Fluorine has played a key role in these materials, since fluorine-containing liquid crystals induce a relatively large dipole moment and little viscosity without loss of liquid crystallinity.

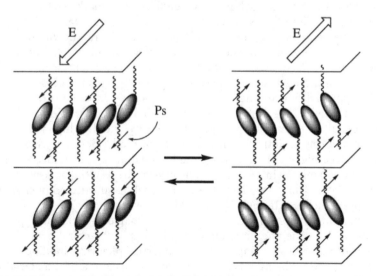

Fig. 6.16. Switching model of ferroelectric liquid crystals

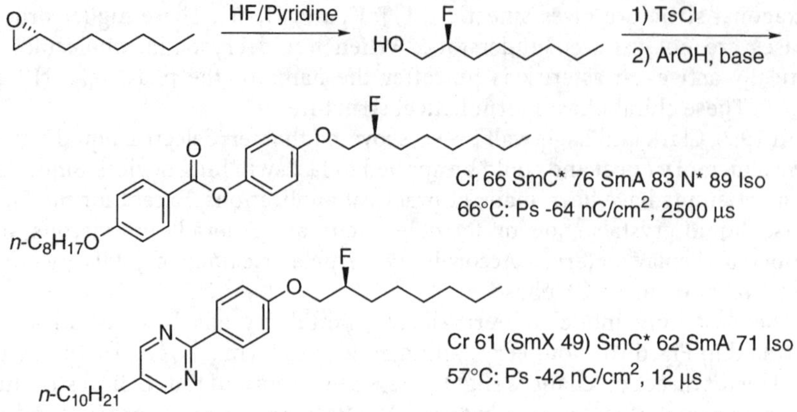

Scheme 6.3

A synthesis of typical ferroelectric liquid crystals is shown in Scheme 6.3. Treatment of optically active 1,2-epoxyoctane with HF/pyridine gives optically active 2-fluoro-1-octanol. Fluorine is introduced at C-2 with inversion of configuration. Tosylation and nucleophilic substitution with a 4-substituted phenol derivative affords the target 2-fluorooctyloxyphenyl benzoate or -pyrimidine [60, 61]. Both the compounds shown in Scheme 6.3 exhibit a SmC* phase with moderate values of Ps: −66 nC/cm^2 (at 66 °C) or −42 nC/cm^2 (at 57 °C), respectively. In addition, the benzoate derivative responds relatively slowly ($\tau = 2500$ μs) due probably to a polar ester functional group. In contrast, the pyrimidine derivative responds very quickly with $\tau = 12$ μs. Both fluorine and a pyrimidine mesogen apparently contribute to the fast switching. This compound, and its derivatives, are used for the ferroelectric liquid crystalline displays marketed by Canon.

Ferroelectric liquid crystals containing a 3- or 4-fluoroalkoxyl group also respond fast and have a high Ps value. Examples are shown in Fig. 6.17. Generally, the Ps value of such a ferroelectric liquid crystalline molecule changes inversely proportionally to the distance between a fluorine-substituted carbon and a mesogen or an ethereal oxygen [62]. This trend is seen also in bis(alkoxyphenyl)pyrimidines but less obviously. The noteworthy effect of fluorine at a position remote from an ethereal oxygen is characterized by a wide liquid crystalline temperature range, low viscosity, and thus high practicability.

When fluorine is introduced at C-3 of an alkyl side chain, the resulting materials (see Fig. 6.18) show less liquid crystallinity and induce smaller Ps. However, upon mixing in a smectic liquid crystal, the resulting mixture responds equally or faster. Thus, the viscosity of 3-fluoroalkyl-substituted phenylpyrimidines appears to be extremely low. In addition, when the sign of Ps of 3-fluoroalkyl-substituted phenylpyrimidine and 2-fluoroalkoxy-substituted phenylpyrimidine is the same, the direction of the helices in the chiral nematic phases compensates. Taking advantage of these properties, phenylpyrimidine with both of the fluorinated substituents has been prepared as a practical material which allows the control of the helical pitch of nematic phases [63, 64].

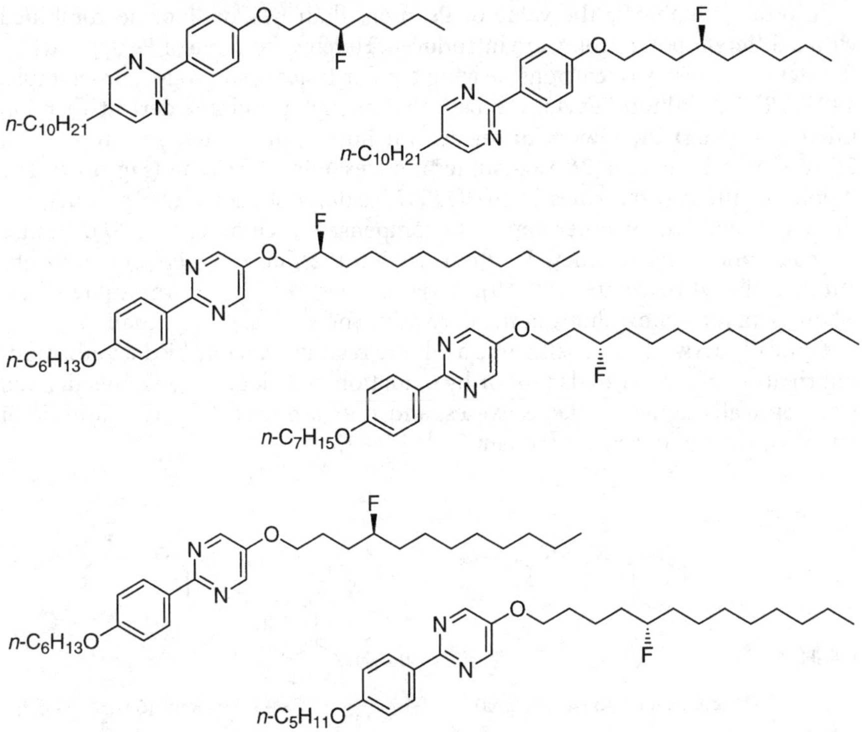

Fig. 6.17. Additional fluorine-containing ferroelectric liquid crystals

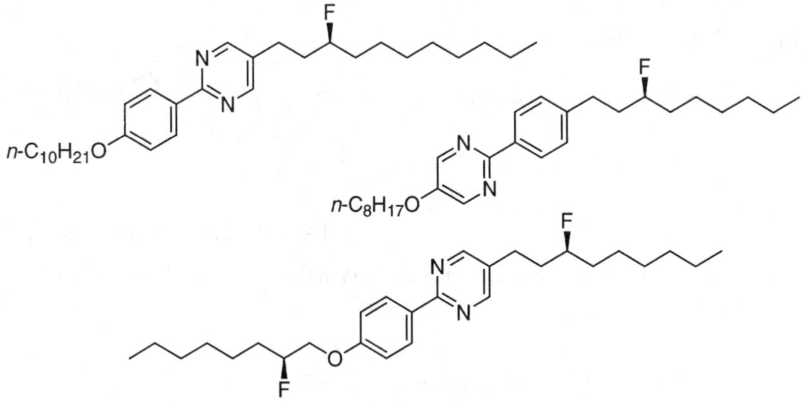

Fig. 6.18. Ferroelectric liquid crystals with a fluoroalkyl group

In order to enhance the value of Ps, more fluorines, or fluorine combined with a different polar group, are introduced. Hereby, the value of Ps depends on the stereochemistry of carbons bearing a polar functional group. For example, a (2R,3R)-2,3-difluorohexyloxy-substituted phenylpyrimidine derivative, upon mixing (doping) by 5 wt% in an achiral host, induces a fairly large Ps of 8.9 nC/cm^2, whereas its (2S,3R)-isomer induces only 1.5 nC/cm^2 (Fig. 6.19). The dipoles of the two fluorines in the (2R,3R)-isomer should direct parallel, but those of the (2S,3R)-isomer appear to compensate each other [65–67]. Similar stereochemical dependency is observed with 2-fluoro-3,4-epoxynonyl-substituted phenyl benzoates [68, 69]. A *syn*-isomer exhibits a large value of Ps, whereas an *anti*-epoxyfluorononyl derivative shows a relatively small Ps.

Optically active 2-fluoroalkanoic acids are readily available by diazonization-fluorination of amino acids [70] or by oxidation of 2-fluoro-1-alkanols derived from optically active 1-alkene oxides, and thus are used for the synthesis of ferroelectric liquid crystals (Scheme 6.4) [71–73].

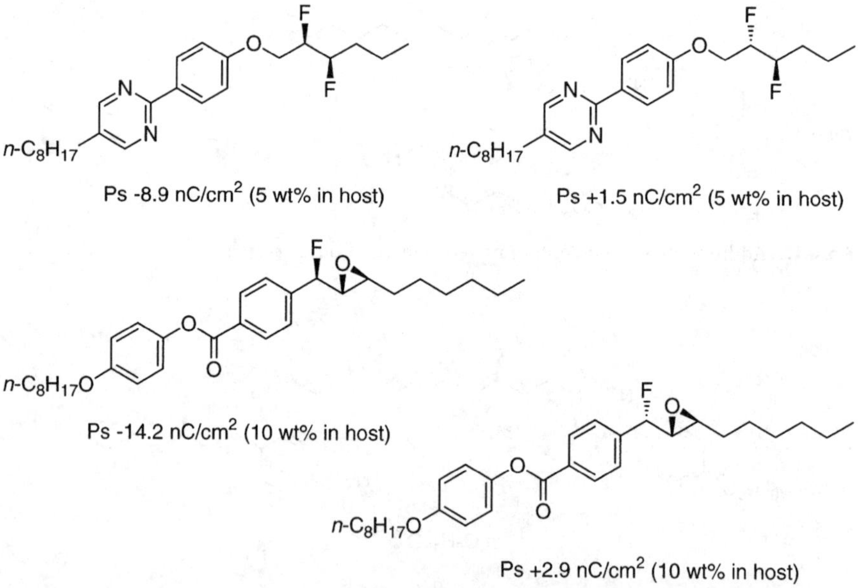

Ps -8.9 nC/cm^2 (5 wt% in host)

Ps +1.5 nC/cm^2 (5 wt% in host)

Ps -14.2 nC/cm^2 (10 wt% in host)

Ps +2.9 nC/cm^2 (10 wt% in host)

Fig. 6.19. Fluorine-containing ferroelectric liquid crystalline compounds with large Ps

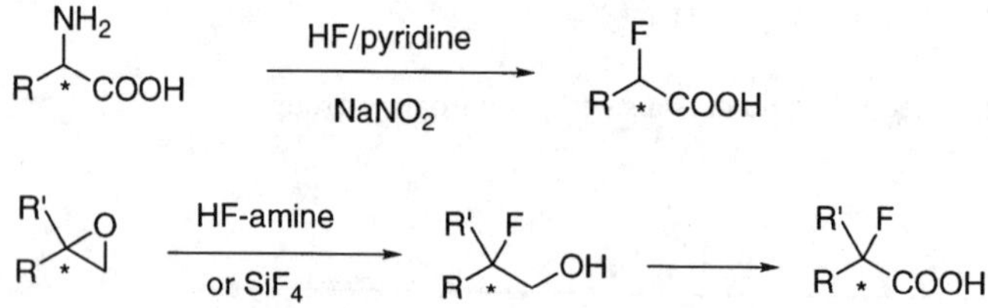

Scheme 6.4

6.3 Fluorine-Containing Liquid Crystals

Phenyl 2-fluorononanoate responds, in spite of a viscous polar ester group, as fast as the 2-fluorooctyloxy-substituted phenylpyrimidine in Scheme 6.3. The effect of an additional methyl group is positive in enhancing Ps as seen in (4-biphenylcarboxy)phenyl esters (Fig. 6.20): a 2-fluorohexanoate ester showed 56 nC/cm^2, and 2-fluoro-2-methylhexanoate ester 90 nC/cm^2, much larger than

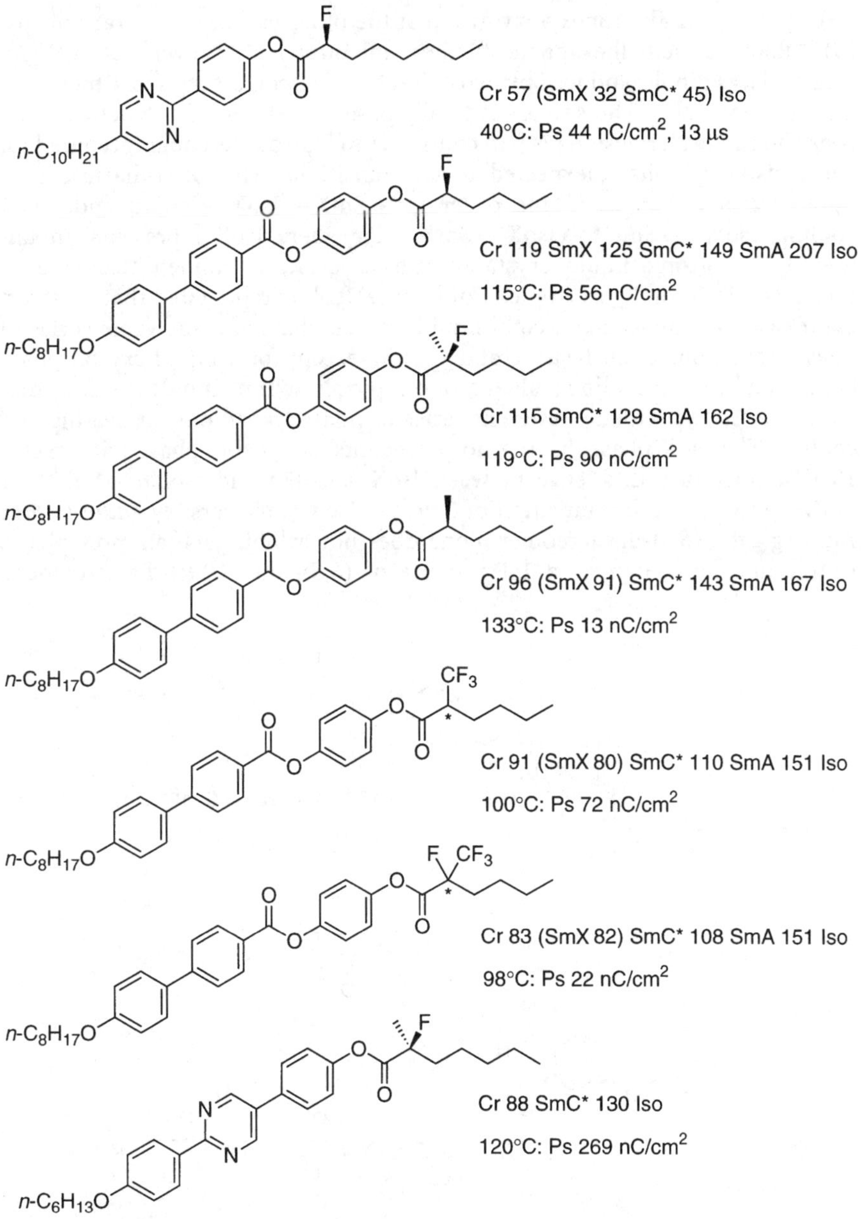

Cr 57 (SmX 32 SmC* 45) Iso
40°C: Ps 44 nC/cm^2, 13 μs

Cr 119 SmX 125 SmC* 149 SmA 207 Iso
115°C: Ps 56 nC/cm^2

Cr 115 SmC* 129 SmA 162 Iso
119°C: Ps 90 nC/cm^2

Cr 96 (SmX 91) SmC* 143 SmA 167 Iso
133°C: Ps 13 nC/cm^2

Cr 91 (SmX 80) SmC* 110 SmA 151 Iso
100°C: Ps 72 nC/cm^2

Cr 83 (SmX 82) SmC* 108 SmA 151 Iso
98°C: Ps 22 nC/cm^2

Cr 88 SmC* 130 Iso
120°C: Ps 269 nC/cm^2

Fig. 6.20. Fluorocarboxylate-type ferroelectric liquid crystals

a fluorine-free derivative or a trifluoromethyl counterpart. In a derivative having both fluorine and a trifluoromethyl group, a dipole of each group is offset to induce a net small Ps. The type of core mesogen also influences the value of Ps. For example, a diazaterphenyl ester of 2-fluoro-2-methylhexanoic acid induces 269 nC/cm², one of the top class data among fluorine-containing liquid crystals [74].

Use of a (S)-2-fluorohexyloxy group at the other end of the diazaterphenyl (S)-2-fluoro-2-methylhexanoate further enhances Ps to over 600 nC/cm² (Fig. 6.21), as dipoles induced by two chiral fluorine-containing substituents are arranged favorably. The success is totally based on systematic structural analysis and molecular design [74]. In contrast, its (R,S)-diastereomer prepared for comparison exhibits unexpected phase transitions. The (R,S)-diastereomer, upon heating, changes in the order Cr→SmX→IsoX→IsoLiq and, upon cooling, IsoLiq→SmC*→IsoX→SmX→Cr, where IsoX represents an unidentified isotropic liquid crystalline phase; SmX, an unidentified smectic phase, IsoLiq, a totally isotropic liquid phase. It should be noted that, in sharp contrast to low molecular weight liquid crystals that show the same series of phase transitions upon heating and cooling except for a liquid crystal phase appearing by supercooling (called monotropic phase transition), the (R,S)-diastereomer shows a totally different series of phase transitions on heating and cooling. The IsoX phase, having no precedents, is a stable phase with over a 10°C temperature difference between IsoX→IsoLiq and IsoLiq→SmC* or SmC*→IsoX, and the transition of SmC*→IsoX is irreversibly *endothermic*. Although the (S,S)-diastereomer alone does not exhibit such an IsoX phase, its racemic form, namely a 1:1 mixture of (S,S)- and (R,R)-diastereomers,

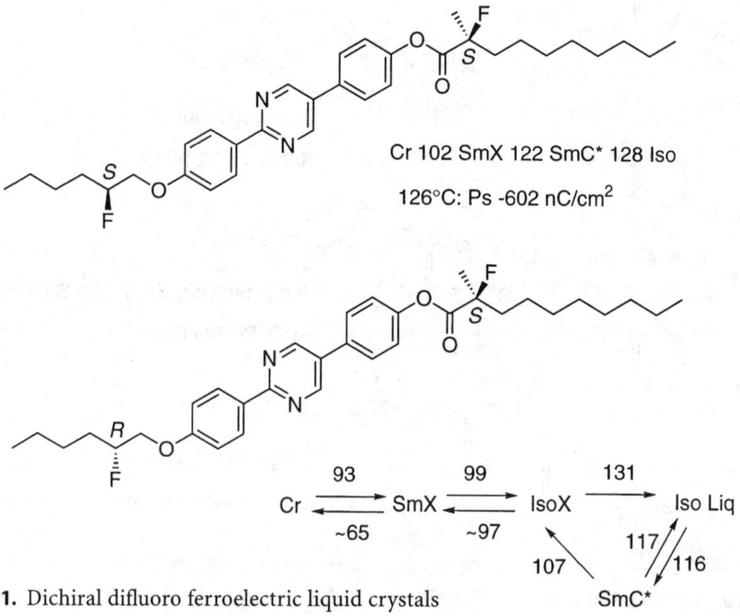

Cr 102 SmX 122 SmC* 128 Iso

126°C: Ps -602 nC/cm²

Cr ⇌ SmX ⇌ IsoX → Iso Liq
93/~65   99/~97   131   117/116
                    ↘107 ↗ SmC*

Fig. 6.21. Dichiral difluoro ferroelectric liquid crystals

does show the IsoX phase. Thus, chiral recognition by two fluorine atoms on two substituents is apparently operating to behave thermally in a quite different way.

6.3.3
Antiferroelectric Liquid Crystals

Liquid crystals that show antiferroelectricity respond quickly enough to be promising materials for future displays. In 1989, Fukuda and Takezoe observed that (R)-4-(1-*m*ethyl*h*exyloxycarbonyl)*p*henyl 4'-*o*ctyloxy*b*iphenyl-4-*c*arboxylate (MHPOBC) (Fig. 6.22) switched in a manner different from ferroelectric liquid crystals, disclosed that it was antiferroelectric, and proposed a display device using antiferroelectric liquid crystals [75].

Antiferroelectric liquid crystals form a layer structure like ferroelectric liquid crystals, but with alternate tilting, as illustrated in Fig. 6.23. This layer structure is called anticlinic and offsets the dipole moment of each layer. When

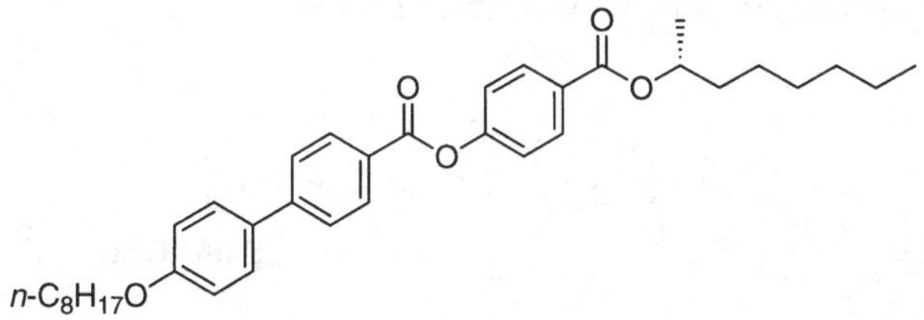

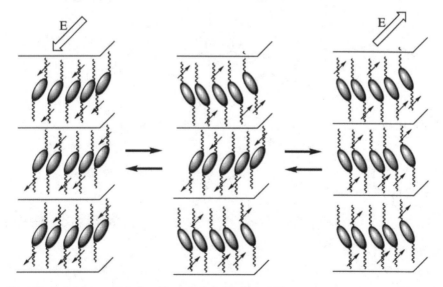

Cr 73 SmC$_A$* 118 SmCγ 119 SmC* 121 SmCα 122 SmA 148 Iso

Fig. 6.22. MHPOBC

Fig. 6.23. Switching model of antiferroelectric liquid crystals

an electric field is applied, however, the liquid crystalline molecules align unidirectionally as do ferroelectric liquid crystals. Thus, the molecules respond as fast as ferrelectric liquid crystals. However, because compounds showing an antiferroelectric phase (SmC$_A^*$) are limited to those in Fig. 6.23 and their derivatives, and no appropriate host liquid crystals of low viscosity are available, the dopant method common with ferroelectric liquid crystals is not applicable. Moreover, as no principle for molecular design is available for antiferroelectric liquid crystals, synthetic studies on antiferroelectric liquid crystals are not easy.

Although one or two fluorines introduced at C-1 of a 2-octanol moiety do not induce an SmC$_A^*$ phase, the ester (TFMHPOBC, Fig. 6.24) of 1,1,1-trifluoro-2-octanol exhibits an Sm$_A$ and an SmC$_A^*$ phase but not an SmC* [76]. Probably an electrostatic repulsion between the CF$_3$ and COO groups induces a conformation appropriate to an SmC$_A^*$ phase.

Thus, a conformational change appears to be crucial for inducing an SmC$_A$* phase. Additional introduction of a methyl group at C-3 is remarkable: a (2R,3S)-1,1,1-trifluoro-2-octanol ester exhibits a stable SmC$_A^*$ phase, whereas its (2S,3S)-diastereomer also showed SmC* (Fig. 6.24). The (2R,3S)-isomer stabilizes a bent conformation pertinent to SmC$_A^*$ [77–80].

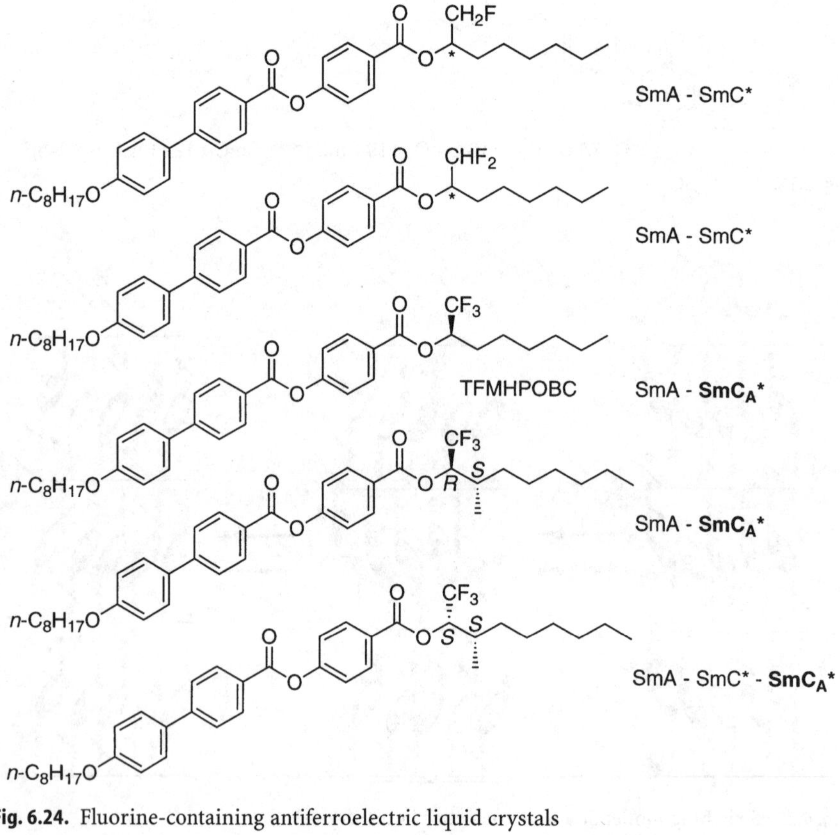

Fig. 6.24. Fluorine-containing antiferroelectric liquid crystals

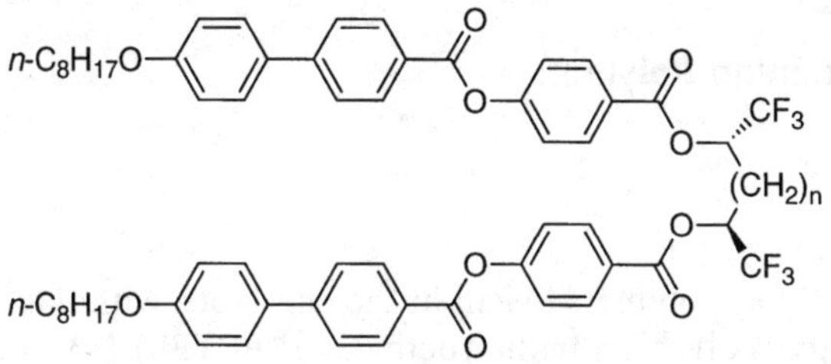

6.3 Fluorine-Containing Liquid Crystals

n	
n = 7	Cr 145 SmC$_A$* 179 Iso
n = 8	Cr 154 SmC* 166 SmA 173 Iso
n = 9	Cr 145 SmC$_A$* 169 Iso
n = 10	Cr 140 SmC* 158 SmA 166 Iso
n = 11	Cr 125 SmC$_A$* 161 Iso

Fig. 6.25. Dimeric models of TFMHPOPC

A steric effect also operates to form antiferroelectric liquid crystals. Thus, dimeric liquid crystals are prepared, each fluorine-functionality being connected by a $(CH_2)_n$ group (Fig. 6.25). Depending on an odd or even number of n, the dimeric liquid crystal stabilizes an SmC$_A^*$ or SmC* phase [81].

Antiferroelectric liquid crystals other than MHPOBC and its derivatives are very rare. However diazaterphenyls, 3-(4-alkoxyphenyl)-4,4,4-trifluorobutyl ethers (Fig. 6.26) (m = 1 – 6), have been recently shown to exhibit a SmC$_A^*$ phase over a wide temperature range [82].

It is obvious that fluorine plays a key role in the fluorine-containing nematic, ferroelectric, and antiferroelectric liquid crystals reviewed above. Future progress in this field will definitely be made by the novel design and synthesis of fluorine-containing chiral molecules.

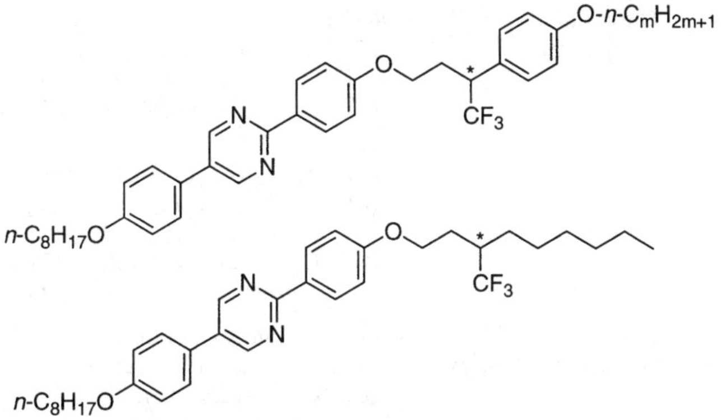

Fig. 6.26. New antiferroelectric liquid crystals

6.4
Fluorine-Containing Polymers

6.4.1
Brief History

The history of fluoropolymers began in the late 1930s with the synthesis of low molecular weight polychlorotrifluoroethylene (PCTFE) [83] and with the accidental discovery of high molecular weight polytetrafluoroethylene (PTFE) [84]. The discovery of PTFE, i.e. Teflon, by Plunkett has been frequently cited as scientific serendipity. When he experimented the synthesis of a new refrigerant with tetrafluoroethylene (TFE), he found a white solid was produced unexpectedly in the steel cylinder and within a few days recorded in his notebook remarkable properties such as thermal behavior and chemical stability in addition to basic ones including elemental analysis.

Fluorinated polymers have been widely used in industry as thermoplastics, elastomers, membranes, textile finishes, coatings, and functional materials based on their unique combinations of properties [85, 86]. In particular, highly fluorinated polymers exhibit high thermal stability, low dielectric constant, low moisture absorption, excellent weatherability, low flammability, low surface energy, and outstanding resistance to most chemicals. A brief history of the commercialization of fluoropolymers is shown in Table 6.9 [87].

Table 6.9. Chronology of fluoropolymers

Period		Major development
1930–1940	The era of homopolymers	$-(CF_2-CF_2)_n-$ (PTFE), $-(CF_2-CFCl)_n-$ (PCTFE) $-(CH_2-CF_2)_n-$ (PVdF), $-(CH_2-CHF)_n-$ (PVF)
1950–mid 1970s	The era of copolymers	Resins $-(CF_2-CF_2)_x-(CF(CF_3)-CF_2)_y-$ (FEP) $-(CF_2-CF_2)_x-(CH_2-CH_2)_y-$ (ETFE) $-(CF_2-CF_2)_x-(CF_2-CF(OC_3F_7))_y-$ (PFA)
		Elastomers $-(CH_2-CF_2)_x-(CF(CF_3)-CF_2)_y-$ (FKM) $-(CH_2-CF_2)_x-(CF(CF_3)-CF_2)_y-(CF_2-CF_2)_z-$ (FKM) $-(CF_2-CF_2)_x-(CH(CH_3)-CH_2)_y-$ (TFE-P) $-(CF_2-CF_2)_x-(CF_2-CF(OCF_3))_y-$ (TFE-PMVE) Rf-containing acrylate copolymers
mid 1970s	The era of functional polymers	Ion-exchange membranes Ambient-curable paint resins $-(CF_2-CFX)_x-(CH_2-CH(OR))_y-$ Thermoplastic elastomers Cyclopolymers

6.4 Fluorine-Containing Polymers

The majority of commercial homopolymers are produced from only four monomers: tetrafluoroethylene (TFE), chlorotrifluoroethylene (CTFE), vinyl fluoride (VF), and vinylidene fluoride (VdF) (abbreviations are listed in Table 6.10) [88, 89]. Perfluorinated polymers are prepared by a free-radical polymerization reaction in water or in a fluorinated solvent. The physical properties of the fluoromonomers are summarized in Table 6.11. Fluorinated monomers, with the exception of perfluoro(propyl vinyl ether) (PPVE), are colorless and show relatively low toxicity with lethal concentrations ranging from about 1000 ppm for CTFE to 800,000 ppm for VF and VdF. In contrast, perfluoroisobutylene is extremely toxic (several ten times more toxic than phosgene), and thus even a trace amount of such a contaminant must not be neglected. The fluoromonomers are in general flammable except for hexafluoropropylene (HFP). In particular, TFE homopolymerizes exothermically

Table 6.10. Abbreviations

CTFE	chlorotrifluoroethylene
ECTFE	ethylene-chlorotrifluoroethylene copolymer
ETFE	ethylene-tetrafluoroethylene copolymer
FEP	fluorinated ethylene-propylene; tetrafluoroethylene-hexafluoropropylene copolymer
HFP	hexafluoropropylene
HFPO	hexafluoropropylene oxide
PCTFE	poly(chlorotrifluoroethylene)
PFA	perfluoroalkoxy copolymer; tetrafluoroethylene-perfluoro(propyl vinyl ether) copolymer
PPVE	perfluoro(propyl vinyl ether)
PTFE	poly(tetrafluoroethylene)
PVdF	poly(vinylidene fluoride)
PVF	poly(vinyl fluoride)
TFE	tetrafluoroethylene
VdF	vinylidene fluoride
VF	vinyl fluoride

Table 6.11. Properties of fluorinated monomers

Monomer	MW	Boiling point (°C)	Melting point (°C)	Critical temp (°C)	Critical pressure (mPa)
TFE	100.02	−75.6	−142.5	33.3	3.82
HFP	150.02	−29.4	−156.2	86.2	2.75
PPVE	266.04	36.0		432.6	1.9
CTFE	116.47	−28.4	−158.2	105.8	3.93
VF	46.04	−72.2	−160.5	54.7	5.43
VdF	64.03	−82.0	−144.0	30.1	4.29

and sometimes undergoes a violent disproportionation reaction to afford CF_4 and carbon. Since it can form a peroxide by reacting with oxygen violently, it must usually be stored with an inhibitor that is removed before the polymerization reaction. CTFE is less reactive than TFE but explosively forms peroxides and oxygenated products. VF and VdF can be handled and stored in the absence of an inhibitor. HFP and PPVE homopolymerize only under forcing conditions.

6.4.2
Monomer Synthesis

The chemistry of fluorinated olefins originates from the pioneering work of Swarts near the end of the 19th century. He reacted perchlorinated hydrocarbons with HF and obtained variously chlorine- and fluorine-substituted carbons [90, 91]. The substitution reaction not only allowed the synthesis of fluorinated alkanes but also provided the precursors of fluorinated alkenes [92–94].

$$C_nCl_{2n+2} \xrightarrow[\text{Sb cat.}]{\text{HF}} C_nCl_{2n+2-a}F_a \qquad (6.1)$$

Tetrafluoroethylene (TFE). TFE is one of the most important raw materials for the production of fluorinated carbon polymers. TFE is prepared by dehalogenation of 1,2-dichlorotetrafluoroethane or 1,2-dibromotetrafluoroethane. TFE is manufactured by the pyrolysis of chlorodifluoromethane at 600–900 °C (Eq. 6.2) in the gas phase neat or by using steam or an inert gas as a diluent [95], or by pyrolysis with steam and alumina at 150 °C (Eq. 6.3) [96]. The reaction probably proceeds through the formation and dimerization of difluorocarbene, giving TFE in yields of more than 95%. TFE should be stored with an oxygen contamination of less than 0.002%. Addition of an inhibitor like limonene allows storage for several months at room temperature.

$$2\ CHClF_2 \xrightarrow{650\ °C} F_2C=CF_2 + HCl \qquad (6.2)$$

$$CHClF_2 + H_2O \xrightarrow[\text{Al}_2\text{O}_3]{150\ °C} \qquad (6.3)$$

$$F_2C=CF_2 \ + \ F_2C=CFCF_3 \ + \ \text{(octafluorocyclobutane)}$$
$$\text{TFE} \qquad\qquad \text{HFP}$$

Hexafluoropropene (HFP). HFP is co-produced during the synthesis of TFE using steam and an alumina catalyst. HFP is also prepared by pyrolysis of TFE at 700–900 °C under reduced pressure [97].

$$3\ F_2C=CF_2 \xrightarrow[\text{700-900 °C}]{\Delta} 2\ F_2C=CFCF_3 \qquad (6.4)$$

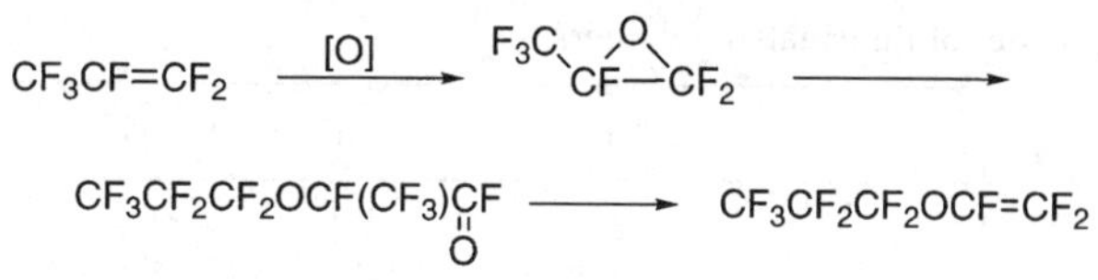

Scheme 6.5

$$CF_3CF=CF_2 \xrightarrow{[O]} F_3C-\overset{O}{\underset{CF-CF_2}{\diagup\diagdown}} \longrightarrow$$

$$CF_3CF_2CF_2OCF(CF_3)\underset{\underset{O}{\|}}{C}F \longrightarrow CF_3CF_2CF_2OCF=CF_2$$

Perfluoro(propyl vinyl ether) (PPVE). PPVE is prepared in three steps from HFP as illustrated in Scheme 6.5 [86]. HFP is oxidized to hexafluoropropylene oxide (HFPO) by reaction with oxygen, hypochlorite, or hydrogen peroxide in an alkaline media or, alternatively, by electrochemical oxidation. HFPO rearranges and then dimerizes in the presence of an alkali metal fluoride catalyst to give acid fluoride, $CF_3CF_2CF_2OCF(CF_3)COF$, which is converted into PPVE by passing it through a pad of alkali metal carbonate or phosphate at 75–300 °C.

Chlorotrifluoroethylene (CTFE). CTFE is prepared by dehalogenation of 1,1,2-trichloro-1,2,2-trifluoroethane (CFC-113) or 1,2-dibromo-1-chloro-1,2,2-trifluoroethane. Dehalogenation of CFC-113 is achieved using zinc in methanol [98], or in the vapor phase using aluminum fluoride/nickel phosphate [99] or a metal oxide in the presence of molecular hydrogen [100].

$$CF_2ClCFCl_2 \xrightarrow{Zn} F_2C=CFCl + ZnCl_2 \qquad (6.5)$$

Vinyl Fluoride (VF). VF, one of the fluorinated olefins with hydrogen atoms, is prepared directly from acetylene in the presence of a mercury catalyst [101], or by the addition reaction of two moles of HF to acetylene to give 1,1-difluoroethane and subsequent elimination of HF (Scheme 6.6) [102].

Vinylidene Fluoride (VdF). VdF is produced by the dehydrochlorination reaction of 1-chloro-1,1-difluoroethane at temperatures of 700–900 °C in the gas phase [103]. The elimination reaction can also be carried out at high temperatures in the presence of a copper catalyst.

$$CH_3CF_2Cl \longrightarrow H_2C=CF_2 + HCl \qquad (6.6)$$

General Properties. The functional properties of fluorinated polymers are strongly related to the number of fluorine atoms. PTFE, PVdF, and PVF are fluorinated forms of polyethylene (PE, $-(CH_2\text{-}CH_2)_n-$). Typical properties of

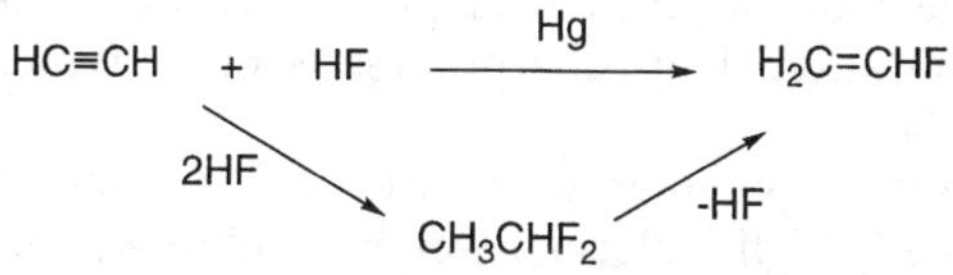

Scheme 6.6

$$HC\equiv CH + HF \xrightarrow{Hg} H_2C=CHF$$

$$\underset{2HF}{\searrow} \quad CH_3CHF_2 \quad \underset{-HF}{\nearrow}$$

Table 6.12. Properties of fluorinated polymers

	PTFE	PVdF	PVF	PE
Fluorine content (%)	76	59.3	41.3	0
Refractive index	1.35	1.42	1.46	1.51
Critical surface tension (mN/m)	18.5	25	28	31
Friction coefficient	0.04	0.30	0.30	0.33
Dielectric constant	2.1	8	9	2.3

these polymers are summarized in Table 6.12. Their refractive index and critical surface tension, that correspond closely to both the hydrophobic and oleophobic character, are roughly inversely proportional to the fluorine content. Worthy of note is that the friction coefficient of PTFE is characteristically low.

Also noteworthy is that polymethacrylate containing a $CH_2CH_2C_7F_{15}$ alcoholic part has a critical surface tension of 11 mN/m, although its fluorine content is 59%, almost the same as PVdF which exhibits a critical surface tension of 25 mN/m (Fig. 6.27). Thus, the properties of fluorinated polymers depend not only on the fluorine content but also on the molecular structure.

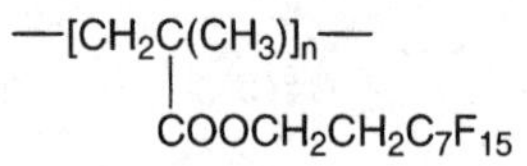

$$-[CH_2C(CH_3)]_n-$$
$$|$$
$$COOCH_2CH_2C_7F_{15}$$

Fig. 6.27

6.4.3
Fluoroplastics

PTFE: $-(CF_2-CF_2)_n-$. PTFE with a high molecular weight was prepared for the first time by the polymerization of TFE in the presence of $ZnCl_2$ or $AgNO_3$. Polymerization of TFE for commercial production is typically done under the conditions of 1–6 MPa at 70–120 °C with a water-soluble radical initiator, e.g. ammonium persulfate [104]. There are two different polymerization processes, depending on how much dispersing agent is used and to what extent agitation is applied. The first one is the suspension polymerization process with a little or no dispersing agent and vigorous agitation to produce a coagulated granular resin. The second one is dispersion polymerization, which uses an emulsifying agent, such as ammonium perfluorooctanoate, under gentle agitation. This process gives a stable aqueous dispersion of small polymer particles which is prevented from coagulation during the polymerization by addition of a hydrocarbon wax. The amount of emulsifying agent in the dispersion process is usually less than its critical micelle concentration. Thus, the process is not a typical emulsion polymerization [86, 105].

PTFE is a linear homopolymer of TFE with a repeat unit of $-(CF_2CF_2)-$. The one with an extremely high molecular weight exhibits a high degree of crystallinity: melting point, 327 °C and glass transition temperature, 130 °C. PTFE has a helical conformation with a torsion period of thirteen carbons each at a 180° angle below 19 °C, in contrast to the planar zigzag structure of polyethylene (cf. Chap. 6.1).

Amongst all solid materials PTFE is an excellent insulator with a high dielectric strength and the lowest dielectric constant of 2.1. It exhibits a low surface energy and low friction coefficient against other materials. Thus, PTFE is used for anti-stick. The chemical stability of PTFE is extremely high: PTFE is inert to almost all chemicals and solvents. It even tolerates fuming H_2SO_4, conc. H_2SO_4 at 300 °C, and 50% NaOH at 100 °C.

Because of the extremely high viscosity of the melt (ca. 10^{11} poise at 380 °C), the conventional melt processing technique is not applicable to PTFE. Three methods have been developed for preparing shaped objects: sintering for granular resins from suspension polymerization, dispersion coating, or paste extrusion of fine powders produced by the dispersion polymerization.

PCTFE: $-(CClF-CF_2)_n-$. Polymerization of CTFE was patented in 1936 [106]. PCTFE is a linear homopolymer prepared by a free-radical polymerization of CTFE [86]. Its melting point is 210 °C, and its relative density falls in the range 2.08–2.19. Although it exhibits better mechanical properties and melt processibility than PTFE, PCTFE is inferior in thermal, electrical, and chemical properties. PCTFE is stable to most chemicals, but is attacked by some organic amines, powerful nucleophiles, and strong reducing agents. PCTFE is used for molded parts in chemical process equipment, and cryogenic applications, such as gaskets, seals, tubes, linings, and electrical insulators. Its films are applied to package air- and moisture-sensitive materials, such as pharmaceuticals and electronic equipment.

PCTFE is manufactured by an aqueous suspension process or an emulsion process using an organic or water-soluble initiator. The aqueous suspension polymerization needs a redox initiator such as an alkali metal persulfate or bisulfite in combination with an accelerator like iron or copper. A perfluorooctanoic acid salt is also used as an emulsifying agent. Polymerization conditions are similar to those for PTFE.

PVF: $-(CH_2-CHF)_n-$. Since VF is a less reactive monomer, its polymerization requires severe conditions. PVF was first prepared in a toluene solvent at a pressure as high as 600 MPa [107]. The polymerization conditions were later improved and the process is now carried out in an aqueous medium at temperatures of 50–165 °C and pressures of 100 MPa or below by UV irradiation or with a free-radical initiator including an acyl peroxide, peroxyammonium sulfate, or a water-soluble azo compound. Trialkylborane-oxygen or a modified vanadium-aluminum Ziegler-Natta catalyst is also used as the initiator [108, 109].

PVF normally has a melting point of 185–210 °C and forms a semicrystalline polymer with an atactic and planner zigzag structure [86]. PVF is insoluble in organic solvents, resistant to acids and bases below 100 °C, but becomes soluble in polar solvents at higher temperatures [87]. A melt technique such as injection molding for the processing cannot be applied, as PVF is thermally labile above its melting point. Films are prepared by an organosol technology that involves mixing with a solvent, pigments, and stabilizers. PVF is mainly applied to films, including PVF-clad metal and plastic laminates such as the exterior cladding of tall buildings.

PVdF: $-(CH_2-CF_2)_n-$. PVdF, prepared by a free-radical initiated polymerization in an aqueous suspension or emulsion, is a linear, semicrystalline polymer with a repeat structure consisting mostly of a head-to-tail monomer unit, and has excellent physical and chemical properties based on an intermolecular force through the H–F hydrogen bonds. Initially, a high pressure of 30 Mpa was required for the polymerization, but current processes employ moderate pressures (1–10 MPa) and temperatures (10–50 °C). PVdF melts in the range 155–192 °C and has four crystal structures, such as the α (phase II) and the β (phase I) [86]. The α-form is thermodynamically the most stable phase and its main chain conformation is *trans-gauche* to minimize the steric interaction between fluorine and adjacent hydrogen atoms. The β-phase is attributed to the mechanical deformation (stretching or drawing) of the melt crystallized polymer and exhibits an all-*trans* planar zigzag conformation with fluorine atoms on one side and hydrogen atoms on the other.

PVdF is resistant to inorganic acids and bases, alcohols, and hydrocarbons, but is soluble in polar solvents, such as ketones and diethyl ether, and highly miscible in N,N-dimethylformamide, dimethyl sulfoxide, and acetone.

The mechanical properties of PVdF are superior to PTFE and PCTFE. A major use of PVdF is as a weather-resistant coating for metal in architectural applications. The piezoelectric and pyroelectrical properties of PVdF films in the β-form have recently found more applications [110].

Perfluorinated Copolymers. Although PTFE has many desirable properties as a material, its high melt viscosity prevents wide application. Endowment with a melt viscosity low enough for conventional processing is accomplished by copolymerization with a perfluorinated comonomer. Thus melt processability and degree control of crystallinity for desirable mechanical properties are improved to allow injection molding and extrusion. HFP and PPVE are comonomers and their random copolymers with TFE are called FEP (fluorinated ethylene propylene) and PFA (perfluoroalkoxy copolymer), respectively (Fig. 6.28). The comonomer content in FEP is about 10–12% and that of PFA is 2–4%. Thus the large perfluoropropoxy group in the PPVE unit is efficient for reducing the crystallinity of PTFE.

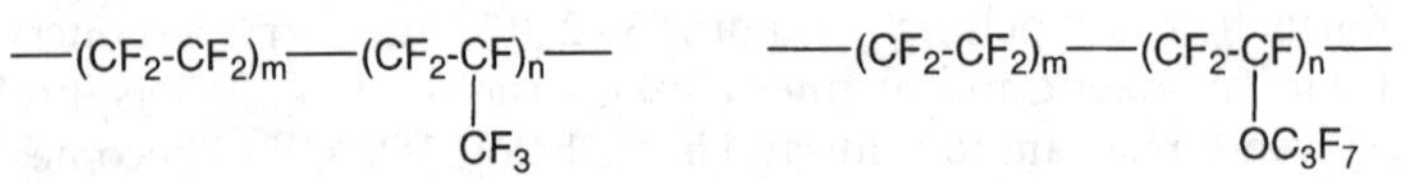

$$-(CF_2-CF_2)_m-(CF_2-CF)_n- \qquad\qquad -(CF_2-CF_2)_m-(CF_2-CF)_n-$$
$$\qquad\qquad\quad | \qquad\qquad\qquad\qquad\qquad\qquad | $$
$$\qquad\qquad\quad CF_3 \qquad\qquad\qquad\qquad\qquad\qquad OC_3F_7$$

Fig. 6.28   FEP   PFA

6.4 Fluorine-Containing Polymers

TFE and HFP or PPVE are copolymerized in aqueous or nonaqueous media [86]. The aqueous process is similar to the dispersion polymerization of TFE using a dispersing agent such as ammonium perfluorooctanoate, and a water-soluble free-radical initiator like ammonium persulfate. The nonaqueous polymerization employs the initiator perfluoropropionyl peroxide in a fluorinated solvent at relatively low temperatures (30–60 °C). FEP is generally produced in an aqueous system; PFA is prepared either in aqueous or nonaqueous media.

PFA has a reactive end group such as –COOH or –COF when prepared in aqueous or nonaqueous media, respectively. The end group is usually converted to an ester or amide functionality. Otherwise, bubble formation and/or dark coloration may occur during the high-temperature processing. For use in electronics, the reactive group can be transformed to a stable CF_3 group by treatment with elemental fluorine [111, 112].

Partially Fluorinated Copolymers. ECTFE (ethylene-chlorotrifluoroethylene copolymer) is prepared from the corresponding two monomers, in a nearly alternating and 1:1 molar ratio, usually under free-radical conditions in an aqueous, organic, or mixed system. A peroxide is used as the initiator. A combination of trialkylborane and oxygen improves the degree of alternation and allows the copolymerization to be performed at very low temperatures (<10 °C). ECTFE is a semicrystalline polymer with a zigzag structure which melts at 240 °C. The mechanical properties of ECTFE are slightly better than ETFE. Salient features of ECTFE are excellent weatherability and barrier properties against water vapor and various gases.

ETFE (ethylene-tetrafluoroethylene copolymer), a linear copolymer consisting of a nearly 1:1 molar ratio of the two monomers, is produced by a free-radical suspension polymerization in an aqueous or mixed medium. ETFE is a semicrystalline polymer with a zigzag structure and a melting point of 275 °C. ETFE is mainly used for the jacketing and insulation of electric cables. In particular, high viscosity grades of ETFE, crosslinked materials made by radiation to cause simultaneous cleavage of polymer chains and intermolecular crosslinking, are applied to insulation against heat sources. ETFE films are used to provide tough and flexible windows in greenhouses due to their high transparency to both UV and visible light.

Amorphous Perfluoroplastics. Amorphous perfluoroplastics, which conserve the special advantages in electrical, chemical, and thermal properties of perfluoropolymers, provide high optical clarity. Examples are Cytop and Teflon AF (Fig. 6.29). Cytop is produced by the cyclopolymerization of a linear

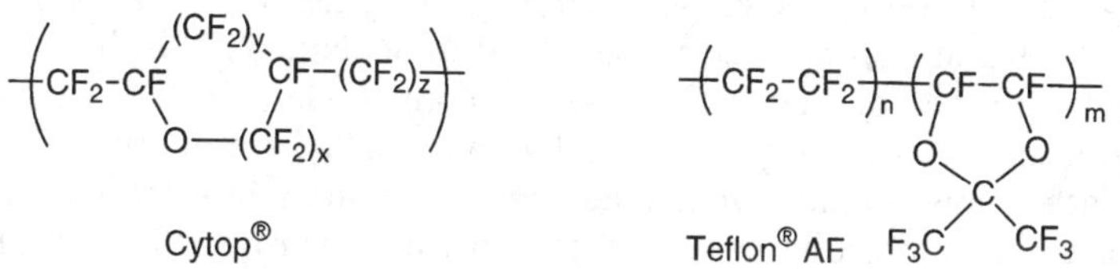

Fig. 6.29   Cytop®   Teflon® AF

Fig. 6.30

perfluorodiene of the general structure $CF=CFO(CF_2)_nCF=CF_2$ [113]. Teflon AF is a copolymer of TFE and a cyclic monomer, perfluoro-(2,2-dimethyl-1,3-dioxole), available in four steps from hexafluoroacetone and ethylene oxide [114]. The polymer has found applications in antireflective coatings, dielectric interlayers, and optical fibers.

A similar cyclic monomer and its trifluoromethoxy-substituted derivative have recently been reported (Fig. 6.30) [115].

6.4.4
Fluoroelastomers

Of all the elastomers, fluoroelastomers have the highest thermal stability, resistance to fluids, chemical stability, and weather resistance, and are used primarily as sealants in aircraft, aerospace, automotives, and for chemical, petroleum, and energy plants. Fluoroelastomers lack crystallinity but have permanent three-dimensional networks by a curing or vulcanization cross-linking process, preserving a rubber-like elasticity [116, 117].

Fluoroelastomers are classified into three groups: fluorocarbon elastomers, fluorosilicones, and fluoroalkoxylated poly(phosphazenes) (Fig. 6.31). While fluorocarbon elastomers are carbon-chain polymers, fluorosilicones and fluoropoly(phosphazenes) are constructed from inorganic backbones.

VdF-HFP Copolymers. VdF-HFP copolymers are prepared by a free-radical emulsion polymerization using potassium persulfate or ammonium persulfate as initiator. Aqueous emulsion polymerizations are generally operated at 5–15 MPa and 60–120 °C. When necessary, a perfluorinated carboxylate salt inert to the highly reactive fluorocarbon radicals in growing polymer chains is used as the emulsifying agent. Commercial elastomers made from VdF and HFP consist of an 80:20 molar ratio of the two key monomers. Elastomers of VdF/HFP/TFE are made in a molar ratio of 60:15–25:25–15. About one HFP unit for every 2–4 moles of the other monomers is required to prevent crystallization.

In order to obtain useful properties, fluoroelastomers need to be crosslinked by curing (vulcanization) with bisphenols, diamines, or peroxides. A typical example is Viton that is crosslinked with bisphenol AF [2,2-bis(4-hydroxyphenyl)hexafluoropropane], as described in Scheme 6.7. Bisphenol AF does not react with the polymer in the absence of an accelerator. The accelerator is, in general, an organic phosphonium salt or a tetraalkyl ammonium salt as combined with a metal oxide, or magnesium oxide/potassium hydroxide. A bisphenolate ion is a strong base that can abstract hydrogen fluoride from the

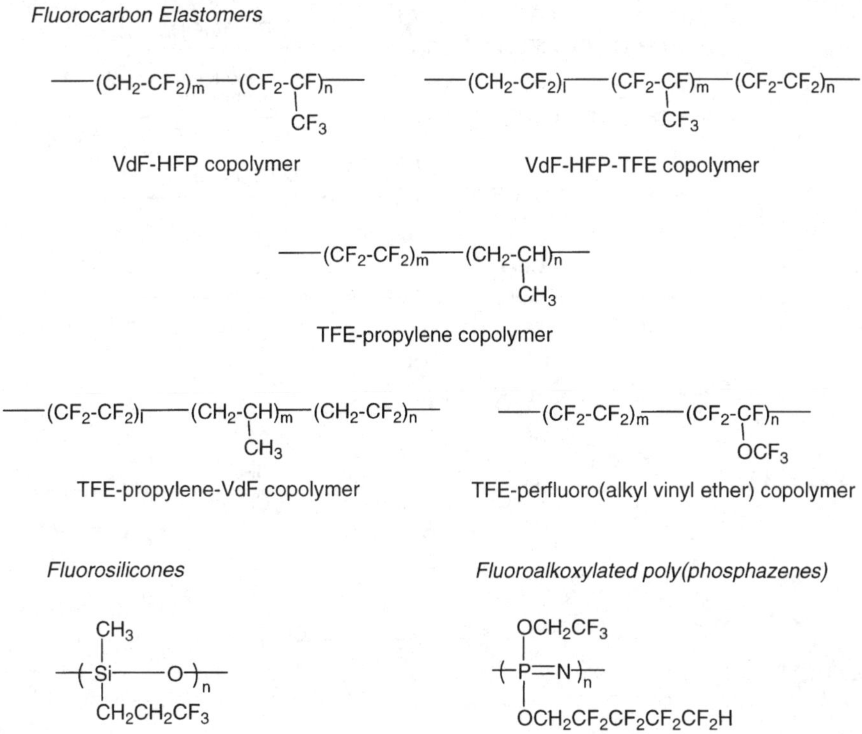

6.4 Fluorine-Containing Polymers

Fluorocarbon Elastomers

$-(CH_2\text{-}CF_2)_{\overline{m}}-(CF_2\text{-}CF)_{\overline{n}}-$
                     |
                    CF_3

VdF-HFP copolymer

$-(CH_2\text{-}CF_2)_{\overline{l}}-(CF_2\text{-}CF)_{\overline{m}}-(CF_2\text{-}CF_2)_{\overline{n}}-$
                                        |
                                       CF_3

VdF-HFP-TFE copolymer

$-(CF_2\text{-}CF_2)_{\overline{m}}-(CH_2\text{-}CH)_{\overline{n}}-$
                                  |
                                 CH_3

TFE-propylene copolymer

$-(CF_2\text{-}CF_2)_{\overline{l}}-(CH_2\text{-}CH)_{\overline{m}}-(CH_2\text{-}CF_2)_{\overline{n}}-$
                        |
                       CH_3

TFE-propylene-VdF copolymer

$-(CF_2\text{-}CF_2)_{\overline{m}}-(CF_2\text{-}CF)_{\overline{n}}-$
                                   |
                                  OCF_3

TFE-perfluoro(alkyl vinyl ether) copolymer

Fluorosilicones

CH_3
$|$
$-(-Si-O-)_n-$
$|$
$CH_2CH_2CF_3$

Fluoroalkoxylated poly(phosphazenes)

OCH_2CF_3
$|$
$-(-P=N-)_n-$
$|$
$OCH_2CF_2CF_2CF_2CF_2H$

Fig. 6.31

polymer backbone to afford finally a diene, which then reacts with the bisphenol to give the crosslinked network successively. A suggested mechanism is also illustrated in Scheme 6.7 [117].

Proposed mechanisms for the curing with diamines and peroxides are shown in Schemes 6.8 and 6.9, respectively. Diamine curing to afford elastomers of high strength was a predominant method for crosslinking before bisphenol curing was introduced in the late 1960s. Hexamethylenediamine and a metal oxide (MgO, CaO) as an acid acceptor to neutralize coproduced hydrogen fluoride are commonly used. Initially, hydrogen fluoride is eliminated from the polymer chain by the amine to give a fluoroolefin moiety which is attacked by the diamine to form a highly crosslinked network.

The peroxide curing needs a third monomer (often called a cure-site monomer) such as 1-bromo-2,2-difluoroethylene. The bromine atom of the resulting elastomeric terpolymer is pulled off by an oxy radical to give a free-radical site in the polymer backbone. The radical then reacts with a crosslinking co-agent trially isocyanurate (TAIC) to form a network (Scheme 6.9). The resulting elastomer is superior in fluid resistance.

TFE-Propylene Copolymers. TFE copolymerizes with propylene in an almost alternating manner to give a fluoroelastomer (T_g of $-2\,°C$) that possesses

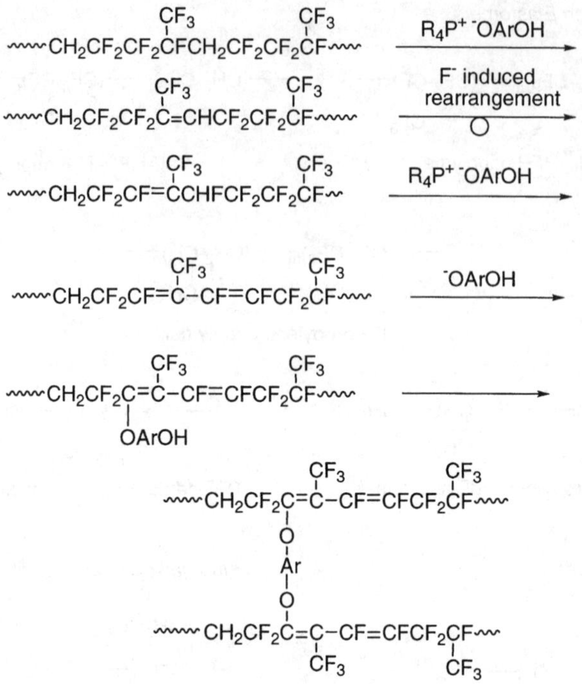

Scheme 6.7

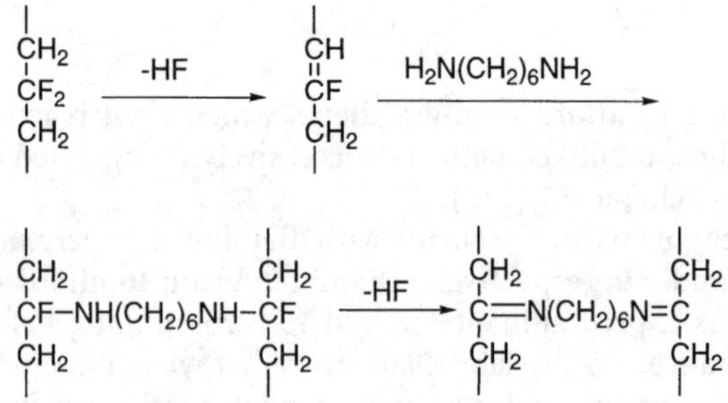

Scheme 6.8

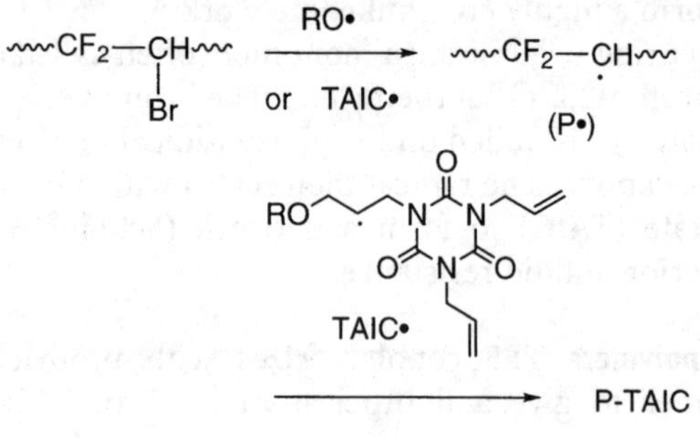

Scheme 6.9

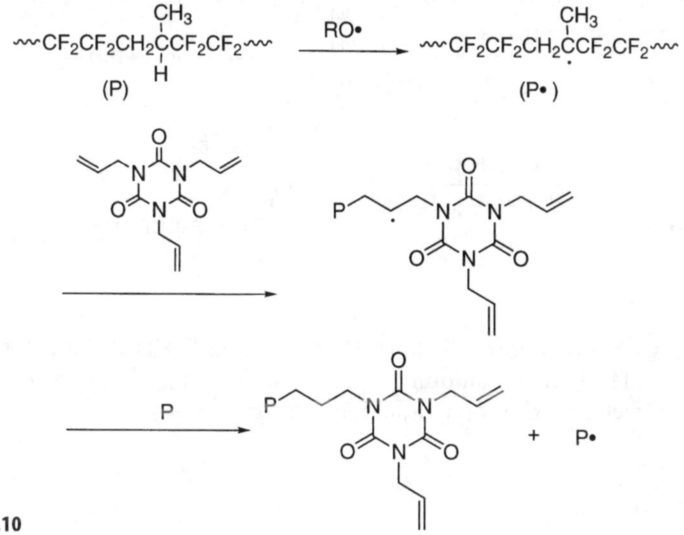

6.4 Fluorine-Containing Polymers

Scheme 6.10

good electrical, thermal, and chemical properties. The TFE-propylene copolymers of high molecular weight (M_n 100,000–180,000) are manufactured by an emulsion polymerization with a modified redox system consisting of ammonium persulfate, ferrous sulfate, ethylenediamine tetraacetic acid, and a sodium hydroxymethanesulfinate initiator at room or slightly higher temperatures [118]. The polymers are cured by a peroxide-radical trap system similar to the curing of VdF-HFP copolymers but without a bromide or iodide agent. A proposed mechanism for the curing process is illustrated in Scheme 6.10.

Fluorosilicones. Fluorosilicones $-[Si(CH_3)(CH_2CH_2CF_3)-O-]_n-$ have Si-O bonds in the main chain and a 3,3,3-trifluoropropyl substituent. They exhibit high thermal stability and weatherability and can be used over a wide temperature range of between −60 and 200 °C (see Fig. 6.31) [116]. The polymers are generally prepared by a ring-opening polymerization of a cyclic oligomer of trifluoropropyl methyl cyclosiloxane in the presence of a base catalyst. A small amount of methyl vinyl siloxane is added as a third comonomer for the curing with a peroxide. Molecular weights of commercial fluorosilicones fall in the range of several 10,000–100,000. The polymers are inferior in thermal, mechanical, and chemical properties to other perfluoroelastomers, but superior in resistance to fuel oils.

Fluoropoly(phosphazenes). Fluorophosphazene elastomers are made of a backbone of P=N bonds, exhibit good weatherability and resistance to fluids, and thus can be used over a wide temperature range −60 to 175 °C [88, 116]. The elastomers are prepared by a ring-opening polymerization of cyclic phosphazene $[PCl_2=N]_3$ at high temperatures followed by substitution

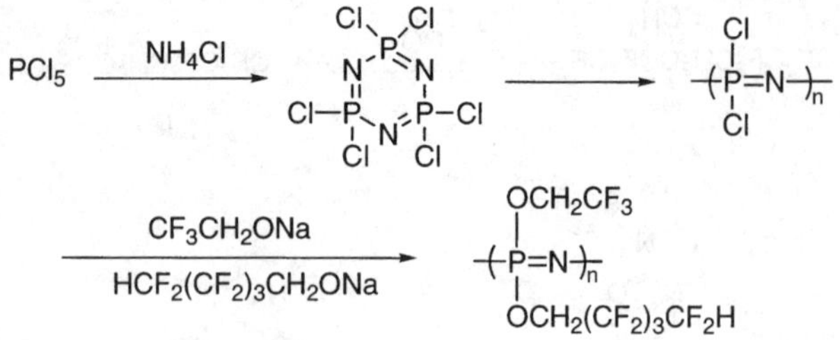

Scheme 6.11

$$PCl_5 \xrightarrow{NH_4Cl} \text{[cyclic trimer } (NPCl_2)_3\text{]} \longrightarrow \{P(Cl)_2=N\}_n$$

$$\xrightarrow[HCF_2(CF_2)_3CH_2ONa]{CF_3CH_2ONa} \{P(OCH_2CF_3)(OCH_2(CF_2)_3CF_2H)=N\}_n$$

reactions with two fluoro alcohols, $HOCH_2CF_3$ and $HOCH_2CF_2CF_2CF_2CF_2H$ (Scheme 6.11). A small amount of an unsaturated alcohol may be introduced to effect peroxide or irradiation curing.

6.4.5
Fluoropolymer Coatings

Because of their special surface properties towards a variety of materials, fluoropolymers offer several advantages as coatings [116, 119]. Those of high crystallinity, particularly PTFE, PVdF and ETFE, are the sole polymers employed for aqueous or nonaqueous dispersion coatings and powder coatings mainly for the purpose of anti-stick and anti-corrosive application. The disadvantages of fluoropolymers in paints and coatings are their poor solubility in common organic solvents and their high baking temperature (>250 °C).

PTFE coatings furnish the optimum effect in relation to such special properties as nonstick and good lubricity and thus are extensively used in housewares (nonstick cookware), food processing (bread and confectioneries), textiles (starching roll), and rubber and plastics machinery (molds and lamination rolls). The major drawback of PTFE coating is caused by its very high melt viscosity at the baking (sintering) temperature (380 °C).

PVdF, a crystalline polymer, is soluble in highly polar solvents and can be dispersed in a polar solvent such as dimethylformamide or dimethylacetamide. The resultant dispersion is baked at a temperature of 300 °C. PVdF paint is formulated by blending in 20–30 wt% of an acrylic resin to improve both melt flow behavior at the baking temperature and adhesion characteristics. The weather resistance of PVdF coating has been shown by long-term exposure tests to last for over 20 years.

To overcome the drawbacks of the use of these polymers as coatings and paints, crosslinkable fluoropolymers have been developed which contain suitable cure sites. One commercialized example is an excellent hard coating on plastic objects like acrylic sheets, based on crosslinkable liquid mixtures composed of TFE-hydroxyalkyl vinyl ether copolymers, hexa(methoxymethyl)-melamine, and silica [120].

Soluble fluoroolefin-vinyl ether copolymers, comprising an alternating sequence of a fluoroolefin and several specific vinyl ether monomer units, have been developed. A typical structure is shown in Fig. 6.32 [121]. The hydroxyl

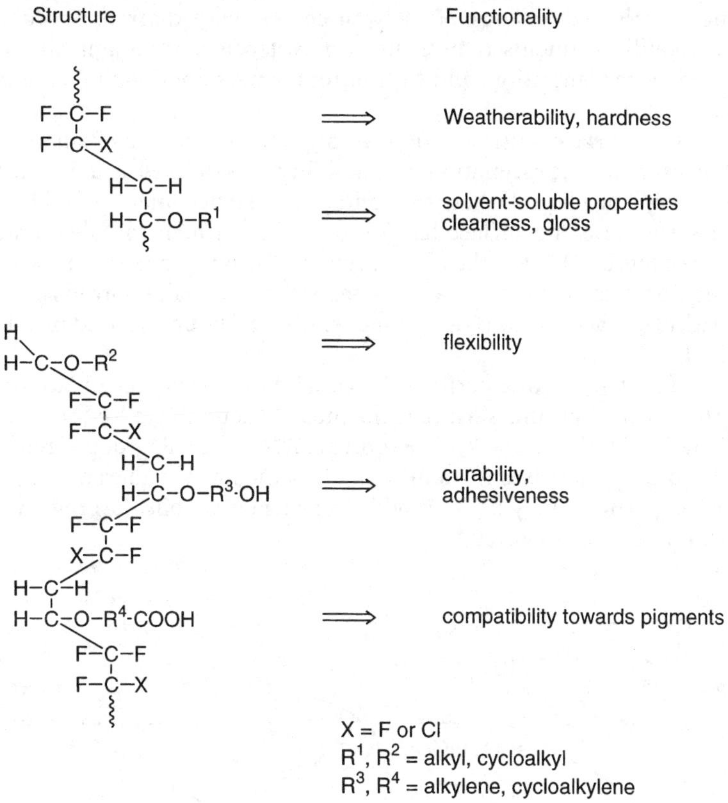

Structure	Functionality
F–C–F F–C–X	⟹ Weatherability, hardness
H–C–H H–C–O–R¹	⟹ solvent-soluble properties, clearness, gloss
H H–C–O–R²	⟹ flexibility
F–C–F F–C–X H–C–H H–C–O–R³·OH	⟹ curability, adhesiveness
F–C–F X–C–F H–C–H H–C–O–R⁴·COOH F–C–F F–C–X	⟹ compatibility towards pigments

X = F or Cl
R^1, R^2 = alkyl, cycloalkyl
R^3, R^4 = alkylene, cycloalkylene

Fig. 6.32. Structural features of Lumiflon polymer

group in the polymer acts as a cure site with polyisocyanates at room temperature, and also as a cure site with melamine resins or blocked isocyanates at high temperatures. Thus the polymer makes it possible to achieve both on-site coating and thermoset coating in factories.

6.4.6
Fluorosurfactants

The characteristics of the surface of materials carrying perfluoroalkyl or polyfluoroalkyl groups are ascribed to the unique properties of the fluorine atom. Such materials are, due to a small intermolecular cohesion, closely packed to form a rod-like structure [116, 122]. The characteristics of the surface of fluorochemicals are summarized by the pioneering work of Zisman [123]: (1) the contact angle (θ) of most solvents on a surface covered by a -CF_3 or -CF_2- group is larger than that on a surface with a -CH_3 or -CH_2- group; (2) polyfluoro- or perfluoroalkanes endowed with such a polar substituent as a carboxy, sulfo, or hydroxyl group become much more surface active in aqueous solutions than their nonfluorinated analogs. Namely, molecules with discrete perfluoroalkyl *and* hydrophilic segments reduce the surface tension of an aqueous solution even at a low concentration; and (3) fluoropolymers are lipophobic in organic solvents.

The critical surface tension (γ_c) of wetting was proposed by Zisman and is a useful measure for the estimation of wettability. γ_c is defined as the value for the surface tension of a liquid where the contact angle θ becomes zero ($\cos\theta = 1$) on a given surface, i.e. the surface tension of a liquid which completely wets the surface (see Fig. 6.33). Since the critical surface tension γ_c of PTFE is 18.5 mN/m, only a liquid whose surface tension is smaller than this value spreads on the surface. Critical surface tensions of polymer surfaces are summarized in Table 6.13 [116, 123].

The surface tension of a perfluorolauric acid monolayer on a platinum plate is only 6 mN/m. Thus, this surface is the most difficult to get wet, since the surface is covered by closely packed CF_3 groups. When the CF_3 group is replaced by a CF_2H group, γ_c increases to 15 mN/m. Thus, the low γ_c values of the surfaces containing perfluoropolymers provide repellency to oils (average γ_c is ca. 30 mN/m) as well as to water.

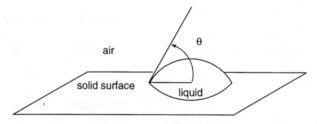

Fig. 6.33. Definition of contact angle θ

Table 6.13. Critical surface tensions γ_c of various polymer surfaces

Polymer	Surface	γ_c (mN/m)	Compound
Fluoro- polymer	$-(CF_2)_nCF_3$	6	perfluorolauric acid monolayer
	$-(CF_2)_nCF_2H$	15	ω-hydroperfluoroundecanoic acid monolayer
	$-(CF_2CF_2)_n-$	18.5	PTFE
	$-(CF_2CFCl)_n-$	31	PCTFE
	$-(CF_2CH_2)_n-$	25	PVdF
Hydro- carbon polymer	$-(CH_2)_nCH_3$	24	aliphatic-amine-monolayer
	$-(CH_2CH_2)_n-$	31	PE
	$-[NHCO(CH_2)_6CONH]_n$	46	poly(hexamethylene-adipamide)
	$-(CH_2CH_2O_2C-C_6H_4CO_2CH_2CH_2)_n-$	43	poly-(ethylene-terephthalate)
Silicone	$-[Si(CH_3)_2-O]_n-$	24	poly(dimethylsiloxane)
Chloro- polymer	$-(CClHCH_2)_n-$	39	poly(vinyl chloride)
	$-(CCl_2CH_2)_n-$	40	poly(vinylidene chloride)

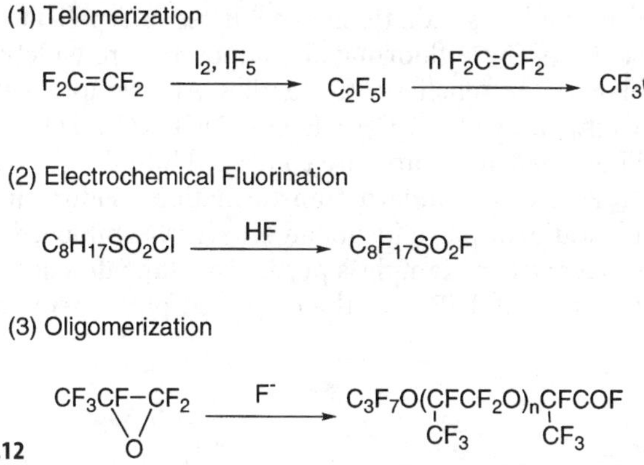

6.4 Fluorine-Containing Polymers

(1) Telomerization

$$F_2C{=}CF_2 \xrightarrow{I_2,\ IF_5} C_2F_5I \xrightarrow{n\ F_2C{=}CF_2} CF_3CF_2(CF_2CF_2)_nI$$

(2) Electrochemical Fluorination

$$C_8H_{17}SO_2Cl \xrightarrow{HF} C_8F_{17}SO_2F$$

(3) Oligomerization

$$CF_3\underset{\diagdown\ \diagup}{CF{-}CF_2} \xrightarrow{F^-} C_3F_7O(\underset{CF_3}{\underset{|}{C}}FCF_2O)_n\underset{CF_3}{\underset{|}{C}}FCOF$$
$$O$$

Scheme 6.12

It is worth noting that fluoroalkyl groups exhibit both hydrophobic and lipophobic properties. Consequently, fluoropolymers are surface active even in organic solvents, acting as an amphiphilic agent by changing the hydrophilic part of a surfactant to a lipophilic one. Fluorocarbon surfactants are prepared via functionalized fluoroalkane intermediates, which are available by telomerization, electrochemical fluorination, or oligomerization (Scheme 6.12).

TFE can be fluoroiodinated to give pentafluoroiodoethane, which is then used for the telomerization of TFE in the presence of strong Lewis acids such as antimony or tantalum pentafluoride or in the presence of a radical initiator [124, 125]. Perfluoroalkanesulfonyl fluorides are manufactured by electrochemical fluorination of the corresponding chlorides in anhydrous hydrogen fluoride [122]. Similarly, perfluoroalkanoic acids are converted into perfluoroalkanoyl fluorides. Perfluoro acid fluorides are also available by oligomerization of HFPO in a polar solvent.

Perfluoroalkyl iodides (R_fI) are prepared by the telomerization of lower homologs with TFE and are successively converted into various intermediates, as shown in Scheme 6.13. Perfluoroalkyl iodides (R_fI) add to ethylene to give $R_fCH_2CH_2I$ under radical conditions or transition metal catalysis. $R_fCH_2CH_2I$

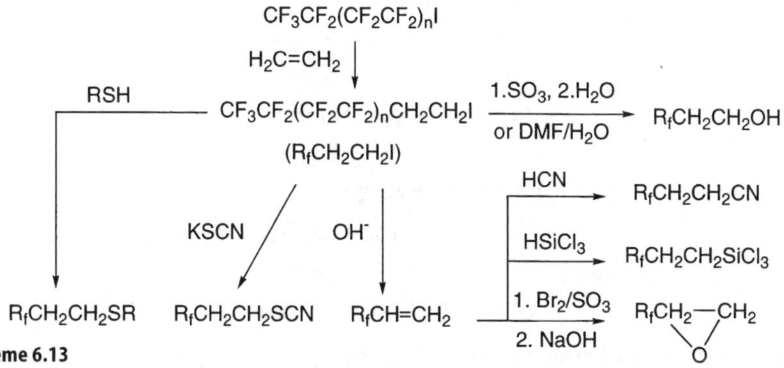

Scheme 6.13

are then transformed to the key synthetic intermediates of repellent products and fluorosurfactants [126]. These fluorinated compounds are widely used as low and high molecular weight repellents for textiles. An example is the polymer of perfluoroalkylethyl acrylate, $R_fCH_2CH_2OC(O)CH=CH_2$. On the other hand, the fluoroalkyl intermediates above are converted into fluorosurfactants, $R_fCH_2CH_2SCH_2CH_2N^+R_3X^-$, by standard transformations. Fluorosurfactants are available in a solid, liquid, or paste form and are widely employed as emulsifying and dispersing agents. For example, a perfluorooctanoate salt is used for the emulsion polymerization of TFE with the dispersed phase being a fluorocarbon.

6.4.7
Fluorinated Membranes

Generally, the term fluorinated membranes means ion-exchange membranes with a perfluorinated polymer backbone. Perfluorinated membranes exhibit excellent thermal and chemical stability with retention of mechanical properties in corrosive and oxidative environments. A typical example is Nafion, membranes made of the perfluorinated sulfonic acid ionomer called XR resin and commercialized in the early 1970s [127]. Nafion was first employed as a separator in fuel cells, in particular for space exploration, and then as ion-exchange membranes that opened the way to the innovative electrolytic process for a chlor-alkali production. Flemion, a perfluorinated carboxylic acid membrane, was later developed and has been commercially produced since 1978 to improve the physical and electrochemical properties of the membranes [128].

The general structures of the perfluorinated membranes manufactured at present are shown in Fig. 6.34. Those with both $X=SO_2F$ and $Y=COOCH_3$ are melt-processable and can be fabricated into films by extrusion. The obtained films are easily converted into the corresponding ion-exchange membranes ($X=SO_3Na$ and $Y=COONa$) by alkaline hydrolysis [129].

Perfluorinated ionomer membranes are generally produced by the following sequence of reactions: preparation of a functionalized perfluorovinyl ether, its

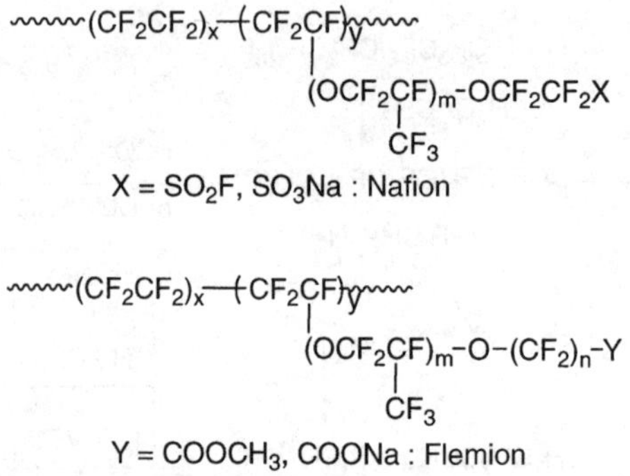

$$\sim\sim(CF_2CF_2)_x-(CF_2CF)_y\sim\sim$$
$$|$$
$$(OCF_2CF)_m-OCF_2CF_2X$$
$$|$$
$$CF_3$$

$X = SO_2F, SO_3Na$: Nafion

$$\sim\sim(CF_2CF_2)_x-(CF_2CF)_y\sim\sim$$
$$|$$
$$(OCF_2CF)_m-O-(CF_2)_n-Y$$
$$|$$
$$CF_3$$

$Y = COOCH_3, COONa$: Flemion

Fig. 6.34. Perfluorinated membranes manufactured

copolymerization with tetrafluoroethylene (TFE), and fabrication of a membrane from the resulting copolymer. Usually, membranes are reinforced by fabrics or other materials to improve the reliability of their mechanical properties for long-term use. The key component of the membranes is hexafluoropropene oxide (HFPO) which is often used for the commercial production of various perfluoro(vinyl ethers). The synthetic schemes of key vinyl ether monomers for the manufacture of Nafion and Flemion are shown in Schemes 6.14 and 6.15, respectively.

The functionalized vinyl ethers are then polymerized with TFE in the presence of a radical initiator in an aqueous medium or in an inert organic solvent. Amorphous or partially crystalline copolymers are fabricated into films

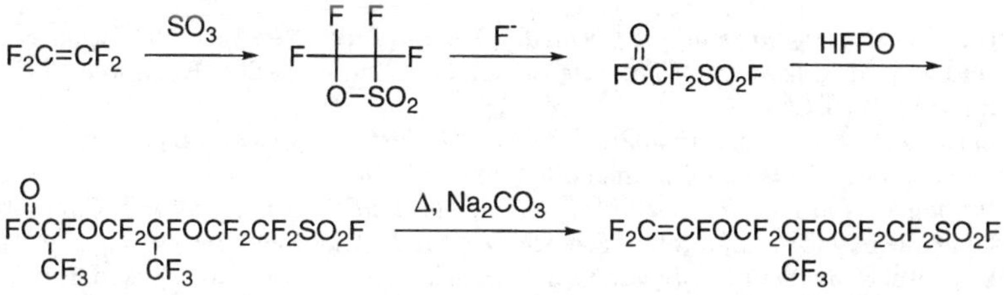

$$F_2C=CF_2 \xrightarrow{SO_3} F-CF_2-CF_2-F \text{ (O-SO}_2\text{)} \xrightarrow{F^-} FCCF_2SO_2F \xrightarrow{HFPO}$$

$$FCCFOCF_2CFOCF_2CF_2SO_2F \text{ (CF}_3\text{, CF}_3\text{)} \xrightarrow{\Delta,\ Na_2CO_3} F_2C=CFOCF_2CFOCF_2CF_2SO_2F \text{ (CF}_3\text{)}$$

Scheme 6.14

(1) Conversion of TFE with Iodine

$$F_2C=CF_2 \xrightarrow{I_2} I(CF_2CF_2)_2I \xrightarrow{oleum} \text{[lactone]} \xrightarrow{CH_3OH}$$

$$FCCF_2CF_2COOCH_3 \xrightarrow{(x+1)\ HFPO} FC(CFOCF_2)_{x+1}CF_2CF_2COOCH_3 \text{ (CF}_3\text{)}$$

$$\xrightarrow[base]{\Delta} F_2C=CFO(CF_2CFO)_xCF_2CF_2CF_2COOCH_3 \text{ (CF}_3\text{)}$$

(2) Conversion of TFE with Dimethyl carbonate

$$CH_3OCOCH_3 \xrightarrow[NaOCH_3]{F_2C=CF_2} CH_3O(CF_2)_2COOCH_3 \xrightarrow{oleum}$$

$$FCCF_2COOCH_3 \xrightarrow{(m+1)\ HFPO} FC(CFOCF_2)_{m+1}CF_2COOCH_3 \text{ (CF}_3\text{)}$$

$$\xrightarrow[base]{\Delta} F_2C=CFO(CF_2CFO)_mCF_2CF_2COOCH_3 \text{ (CF}_3\text{)}$$

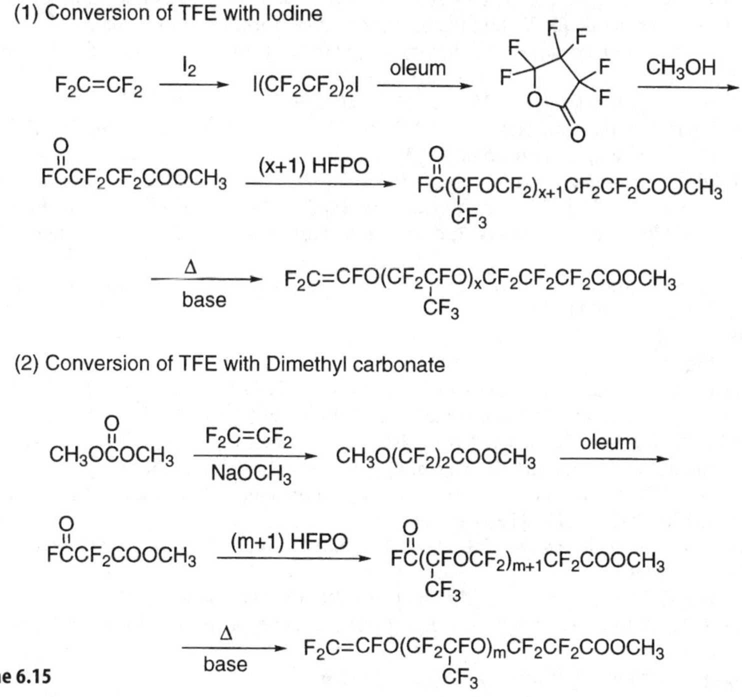

Scheme 6.15

with a thickness of 100–250 mm by a conventional extrusion technique. The membranes are widely used as separators in electrochemical processes such as brine electrolysis, water electrolysis, and fuel cells. Electrochemically basic characteristics, such as the electric conductivity and current efficiency of the membrane, depend mainly on the inherent properties of the polymers like ion-exchange capacities, water content, and fixed-ion concentrations.

Recently, phosphonate-type monomers, $CF_2=CF(CF_2)_nP(O)(OMe)_2$, have been used for copolymerization with TFE. It has been suggested that perfluorinated phosphonic acids may be useful as a substitute for, or an additive to, H_3PO_4 electrolytes in fuel cells [130].

References

1. Hudlicky, M. *Organic Fluorine Chemistry*, Plenum Press: New York, 1971, Chap 6
2. Hudlicky, M. *Chemistry of Fluorine Compounds*, 2nd edn., Ellis Horwood: Chichester, England, 1976; Chap 7
3. Banks, R. E., Ed., *Preparation, Properties, and Industrial Applications of Organic Fluorine Compounds*, Ellis Horwood: Chichester, England, 1982
4. *Introduction to Fluorine Chemistry – Principles and Experiments*, 155th Committee of Japan Society for Promotion of Science, Daily Industrial News, 1997; Chap 3, p 293
5. Ishikawa, N. *Syntheses and Functions of Fluorine Compounds*, CMC: Tokyo, 1987
6. Yamabe, M.; Matsuo, M. *New Aspects of Fluoro Functional Material*, CMC: Tokyo, 1994
7. Manzar, L. E. *Science* **1990**, *249*, 31
8. Rao, V. N. M. In: *Organofluorine Chemistry: Principles and Commercial Applications*; Banks, R. E., Ed.; Plenum Press: New York, 1994; Chap 7
9. Powell, R. L.; Steven, J. H. In *Organofluorine Chemistry: Principles and Commercial Applications*; Banks, R. E., Ed.; Plenum Press: New York, 1994; Chap 28
10. Ishikawa, I.; Kobayashi, Y. In *Fusso no Kagoubutsu* (Japanese); Kodansha: Tokyo, 1979; Chap 4
11. Molina, M. J.; Rowland, F. S. *Nature* **1974**, *810*, 249
12. (a) Swart, F. *Bull. Acad. Roy. Belg.* **1892**, *24*, 474; (b) Swart, F. *Bull. Acad. Roy. Belg.* **1895**, *29*, 874; (c) Swart, F. *Centralblat*, **1903**, *1*, 3
13. Midgley, T.; Henne, A. L. *Ind. Eng. Chem.* **1930**, *22*, 542
14. (a) Clayton, J. W. Jr., *Fluorine Chem. Rev.* **1967**, *1*, 197; (b) Hudlicky, M. *Chemistry of Organic Fluorine Compounds*, 2nd edn., Ellis Horwood Ltd.: Chichester, England, 1976, p 557
15. Yamabe, M.; Matsuo, M. In: *Fusso kei Zairyo no Saishin Doukou* (Japanese), Shi-Emu-Shi (CMC): Tokyo, 1994; Chap 1
16. Patents cited in refs 2 and 9
17. Feiring, A. E. *J. Fluorine Chem.* **1979**, *14*, 7
18. Marangoni, L.; Rasia, G.; Gervasutti, C.; Colombo, L. *Chim. Ind.* (Milan) **1982**, *64*, 135
19. Knaak, J, *US Patent Application* 328127 [*CA 85*, 142607 (1976)]
20. Miller, E. T. *J. Am. Chem. Soc.* **1940**, *62*, 993
21. Gervasutti, C.; Marangoni, L.; Marra, W. *J. Fluorine Chem.* **1981/82**, *19*, 1
22. Morikawa, S.; Samejima, S.; Yoshitake, M.; Tatematsu, S. *European Patent Application* 347830 [*CA 112*, 234785 (1989)]
23. Morikawa, S.; Yoshitake, M.; Tatematsu, S. *Japanese Patent Application* 1-319443 [*CA 112*, 197613 (1989)]
24. Kellner, C. S.; Rao, V. N. M. *US Patent* 4,873,381 [*CA 112*, 197608 (1990)]
25. Lerot, L.; Wilmet, V.; Pirotton, J. *European Patent Application* 355907 [*CA 113*, 58473 (1990)]
26. *Japanese Patent Application* 4-117333 (1992)

27. *Japanese Patent* 62-23728 (1987)
28. Gumprecht, W. H.; Schindel, W. H.; Felix. V. M. *PCT International Application* 8,912,615 [*CA* **112**, 234788 (1989)]
29. Manzer, L. E.; Rao, V. N. M. *US Patent* 4,766,260 [*CA* **110**, 97550 (1988)]
30. Furutaka, Y.; Aoyama, H.; Yomoto, H. *Japanese Patent Application* 1-149739 [*CA* **111**, 232067 (1989)]
31. Morikawa, S.; Yoshitake, M.; Tatematsu, S. *Japanese Patent Application* 1-319440 [*CA* **112**, 216213 (1989)]
32. Torii. S.; Tanaka, H.; Yamashita, S.; Hotate, M.; Suzuki, A.; Okuma, Y. *Japanese Patent Application* 2-1414 [*CA* **112**, 197614 (1990)]
33. Kondo, T.; Yoshikawa, S. *German Patent* 3,834, 038 [*CA* **111**, 194095 (1989)]
34. Takamatsu, S.; Misaki, S. *Japanese Patent Application* 58-222036 [*CA* **100**, 156218 (1983)]
35. Furutaka, Y.; Homoto, Y.; Honda, T. *Japanese Patent Application* 1-290638 [*CA* **112**, 178061 (1989)], 1-290639 [*CA* **112**, 178602 (1989)]
36. Paleta, O.; Posta, A.; Tesarik, K. *Coll. Czech. Chem. Commun.* **1971**, *36*, 1867
37. Morikawa, S.; Yoshitake, M.; Tatematsu, S.; Tanuma, T. *Japanese Patent Application* 2-17134 [*CA* **113**, 5704 (1990)]
38. Fineganm C. E.; Rusch, G. M. *Presented at International CFC & Halon Alternatives Conference*, 1993, Washington D. C., USA
39. *Scientific Assessment of Stratospheric Ozone 1992; United Nations Environment Programme/World Meteorological Global Ozone Research and Monitoring Project* UNEP (1992)
40. Bertnthal, F. Ed., *Climate Change; the Intergovernmental Panel for Climate Change Response Strategies, United Nations Environment Programme/World Meteorological Organization* IPCC (1990)
41. Wigley. T. M. L; Raper, S. C. B. *STUGE (an Interactive Greenhouse Model) User's Manual*, Climate Research Unit, Norwich, UK.,1991
42. Sekiya, A.; Misaki, S. *International Conference on Ozone Protection Technologies*, 1997, Baltimore, USA; p 26
43. Williams, R. *Nature*, **1963**, *199*, 273
44. Schadt, M.; Helfrich, W. *Appl. Phys. Lett.*, **1971**, *18*, 127
45. Schadt, M.; Gerber, P. R. *Z. Naturforsch.*, **1982**, *37a*, 165
46. Gerber, P. R.; Schadt, M. *Z. Naturforsch.*, **1982**, *37a*, 179
47. Osman, M. A.; Huynh-Ba, T. *Mol. Cryst. Liq. Cryst.*, **1983**, *92*, 57
48. Inukai, T.; Furukawa, K.; Inoue, H.; Terashima, K. *Mol. Cryst. Liq. Cryst.*, **1983**, *94*, 109
49. Reiffenrath, V.; Krause, J.; Plach, H. J.; Weber, G. *Liq. Cryst.*, **1989**, *5*, 159
50. Gray, G. W.; Hogg, C.; Lacey, D. *Mol. Cryst. Liq. Cryst.*, **1981**, *67*, 1
51. Takatsu, H.; Takeuchi, K.; Tanaka, Y.; Sasaki, M. *Mol. Cryst. Liq. Cryst.*, **1986**, *141*, 279
52. Inoue, S.; Machida, K. *Senryo to Yakuhin*, **1996**, *41*, 57
53. Inukai, T.; Miyazawa K. *Ekisho*, **1997**, *1*, 9
54. Goto, Y.; Ogawa, T.; Sawada, S.; Sugimori, S. *Mol. Cryst. Liq. Cryst.*, **1991**, *209*, 1
55. Demus, D.; Goto, Y.; Sawada, S.; Nakagawa, E.; Saito, H.; Tarao, R. *Mol. Cryst. Liq. Cryst.*, **1995**, *260*, 1
56. Greenfield, S.; Coates, D.; Goulding, M.; Clemiton, R. *Liq. Cryst.*, **1995**, *18*, 665
57. Yamamoto, H.; Takeshita, F.; Terashima, K.; Kubo, Y.; Goto, Y.; Sawada, S.; Yano, S. *Mol. Cryst. Liq. Cryst.*, **1995**, *260*, 57
58. Meyer, R. B.; Liebert, L.; Strazelecki, L.; Keller, P. *J. Phys. (France)*, **1975**, *36*, L69
59. Clark, N. A.; Lagerwall, S. T. *Appl. Phys. Lett.*, **1980**, *36*, 899
60. Nohira, H.; Nakamura, S.; Kamei, M. *Mol. Cryst. Liq. Cryst.*, **1990**, *180B*, 379
61. Nakamura, S.; Nohira, H. *Mol. Cryst. Liq. Cryst.*, **1990**, *185*, 199
62. Nagashima, Y.; Ichihashi, T.; Noguchi, K.; Iwamoto, M.; Aoki, Y.; Nohira, H. *Liq. Cryst.*, **1997**, *23*, 537
63. Kusumoto, T.; Ogino, K.; Sato, K.; Hiyama, T.; Takehara, S. Nakamura, K. *Ferroelectrics*, **1993**, *148*, 147

64. Kusumoto, T.; Ogino, K.; Sato, K.; Hiyama, T.; Takehara, S. Nakamura, K. *Chem. Lett.*, **1993**, 1243
65. Thurmes, W. N.; Wand, M. D.; Vohra, R. T.; Walba, D. M. *Ferroelectrics*, **1991**, *121*, 213
66. Wand, M. D.; Thurmes, W. N.; Vohra, R. T.; More, K. Walba, D. M. *Ferroelectrics*, **1991**, *121*, 219
67. Wand, M. D.; Vohra, R. T.; Walba, D. M.; Clark, N. A.; Shao, R. *Mol. Cryst. Liq. Cryst.*, **1991**, *202*, 183
68. Walba, D. M.; Razavi, H. A.; Clark, N. A. Parmae, D. S. *J. Am. Chem. Soc.*, **1998**, *110*, 8686
69. Walba, D. M.; Razavi, H. A.; Horiuchi, A.; Eidman, K. F.; Otterholm, B.; Haltiwanger, R. C.; Clark, N. A.; Shao, R.; Parmer, D. S.; Wand, M. D.; Vohra, R. T. *Ferroelectrics*, **1991**, *113*, 21
70. Boemelbutg, J.; Heppke, G.; Ranft, A. *Z. Naturforsch., B: Chem. Sci.*, **1989**, *44*, 1127
71. Shiratori, N.; Nishiyama, I.; Yoshizawa, A.; Hirai, T. *Jpn. J. Appl. Phys.*, **1990**, *29*, L2086
72. Shiratori, N.; Yoshizawa, A.; Nishiyama, I.; Fukumasa, M.; Yokoyama, A.; Hirai, T.; Yamane, M. *Mol. Cryst. Liq. Cryst.*, **1991**, *199*, 129
73. Yokoyama, A.; Yoshizawa, A.; Hirai, T. *Ferroelectrics*, **1991**, *121*, 225
74. Yoshizawa, A.; Umezawa, J.; Ise, N.; Sato, R.; Soeda, Y.; Kusumoto, T.; Sato, K.; Hiyama, T.; Takanishi, Y.; Takezoe, H. *Jpn. J. Appl. Phys.*, **1998**, *37*, L942
75. Chandani, A. D. L.; Gorecka, E.; Ouchi, Y.; Takezoe, H.; Fukuda, A. *Jpn. J. Appl. Phys.*, **1989**, *28*, L1265
76. Suzuki, Y.; Nonaka, O.; Koide, Y.; Okabe, N.; Hagiwara, T.; Kawamura, I.; Yamamoto, N.; Yamada, Y.; Kitazume, T. *Ferroelectrics*, **1993**, *147*, 109
77. Mikami, K.; Yajima, T.; Siree, N.; Terada, M.; Suzuki, Y.; Kobayashi, I. *Synlett*, **1996**, 837
78. Mikami, K.; Yajima, T.; Terada, M.; Kawauchi, S.; Suzuki, Y.; Kobayashi, I. *Chem. Lett.*, **1996**, 861
79. Mikami, K.; Yajima, T.; Terada, M.; Suzuki, Y.; Kobayashi, I. *J. Chem. Soc., Chem. Commun.*, **1997**, 57
80. Kobayashi, I.; Suzuki, Y.; Yajima, T.; Kawauchi, S.; Terada, M.; Mikami, K. *Mol. Cryst. Liq. Cryst.*, **1997**, *303*, 165
81. Suzuki, Y.; Isozaki, T.; Kusumoto, T.; Hiyama, T. *Chem. Lett.*, **1995**, 719
82. (a) Aoki, Y.; Nohira, H. *Liq. Cryst.*, **1995**, *19*, 15; (b) Aoki, Y.; Nohira, H. *Ferroelectrics*, **1996**, *178*, 213
83. *British Patent* 465,520 (1937) [*CA* **31**, 7145 (1937)]
84. Plunkett, R. J. *US Patent* 2,230,654 (1941) [*CA* **35**, 3365 (1941)]
85. (a) Yamabe, M. *Makromol. Chem., Macromol. Symp.* **1992**, *64*, 11; (b) Scheirs, J. *Modern Fluoropolymers*; John Wiley & Sons: Chichester, 1997
86. Feiring, A. E. In: *Organofluorine Chemistry: Principles and Commercial Applications*; Banks, R. E.; Smart, B. E.; Tatlow, J. C. Eds.; Plenum Press: New York, 1994; Chap 15
87. (a) Swart, F. *Bull. Acad. Roy. Belg.* **1892**, *24*, 474; (b) Swart, F. *Bull. Acad. Roy. Belg.* **1895**, *29*, 1895; (c) Swart, F. *Zentralblatt*, **1903**, *1*, 3
88. Ishikawa, N.; Kobayashi, Y. In: *Fusso no Kagobutsu; Kodansha*: Tokyo, 1979; Chap 4
89. Satokawa, T. In: *Kinousei Fusso Kobunshi; Nikkan Kogyo Shinbunsha*: Tokyo, 1982; Chap 1
90. Yamabe, M.; Matsuo, M.; Miyake, H. In *Macromolecular Design of Polymeric Materials*; Hatada, K.; Kitayama, T.; Vogel, O. Eds.; Marcel Dekker, Inc.: New York, 1997; Chap 25
91. Fearn, J. E. In: *Fluoropolymers*; Wall, L. A. Ed.; Wiley-Interscience: New York, 1972; Chap 1
92. Midgley, T. Jr.; Henne, A. L. *Ind. Eng. Chem.* **1930**, *22*, 542
93. Booth. H. S.; Mong, W. L.; Burchfield, P. E. *Ind. Eng. Chem.* **1932**, *24*, 328
94. Booth, H. S.; Burchfield, P. E.; Mckelvey, J. B. *J. Am. Chem. Soc.* **1931**, *55*, 2231
95. Hamilton, J. M. Jr. In: *Advances in Fluorine Chemistry*, Vol. 3; Stacey, M.; Tatlow, J. C. Sharpe, A. G. Eds.; Butterworth: London, 1963
96. Ukihashi, H.; Hisasue, M. *US Patent* 3,459,818 [*CA* **71**, 123506 (1969)]
97. Nelson, D. A. *US Patent* 2,758,138 [*CA* **51**, 3654 (1957)]
98. Blum, O. A. *US Patent* 2,590,433 [*CA* **47**, 1180 (1953)]
99. Gardner, L. E. *US Patent* 3,789,016 [*CA* **80**, 82052 (1974)]

100. Ohira, K.; Yoneda, S.; Goto, I. *Japanese Patent* 01 29,328 [*CA* **111**, 173591 (1989)]
101. Coffman, D. D.; Cramer, R. D. *US Patent* 2,461,523 [*CA* **43**, 34372 (1949)]
102. Harmon, J. *US Patent* 2,599,6316 [*CA* **47**, 1725 (1953)]
103. (a) Downing, F. B.; Benning, F. A.; McHarness, R. C. *US Patent* 2,551,573 [*CA* **45**, 9072 (1951)]; (b) Miller, C. B. *US Patent* 2,628,989 [*CA* **48**, 1406 (1954)]
104. Gangal, S. V. *US Patent* 4,186,121 [*CA* **92**, 129921 (1980)]
105. Bankoff, S. G. *US Patent* 2,612,484 [*CA* **47**, 3618 (1953)]
106. Schloffer, F.; Sherer, O. *French Patent* 796,026 (1936)
107. Starkweather, H. W. *J. Am. Chem. Soc.* **1934**, *56*, 1870
108. Haszeldine, R. N.; Tait, P. J. T. *Polymer* **1973**, *14*, 221
109. Natta, G.; Allegra, G.; Bassi, I. W.; Sianesi, D.; Caporiccio, G.; Torti, E. *J. Polym Sci. Part A* **1965**, *3*, 4263
110. Furukawa, T. *Kobunshi* **1987**, *36*, 868
111. Imbalzano, J. F; Kerbow, D. L. *US Patent* 4,943,658 [*CA* **107**, 116143 (1987)]
112. (a) Ihara, K.; Ootani, K.; Aomi, H. *Proc.-Inst. Environ. Sci.*, **1991**, *37*, 802; (b) Goodman, J. B. *Proc.-Inst. Environ. Sci.*, **1991**, *37*, 551
113. Matsuo, H.; Kaneko, I.; Kanba, M.; Nakamura, H. *Japanese Patent Application* JP 63, 238,115 [*CA* **110**, 135936 (1989)]
114. (a) Squire, E. N. *US Patent* 4,754,009 [*CA* **110**, 77691 (1989)]; (b) Smart, B. E.; Feiring, A. E.; Krespan, C. G.; Yang, Z.-Y.; Resnick, P. R. *Macromol. Symp.* **1995**, *98*, 753; (c) Hung, M.-H. *Macromolecules* **1993**, *26*, 5829
115. (a) Navarrini, W.; Bragante, L.; Fontana, S.; Tortelli, V.; Zedda, A. *J. Fluorine Chem.* **1995**, *71*, 111; (b) *Japanese Patent Application* 07 70,107
116. Yamabe, M.; Matsuo, M. In *Fusso kei Zairyo no Saishin Douko*: Shi Emu Shi; Tokyo, 1994; Chap 2
117. Logothetis, A. L. L. In: *Organofluorine Chemistry: Principles and Commercial Applications*; Banks, R. E.; Smart, B. E.; Tatlow, J. C. Eds.; Plenum Press: New York, 1994; Chap 16
118. (a) Kijima, G.; Wachi, H. *Rubber Chem. Technol.* **1977**, *50*, 403; (b) Kojima, G.; Kojima, H.; Morozumi, M.; Wachi, H.; Hisasue, M. *Rubber Chem. Technol.* **1981**, *54*, 779
119. Yamabe, M. In: *Organofluorine Chemistry: Principles and Commercial Applications*; Banks, R. E.; Smart, B. E.; Tatlow, J. C. Eds.; Plenum Press: New York, 1994; Chap 17
120. Bechtoldl, M. F. *US Patent* 3,651,003 [*CA* **73**, 78669 (1970)]
121. Yamabe, M.; Higaki, H.; Kijima, G. In: *Organic Coatings Science and Technology*; Parfitt, G. D.; Pastis, A. V. Eds.; Marcel Dekker: New York (1984); Vol. 7, pp 25–39
122. Rao, N. S.; Baker, B. E. In: *Organofluorine Chemistry: Principles and Commercial Applications*; Banks, R. E.; Smart, B. E.; Tatlow, J. C. Eds.; Plenum Press: New York, 1994; Chap 14
123. Zisman, W. A. In: *Advances in Chemistry*, ACS Symposium Series No. 43; Fowker, F. M.; Zisman, W. A. Eds.; American Chemical Society, Washington, D. C., 1963; p 1
124. Haszeldine, R. N. *J. Chem. Soc.* **1949**, 2856
125. Parsons, R. E. *US Patent* 3,234,294 [*CA* **62**, 13045 (1965)]
126. Other examples are shown in ref. 39
127. (a) Vaughan, D. J. *Du Pont Innovation* **1973**, *4*, 10; (b) Grot, W. G. *Macromol. Symp.* **1994**, *82*, 161
128. Ukihashi, H.; Yamabe, M.; Miyake, H. *Prog. Polym. Sci.* **1986**, *12*, 229
129. Yamabe, M.; Miyake, H. In: *Organofluorine Chemistry: Principles and Commercial Applications*; Banks, R. E.; Smart, B. E.; Tatlow, J. C. Eds.; Plenum Press: New York, 1994; Chap 18
130. (a) Pedersen, S. D.; Qiu, W.; Qiu, Z.-M.; Kotov, S. V.; Burton, D. J. *J. Org. Chem.* **1996**, *61*, 8024; (b) Kato, M.; Yamabe, M. *J. Chem. Soc., Chem. Commun.* **1981**, 1173

CHAPTER 7
Fluorous Media

The isolation and purification of a desired organic compound after workup are inevitable but often tedious procedures. Thus, solid-phase synthesis [1] and phase-separation techniques [2] are introduced that allow isolation of the desired compounds in an almost pure form without such time-consuming procedures as column chromatography [3]. Very recently, in order to make the separation and purification of products and catalysts easier, PFCs have been introduced as solvents in organic synthesis.

7.1
Organic Reactions in Perfluorocarbons

The salient features of PFC reaction media were first described in 1993 by Zhu [4], who used FC-77 (bp 97 °C, consisting mainly of isomeric mixtures of perfluoroalkyl ethers), as a reaction medium for esterification, transesterification, acetalization, enamine formation, and suspension polymerization of styrene/vinylbenzenes, as summarized in Table 7.1.

The advantages of PFC reaction media are: (i) The desired compounds are easily separated and obtained in a pure form simply by phase separation. This procedure is particularly suitable for the synthesis of labile methacrylates $CH_2=C(CH_3)CO_2C_{10}H_{21}$. (ii) Reactivity in PFCs is often better than that in toluene. (iii) The nontoxicity and zero ozone-depletion potential of PFCs are friendly to life and environment. (iv) Various PFCs with a wide range of boiling

Table 7.1. Organic reactions in PFC media

Substrate Reagent	Product	Yield (%)
$PhCO_2C_3H_7$ $HO(CH_2)_4OH$	$PhCO_2(CH_2)_4OCOPh$	67
$CH_2=C(CH_3)CO_2H$ $C_{10}H_{21}OH$	$CH_2=C(CH_3)CO_2C_{10}H_{21}$	92
$CH(OEt)_3$ cyclohexanone	1,1-diethoxycyclohexane	41
morpholine cyclohexanone	1-morpholino-1-cyclohexene	53

temperatures from 56 to 220 °C are commercially available. (v) PFCs are stable under most reaction conditions.

In 1990, the use of PFCs for a suspension polymerization of lauryl methacrylate was filed in a patent [5]. This reaction medium was claimed to be appropriate to the polymerization of a labile monomer in aqueous media.

Bromination of alkenes is carried out in perfluorohexane (PFH) [6]. The solvent is recovered easily in a pure form and can be reused without any further purification. PFH is suggested to be an excellent alternative for carbon tetrachloride.

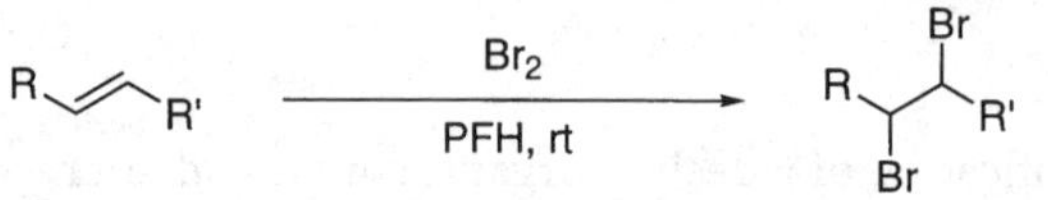

$$R-CH=CH-R' \xrightarrow{Br_2, \text{PFH, rt}} R-CHBr-CHBr-R' \quad (7.1)$$

Since PFH is UV transparent above 185 nm and is relatively inert against radical species, photolysis of bis(perfluorooctyl)diazene is performed in PFH more conveniently than in benzene [7].

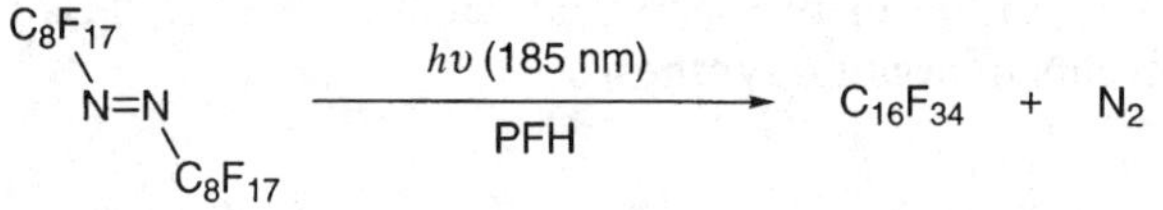

$$C_8F_{17}-N=N-C_8F_{17} \xrightarrow{h\upsilon \text{ (185 nm)}, \text{PFH}} C_{16}F_{34} + N_2 \quad (7.2)$$

Owing to their high oxygen solubility in PFH, organozinc compounds are effectively oxidized in PFH to give hydroperoxides in moderate to high yields. Scheme 7.1 shows a transformation from an olefin to a hydroperoxide [8].

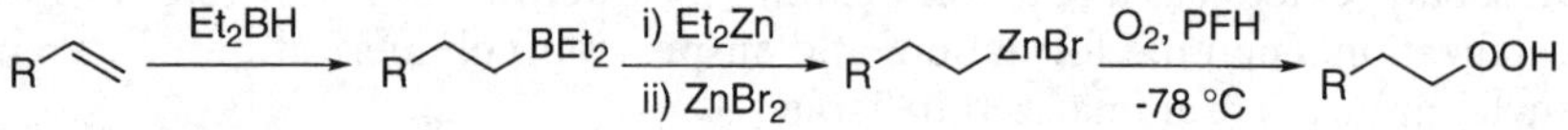

$$R-CH=CH_2 \xrightarrow{Et_2BH} R-CH_2-CH_2-BEt_2 \xrightarrow[\text{ii) ZnBr}_2]{\text{i) Et}_2Zn} R-CH_2-CH_2-ZnBr \xrightarrow[-78\,°C]{O_2, \text{PFH}} R-CH_2-CH_2-OOH$$

Scheme 7.1. Synthesis of hydroperoxides in PFH

The reaction of alkenes with singlet oxygen under photochemical conditions in the presence of a sensitizer produces allylic hydroperoxides in high yields [9].

Generally, partially fluorinated compounds differ from PFCs, especially in their dipole moments and dielectric constants, as described in Chap. 1. Benzotrifluoride (BTF, bp 40 °C) is slightly more polar than THF or ethyl acetate, slightly less polar than chloroform or dichloromethane, and thus has been suggested as a solvent for organic synthesis as an alternative to dichloromethane [10]. BTF is used for such transformations as acylation, tosylation, and silylation of alcohols, Swern oxidation, Dess–Martin oxidation, selenoxide elimination reaction, and Lewis acid catalyzed allylation, Mukaiyama aldol reaction, Friedel-Crafts acylation, and Diels–Alder reaction. Reactivities and yields in BTF are demonstrated to be similar or superior to those in dichloromethane. Representatives are shown below.

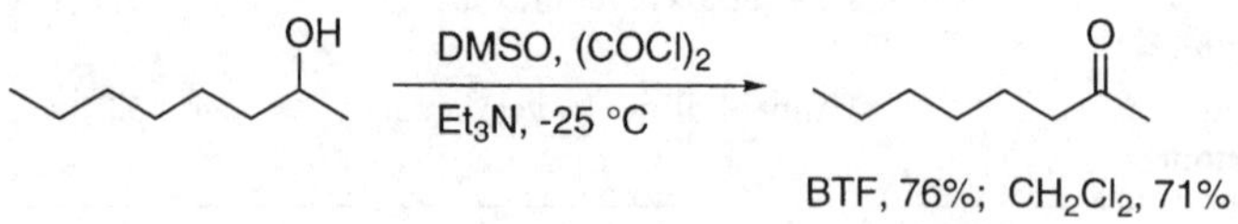

$$\text{R-CH(OH)-R'} \xrightarrow[\text{Et}_3\text{N, -25 °C}]{\text{DMSO, (COCl)}_2} \text{R-CO-R'} \quad (7.3)$$

BTF, 76%; CH$_2$Cl$_2$, 71%

$$\text{Ph-C(=O)-Cl} + \text{4-tBu-2-OMe-C}_6\text{H}_3\text{-H} \xrightarrow{\text{ZnCl}_2} \text{Ph-C(=O)-(2-OMe-4-tBu-C}_6\text{H}_3\text{)} \quad (7.4)$$

BTF, reflux, 1 day, 81%
CHCl$_2$CHCl$_2$, 138 °C, 2 days, 66%

The limitations of BTF are that it freezes at −29 °C and is sensitive at higher temperatures to AlCl$_3$, some reductants and aqueous acids.

7.2
Fluorous Biphase Reactions

A liquid-liquid biphase system consisting of organic and aqueous media has been used chiefly for the separation of inorganic materials from organic reaction mixtures. The phase separation technique is also effective in the separation of acid/base components from organic mixtures by pH control of the aqueous phase.

The term "fluorous" was coined in 1994 by Horváth and Rábai [11] in analogy to the term "aqueous". They used a fluorous phase instead of an aqueous phase for phase separation and demonstrated a method for the facile separation and recycle of an organo transition metal catalyst from an organic reaction mixture using a fluorous biphase system (FBS). The key to this concept lies in the immiscibility of PFCs in organic and aqueous liquids at room temperature and dissolution of an insoluble organo transition metal complex in PFCs.

Another approach to the separation and purification of organic products was suggested in 1997 by Curran, who showed that the residue of a reagent or a reactant having a perfluoroalkyl chain was easily separated from an organic phase containing a product by a simple phase separation using PFCs [12]. The recovered reagent or reactant residue dissolved in the fluorous phase is easily reconverted into an active reagent. The two methods described above will provide new experimental techniques for future organic synthesis [2a, 13].

7.2.1
Hydroformylation

An organic/aqueous biphase system (BS) was first introduced to ease separation and recovery of catalysts using sulfonated triphenylphosphines [14]. Although this method is useful in many homogeneous reactions, it is not applicable to a system containing a water-sensitive component. An FBS employs a PFC in lieu of water in combination with an organic solvent such as toluene, hexane, THF, acetone, or an alcohol. High gas solubility in PFCs accelerates homogeneous reactions of oxygen, hydrogen, and carbon monoxide. A homo-

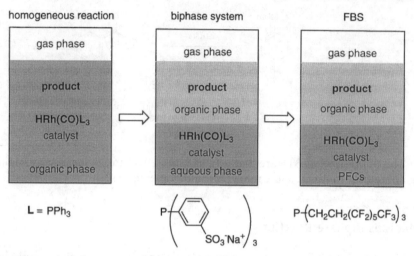

Fig. 7.1. Homogeneous reaction, biphase system, and FBS using a rhodium catalyst

geneous reaction, a biphase system and an FBS using a rhodium catalyst are compared schematically in Fig. 7.1.

For the FBS catalyst, a partially fluorinated phosphine ligand, $P(CH_2CH_2(CF_2)_5CF_3)_3$, is prepared from $CH_2=CH(CF_2)_5CF_3$. The perfluorohexyl moiety allows the metal complex to dissolve in PFCs; a CH_2CH_2 insulator minimizes the electron-withdrawing effect of the n-C_6F_{13} group. In this way, a "fluorous" soluble catalyst which can maintain its catalytic activity is designed and prepared. The $R_fCH_2CH_2$ group has also been applied to a cyclopentadienyl metal complex [15]. A $R_fCH_2CH_2$ group is often called a "ponytail".

A typical example of a transition metal catalyzed organic reaction in an FBS is the rhodium-catalyzed hydroformylation of 1-decene carried out in toluene/perfluoro(methylcyclohexane) using $Rh(CO)_2(acac)$ and $P[CH_2CH_2(CF_2)_5CF_3]_3$.

$$n\text{-}C_{10}H_{21}\text{-CH=CH}_2 \xrightarrow[\substack{Rh(CO)_2(acac) \\ P[CH_2CH_2(CF_2)_5CF_3]_3 \\ P/Rh = 40}]{CO:H_2\ (1:1,\ 75\ psi)} n\text{-}C_{10}H_{21}\text{-CH}_2\text{-CHO} + n\text{-}C_{10}H_{21}\text{-CH(CHO)-} \quad (7.5)$$

normal   iso

The catalyst, $HRh(CO)\{P[(CH_2)_2(CF_2)_5CF_3]_3\}_3$, is generated in situ under an atmosphere of CO/H_2 (1:1) at 5 bar. Hydroformylation of 1-decene, for example, proceeds in high yields at 100 °C with CO/H_2 (1:1) at 5 bar to give a high normal/iso ratio of products. Under the reaction conditions, the reaction mixture becomes homogeneous; after the reaction is complete, the homogeneous phase separates into an organic and a fluorous phase. Almost all the aldehyde and the rhodium catalyst remain in the organic and the fluorous phase, respectively.

Fig. 7.2. An X-ray crystal structure of two models of trans-(IrCl(CO){P[CH$_2$CH$_2$(CF$_2$)$_5$CF$_3$]$_3$}$_2$)

Thus, the recovered fluorous phase can be used for a second reaction. After nine consecutive reaction/separation cycles, a total turnover of more than 35,000 is achieved with a loss of 1.18 ppm of Rh/mol of aldehyde [16].

The estimated enthalpy of the reaction of fluorous phosphine P(CH$_2$CH$_2$(CF$_2$)$_5$CF$_3$)$_3$ with [Rh(CO)$_2$Cl]$_2$ and CpRu(COD)Cl suggests that the phosphine ligand is a poorer donor than PEt$_3$, and similar to that for PPhMe$_2$ [17]. An X-ray structural analysis has disclosed that the three fluorous ponytails of the rhodium complex [RhCl(CO){P(CH$_2$CH$_2$(CF$_2$)$_5$CF$_3$}$_2$] [18] and the corresponding iridium analog of Vaska's complex trans-(IrCl(CO){P[CH$_2$CH$_2$-(CF$_2$)$_5$CF$_3$]$_3$}$_2$) [19] possess two different conformations due to the packing manner of the two phosphines in each couple. Because perfluorocarbons tend to self-aggregate, it is easily understood that four ponytails in each complex align in a parallel and coplanar fashion (Fig. 7.2). Although the most stable conformation of a perfluoroalkyl chain is a helix, the four chains take an *anti* conformation due to aggregation. Indeed, the remaining ponytails take helix conformations, as shown in Fig. 7.2.

Oxidative addition of *trans*-(IrCl(CO){P[CH$_2$CH$_2$(CF$_2$)$_5$CF$_3$]$_3$}$_2$) to an alkyl halide proceeds in PFCs through a free-radical chain mechanism [19, 20], because an S$_N$2-type mechanism is disfavored in nonpolar PFC media.

The FBS technique is also applicable to a rhodium-catalyzed hydroboration [21]. In the presence of RhCl[P(CH$_2$CH$_2$(CF$_2$)$_5$CF$_3$]$_3$, an analog of the Wilkinson catalyst RhCl(PPh$_3$)$_3$, hydroboration of olefins is carried out using catecholborane in a mixture of toluene and c-C$_6$F$_{11}$CF$_3$ in high yields with a high turnover number. Recycle of the catalyst is again achieved easily.

7.2.2
Oxidation

Transition metal catalyzed oxidation of organic substrates with molecular oxygen is advantageous in view of industrial production. An FBS is applied to this process, as illustrated in Fig. 7.3. Upon heating, the two-phase system

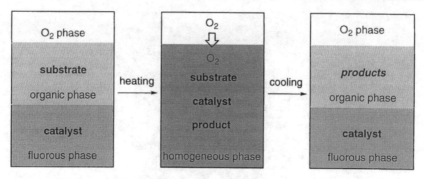

Fig. 7.3. Oxidation with molecular oxygen in an FBS

becomes homogeneous, and oxidation proceeds in an oxygen-rich medium. After the reaction, the reaction mixture is cooled down to again separate into two phases with products in the organic phase and the catalyst in the fluorous phase. Separation of each phase allows isolation of the products and reuse of the catalyst.

By means of this method, aldehydes, sulfides, and alkenes are oxidized [22] using a fluorous soluble catalyst prepared from a perfluoroalkyl-substituted 1,3-diketone, as illustrated in Scheme 7.2. With the nickel catalyst used in Scheme 7.2, aldehydes are oxidized in an FBS to carboxylic acids in high yields. Reuse of the fluorous phase containing the catalyst is readily done without any loss of the catalyst. For example, the yield of p-chlorobenzoic acid from p-chlorobenzaldehyde ranges from 87% (1st cycle) to 70% (6th cycle).

$$\text{RCHO} \xrightarrow[\substack{\text{toluene/perfluorodecalin} \\ 64\,°C,\ 12\,h}]{\text{Ni cat. (3 mol\%), } O_2\ (1\ \text{atm})} \text{RCOOH} \quad\quad (7.6)$$
$$71\text{-}87\%$$

cycle	yield (%)
1	87
2	83
3	75
4	77
5	71
6	70

Oxidation of sulfides with molecular oxygen is performed in the presence of isobutyraldehyde and the nickel catalyst shown in Scheme 7.2 [23]. The amount of isobutyraldehyde is crucial: 1.6 mol are appropriate for sulfoxides but sulfone synthesis needs 5 mol (Scheme 7.3).

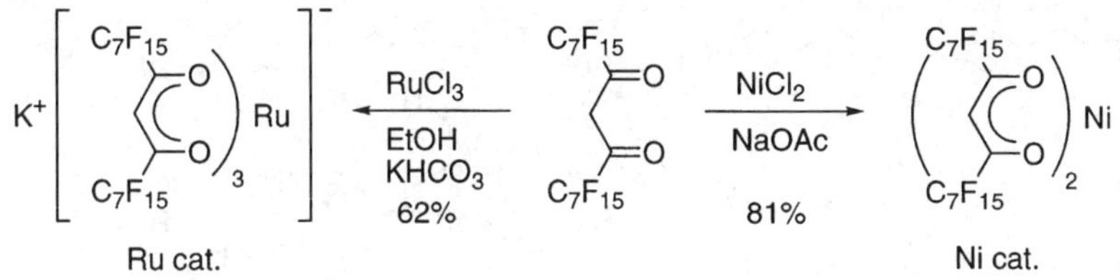

7.2 Fluorous Biphase Reactions

Scheme 7.2. Preparation of perfluorinated complexes

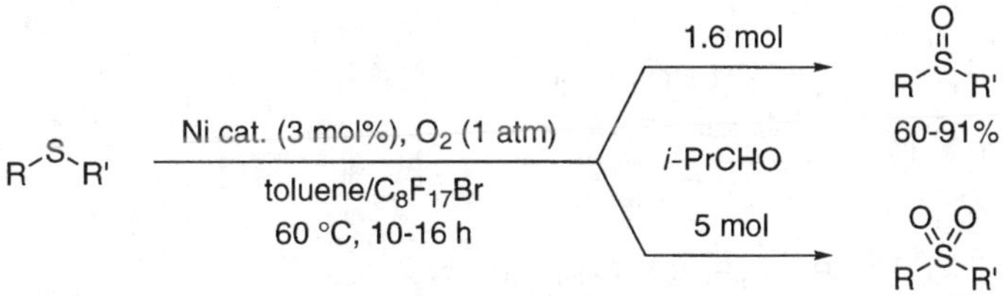

Scheme 7.3. Oxidation of sulfides in an FBS

For the epoxidation of olefins, an FBS with the ruthenium complex from Scheme 7.2 is highly effective. The catalyst is easily separated and can be reused several times for repeated epoxidations.

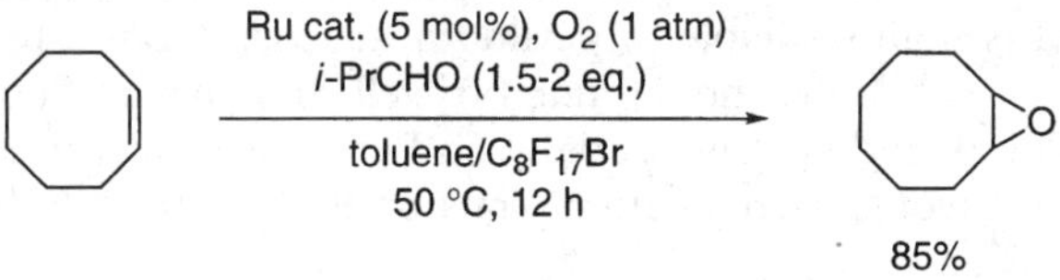

(7.7)

Epoxidation of olefins is alternatively carried out using a cobalt complex of tetraarylporphyrin bearing eight C_8F_{17} substituents [24] in PFH and acetonitrile [25]. The reaction is performed with a substrate/catalyst ratio of 1000:1, much higher than that of the ruthenium system (20:1). Reuse of the catalyst does not deteriorate the conversion or selectivity. Enantioselective epoxidation of indene using a chiral fluorous soluble (salen) Mn complex [26–28] in CH_2Cl_2 and D-100, a mixture that contains mainly perfluorooctane. The desired epoxide is obtained in 83% yield with 92% enantiomeric excess. Applicability of the reaction to other alkenes still remains to be investigated.

Direct oxidation of hydrocarbons with molecular oxygen in an FBS is achieved by use of azamacrocycles R_fTACTD [29] and R_fTACN [30] (Fig. 7.4) which have perfluoro ponytails.

Cyclohexene is oxidized with *tert*-butyl hydroperoxide (TBHP) under an oxygen atmosphere using a fluorous soluble active catalyst generated in situ from $[Mtl(O_2C(CH_2)_2C_8F_{17})_2]$ (Mtl=Mn, Co) and R_fTACN in perfluoroheptane.

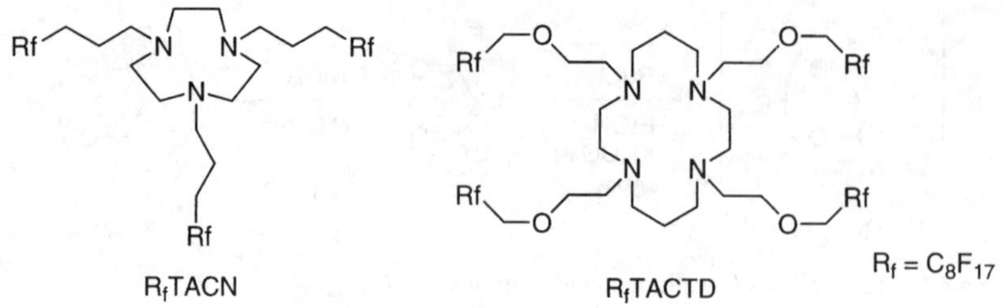

Fig. 7.4. Fluorous soluble azamacrocycles

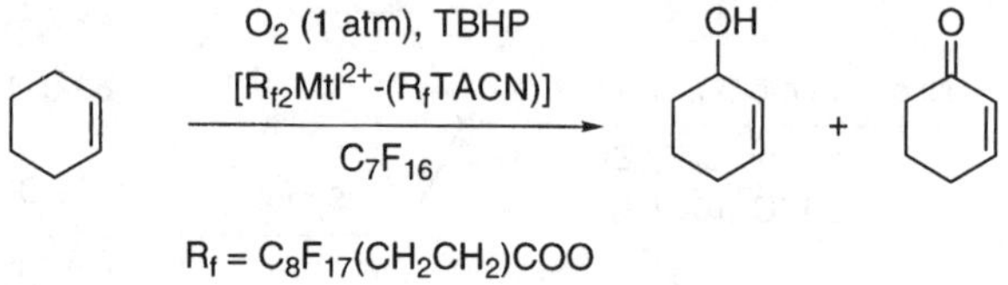

$$\text{cyclohexene} \xrightarrow[C_7F_{16}]{\substack{O_2 \text{ (1 atm), TBHP} \\ [R_{f2}Mtl^{2+}\text{-}(R_fTACN)]}} \text{cyclohexenol} + \text{cyclohexenone} \quad (7.8)$$

$R_f = C_8F_{17}(CH_2CH_2)COO$

For an efficient allylic oxidation, use of O_2 and TBHP in a stoichiometric amount is essential. Some results are summarized in Table 7.2. The Mn catalyst is easily recovered from the reaction mixture as compared with the conventional procedure.

Singlet oxygen is efficiently generated in an FBS using 5,10,15,20-tetrakis-(heptafluoropropyl)porphyrin (TPFPP) as a fluorous soluble photosensitizer [31, 32]. TPFPP tolerates the oxidation conditions better than tetraphenylporphyrin (TPP) and is highly soluble in perfluorocarbons. Because the organic phase dissolves the substrate and product(s) well and the perfluorocarbon phase dissolves TPFPP and O_2 preferentially, an FBS allows an easy separation of the sensitizer and avoids overoxidation. Thus, an FBS is particularly suitable for singlet oxygenation.

Reaction of cyclohexene with singlet oxygen generated in a C_6F_{14}/CD_3CN biphasic system in the presence of 0.2 mol% TPFPP under visible light irradia-

Table 7.2. Oxidation of hydrocarbons with O_2 in FBS[a]

Compound	Oxidant	Product (ratio)	Time (h)	Yield[b] (%)
cyclohexene	TBHP/O_2	cyclohexenol-cyclohexanone (1:2)	3	650
cyclohexene	TBHP	cyclohexenol-cyclohexanone (1:2)	7	trace
cyclohexene	O_2	cyclohexenol-cyclohexanone (1:2)	24	trace
toluene	TBHP/O_2	PhCHO/PhCH$_2$OH (1:2)	24	65

[a] [Mtl($O_2C(CH_2)_2C_8F_{17})_2$] (5 mol%TBHP) and R_fTACN (5 mol%/TBHP) were used.
[b] Based on TBHP added.

7.3 Purification and Isolation by Phase Separation

tion at 0 °C gives 2-cyclohexenyl hydroperoxide in high yields. The lifetime of the sensitizer is definitely prolonged. For example, after 8, 32, and 46 h of reaction, the yield of the hydroperoxide increases to 23, 84, and 96 % with a residue of TPFPP >99, 94, and 83 %, respectively. The sensitizer-containing perfluorohexane solution can be reused directly after separation.

$$\text{cyclohexene} \xrightarrow[\text{C}_6\text{F}_{14}/\text{CD}_3\text{CN, 0 °C}]{\text{O}_2 \text{ (1 atm)}, h\nu \quad \text{TPFPP (0.2 mol\%)}} \text{2-cyclohexenyl hydroperoxide (OOH)} \quad (7.9)$$

7.3
Purification and Isolation by Phase Separation

Separation, isolation, and purification of desired products after workup are important but very often tedious. Although chromatography is powerful, it is necessary to use a column packed with relatively expensive solid gel or resin and a large volume of an eluent. In contrast, solid-phase synthesis allows purification of products simply by filtration. However, additional steps to connect the substrate or the reagent to the solid support and to disconnect the product from the support are necessary.

Since a solid-phase system and an FBS have their own characteristic advantages in the separation and purification of products and catalysts, a combination of both would be suitable for conducting multiple synthesis. In addition, an FBS can compensate for the drawbacks of solid-phase synthesis. For example, the reaction in an FBS can be performed by a procedure similar to conventional organic synthesis. The high chemical and thermal stabilities of perfluoroalkyl ponytails in a fluorous soluble substrate allow the use of various types of reagents even under harsh conditions. Furthermore, because of the relatively low boiling and melting temperatures of the substrates with fluorous ponytails, identification and analysis of products in an FBS are easily performed, whereas it is hard to characterize products on a solid phase.

Three approaches have been introduced to fluorous biphase systems: fluorous synthesis, fluorous phase switch, and fluorous multiphase condensations for a new separation technique, by labeling a substrate and/or a product with perfluoroalkyl-substituted ponytails to dissolve a substrate and/or a product in the fluorous phase [12].

A typical example is the nitrile oxide cycloaddition to a terminal alkene or alkyne, as illustrated in Scheme 7.4. Generally, a large excess of the nitrile oxide is required to obtain the desired heterocycle in high yields because the dimerization of a nitrile oxide often lowers the yield of the cycloadduct. Therefore, chromatographic purification of the desired heterocycle from the resulting complex mixture of products is inevitable.

A fluorous phase separation technique is applied to the cycloaddition reaction using a fluorous soluble substrate with fluorous phase labeling. To dissolve the substrate and product in the fluorous phase, an allyl silyl ether having pefluorocarbon ponytails on silicon is employed for the cycloaddition, as shown

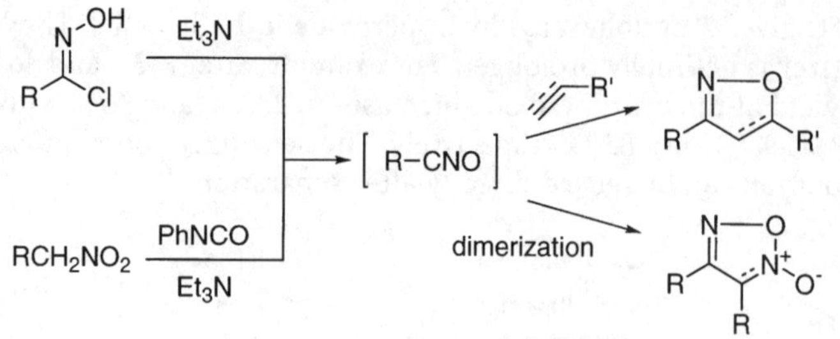

Scheme 7.4. Cycloaddition reaction of nitrile oxide with unsaturated compounds

in Scheme 7.5. A fluorous allyl silyl ether is prepared and allowed to react with an excess of the nitrile oxide. Isolation and purification of the fluorous soluble desired product are achieved by a single fluorous/organic/aqueous three-phase liquid extraction to give only the desired cycloadduct in high purity. To implement the phase separation, the fluorous ponytails are cleaved with HF/-pyridine. After the deprotection and separation of the products in the organic phase, the organosilicon reagent is recovered and reused. As shown in Scheme 7.5, both overall yields and purities of the products are of synthetic meaning.

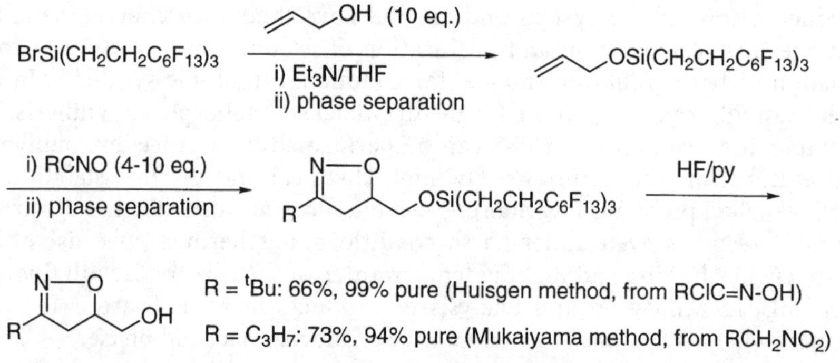

Scheme 7.5. Nitrile oxide cycloaddition with a fluorous labeled allyl alcohol

Organic fluorous phase switch is also effective for the separation and purification of products. The core of the technique is an in situ fluorous labeling of the reaction product, which is transported from the organic phase into the fluorous phase and then separated from the fluorous phase selectively. An example of a Grignard reaction is shown in Scheme 7.6. When an excess amount of aldehyde is applied, the unreacted aldehyde remains in the organic phase, whereas the adduct magnesium alkoxide is connected to fluorous ponytails and extracted into the fluorous phase selectively by a three-phase separation. Thus, finally, the fluorous phase contains only the desired silyl ether; excess bromosilane is recovered by the first separation. Then the fluorous phase is treated with cesium fluoride to cleave a O–Si bond, and an alcohol and a fluorous halosilane result. The alcohol is extracted into the organic phase, the halosilane

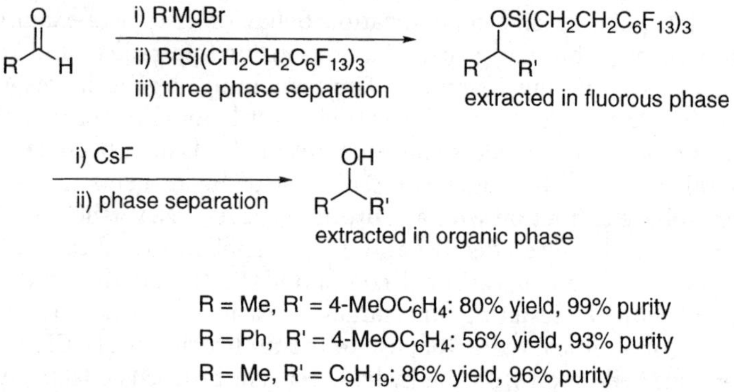

7.3 Purification and Isolation by Phase Separation

$$R-CHO \xrightarrow[\text{iii) three phase separation}]{\text{i) R'MgBr} \atop \text{ii) BrSi(CH}_2\text{CH}_2\text{C}_6\text{F}_{13})_3} \underset{\text{extracted in fluorous phase}}{R-CH(OSi(CH}_2\text{CH}_2\text{C}_6\text{F}_{13})_3)-R'}$$

$$\xrightarrow[\text{ii) phase separation}]{\text{i) CsF}} \underset{\text{extracted in organic phase}}{R-CH(OH)-R'}$$

R = Me, R' = 4-MeOC$_6$H$_4$: 80% yield, 99% purity
R = Ph, R' = 4-MeOC$_6$H$_4$: 56% yield, 93% purity
R = Me, R' = C$_9$H$_{19}$: 86% yield, 96% purity

Scheme 7.6. Grignard reaction and purification by a fluorous phase switch

into the fluorous phase. The yield and purity of the product are satisfactory for an organic synthesis.

Purification using a fluorous phase is applicable to multicomponent condensation reactions; the desired product is isolated from the substrate and two or more reactants in a single step [33]. Examples of the Ugi and the Biginelli reaction in Fig. 7.5 show that the fluorous separation technique is extremely facile in the isolation of a desired condensation product.

The fluorous technique is also applied to the preparation and purification of a disaccharide having a fluorous soluble protecting group [34].

An organic mess in a reaction mixture sometimes raises serious purification problems, particularly when by-products resemble the desired product in

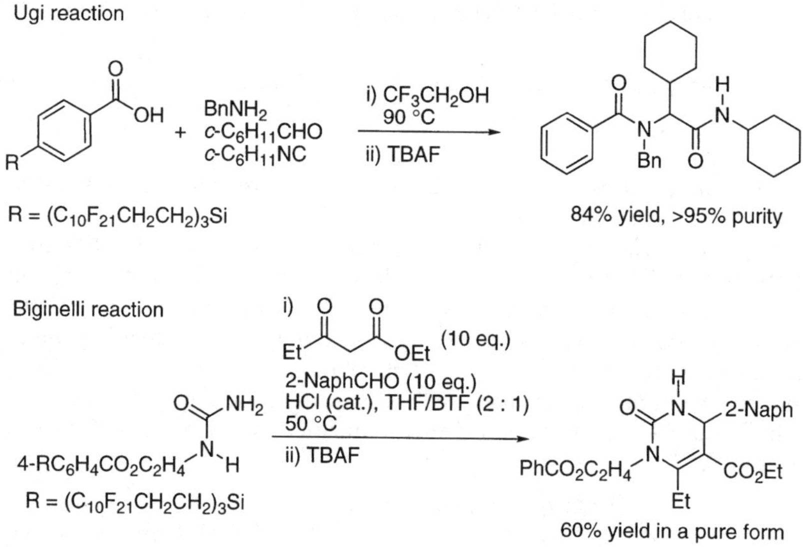

Ugi reaction

4-R-C$_6$H$_4$-CO$_2$H + BnNH$_2$ + c-C$_6$H$_{11}$CHO + c-C$_6$H$_{11}$NC $\xrightarrow[\text{ii) TBAF}]{\text{i) CF}_3\text{CH}_2\text{OH, 90 °C}}$ product

R = (C$_{10}$F$_{21}$CH$_2$CH$_2$)$_3$Si

84% yield, >95% purity

Biginelli reaction

4-RC$_6$H$_4$CO$_2$C$_2$H$_4$-NH-C(O)-NH$_2$ $\xrightarrow[\text{ii) TBAF}]{\text{i) EtO}_2\text{CCH}_2\text{CO}_2\text{Et (10 eq.), 2-NaphCHO (10 eq.), HCl (cat.), THF/BTF (2:1), 50 °C}}$ product

R = (C$_{10}$F$_{21}$CH$_2$CH$_2$)$_3$Si

60% yield in a pure form

Fig. 7.5. Multicomponent condensation and purification by the fluorous technique

boiling point, polarity, and crystallization behavior. A typical example is the reaction with organotin reagents, widely used for free-radical and transition metal catalyzed coupling reactions. The remaining organotin reagents and organotin halide and alkoxide residues frequently hamper the purification [35]. In order to remove the tin mess conveniently, methods have been devised such as immobilization of tin reagents, dissolution of tin reagents in the aqueous phase, and phase extraction with a coordinative agent. Nevertheless, a fluorous extraction is much more effective and can be applied to radical reduction of halides [36], allylation of carbonyl compounds [37], and the Stille coupling reaction [38–40]. The resulting tin residues are easily removed by a single-phase separation. For a radical reduction, use of the tin hydride $(C_6F_{13}CH_2CH_2)_3SnH$ is convenient. Reduction with the hydride reagent proceeds effectively; the tin residue is readily removed by evaporation of the solvent BTF followed by liquid-liquid extraction with perfluoromethylcyclohexane (PFMC) and dichloromethane. This procedure is applicable to the catalytic reduction of halides using 10% tin hydride and 1.3 eq. of $NaCNBH_3$ in a 1:1 mixture of BTF and *tert*-butyl alcohol.

$$\text{Br-Ad} \xrightarrow[\text{AIBN (10\%)}]{\substack{NaBH_3CN\ (1.3\ mol) \\ (C_6F_{13}CH_2CH_2)_3SnH\ (1\ mol\%) \\ \text{BTF-}t\text{-BuOH, reflux, 3 h}}} \text{Ad} \quad (7.10)$$

1 mol → 95%

The stoichiometric procedure is also applicable to the reduction of phenylselenyl, nitro, and methyl dithiocarbonyloxy groups.

The phase separation technique has been successfully employed in a combinatorial synthesis and a reaction in supercritical CO_2 [41].

Another example of the fluorous biphase purification technique is demonstrated by thermal allylation of aldehydes with a fluorous soluble allylstannane [42].

$$RCHO + \diagup\!\!\!\diagup\!\!\!\text{Sn}(CH_2CH_2C_6F_{13})_3 \xrightarrow[\substack{140\ °C \\ 3\ d}]{\text{neat}} R\!\!\diagup\!\!\!\overset{OH}{\diagup}\!\!\!\diagup \quad (7.11)$$

(3 eq.)

Yields and purities of crude products after a fluorous biphase separation or fluorous solid extraction are summarized in Table 7.3.

For a fluorous solid-phase extraction, fluorous reverse-phase (FRP) silica gel is prepared by silylating the surface of silica gel for flash chromatography using dimethyl[(2-perfluorohexyl)ethyl]silyl chloride. The crude reaction mixture is chromatographed on FRP silica gel using acetonitrile and then hexane as an eluent. Concentration of the acetonitrile elutant gives the products listed in Table 7.3. Fluorous tin residues on FRP silica gel are eluted by hexane. The method has been further applied to the purification of organometallic complexes [42].

$$\text{Silica}-OH \xrightarrow{ClSi(Me)_2CH_2CH_2C_6F_{13}} \text{Silica}-OSi(Me)_2CH_2CH_2C_6F_{13} \quad (7.12)$$

FRP silica gel

Table 7.3. Allylation of aldehydes with a fluorous allyl stannane

Aldehyde (R)	Fluorous biphase extraction		Fluorous solid extraction		
	crude yield (%)	purity (%)	crude yield (%)	purity (%)	isolated yield (%)
p-MeOC$_6$H$_4$	61	83	98	76	70
1-Naphthyl	92	85	95	84	70
PhCH$_2$CH$_2$	82	88	80	100	71
c-C$_6$H$_{11}$	62	94	100	100	80
p-NO$_2$C$_6$H$_4$	90	94	98	91	85

The Stille coupling [38–40] with aryl halides proceeds in the presence of a Pd catalyst to give coupled products as shown below. Separation of the products and the tin residue with a fluorous ponytail is easily performed through a three-phase separation: evaporation of the fluorous phase provides in 80–90% yields the tin residue $(C_6F_{13}CH_2CH_2)_3SnCl$ which is easily recycled for the next reaction. From the organic phase, the desired hetero-coupled products and a homo-coupled product are isolated. These, however, need to be separated by flash column chromatography.

$$R-X + (C_6F_{13}CH_2CH_2)_3SnAr \xrightarrow[\text{LiCl, 80 °C, 22 h}]{\substack{\text{PdCl}_2(\text{PPh}_3)_2 \text{ (2 mol\%)} \\ \text{DMF/THF (1/1)}}} \quad (7.13)$$
$$1 \text{ mol} \qquad 1.2 \text{ mol}$$

$$R-Ar + Ar-Ar + (C_6F_{13}CH_2CH_2)_3SnCl$$

R–X	Ar and yields of R-Ar (and Ar-Ar)	
	4-MeOC$_6$H$_4$ (%)	2-Pyridyl (%)
p-CH$_3$COC$_6$H$_4$Br	90 (7)	87 (6)
p-NO$_2$C$_6$H$_4$Br	94 (9)	98 (8)
p-NO$_2$C$_6$H$_4$OTf	82 (9)	86 (8)
PhCH$_2$Br	77 (5)	98 (7)

The salient features of fluorocarbons in relation to chemical and thermal stability, gas solubility, and miscibility with common organic solvents at high temperatures allow novel applications to organic reaction media. It is expected that this new fluorous technique will make many synthetic processes sustainable in the future.

References

1. (a) N. C. Mathur, C. K. Narang, R. E. Williams, *Polymers as Aids in Organic Chemistry*, Academic Press, New York (1980); (b) L. M. Gayo, *Biotechnol. Bioeng. (Combinatorial Chemistry)* **1998**, *61*, 95

2. (a) D. P. Curran, *Angew. Chem. Int. Ed. Engl.* **1998**, *37*, 1174; (b) B. Cornils, *Angew. Chem. Int. Ed. Engl.* **1995**, *34*, 1575; (c) I. T. Horváth, F. Joó, Eds., *Aqueous Organometallic Chemistry and Catalysts*, Kluwer, Dordrecht (1995)
3. (a) W. C. Still, M. Kahn, A. Mitra, *J. Org. Chem.*, **1978**, *43*, 2923; (b) Y. Naruta, H. Uemori, M. Fukumoto, H. Sugiyama, Y. Sakata, K. Maruyama, *Bull. Chem. Soc. Jpn.*, **1988**, *61*, 1815
4. D.-W. Zhu, *Synthesis* **1993**, 953
5. K. Tsubushi, *Jpn. Kokai Tokkyo Koho* JP 2-24362 [90-243602]
6. S. M. Pereira, G. P. Savage, G. W. Simpson, *Synth. Commun.* **1995**, *25*, 1023
7. T. Nalkamura, A. Yabe, *Chem. Lett.* **1995**, 533
8. I. Klement, P. Knochel, *Synlett.* **1995**, 1113
9. R. D. Chambers, G. Sandford, A. Shah, *Synth. Commun.* **1996**, *26*, 1861
10. A. Ogawa, D. P. Curran, *J. Org. Chem.* **1997**, *62*, 450
11. (a) I. T. Horváth, J. Rábai, *Science* **1994**, *266*, 72; (b) I. T. Hováth, *Acc. Chem. Res.*, **31**, 61 (1998)
12. A. Studer, S. Hadida, R. Ferritto, S.-Y. Kim, P. Jeger, P. Wipf, D. P. Curran, *Science* **1997**, *275*, 823
13. (a) J. A. Gladysz, *Science* **1994**, *266*, 55; (b) B. Cornils, *Angew. Chem. Int. Ed. Engl.* **1997**, *36*, 2057
14. E. G. Kuntz, *Chemtech* **1987**, *17*, 570
15. R. P. Hughes, H. A. Trujillo, *Organometallics* **1996**, *15*, 286
16. I. T. Horváth, G. Kiss, R. A. Cook, J. E. Bond, P. A. Stevens, J. Rábai, E. J. Mozeleski, *J. Am. Chem. Soc.* **1998**, *120*, 3133
17. C. Li, S. P. Nolan, I. T. Horváth, *Organometallics* **1998**, *17*, 452
18. J. Fawcett, E. G. Hope, R. D. W. Kemmitt, D. R. Paige, D. R. Russell, A. M. Stuart, D. J. Cole-Hamilton, M. J. Payne, *Chem. Commun.* **1997**, 1127
19. M.-A. Guillevic, A. M. Arif, I. T. Horváth, *Angew. Chem. Int. Ed. Engl.* **1997**, *36*, 1612
20. M.-A. Guillevic, C. Rocaboy, A. M. Arif, I. T. Horváth, J. A. Gladysz, *Organometallics* **1998**, *17*, 707
21. J. J. J. Juliette, I. T. Horváth, J. A. Gladysz, *Angew. Chem. Int. Ed. Engl.* **1997**, *36*, 1610
22. I. Klement, H. Lütjens, P Knochel, *Angew. Chem. Int. Ed. Engl.* **1997**, *36*, 1454
23. T. Yamada, T. Takai, O. Rhode, T. Mukaiyama, *Chem. Lett.* **1991**, 1
24. G. Pozzi, S. Banfi, A. Manfredi, F. Montanari, S. Quici, *Tetrahedron* **1996**, *52*, 11879
25. G. Pozzi, F. Montanari, S. Quici, *Chem. Commun.* **1997**, 69
26. G. Pozzi, F. Cinato, F. Montanari, S. Quici, *Chem. Commun.* **1998**, 877
27. Epoxidation of alkenes using manganese(III) acetate dihydrate in perfluoro-2-butyltetrahydrofuran or perfluoropropanol: K. S. Ravikumar, F. Barbier, J.-P. Bégué, D. Bonnet-Delpon, *Tetrahedron* **1998**, *54*, 7457
28. (a) T. Yamada, K. Imagawa, T. Nagata, T. Mukaiyama, *Chem. Lett.* **1992**, 2231; (b) K. Imagawa, T. Nagata, T. Yamada, T. Mukaiyama, *Chem. Lett.* **1994**, 527
29. G. Pozzi, M. Cavazzini, S. Quici, S. Fontana, *Tetrahedron Lett.* **1997**, *38*, 7605
30. J.-M. Vincent, A. Rabion, V. K. Yachandra, R. H. Fish, *Angew. Chem. Int. Ed. Engl.* **1997**, *36*, 2346
31. S. G. DiMagno, P. H. Dussault, J. A. Schultz, *J. Am. Chem. Soc.* **1996**, *118*, 5312
32. S. G. DiMagno, R. A. Williams, M. J. Therien, *J. Org. Chem.* **1994**, *59*, 6943
33. A. Studer, P. Jeger, P. Wipf, D. P. Curran, *J. Org. Chem.* **1997**, *62*, 2917
34. D. P. Curran, R. Ferritto, Y. Hua, *Tetrahedron Lett.* **1998**, *39*, 4767
35. D. Crich, S. Sun, *J. Org. Chem.* **1996**, *61*, 7200
36. D. P. Curran, S. Hadida, *J. Am. Chem. Soc.* **1996**, *118*, 2531
37. D. P. Curran, S. Hadida, M. He, *J. Org.Chem.* **1997**, *62*, 6714
38. D. P. Curran, M. Hoshino, *J. Org.Chem.* **1996**, *61*, 6480
39. M. Hoshino, P. Degenkolb, D. P. Curran, *J. Org.Chem.* **1997**, *62*, 8341
40. M. Larhed, M. Hoshino, S. Hadida, D. P. Curran, A. Hallberg, *J. Org.Chem.* **1997**, *62*, 5583
41. S. Hadida, M. S. Super, E. J. Beckman, D. P. Curran, *J. Am. Chem. Soc.* **1997**, *119*, 7406
42. N. Spetseris, S. Hadida, D. P. Curran, T. Y. Meyer, *Organometallics*, **1998**, *17*, 1458

CHAPTER 8

Organic Reactions with Fluorinated Reagents

Fluorine-containing compounds involving KF, CsF, R_4NF, and $(R_2N)_3S^+$ $Me_3SiF_2^-$ (TASF) are widely used as fluorinating reagents (see Chap. 2). However, their use is not limited to acting as a fluorine source for the synthesis of organofluorine compounds. These fluorine-containing compounds play various important roles in organic synthesis [1]. This chapter describes organic reactions mediated by fluorine-containing reagents.

8.1
Fluoride Ion in Organic Synthesis

8.1.1
Fluoride Base

Since the bond energy of H–F is strong (~569 kJ/mol, cf. H–Cl ~432 kJ/mol, H–O ~428 kJ/mol, H–N ~314 kJ/mol), a fluoride ion can easily react with a proton [2]. Thus, base-assisted reactions such as alkylation, arylation, esterification, elimination, and condensation proceed cleanly with potassium fluoride, cesium fluoride, or an onium fluoride under mild and essentially neutral conditions. For example, alkylation of phenol with an alkyl halide proceeds by use of Et_4NF as a base [3].

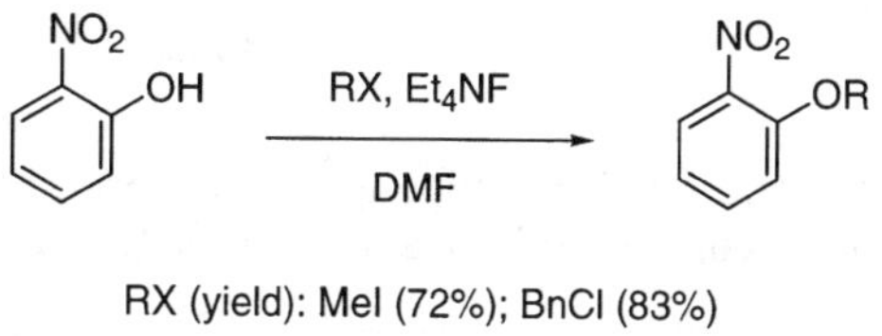

$$\text{2-NO}_2\text{-C}_6\text{H}_4\text{-OH} \xrightarrow{\text{RX, Et}_4\text{NF, DMF}} \text{2-NO}_2\text{-C}_6\text{H}_4\text{-OR} \quad (8.1)$$

RX (yield): MeI (72%); BnCl (83%)

Selective dehydrochlorination from primary chlorides of norbornane derivatives is induced by CsF [4].

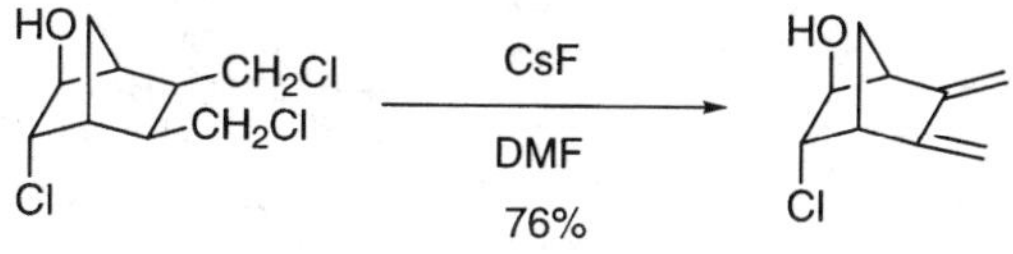

$$\quad (8.2)$$

Enolizable ketones such as pentane-2,4-dione undergo intermolecular condensation and subsequent cyclization in the presence of KF [5].

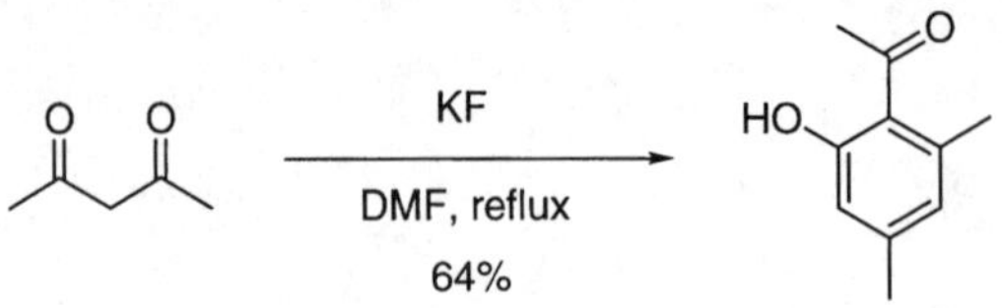

$$\text{MeCOCH}_2\text{COMe} \xrightarrow[\text{64\%}]{\text{KF, DMF, reflux}} \text{3-acetyl-2-hydroxy-4,6-dimethylphenyl derivative} \quad (8.3)$$

Trimerization of isocyanate is catalyzed by CsF or Bu$_4$NF to give an aromatic isocyanurate [6].

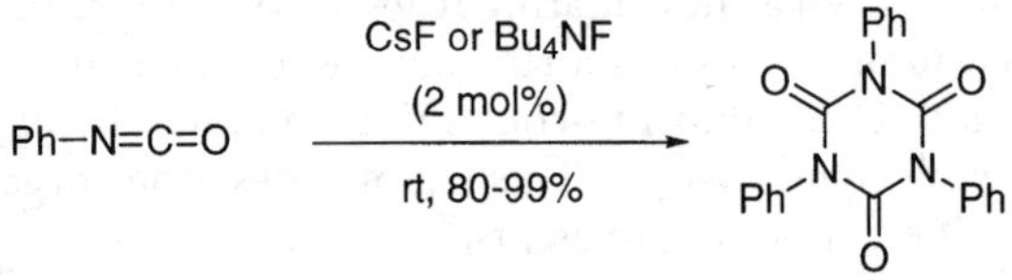

$$\text{Ph–N=C=O} \xrightarrow[\text{rt, 80-99\%}]{\text{CsF or Bu}_4\text{NF (2 mol\%)}} \text{triphenyl isocyanurate} \quad (8.4)$$

8.1.2
Desilylative Elimination and Deprotection

Due to the high energy of a Si–F bond, a fluoride ion displays a strong nucleophilic character towards silicon. Hence, treatment of organosilicon compounds bearing a leaving group at an α- or β-position with a fluoride ion gives vinylidene carbenes, alkenes, alkynes, or allenes via α- or β-elimination, respectively. Alkylidene carbene is conveniently generated from 1-trimethylsilylalkenyl triflate with KF and reacts with cyclohexene to give the corresponding cyclopropanane derivatives [7].

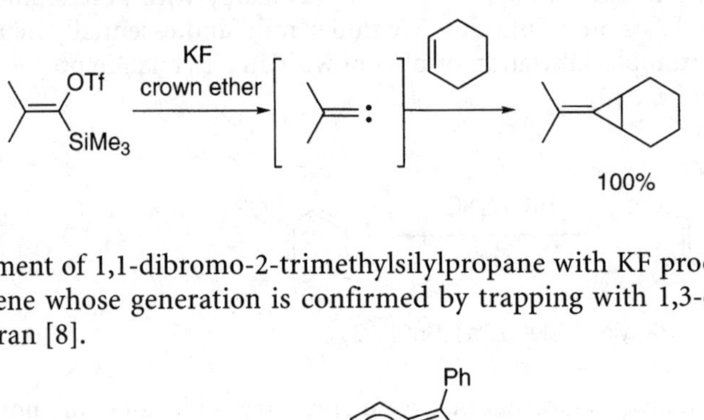

$$\underset{\text{SiMe}_3}{\overset{\text{OTf}}{>\!=\!<}} \xrightarrow{\text{KF, crown ether}} \left[>\!=\!: \quad \square \right] \longrightarrow \text{bicyclic product} \quad (8.5)$$

100%

Treatment of 1,1-dibromo-2-trimethylsilylpropane with KF produces 1-bromopropene whose generation is confirmed by trapping with 1,3-diphenylisobenzofuran [8].

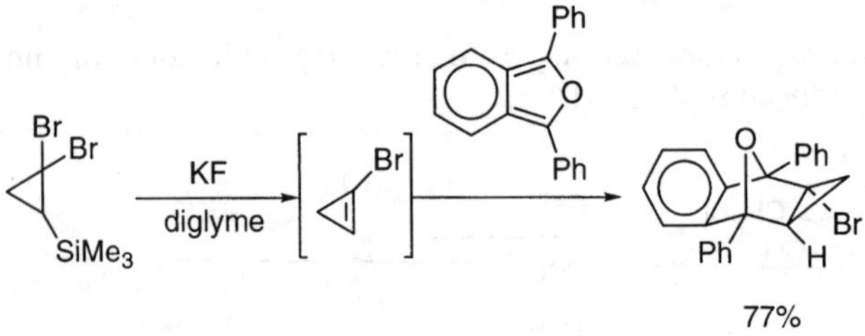

$$\underset{\text{SiMe}_3}{\overset{\text{Br}\;\text{Br}}{\triangleleft}} \xrightarrow[\text{diglyme}]{\text{KF}} \left[\overset{\text{Br}}{\triangleleft} \right] \xrightarrow{\text{1,3-diphenylisobenzofuran}} \text{adduct} \quad (8.6)$$

77%

8.1 Fluoride Ion in Organic Synthesis

Phenylallene is prepared by the reaction of 2-triphenylsilyl-3-chloro-3-phenylpropene with Bu_4NF [9].

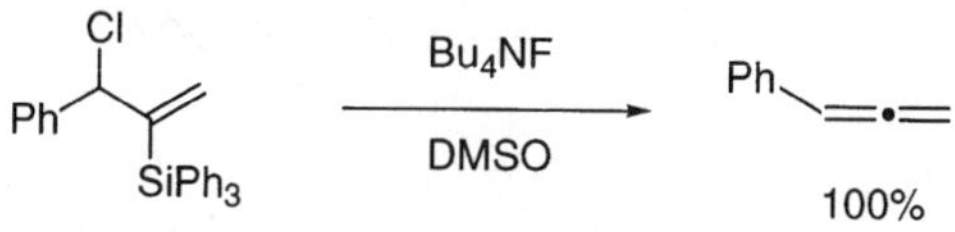

(8.7)

Fluoride-induced 1,3- and 1,4-elimination of silicon-containing compounds are effective for the generation of such unstable species as azomethine ylide and o-quinodimethane. Treatment of benzyl (cyanomethyl) (trimethylsilylmethyl)-amine with silver fluoride produces the corresponding azomethine ylide, which readily undergoes cycloaddition reactions with electron-deficient olefins to give pyrrolidines in good yields [10].

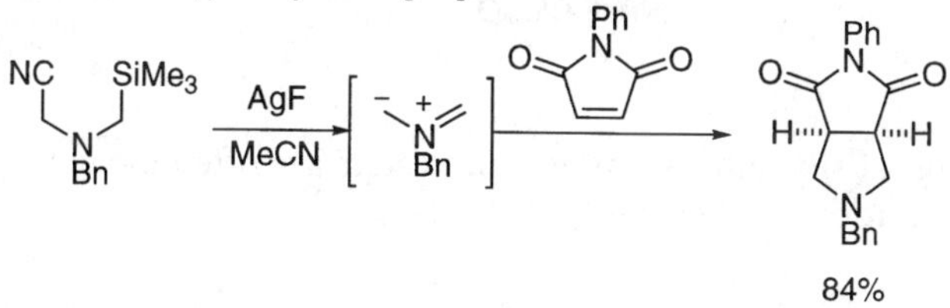

(8.8)

Cesium fluoride reacts with [o-(α-(trimethylsilyl)alkyl)benzyl]trimethylammonium iodide at ambient temperatures to generate an o-quinodimethane intermediate which acts as a diene in the intramolecular Diels-Alder reaction giving rise to the steroidal structure estrone stereoselectively [11].

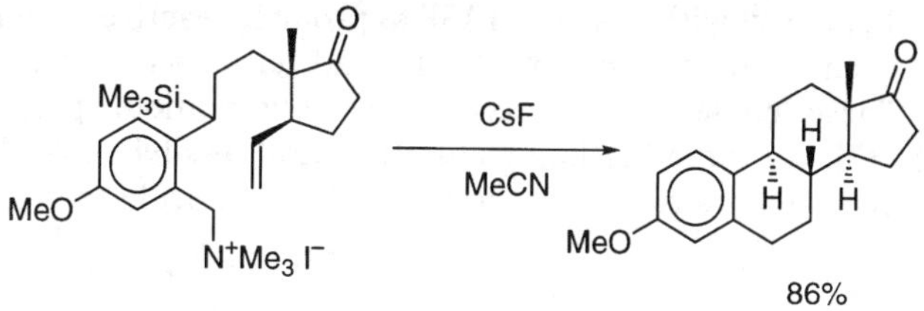

(8.9)

Cleavage of a C–Si bond without a leaving group or an anion-stabilizing group in a suitable position is also effected by use of a fluoride ion. Desilylation of acyl- or alkynylsilanes proceeds with Bu_4NF or KF at ambient temperatures conveniently [12].

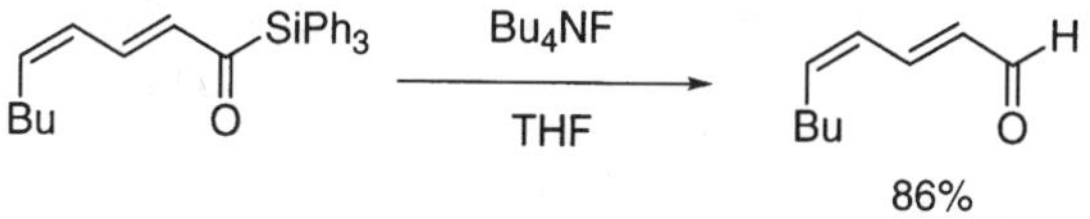

(8.10)

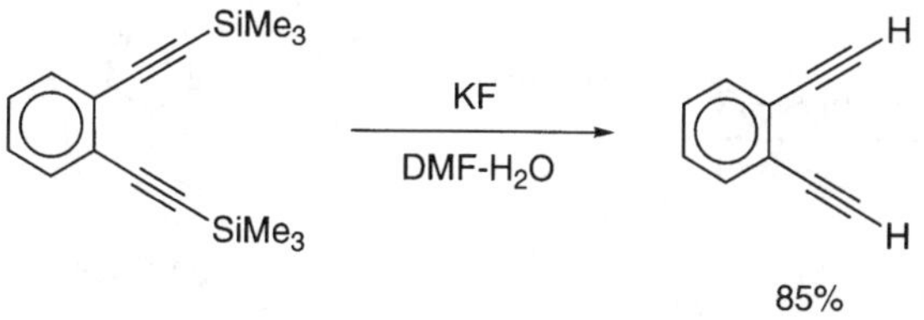

(8.11)

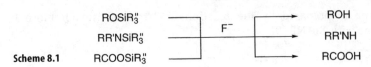

Scheme 8.1

A vinylic C–Si bond is cleaved with Bu_4NF with retention of configuration of the starting alkene [13]. In this reaction, the presence of a phenyl group on the silicon atom plays a critical role.

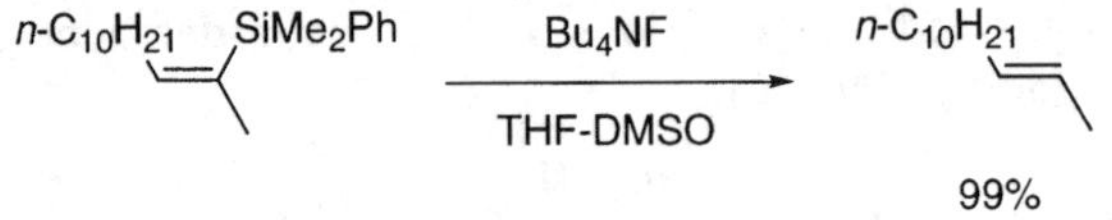

$$n\text{-}C_{10}H_{21}\text{-CH=CH-}SiMe_2Ph \xrightarrow[\text{THF-DMSO}]{Bu_4NF} n\text{-}C_{10}H_{21}\text{-CH=CH}_2 \quad 99\% \tag{8.12}$$

Deprotection of silyl ethers, amines, and esters are often mediated by TBAF under mild conditions (Scheme 8.1) [14].

8.1.3
Naked Anions and Fluorosilicates

A fluoride ion reacts with organosilicon compounds to generate naked anion species or five- or six-coordinated silicates which also serve as versatile sources of anion species [15].

Silyl enol ethers react with TBAF or TASF to produce reactive naked enolates which have been characterized by NMR and electrochemical measurements [16]. These enolates are alkylated cleanly without polyalkylation (Scheme 8.2) [17]. On the other hand, aldol reaction of silyl enol ethers with aldehydes proceeds with a catalytic amount of TBAF or TASF to give silylated aldols [18].

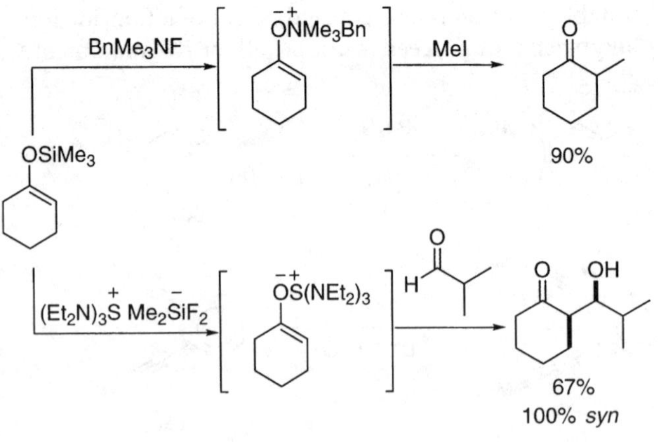

Scheme 8.2

8.1 Fluoride Ion in Organic Synthesis

Treatment of ethyl (trimethylsilyl)acetate with TBAF generates the naked enolate of acetate which does not act as a nucleophile with a ketone but behaves as a base to produce a silyl enol ether of a ketone [19].

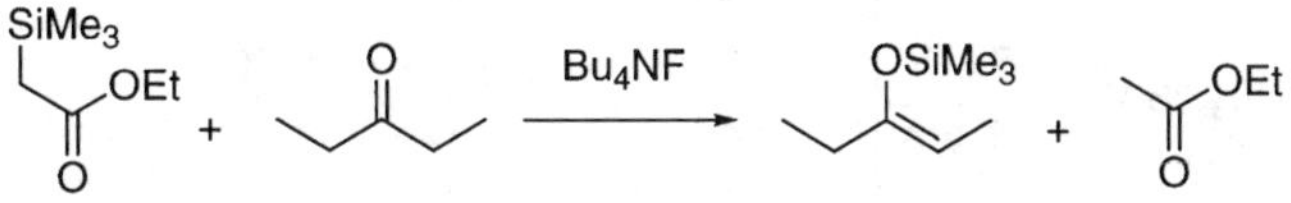

$$\text{Me}_3\text{Si-CH}_2\text{CO}_2\text{Et} + \text{MeCOEt} \xrightarrow{\text{Bu}_4\text{NF}} \text{CH}_2=\text{C(OSiMe}_3\text{)Et} + \text{CH}_3\text{CO}_2\text{Et} \quad (8.13)$$

Ketene silyl acetals undergo conjugate addition to α,β-unsaturated carbonyl compounds in the presence of a catalytic amount of TASF. This reaction has been applied to group transfer polymerization (Scheme 8.3) [20].

Cleavage of a C–Si bond with TASF produces the corresponding naked carbanion when the generated species is stabilized by halogen atoms.

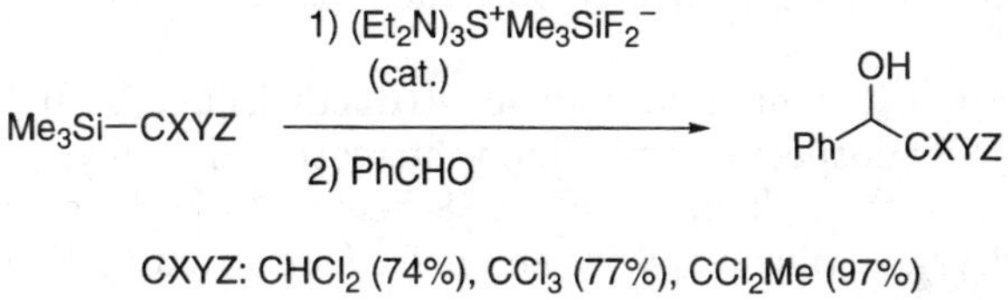

$$\text{Me}_3\text{Si-CXYZ} \xrightarrow[\text{2) PhCHO}]{\text{1) (Et}_2\text{N})_3\text{S}^+\text{Me}_3\text{SiF}_2^- \text{ (cat.)}} \text{Ph-CH(OH)-CXYZ} \quad (8.14)$$

CXYZ: CHCl$_2$ (74%), CCl$_3$ (77%), CCl$_2$Me (97%)

The produced anion species undergo addition reactions to aldehyde and ketone carbonyls smoothly even at room temperature (Scheme 8.4) [21].

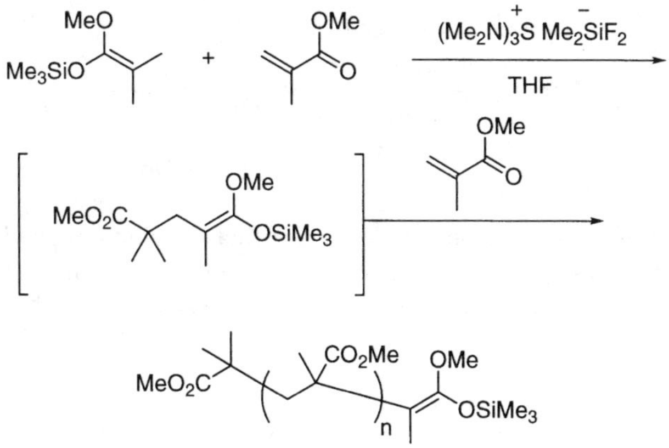

Scheme 8.3

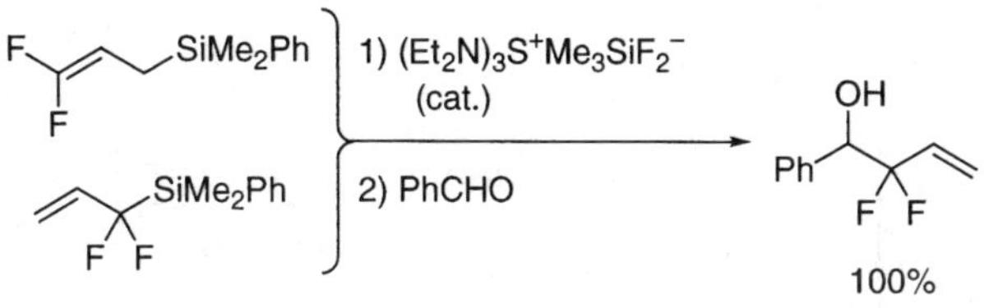

Scheme 8.4

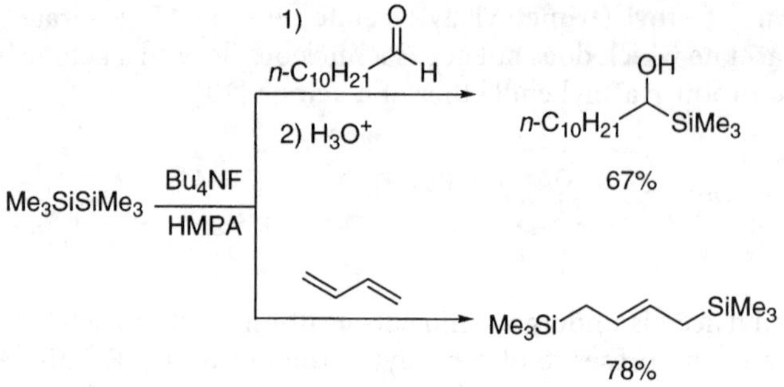

Scheme 8.5

Treatment of disilane with TBAF gives a metal-free silyl anion which undergoes bis-silylation of unsaturated bonds such as C=O and C=C–C=C (Scheme 8.5) [22].

Allylation of carbonyl compounds with allyltrimethylsilane is catalyzed by a fluoride ion [23]. Regiospecific and highly stereoselective allylation of aldehydes is possible with allyltrifluorosilane (Scheme 8.6) [24]. To explain the selectivities, hypervalent silicon species and a six-membered cyclic transition state have been proposed.

Dimethylphenylsilane reduces aldehydes and ketones in the presence of a catalytic amount of TBAF or TASF to give the *threo*-isomers selectively [25]. The kinetic and stereochemical outcome suggests a hexacoordinated fluorosilicate intermediate.

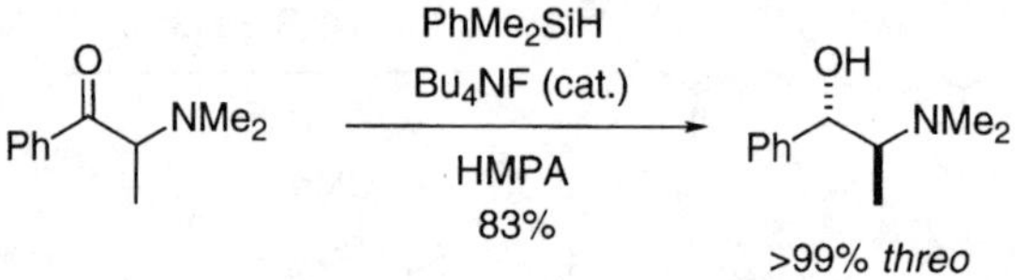

(8.15)

Activation of organosilicon compounds with a fluoride ion produces a transient pentacoordinated silicate which is capable of transmetalation with an organopalladium complex and achieves cross-coupling reaction of organosilicon

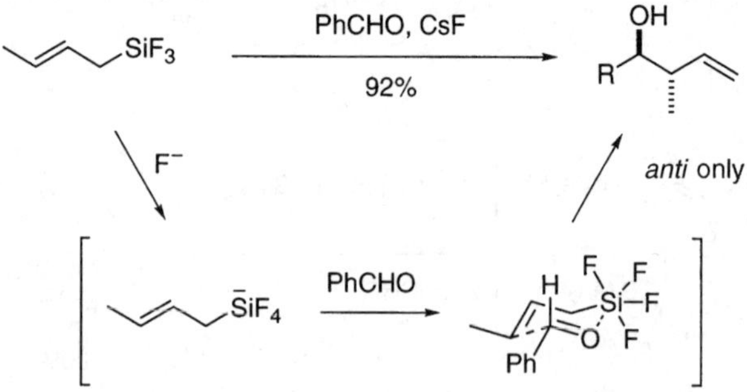

Scheme 8.6

compounds [26]. Alkynyl-, alkenyl-, aryl-, allyl-, and alkylsilanes and disilanes are applicable to this coupling reaction which is tolerant towards a variety of functional groups.

$$\text{R–Si} + \text{R'–X} \xrightarrow[\text{Pd cat.}]{F^-} \text{R–R'} \qquad (8.16)$$

R = alkynyl, alkenyl, aryl, alkyl, allyl, silyl
R' = alkenyl, aryl, allyl, alkyl
X = Br, I, OSO_2CF_3

A C–Si bond with a silyl group substituted by at least one heteroatom such as H, F, Cl, OR, or NR_2 is oxidatively cleaved with hydrogen peroxide and KF and is converted into a C–O bond stereospecifically [27].

$$-\overset{|}{\underset{|}{C}}-SiR_2X \xrightarrow[\text{DMF or MeOH}]{H_2O_2,\ KF\ or\ KHF_2} -\overset{|}{\underset{|}{C}}-OH \qquad (8.17)$$

X = H, F, Cl, OR', NR'$_2$

8.2
Trifluoroacetic Acid and Trifluoroperacetic Acid

8.2.1
Trifluoroacetic Acid

Trifluoroacetic acid (TFA) is a general catalyst for most acid-catalyzed rearrangements and has advantages over conventional acids such as sulfuric or *p*-toluenesulfonic acid in product isolation because TFA is volatile (bp 72°C) and easily removed by evaporation without neutralization or extractive workup. For example, TFA is commonly used in acid-induced diazoketone cyclization which is useful for the construction of spiro frameworks [28], while Cope rearrangement of 2-acyl-1,5-dienes is strongly accelerated by TFA [29].

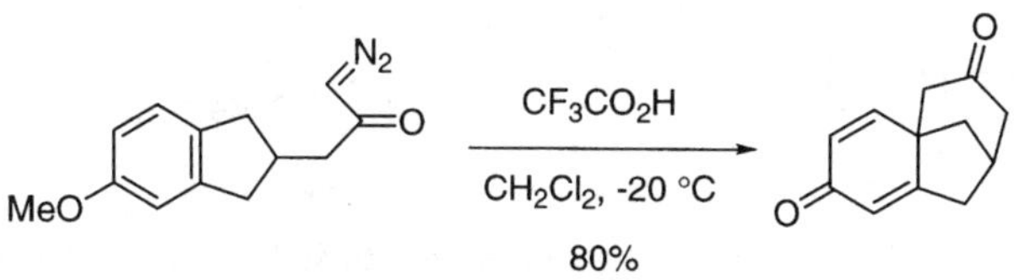

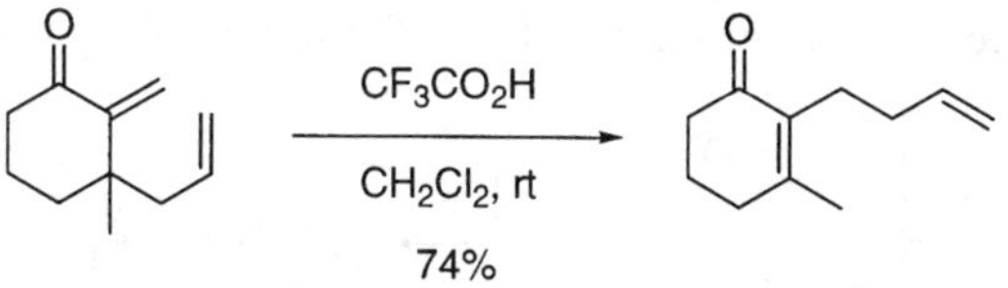

Cleavage of acid-labile nitrogen- and oxygen-protecting groups, such as *N*-Boc, *N*-benzyloxymethyl, benzyl ether, *p*-methoxybenzyl ether, *tert*-butyl ether, tert-butyloxymethyl ether, triphenylmethyl ether, and dimethyl acetal, is performed with TFA under mild conditions [14].

Oxidation of ketones with m-chloroperbenzoic acid in the presence of TFA gives the corresponding Bayer-Villiger products in much shorter reaction times and in higher yields than the reaction with m-chloroperbenzoic acid alone [30].

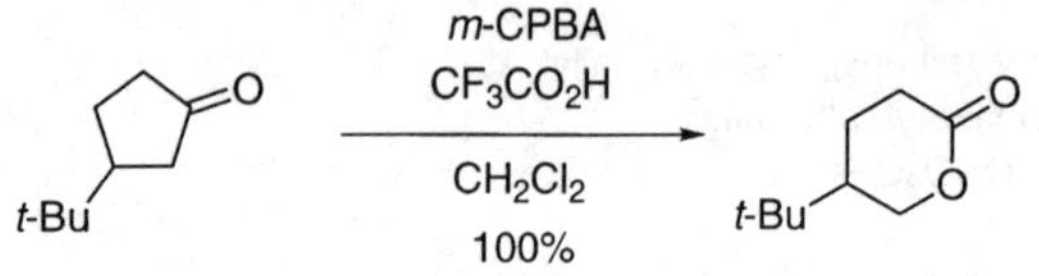

$$\text{(8.20)}$$

A combination of TFA and sodium percarbonate is also effective for Baeyer–Villiger oxidation [31].

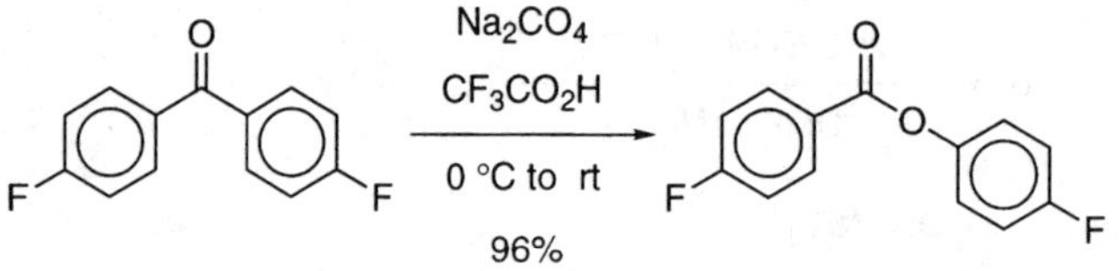

$$\text{(8.21)}$$

8.2.2
Trifluoroperacetic Acid

Trifluoroperacetic acid is prepared from 90% hydrogen peroxide and trifluoroacetic anhydride or, alternatively, 30% hydrogen peroxide and TFA, and is the most reactive organic peroxy acid due to the presence of an electron-withdrawing trifluoromethyl group. Usually, electron-poor alkenes such as monosubstituted alkenes or α,β-unsaturated esters and ketones react sluggishly with standard organic peroxy acids. In sharp contrast, trifluoroperacetic acid oxidizes those compounds smoothly to give the corresponding epoxides [32].

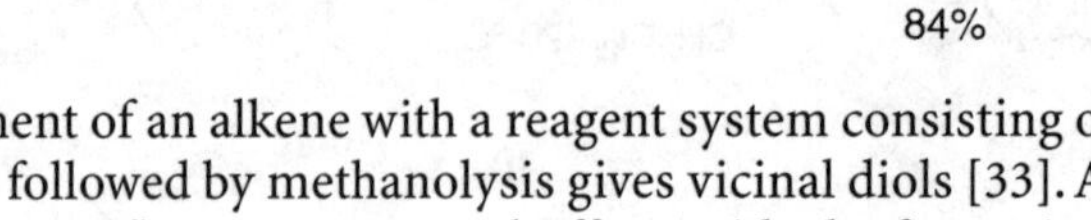

$$\text{(8.22)}$$

Treatment of an alkene with a reagent system consisting of trifluoroperacetic acid/TFA followed by methanolysis gives vicinal diols [33]. Addition of triethylammonium trifluoroacetate as a buffer avoids the formation of by-products.

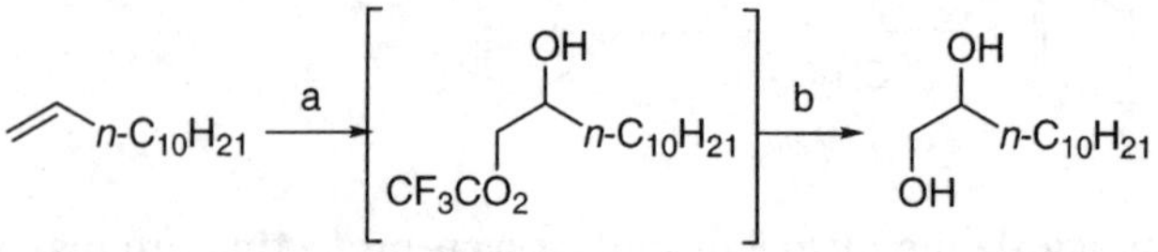

$$\text{(8.23)}$$

a) CF_3CO_3H, $CF_3CO_2^\ominus$ $^\oplus NHEt_3$, CH_2Cl_2 b) HCl, MeOH (95%)

Baeyer–Villiger oxidation with trifluoroperacetic acid proceeds in a manner similar to other peroxy acids [34]. Na$_2$HPO$_4$ buffer is used to prevent reaction between trifluoroacetic acid and the product.

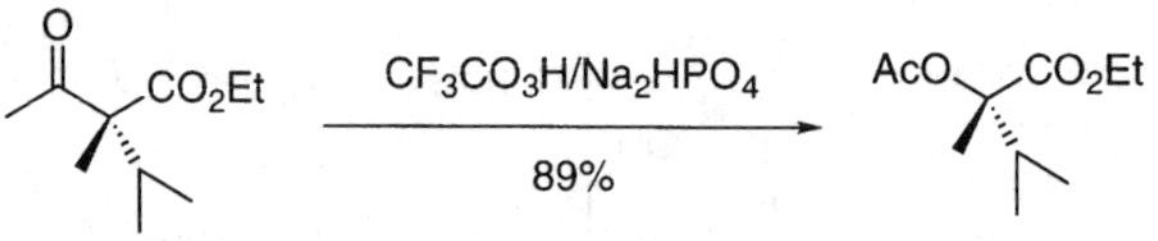

$$\text{ketone substrate} \xrightarrow[89\%]{\text{CF}_3\text{CO}_3\text{H/Na}_2\text{HPO}_4} \text{acetate product} \quad (8.24)$$

Primary amines, nitroso compounds, and sulfides are oxidized efficiently with trifluoroperacetic acid to the corresponding nitro compounds and sulfoxides or sulfones, respectively [35].

8.3 Trifluoromethanesulfonic Acid and Derivatives

8.3.1 Trifluoromethanesulfonic Acid

Trifluoromethanesulfonic acid (TfOH), known to be one of the strongest organic acids, is extremely stable and resistant to oxidation and reduction. TfOH mediates many types of acid-induced reactions such as oligomerization/polymerization of alkenes, Friedel–Crafts reaction, and cationic rearrangements better than a conventional Brønsted acid [36]. For example, intramolecular ionic Diels–Alder reaction of tetraenes is catalyzed by TfOH at ambient temperatures [37].

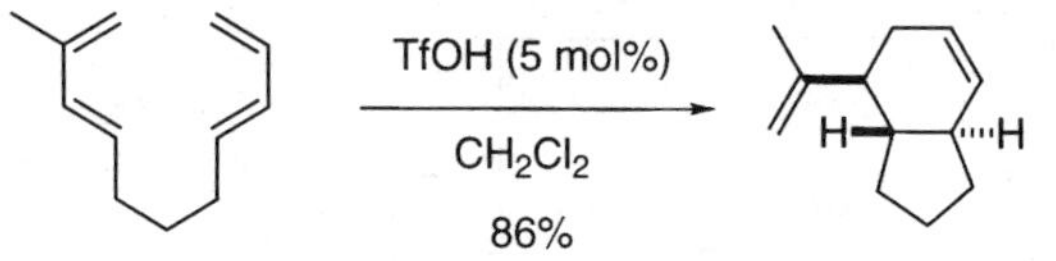

$$\text{tetraene} \xrightarrow[\text{CH}_2\text{Cl}_2,\ 86\%]{\text{TfOH (5 mol\%)}} \text{bicyclic product} \quad (8.25)$$

Protonation of an alkyne with TfOH in the presence of a nitrile gives a pyrimidine derivative conveniently [38].

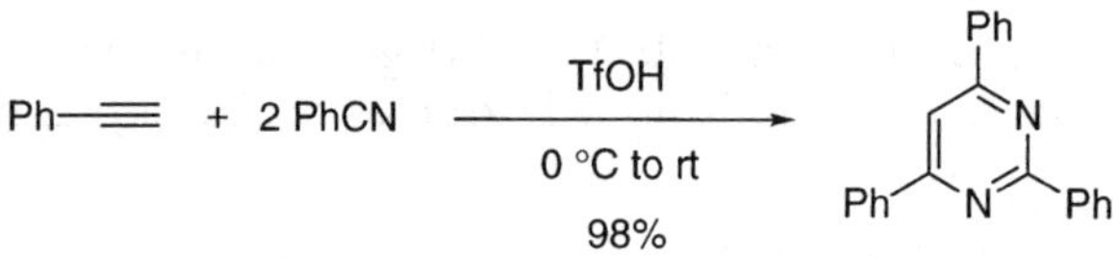

$$\text{Ph}{-}{\equiv}\ +\ 2\ \text{PhCN} \xrightarrow[0\ ^\circ\text{C to rt},\ 98\%]{\text{TfOH}} \text{2,4,6-triphenylpyrimidine} \quad (8.26)$$

Intramolecular Friedel–Crafts acylation of an arylalkanoic acid to a cyclic ketone is effected by TfOH in one pot via in situ preparation of an acid chloride [39].

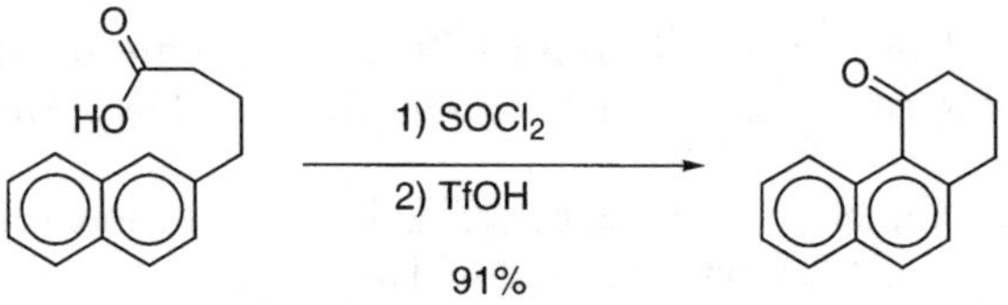

$$\text{arylalkanoic acid} \xrightarrow[91\%]{1)\ \text{SOCl}_2;\ 2)\ \text{TfOH}} \text{cyclic ketone} \quad (8.27)$$

A nitroalkene is protonated by TfOH to produce a *N,N*-dihydroxyiminium carbenium ion which reacts with arene to give finally an α-arylated ketone [40].

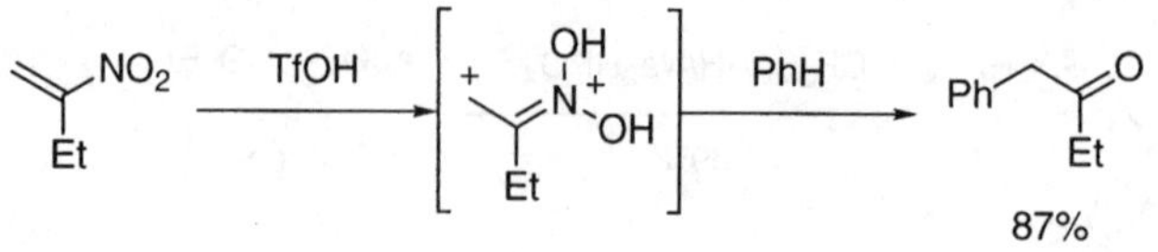

$$\underset{Et}{\diagdown}NO_2 \xrightarrow{TfOH} \left[\underset{Et}{\diagdown}\overset{+}{N}\diagup\underset{OH}{\overset{OH}{|}} \right] \xrightarrow{PhH} \underset{Et}{Ph\diagdown\diagup}\overset{O}{\diagdown} \qquad (8.28)$$

87%

8.3.2
Trimethylsilyl Trifluoromethanesulfonate

Trifluoromethanesulfonic acid (TfOH) reacts with chlorotrimethylsilane or tetramethylsilane to give trimethylsilyl trifluoromethanesulfonate (TMSOTf) which is versatile both as a silylating reagent and as a strong Lewis acid [41].

$$TfOH + Me_3SiCl \text{ or } Me_4Si \longrightarrow Me_3SiOTf + HCl \text{ or } CH_4 \qquad (8.29)$$

Carbonyl compounds are converted into the corresponding silyl enol ethers with a reagent system consisting of TMSOTf/NEt$_3$ (Scheme 8.7). The silylating reactivity of TMSOTf is 6.7×10^8 higher than that of TMSCl. On the other hand, the reaction of an α-silyl ketone and a catalytic amount of TMSOTf also affords a silyl enol ether [42].

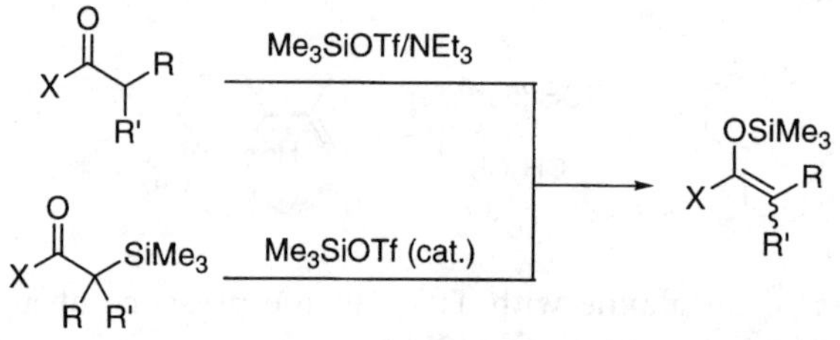

Scheme 8.7

Treatment of an epoxide with TMSOTf/DBU gives an isomeric allyl silyl ether [43]. Tetra-, tri-, and 2,2-disubstituted oxiranes can be used in this transformation.

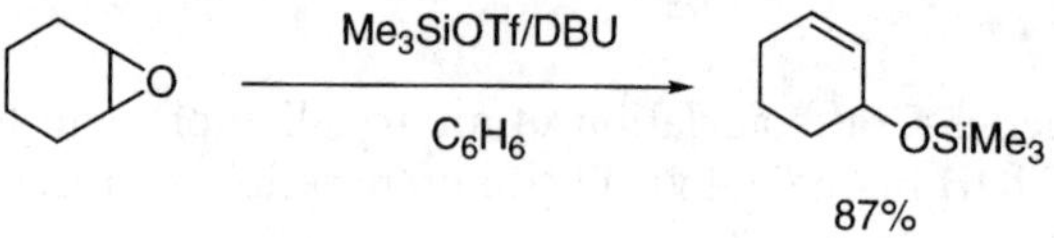

$$\text{(epoxide)} \xrightarrow[C_6H_6]{Me_3SiOTf/DBU} \text{(allyl silyl ether, OSiMe}_3\text{)} \qquad (8.30)$$

87%

Acetalization of aldehydes and ketones with a stoichiometric amount of alkoxysilane is efficiently catalyzed by TMSOTf at low temperatures (Scheme 8.8) [44].

Silyl enol ethers undergo stereoselective aldol-type condensation with acetals in the presence of a catalytic amount of TMSOTf [45].

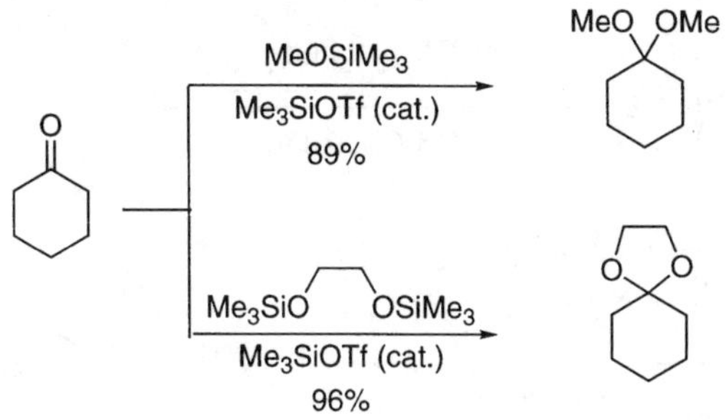

8.3 Trifluoromethanesulfonic Acid and Derivatives

Scheme 8.8

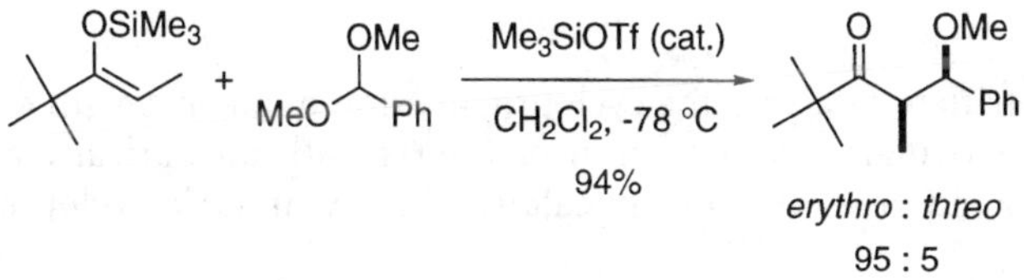

$$\text{(8.31)}$$

Similarly, allylation of acetals with allyltrimethylsilane is effected by TMSOTf catalytically [46].

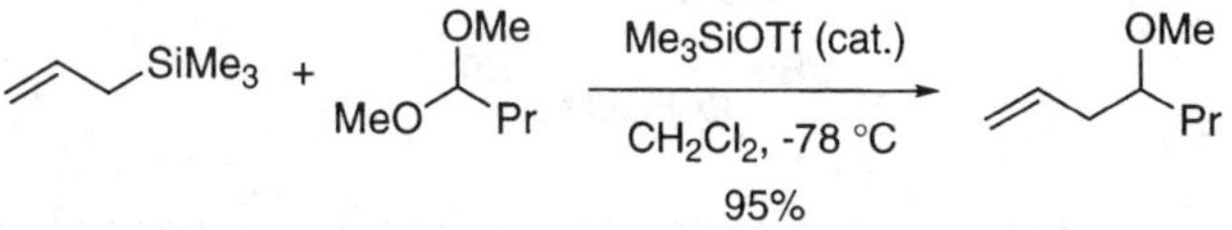

$$\text{(8.32)}$$

8.3.3
Metal Trifluoromethanesulfonates

A variety of metal salts of TfOH are used in various types of reactions because a triflate anion has low nucleophilicity and thus low coordinating ability. Some selected examples are shown below.

Dialkylboryl triflate reacts smoothly with enolizable ketones in the presence of tertiary amines to generate stereoselectively boron enolates which undergo aldol condensation with high stereoselectivity [47]. In addition, various types of chiral boron enolates which play important roles in asymmetric aldol reactions are prepared with boron triflates.

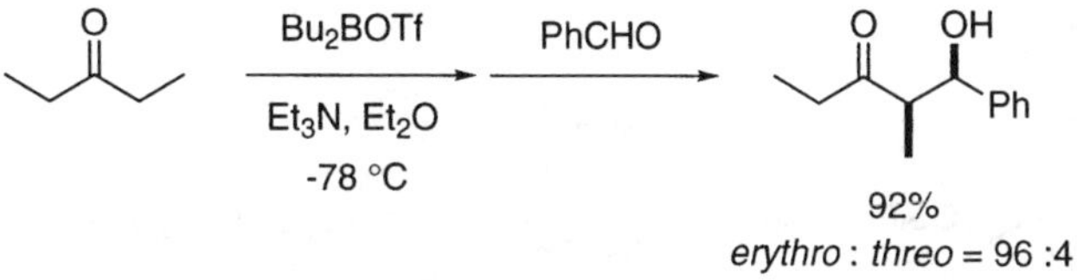

$$\text{(8.33)}$$

Another versatile metal triflate for enolate formation is tin triflate [48]. A reagent system consisting of tin triflate and a chiral diamine derived from (S)-proline performs enantioselective aldol-type reactions (Scheme 8.9) [49]. A catalytic version of such a reaction has also been developed [50].

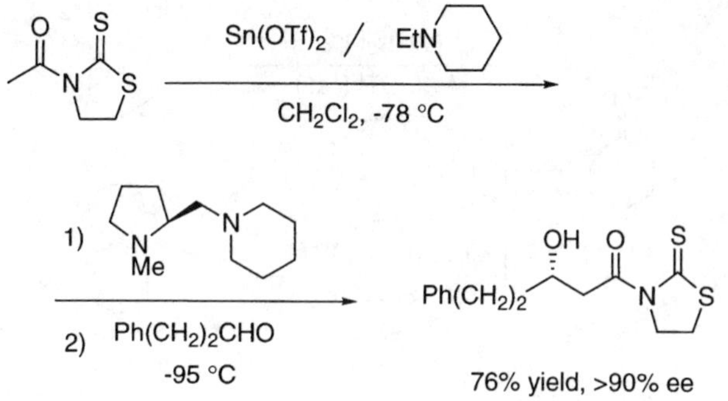

Scheme 8.9

Copper(I) triflate catalyzes the cyclopropanation of alkenes with diazo compounds. In particular, a combination of CuOTf with an optically active bis-(oxazoline) ligand gives an efficient catalyst for asymmetric cyclopropanation [51].

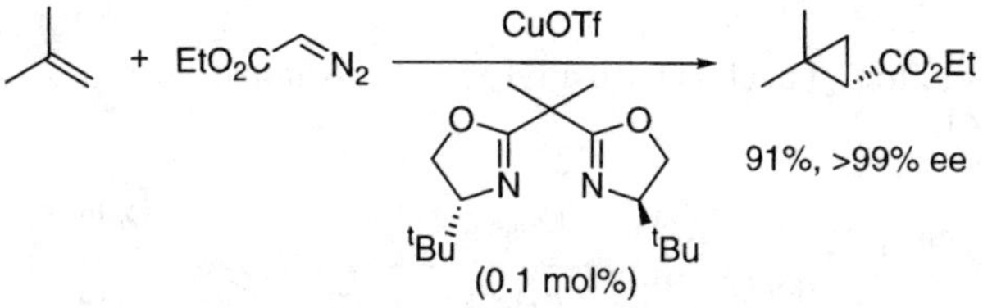

$$\text{(8.34)}$$

Synthesis of 1,4-diketones via oxidative coupling of ketone enolates or silyl enol ethers is effected using copper(II) triflate. For example, treatment of acetophenone with LDA followed by the addition of Cu(OTf)$_2$ gives 1,4-diphenylbutane-1,4-dione in good yield [52].

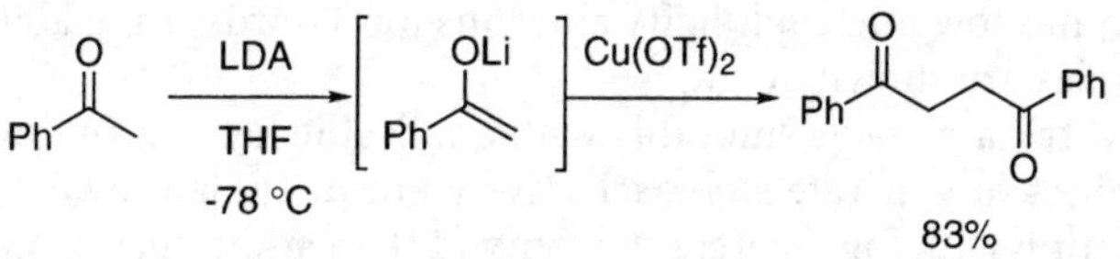

$$\text{(8.35)}$$

Thioacetalization of ketones with 1,2-ethanedithiol proceeds efficiently by use of zinc triflate as a Lewis acid catalyst under mild conditions [53].

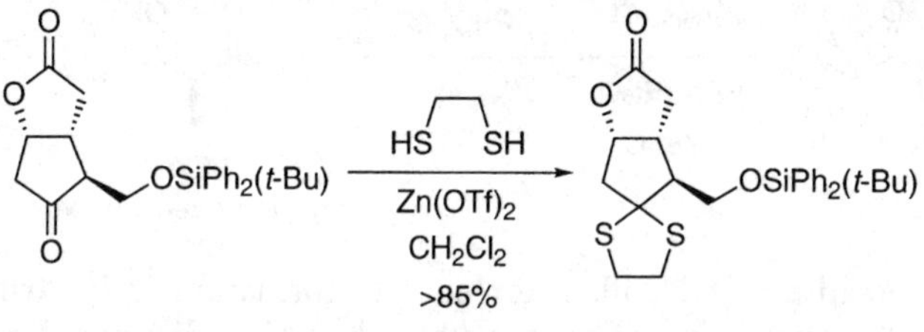

$$\text{(8.36)}$$

The Friedel–Crafts acylation reaction is promoted by silver triflate. For example, treatment of benzoyl chloride with AgOTf generates the highly electrophilic mixed anhydride PhC(O)OTf that reacts with benzene rapidly to give benzophenone in good yield [54].

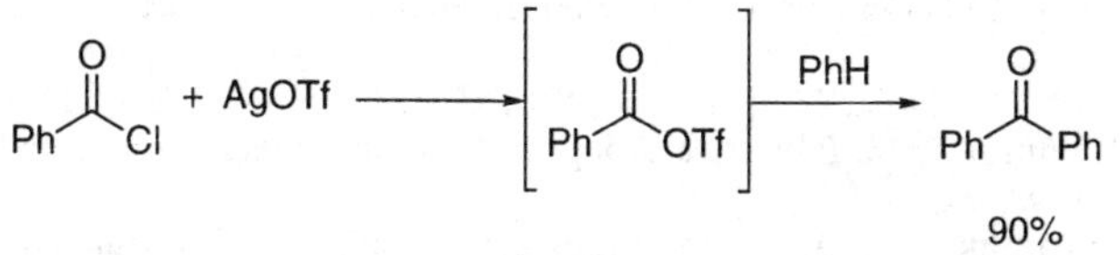

$$\text{Ph-C(O)-Cl} + \text{AgOTf} \longrightarrow [\text{Ph-C(O)-OTf}] \xrightarrow{\text{PhH}} \text{Ph-C(O)-Ph} \quad (8.37)$$
$$90\%$$

Ytterbium triflate catalyzes an aldol-type reaction of enol silyl ethers with aldehydes efficiently in aqueous media [55]. The catalyst, ytterbium triflate, can be recovered and reused.

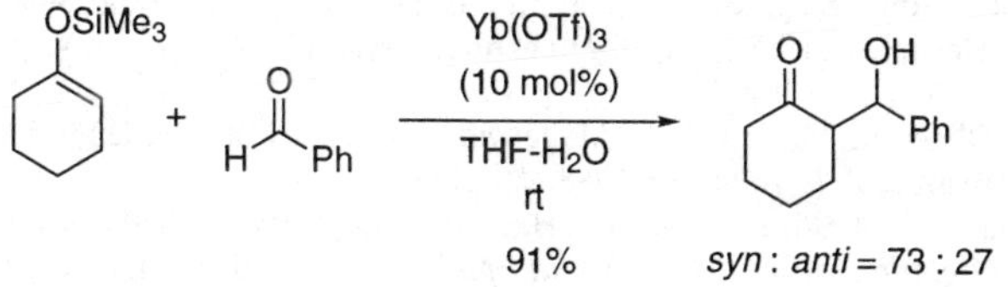

$$(8.38)$$
$$91\% \quad syn : anti = 73 : 27$$

Aldol reactions, Michael additions, allylation reactions, Diels–Alder reactions, Friedel–Crafts reactions, acylations, esterifications, and lactonizations are catalyzed by scandium triflate under mild conditions [55, 56].

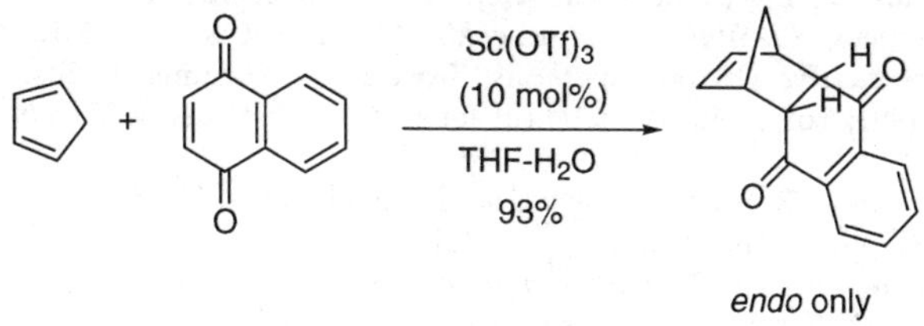

$$(8.39)$$
$$93\% \quad endo \text{ only}$$

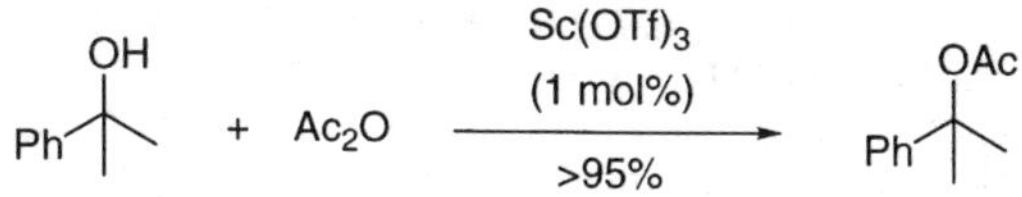

$$(8.40)$$
$$>95\%$$

References

1. Stang, P. J.; Zhdankin, V. V. In: *Chemistry of Organic Fluorine Compounds II. A Critical Review*; Hudlicky, M.; Pavlath, A. E. Eds.; American Chemical Society: Washington, DC, 1995; pp 941–975
2. Clark, J. H. *Chem. Rev.* 1980, *80*, 429–452
3. Miller, J. M.; So, K. H.; Clark, J. H. *Can. J. Chem.* 1979, *57*, 1887
4. Chollet, A.; Hagenbuch, J.-P.; Vogel, P. *Helv. Chim. Acta* 1979, *62*, 511
5. Clark, J. H.; Miller, J. M. *J. Chem. Soc., Perkin Trans. 1* 1977, 2063
6. Nambu, Y.; Endo, T. *J. Org. Chem.* 1993, *58*, 1932
7. Stang, P. J.; Fox, D. P. *J. Org. Chem.* 1977, *42*, 1667
8. Chan, T. H.; Massuda, D. *Tetrahedron Lett.* 1975, 3383
9. Chan, T. H.; Mychajlowskij, W.; Ong, B. S.; Harpp, D. N. *J. Org. Chem.* 1978, *43*, 1526

10. Padwa, A.; Chen, Y.-Y. *Tetrahedron Lett.* 1983, *24*, 3447
11. Ito, Y.; Nakatsuka, M.; Saegusa, T. *J. Am. Chem. Soc.* 1982, *104*, 7609
12. (a) Degl'Innocenti, A.; Stucchi, E.; Capperucci, A.; Mordini, A.; Reginato, G.; Ricci, A. *Synlett* 1992, 329; (b) Semmelhack, M. F.; Neu, T.; Foubelo, F. *Tetrahedron Lett.* 1992, *33*, 3277
13. Oda, H.; Sato, M.; Morizawa, Y.; Oshima, K.; Nozaki, H. *Tetrahedron* 1985, *41*, 3257
14. Greene, T. W.; Wuts, P. G. M. *Protective Groups in Organic Synthesis*; 2nd Edn; John Wiley & Sons, Inc.: New York, 1991
15. Furin, G. G.; Vyazankina, O. A.; Gostevsky, B. A.; Vyazankin, N. S. *Tetrahedron* 1988, *44*, 2675
16. Noyori, R.; Nishida, I.; Sakata, J. *J. Am. Chem. Soc.* 1983, *105*, 1598
17. Kuwajima, I.; Nakamura, E.; Shimizu, M. *J. Am. Chem. Soc.* 1982, *104*, 1025
18. (a) Nakamura, E.; Shimizu, M.; Kuwajima, I.; Sakata, J.; Yokoyama, K.; Noyori, R. *J. Org. Chem.* 1983, *48*, 932; (b) Kuwajima, I.; Nakamura, E. *Acc. Chem. Res.* 1985, *18*, 181
19. Nakamura, E.; Hashimoto, K.; Kuwajima, I. *Bull. Chem. Soc. Jpn.* 1981, *54*, 805
20. Webster, O. W.; Hertler, W. R.; Sogah, D. Y.; Farnham, W. B.; RajanBabu, T. V. *J. Am. Chem. Soc.* 1983, *105*, 5706
21. (a) Hiyama, T.; Obayashi, M.; Sawahata, M. *Tetrahedron Lett.* 1983, *24*, 4113; (b) Fujita, M.; Obayashi, M.; Hiyama, T. *Tetrahedron* 1988, *44*, 4135
22. Hiyama, T.; Obayashi, M.; Mori, I.; Nozaki, H. *J. Org. Chem.* 1983, *48*, 912
23. Majetich, G. In *Organic Synthesis: Theory and Applications*; Hudlicky, T. Ed.; JAI Press Inc.: Greenwich, 1989; Vol. 1; pp 173–240
24. Kira, M.; Hino, T.; Sakurai, H. *Tetrahedron Lett.* 1989, *30*, 1099
25. Fujita, M.; Hiyama, T. *J. Org. Chem.* 1988, *53*, 5405
26. Reviews: (a) Hatanaka, Y.; Hiyama, T. *Synlett* 1991, 845; (b) Hiyama, T.; Hatanaka, Y. *Pure Appl. Chem.* 1994, *66*, 1471; (c) Hiyama, T. In *Metal-catalyzed Cross-coupling Reactions*; Diederich F.; Stang, P. J. Eds.; Wiley-VCH: Weinheim, 1998; pp 421–453
27. Reviews: (a) Tamao, K. *Yuuki Gosei Kagaku Kyokaishi* 1988, *46*, 861; (b) Colvin, E. W. In *Comprehensive Organic Synthesis*; Trost B. M.; Fleming, I. Eds.; Pergamon Press: Oxford, 1991; Vol. 7; pp 641–651; (c) Jones, G. R.; Landais, Y. *Tetrahedron* 1996, *52*, 7599
28. Nicolaou, K. C.; Zipkin, R. E. *Angew. Chem. Int. Ed. Engl.* 1981, *20*, 785
29. Dauben, W. G.; Chollet, A. *Tetrahedron Lett.* 1981, *22*, 1583
30. Koch, S. S. C.; Chamberlin, A. R. *Synth. Commun.* 1989, *19*, 829
31. Olah, G. A.; Wang, Q.; Trivedi, N. J.; Prakash, G. K. S. *Synthesis* 1991, 739
32. Emmons, W. D.; Pagano, A. S. *J. Am. Chem. Soc.* 1955, *77*, 89
33. Emmons, W. D.; Pagano, A. S.; Freeman, J. P. *J. Am. Chem. Soc.* 1954, *76*, 3472
34. Wetter, H. *Helv. Chim. Acta* 1981, *64*, 761
35. (a) Pagano, A. S.; Emmons, W. D. *Org. Synth.* 1969, *49*, 47; (b) Venier, C. G.; Squires, T. G.; Chen, Y.-Y.; Hussmann, G. P.; Shei, J. C.; Smith, B. F. *J. Org. Chem.* 1982, *47*, 3773
36. Reviews of TfOH: (a) Howells, R. D.; McCown, J. D. *Chem. Rev.* 1977, *77*, 69; (b) Stang, P. J.; Hanack, M.; Subramanian, L. R. *Synthesis* 1982, 85
37. Gorman, D. B.; Gassman, P. G. *J. Org. Chem.* 1995, *60*, 977
38. Martinez, A. G.; Fernandez, A. H.; Alvarez, R. M.; Losada, M. C. S.; Vilchez, D. M.; Subramanian, L. R.; Hanack, M. *Synthesis* 1990, 881
39. Hulin, B.; Koreeda, M. *J. Org. Chem.* 1984, *49*, 207
40. Okabe, K.; Ohwada, T.; Ohta, T.; Shudo, K. *J. Org. Chem.* 1989, *54*, 733
41. Reviews of TMSOTf: (a) Noyori, R.; Murata, S.; Suzuki, M. *Tetrahedron* 1981, *37*, 3899; (b) Emde, H.; Dornsch, D.; Feger, H.; Frick, U.; Gotz, A.; Hergott, H. H.; Hofmann, K.; Kober, W.; Krogelsch, K.; Oesterle, T.; Stepp, W. *Synthesis* 1982, 1
42. Emde, H.; Gotz, A.; Hofmann, K.; Simchen, G. *Liebigs Ann. Chem.* 1981, 1643
43. Murata, S.; Suzuki, M.; Noyori, R. *J. Am. Chem. Soc.* 1979, *101*, 2738
44. Tsunoda, T.; Suzuki, M.; Noyori, R. *Tetrahedron Lett.* 1980, *21*, 1357
45. Murata, S.; Suzuki, M.; Noyori, R. *J. Am. Chem. Soc.* 1980, *102*, 3248
46. Tsunoda, T.; Suzuki, M.; Noyori, R. *Tetrahedron Lett.* 1980, *21*, 71

47. Reviews of boron enolates: (a) Kim, B. M.; Williams, S. F.; Masamune, S. In: *Comprehensive Organic Synthesis*; Trost B. M.; Fleming I. Eds.; Pergamon Press: Oxford, 1991; Vol. 2; pp 239-275; (b) Cowden, C. J.; Paterson, I. In: *Organic Reactions*; John Wiley & Sons, Inc.: New York, 1997; Vol. 51; pp 1-200; (c) Evans, D. A.; Vogel, E.; Nelson, J. V. *J. Am. Chem. Soc.* 1979, *101*, 6120; (d) Inoue, T.; Mukaiyama, T. *Bull. Chem. Soc. Jpn.* 1980, *53*, 174
48. Review of tin enolates: Mukaiyama, T.; Kobayashi, S. In: *Organic Reactions*; John Wiley & Sons, Inc.: New York, 1994; Vol. 46; pp 1-103
49. Mukaiyama, T.; Iwasawa, N.; Stevens, R. W.; Haga, T. *Tetrahedron* 1984, *40*, 1381
50. (a) Iwasawa, N.; Yura, T.; Mukaiyama, T. *Tetrahedron* 1989, *45*, 1197; (b) Kobayashi, S.; Uchiro, H.; Fujishita, Y.; Shiina, I.; Mukaiyama, T. *J. Am. Chem. Soc.* 1991, *113*, 4247
51. Evans, D. A.; Woerpel, K. A.; Hinman, M. M.; Faul, M. M. *J. Am. Chem. Soc.* 1991, *113*, 726
52. Kobayashi, Y.; Taguchi, T.; Tokuno, E. *Tetrahedron Lett.* 1977, 3741
53. Corey, E. J.; Shimoji, K. *Tetrahedron Lett.* 1983, *24*, 169
54. Effenberger, F.; Epple, G. *Angew. Chem. Int. Ed. Engl.* 1972, *11*, 299
55. Reviews of rare earth triflates: (a) Kobayashi, S. *Synlett* 1994, 689; (b) Marshman, R. W. *Aldrichimica Acta* 1995, *28*, 77; (c) Kobayashi, S. *Eur. J. Chem.* **1999**, 15.
56. (a) Ishihara, K.; Kubota, M.; Kurihara, H.; Yamamoto, H. *J. Org. Chem.* 1996, *61*, 4560; (b) Tsuchimoto, T.; Tobita, K.; Hiyama, T.; Fukuzawa, S.-i. *J. Org. Chem.* 1997, *62*, 6997

Subject Index

ab initio calculation 131–132
ACE see angiotension converting enzyme
acetalization 258
acetyl CoA 22
acetyl hypofluorite 31–33
acifluorfen 173
aconitic acid 22
acquired immune deficiency syndrome 153
acylation 261
addition-elimination 98
AFEAS see Alternative Fluorocarbons Environmental Acceptability Study
AIDS see acquired immune deficiency syndrome
aldol reaction 93, 100, 252, 258, 261
– asymmetic 259
N-alkyl-N-fluoro-p-toluenesufonamides 37
N-alkyl-N-fluorobenzenesulfonamdies 37
alkylmetals 77, 91
allylation 261
allytrifluorosilane 254
alternatives
– of CFCs 143, 192
– to the third generation 196
Alternative Fluorocarbons Environmental Acceptability Study 195
Alzheimer's disease 149
amidocarbonylation 147
α-amino acid 146–148
Ampére 1
amphiphilic agent 227
anesthetics 164
angiotensin converting enzyme 143
angiotensin I 148
angiotensinogen 148
Anh-Eisenstein model 132
anti-corrosive 224
antiferroelectric liquid crystals 197, 209–211
antifungal 163

antimony pentachloride 191, 192
antimony pentafluoride 192
anti-stick 224
arachidonic acid 154
aromatic nucleophilic substitution 121
artificial blood substitutes 165, 166
3'-azidothymidine 153
azlactone synthesis 146
azomethine ylide 251
AZT see 3'-azidothymidine

Bayer-Villiger product 256
benzotrifluoride 39, 236
bifenthrin 81, 169
Biginelli reaction 245
biphase system 237
bis(fluoroxy)difluoromethane 29, 31
bis(perfluorooctyl)diazene 236
bis(trifluoromethyl)mercury 78
blood-brain barrier 160
boiling points 184
bond energy 140
BrF- see halogen fluoride
bromodifluoromethylation 109
bromofluorocarbene 111
BS see biphase system
BTF see benzotrifluoride
building block 77, 113, 143

calcium fluoride 50
carbohydrates 150, 151
central nervous system 143, 160, 161
cesium fluoride 48, 254
cesium fluorosulfate 33, 34
C-F bond, activation 126–128
CFC 188
CFC-11 191, 192
CFC-1112a 193
CFC-1113 193, 194
CFC-113 191, 193, 194, 215
CFC-113a 79, 171, 172, 193, 194
CFC-114 193

CFC-114a 193, 194
CFC-12 191, 192
chemical stability of CFC 191
chiral nematic phase 204
chiral smectic C phase 204
chiral smectic C_A phase 210
chitin biosynthesis 167
chlodinafop-propargyl 177
chlorodifluoromethane 143
chlorofluorocarbene 110
1-chloromethyl-4-fluoro-1,4-diazoni-
  abicyclo[2,2,2]octane bis(tetrafluoroborate) 36, 151
chlorotrifluoroethylene 143, 213, 214
ciprofloxacin 161
Claisen rearrangement, ester enolate 98
Clark 165, 203
cobaltocenium fluoride 125
code number 189
combinatorial synthesis 246
configuration, anomeric 120
conformation
– antiperiplanar 12
– zig zag 183, 184
conjugate addition 253
connecting group 197
contact angle 226
Cope rearrangement 255
Corey lactone 155
critical surface tension 226
cross-coupling 92
18-crown-6 49, 108
CTFE see chlorotrifluoroethylene
cure site 225
cure-site monomer 220
curing 220, 223
Curran 237
cyclic monomer 220
cyhalopfop-butyl 177
cypermethrin 169
cystic fibrosis 148

DAST see diethylaminosulfur trifluoride
DBH see 3,3-dimethyl-1,3-dibromo-
  hydantoin
dehydrochlorination 249
3′-deoxy-3′-fluorothymidine 153
deprotection 250
desflurane 164
DesMarteu reagent 38
Dess-Martin oxidation 236
dialkylaminosulfur trifluoride 44
diastereomer 208
diazepam 161
dibromofluoromethyllithium 82

2′,3′-dideoxy-5-fluoro-3′-thiacytidine
  153
2′,3′-dideoxy-3′-fluorouridine 153
dielectric constant 14, 217
Diels-Alder reaction 236, 261
diethylaminosulfur trifluoride 43, 150
diethyl difluoromethylphosphonate
  83
1-N,N-diethylcarbamoyl-2,2-difluorovinyl-
  lithium 86
diflubenzuron 167
difluorination 43, 44, 61–65
1,2-difluoroalkenyllithium 84
2,6-difluorobenzoylurea 168
difluorocarbene 109, 214
difluorocyclopropanes 108
α,α-difluoro ethers 46
2,2-difluoro-1-methoxyethoxymethoxyvinyl-
  lithium 86
difluoromethoxy group 173
difluoromethyl
– group 145
– ketones 86, 95, 96
difluoromethyl(dimethyl)phenylsilane
  84
2,2-difluoro-1-tosyloxyvinyllithium 86
2,2-difluorovinyllithium 84
3,3-dimethyl-1,3-dibromohydantoin 59
direct metalation 88
director film 197, 203
disproportionation 214
dTMP see thymidylic acid

eddy diffusion 192
electric field 203
electronegativity 2
electron shuttle 125
electron transfer 125
elemental fluorine 25
elimination
– α- 250
– β- 96, 250
– E2 51
emphysema 148
emulsifying agent 214
enolates perfluoro ketone 98
enoxacin 161
epoxidation 241
Erlenmeyer synthesis 146
esterification 261
ethoxyfen-ethyl 177
ethyl bromodifluoroacetate 95
ethyl bromofluoroacetate 94
ethyl difluoroiodoacetate 95
ethyl trifluoroacetate 101

FAR 43, 46
Favorskii rearrangement 171
FBS *see* fluorous biphase system
FdUMR *see* 5-fluoro-2'-deoxyuridylate
FdUR *see* 5-fluoro-2'-deoxy-β-uridine
fenvalerate 173
FEP *see* fluorinated ethylene propylene
ferroelectric liquid crystals 43, 197, 202–209
ferroelectricity 202, 203
field effect 140
first-aid treatment 20–21
Flemion 228, 229
flomoxef 164
FLT *see* 3'-deoxy-3'-fluorothymidine
flucythrinate 161
fluoride base 249
fluoride reagents 48–56
flucloxacillin 161, 164
fluconazole 163
flucytosine 163
flumiclorac-pentyl 177
flumioxazin 177
fluorinated ethylene propylene 218
fluorination
- asymmetric 37
- chlorinative 169
- diastereoselective 38
- electrochemical 66–70
- electrophilic 25–39
- *gem*-di- 45, 46
- nucleophilic 39–66
- oxidative 59–61, 65, 66
- oxidative desulfurization- 61–65
- radical 29, 30
fluorine effect
- block effect 10, 22
- mimic effect 10, 22
- polar effect 10
fluorine functionality
- CF_3 201
- CF_3O 201
- CHF_2O 201
- difluoromethyene 44
- trifluoromethyl 44
fluorine gas 18, 25–28, 56
fluorine resources
- cryolite 1
- fluorite 1
- phosphorite 1
fluoroacetic acid 21, 132, 141, 142
fluoroacetyl-CoA 141
2-fluoroalkanoic acid 204
2-fluoro-1-alkanols 40
fluoroalkoxylated poly(phosphazene) 220

β-fluoro-D-alanine 161
fluoroalkenyl
- copper 89
- lithium 88
- metals 84–92
- silanes 89
- stannanes 89
- zinc 87, 89
fluoroalkyl amine 43
fluoroalkylmetals 77–84
fluorobenzenes 185
N-fluorobis(sulfon)imide 38
N-fluorobis(trifluoromethylsulfon)imide 38
N-fluoro-2,10-camphorsultam 37
fluorocarbenes 107–111
fluorocarbon elastomer 220
(2R,3R)-2-fluorocitric acid 141
fluorocyclopropanes 107–111
α-fluoro enamines 47
5-fluoro-2'-deoxyuridylate 138
5-fluoro-2'-deoxy-β-uridine 138, 151
fluoroelastomer 220, 221
ω-fluorofatty acid 132
4-fluoroglutamic acid 147
fluorohydrins 53
9α-fluorohydrocortisone acetate 137
fluoromethyl group 145
fluoromonomers 213
fluoroolefin 44, 82
fluoroolefin-vinyl ether copolymer 224
fluorophosphazene elastomer 223
fluoropolymer coating 224
4-fluoroproline 147
fluoro-Pummerer reaction 28, 35, 46, 60, 69
N-fluoropyridinium fluoride 34–36
N-fluorosaccharin 37
fluorosilicone 220
fluoro sugars 45, 50, 119–121
fluorosulfonyldifluoroacetate 143
N-fluorosultam 37
4-fluorothreonine 137
3'-fluorotyrosine 147
5-fluorouracil 137, 138, 151
fluorous biphase purification 246
fluorous biphase reaction 237
fluorous biphase system 237, 243
fluorous multiphase condensation 243
fluorous phase 243
fluorous phase switch 243, 244
fluorous phosphine 239
fluorous ponytail 239, 243, 244, 247
fluorous reverse-phase 246
fluorous synthesis 243
fluoroxene 164

fluoroxytrifluoromethane 29–31
fluoxetine 161
flusilazole 177
fluthiacetmethyl 177
flutolanil 178
flutrimazole 163
fluvalinate 173
fomesafen 173
free-radical polymerization 213
friction coefficient 217
Fried 137
Friedel-Crafts acylation 236
Friedel-Crafts reaction 192, 261
FRP see fluorous reverse-phase
FTC see 2′,3′-dideoxy-5-fluoro-3′-thia-cytidine
5-FU see 5-fluorouracil
fuel cell 228, 230
Fukuda 209
functional group transformation 43–48
fungicides 177, 178

gemcitabine 152
global warming 189, 195
global warming potentials 195
Gloud-Jacobs reaction 162
glycosidation 119–121, 143
glycosyl fluorides 119, 120
group transfer polymerization 253
GWPs see global warming potentials

halfenprox 173
halocarbon global warming potentials 195
halofluorination 56–58
halogen fluoride 56, 65
N-haloimide 56
Halons 189
halothane 164
haloxyfop-R-methyl 177
Hammet
– acidity function 5
– σ constant 3
Haszeldine 104
HCF see hydrochlorofluorocarbon
HCFC see hydrochlorofluorocarbon
HCFC-1123 194
HFC-113a 192
HCFC-122 193
HCFC-123 193–195
HCFC-124 193, 195
HCFC-125 195
HCFC-133a 192–194
HCFC-134a 193–195
HCFC-141b 194–195
HCFC-21 194

HCFC-22 191, 192
HCHC-225ca 194, 195
HCFC-225cb 194, 195
HFC-32 195
helical pitch 204
hemiacetal 100
Henry reaction 148
hexacoordinated silicate 252, 254
hexaflumuron 168
hexafluoropropene 143, 213, 214
hexafluoropropene oxide 143, 229
HF see hydrogen fluoride
HF/amine 40
HF/melamine 40
HF/pyridine 40–43, 150
HF/triethylamine 40, 60, 68, 69
HFC see hydrofluorocarbon
HFE-227me 196
HFE-245mc 196
HFE-347mcc 196
HFE-347mmy 196
HFPO see hexafluoropropene oxide
HGWPs see halocarbon global warming potentials
HIV protease inhibitor 143
HLE see human leukocyte elastase
homogeneous phase 238
Horner-Emmons reagent 153
Horváth 237
human leukocyte elastase 148
hydantoin synthesis 146
hydroboration 239
hydrochlorofluorocarbon 143, 189, 192
hydrofluoric acid 18
hydrofluorocarbon 143, 189, 192
hydroformylation 147, 237, 238
hydrogen bond 11, 12, 218
hydrogen bonding 10, 40, 128, 184
hydrogen fluoride
– nucleophilic fluorination 39, 40, 68, 191
– resources 18
hypervalent silicon 254

IF see halogen fluoride
IGRs see insect growth regulators
immobilization 246
insect growth regulators 167
insecticides 167–173
interaction
– fluorine-hydrogen 128, 129
– fluorine-lithium 131–133
– fluorine-metal 101, 129–133
iodofluorination 57
iodosotoluene difluoride 61, 64

Subject Index

Ishikawa reagent   47
isostere   10
isotope   2, 139, 166
isotropic liquid phase   208

KF   *see* potassium fluoride

β-lactam   143, 163
lactofen   173
lactonization   261
Lagerwall   203
La Mar process   26
leaving group   121
lethal synthesis   141
lipophilicity   140, 143
liquid crystallinity   201
liquid crystals   187, 196
lufenuron   169

magic acid   5
mass spectroscopy   17, 18
mechanism
– anion chain   109
– radical   104, 105, 127, 239
– S_Ni 59
mefloquine   163
melting point   184
memory effect   203
mesogen   197, 204
metal fluoride
– antimony fluoride   1
– uranium fluoride   1, 2
metal fluoro enolate   93–99
metal trifluoromethanesulfonate   259
methyl 2-bromo-3,3,3-trifluoropropanoate   97
methyl 3,3,3-trifluoropanoate   97
Meyer   202
MHPOBC   209, 211
Michael addition   103, 261
Michael-type acceptor   144
Moissan   1, 150
molecular fluorine   18, 19, 25
molecular oxygen   240
Molina   188
Montreal Protocol   189
morph-DAST   44
morpholine-based human NK-1 antagonist   144
morpholinosulfur trifluoride   44
Mukaiyama aldol reaction   236

N*   *see* chiral nematic phase
Nafion   228, 229
naked anion   252

naked enolate   252, 253
negative dielectric anisotropy   197
negative hyperconjugation   8
nematic liquid crystals   197
nematic phase   198, 200
nitrile oxide cycloaddition   243
nitrofen   173
nitrofluorfen   173
nitrofluorination   58
Nn   *see* negative dielectric anisotropy
nomenclature   189
norfloxacin   161
novaluron   169
Np   *see* positive dielectric anisotropy
nuclear magnetic resonance spectroscopy
– chemical shift   15, 16
– coupling constant   15, 16
nucleocidin   137, 151

odd-even rule   185, 211
ODPs   *see* ozone-depletion potentials
Olah reagent   40, 57
oligopeptides fluorinated   147
optical fibers   186
orientation film   197
orthothioesters   62
oxidation potential   25
oxidation
– β-   141
– electrochemical   106
– fluorous biphase system   239, 240
– transition metal catalyzed   239, 240
oxidative addition   126
ozone   189
ozone-depletion potentials   189, 195, 235

paroxetine   161
PCE   *see* tetrachloroethene
PCTFE   *see* polychlorotrifluoroethylene
pentacoordinated silicate   252, 254
pentaoxazone   177
perchloryl fluoride   33, 155
perfluorinated copolymers   218
perfluoroadamantane   67
perfluoroalkanoyl fluorides   67
perfluoroalkoxy copolymer   218
perfluoroalkyl
– aryliodonium salts   111
– copper   78
– ethyne   91
– iodide   227
– lithium   77
perfluoroalkyl
– magnesium reagent   78
– radicals   103, 104, 106

(perfluoroalkyl)-*p*-fluorophenyliodonium triflate 112
perfluoroalkylmethylation 112
(perfluoroalkyl)phenyliodonium tosylate 112
(perfluoroalkyl)-*p*-tolyliodonium chloride 111, 112
perfluoroalkynyl, zinc 92
perfluorocarbons 165, 235
perfluorodecalin 125, 166
perfluorodecane 183
perfluoroheptane 187, 241
perfluorohexane 236
perfluoroisobutylene 213
perfluoropolymethylene chain 183
perfluoro(propyl vinyl ether) 213, 214, 218
permethrin 169
PET *see* positron emission tomography
Peters 137
Peterson olefination 93
PFA *see* perfluoroalkoxy copolymer
PFC *see* perfluorocarbons
PFH *see* perfluorohexane
PGs *see* prostaglandins
phase extraction 246
phase transfer catalyst 57, 109
phenylsulfenyl chloride 58
photochemical processes 104
Plunkett 212
polychlorotrifluoroethylene 212
polyethylene 214
polymerization
- dispersion 214
- emulsion 227
- suspension 214, 236
polymethacrylate 216
polytetrafluoroethylene 125, 212, 216, 226
polyvinylidene fluoride 218, 224
ponytail 238
positive dielectric anisotropy 197, 99–201
positron emisson tomography 166
potassium difluoride 50
potassium fluoride
- for desilylation 250, 251
- freeze-dried 48
- spray-dried 48
potassium hydrogen fluoride 50
PPDA 47
PPVE *see* perfluoro(propyl vinyl ether)
Prince reaction 194
prostacyclin 143
prostaglandins 154
- 16,16-difluoro-PGE$_1$ 155
- 16,16-difluoro-PGF$_{2\alpha}$ 156
- 7,7-difluoro-PGI2 155

- 10,10-difluoro-13,14-dehydro-PGI$_2$ 155
- 7,7-difluoro-18,19-dehydro-16,20-dimethyl-PGI$_2$ 155
- PGF$_{2\alpha}$ 154
- PGH$_2$ 154
- PGI$_2$ 154
prostanoids 154–157
protease inhibitors 148–150
PTFE *see* polytetrafluoroethylene
PVdF *see* polyvinylidene fluoride
pyramidalization angle 6
pyrethrin I 169
pyrethrin II 169
pyrethroids 169

o-**q**uinodimethane 251
quinolone carboxylic acid 143
- fluoro 161

Rábai 237
radical initiator 216–218
radical polymerization 218
radioisotope 2
reductive amination 146
Reformatsky reaction 95, 97, 152, 155
renin 148
renin inhibitor 149
ring-opening polymerization 223
rheumatoid arthritis 148
Rowland 188

Sanger method 123
SAR *see* structure-activity relationship
Selectfluor 151
selectivity
- diastereo 94, 97
- peripheral activity 145
selenofluorination 58
serine protease 148
sevoflurane 164
Shiemann reaction 42
silaflulofen 173
SmA *see* smectic A phase
SmC* *see* chiral smectic C phase
SmC$_A$* *see* chiral smectic C$_A$ phase 210
smectic A phase 202
smectic C phase 202
S$_N$2 reaction 8
S$_N$Ar 121, 122, 124
solubility
- of CFCs 191
- of PFC 186
solvent polarity 14
sparfloxacin 161
specific gravity 184

spontaneous polarization 203
stereochemistry
- acetylene addition 53
- fluorination 27
- inversion 40, 43, 44, 47, 203
- polar functional group 204
- retention 43, 59, 91, 103
- trifluoromethoxylation 30
steric effect 211
steric effect constant 2
steric interaction 218
1,3-steric repulsion 183
Stille coupling 246, 247
STN see supertwisted nematic
stratosphere 188, 189
Strecker synthesis 146
structure-activity relationship 143
substituent effect
- +Ip effect 3, 4, 8, 140
- -Is effect 3, 4, 7, 8, 140
- +R effect 3, 4, 7
- resonance effect 3
suicide inhibitor 141, 144
sulfur tetrafluoride 43-45, 61
super acid 5
supertwisted nematic 197, 199
surface engergy 217
surface tension 186
Swarts 191, 192, 214

TAIC see triallyl isocyanurate
Takezoe 209
TASF see tris(dialkylamino)sulfur difluorotrimethylsilicate
TASF(Et) see tris(diethylamino)sulfur difluorotrimethylsilicate
TASF(Me) see tris(dimethylamino)sulfur difluorotrimethylsilicate
taxoids 143
TBAF see tetrabutylammonium fluoride
TCE see trichloroethene
Teflon 2, 212
teflulthin 169
telomerization 227
tetraalkylammonium fluoride 51
tetrabutylammonium difluorotriphenyl-silicate 54, 55
tetrabutylammonium difluorotriphenyl-stannate 55
tetrabutylammonium dihydrogen trifluoride 53, 57, 60
tetrabutylammonium fluoride 51
- for desilylation 251
tetrabutylammonium hydrogen fluoride 52

tetrachloroethene 193, 194
tetrafluoroethylene 143, 194, 213, 214, 227-229
tetrahedral intermediate 148
5,10,15,20-tetrakis(heptafluoropropyl)-porphyrin 242
tetramethylammonium fluoride 51, 241, 253-254
tetraphenylphosphonium bromide 48
TFA see trifluoroacetic acid
TFE see tetrafluoroethylene
TFE-hydroxyalkyl vinyl ether copolymer 224
TFE-propylene copolymer 221
TFMHPOBC 210
TfOH see trifluoromethanesulfonic acid
TFT see thin film transistor
TFT display 200, 201
thermal stability 223
thin film transistor 197, 199
thiofluorination 58
thromboxanes 154
- TXA_2 154
- 10,10-difluoro-TXA_2 155
thymidylate synthetase 138
thymidylic acid 138
TMSOTf see trimethylsilyl trifluoromethanesulfonate
TNS-Tf see N-trifluoromethyl-N-nitrosotrifluoromethanesulfonamide
TolIF$_2$ see iodosotoluene difluoride
tosufloxacin 161
toxicity
- of CFCs 191
- of fluorine gas 19
- of fluoromonomers 213
- of hydrogen fluoride 19
TPFPP see 5,10,15,20-tetrakis(heptafluoropropyl)porphyrin
transmetalation 92
triallyl isocyanurate 221-223
trichloroethene 193, 194
triethylammonium trifluoroacetate 256
triflate
- copper(I) 260
- copper(II) 260
- dialkylboryl 259
- metal 259
- scandium 261
- silver 260
- tin 259
triflate
- ytterbium 261
triflumuron 167
trifluoroacetaldehyde 99

trifluoroacetic acid 255
trifluoroacetimidoyllithium 86
trifluoroethene 194
2,2,2-trifluoroethyl tosylate 86
trifluoromethanesulfonic acid 257, 258
trifluoromethyl
- amines 64
- chalcogenium salts 113
- copper 78, 79
- ethers 64
- iodide 152
- ketones 81, 101, 148
- radical 9
- zinc 79
trifluoromethylation 63–65, 78, 81, 105, 113, 144, 148
S-, Se-, or Te-(trifluoromethyl)dibenzochalcogenophenium salts 113
N-trifluoromethyl-N-nitrosotrifluoromethanesulfonamide 78
trifluoromethyl(trimethyl)silane 81, 148
5-trifluoromethyluridine 152
5,5,5-trifluoronorvaline 147
1,1,1-trifluoro-2-octanol 210
trifluoroperacetic acid 256
3,3,3-trifluoropropene 147
3,3,3-trifluoropropene oxide 103
trifluoropropyl methyl cyclosiloxane 223
trifluoropropynyl
- lithium 91
- magnesium 91
- zinc 92
trifluoro(trifluoromethyl)silane 109
4,4,4-trifluorovaline 146, 147
trifluorovinyl
- lithium 84
- magnesium 89
- silanes 89, 90
- stannanes 91
trimethylsilyl trifluoromethanesulfonate 258

tris(dialkylamino)sulfur dilfuorotrimethylsilicate 54
tris(diethylamino)sulfur difluorotrimethylsilicate 54, 252, 253
tris(dimethylamino)sulfur difluorotrimethylsilicate 54, 253
troposphere 188
turnover 239
two-phase system 57
TXs see thromboxanes

Ugi reaction 245

vancomycin 122
van der Waals radius 2, 139, 184
VdF see vinylidene fluoride
VdF-HFP copolymer 220
VF see vinyl fluoride
Vienna Conference 189
vinyl fluoride 213, 214
vinylidene fluoride 213, 214
viscosity 187
- absolute 187
- of ferroelectric liquid crystals 203
- kinetic 187
vitamin D_3 143, 157–160
- 26-F_3,27-F_3-1,15-$(OH)_2$ 158
- 26-F_3,27-F_3,1,25-$(OH)_2$ 158
- 1,25-$(OH)_2$ 157
- 25-(OH) 157
Viton 220
voltage holding ratio 199

Wakamatsu reaction 147
weatherability 223
Wilkinson catalyst 239

xenon difluoride 28

Yarovenko reagent 46

Zisman 225, 226

New Approaches and Methods

G. Furin

New Aspects in the Synthesis of Fluoroorganic Compounds

2000. Approx. 400 pp.
410 figs., 121 tabs.
Hardcover
ISBN 3-540-66399-1

Interest in fluorine-containing organic compounds is based on their unique properties, leading to successful applications in a wide range of academic and industrial areas, including chemical, biochemical, biomedical, pharmaceutical and materials research. The purpose of this book is to present new approaches and methods developed in the last 10-15 years for the introduction of fluorine-containing functional groups and moieties into organic compounds. The detailed presentation and analysis of synthetic methodology allows the optimization of the methods involved and the development of new approaches. An essential tool not only for specialists in the field of organic synthesis and organofluorine chemistrybut also for all those who prepare organofluorine compounds, whether for academic, biomedical or industrial applications.

Please order from
Springer · Customer Service
Haberstr. 7
69126 Heidelberg, Germany
Tel: +49 6221 345200
Fax: +49 6221 300186
e-mail: orders@springer.de

Prices and other details are subject to change without notice.
In EU countries the local VAT is effective. d&p · 66689/MNT/1 · Gha

Topics in Current Chemistry

Volume 192
R.D. Chambers (Ed.)

Organofluorine Chemistry

Fluorinated Alkenes and Reactive Intermediates

1997. X, 244 pp.
Hardcover *DM 218
£ 84 / FF 822 / Lit. 240.760;
ISBN 3-540-63171-2

Volume 193
R.D. Chambers (Ed.)

Organofluorine Chemistry

Techniques and Synthons

1997. X, 252 pp.
Hardcover *DM 218
£ 84 / FF 822 / Lit. 240.760
ISBN 3-540-63170-4

Please order from
Springer · Customer Service
Haberstr. 7
69126 Heidelberg, Germany
Tel: +49 6221 345200
Fax: +49 6221 300186
e-mail: orders@springer.de

The excitement of the chemistry of organo fluorine compounds stems from the unique reactions that arise and the "special effects" that introduction of fluorine impart on a molecule. Indeed, these effects are now exploited in a remarkable array of applications the whole of the chemical, pharmaceutical, and plant-protection industries. In this two-volume set, we have gathered authors with immense experience in various aspects of their fields and each is a world-authority on the important topics which they have described.
The first volume treats the chemistry of fluorinated alkenes, which are important "building-blocks" for the synthesis of a range of fluorinated systems and are used widely in industry, while the second volume is directed to techniques and synthons for obtaining fluorinated compounds.

* Recommended retail prices. Prices and other details are subject to change without notice. In EU countries the local VAT is effective. d&p · 66689/MNT/2 · Gha

Druck: Saladruck, Berlin
Buchbinderei: Lüderitz & Bauer, Berlin